Springer Advanced Texts in Life Sciences

David E. Reichle, Editor

David M. Gates

Biophysical Ecology

With 163 Illustrations

Springer-Verlag
New York   Heidelberg   Berlin

David M. Gates
Biological Station
University of Michigan
Ann Arbor, Michigan 48109
USA

Series Editor:
David E. Reichle
Environmental Science Division
Oak Ridge National Laboratory
Oak Ridge, Tennessee 37830
USA

Library of Congress Cataloging in Publication Data

Gates, David Murray, 1921–
 Biophysical ecology.

 (Springer advanced texts in life sciences)
 Bibliography: p.
 Includes index.
 1. Ecology. 2. Biological physics.
 3. Bioenergetics. I. Title.
 QH541.G39   574.5′01′574191   79-14509

Printed in the United States of America.

9 8 7 6 5 4 3 2 1

ISBN 0-387-90414-X Springer-Verlag New York
ISBN 3-540-90414-X Springer-Verlag Berlin Heidelberg

*This book is dedicated to my family
for their enduring patience.*

*Marian Penley Gates and our children,
Murray, Julie, Heather, and Marilyn*

Preface

The objective of this book is to make analytical methods available to students of ecology. The text deals with concepts of energy exchange, gas exchange, and chemical kinetics involving the interactions of plants and animals with their environments. The first four chapters are designed to show the applications of biophysical ecology in a preliminary, simplified manner. Chapters 5–10, treating the topics of radiation, convection, conduction, and evaporation, are concerned with the physical environment. The spectral properties of radiation and matter are thoroughly described, as well as the geometrical, instantaneous, daily, and annual amounts of both shortwave and longwave radiation. Later chapters give the more elaborate analytical methods necessary for the study of photosynthesis in plants and energy budgets in animals. The final chapter describes the temperature responses of plants and animals.

The discipline of biophysical ecology is rapidly growing, and some important topics and references are not included due to limitations of space, cost, and time. The methodology of some aspects of ecology is illustrated by the subject matter of this book. It is hoped that future students of the subject will carry it far beyond its present status. Ideas for advancing the subject matter of biophysical ecology exceed individual capacities for effort, and even today, many investigators in ecology are studying subjects for which they are inadequately prepared. The potential of modern science, in the minds and hands of skilled investigators, to advance our understanding of the interactions of organisms with their environment is enormous.

My students have contributed significantly to the development of this subject matter. They have helped to formulate many of the techniques and have constructively criticized much of the material. I am much indebted to Drs. George Bakken, John Tenhunen, and James Weber and to Peter Harley. My secretary, Meredy Gockel, typed and retyped chapter after chapter with infinite patience. All the early versions, done while I was Director of the Missouri Botanical Garden, were typed by Florence Guth. My administrative assistant, Mark Paddock, has suffered loyally through this arduous task and made it possible for me to devote the necessary time to it. I am deeply grateful to all who have helped. If the science of ecology advances from this effort, I will feel generously rewarded.

Contents

Symbols

Roman Symbols

A surface or cross-sectional area (m²); subscripts 1, 2, 3, 4, 5 for surface area receiving particular streams of radiation) a constant in the respiration-temperature equation (dimensionless); activation energy of reacting molecules in chemical reactions.

A_i increment of surface area exposed to a particular radiation source (m²) (subscripts; a, atmospheric radiation; g, ground radiation; S, solar; s, skylight; rS, reflected solar; rs, reflected skylight).

A_g animal surface area in contact with the ground (m²).

A_p projected area (m²) (subscripts: c, cylinder; s, sphere).

A_r effective emitting area for thermal radiation from an animal surface (m²).

A_0, A_j coefficients in Fourier series expression of environmental temperature (°C).

a radius of the earth (m); acceleration (m s⁻²); absorptance (decimal fraction); altitude angle of the sun (degrees) (subscripts: s, for normal to a slope; c, for the axis of a cylinder).

a_i absorptance of a surface to a particular source i of incident radiation (decimal fraction) (subscripts 1, 2, 3, 4, 5 for direct solar, skylight, reflected sunlight, atmospheric, and ground radiation, respectively).

$_ia_j$ absorptance of the ith incremental element of animal surface

	to the jth source of radiation (j = S, s, rS, rs, and t for direct solar, skylight, reflected direct solar, reflected skylight, and thermal radiation).
$_d a_1$, $_u a_1$	absorptance of a downward- or upward-facing surface to direct sunlight (decimal fraction).
$_\lambda a$, a_λ	monochromatic absorption, coefficient decimal fraction.
$\mathscr{B}$	radiance (W m^{-2} sr^{-1}) (subscript o, in direction of a normal).
B	proportionality constant in chemical rate equation.
B_j	coefficients in Fourier series expression of environmental temperature (°C).
BR	breathing rate (breaths min^{-1} or breaths s^{-1}).
C	quantity of heat exchange by convection (W m^{-2}); heat capacity, (J m^{-2} °C^{-1}); carbon dioxide concentration (mmole m^{-3}) (subscripts: a, in the air; c, chloroplasts; i, intercellular air space; w, sites of respiration).
$\overline{C}$	monthly average amount of sky covered by clouds (decimal fraction).
c	velocity of light (m s^{-1}).
c_b	specific heat of blood (J kg^{-1} °C^{-1}).
c_p	specific heat at constant pressure (J kg^{-1} °C^{-1}).
c_1	a constant in the light response of photosynthesis.
c_2	a constant in the temperature function for photosynthesis.
D	characteristic dimension (m).
$\mathscr{D}_i$	diffusion coefficient (m^2 s^{-1}).
$_h D$, $_h \overline{D}$	average daily diffuse radiation incident on a horizontal surface (J m^{-2} day^{-1}).
D_h, D_c	fraction of total active time spent in the hot or cold microclimate parts, respectively, of a duty cycle when shuttling (s).
d, $\overline{d}$	instantaneous and average distance between earth and sun (m).
d	thickness of material (m) (subscript f for fat); diameter of a cylinder (m).
$_h d$, $_h \overline{d}$	instantaneous and average hourly diffuse radiation incident on a horizontal surface at the ground (W m^{-2}) (subscript λ monochromatic intensity W m^{-2} nm^{-1}).
E	evaporation rate (kg m^{-2} s^{-1}) (subscripts b, by respiration; r, from skin or pelage); energy (J) elastic modulus (kg m^{-2}); apparent activation energy (J mole^{-1}); internal energy of a molecule (J mole^{-1}) (subscripts: e, electronic; v, vibrational; r, rotational); $\mathscr{E}$.
$_\lambda E$	flux density per unit of wavelength (W m^{-2} nm^{-1}).

$_\nu E$ flux density per unit of wavenumber (W m^{-2} cm^{-1}).

ETL   eating time limit (min).

e water vapor pressure (N m^{-2}, mbar) (subscripts: a, of the air, s, at saturation); Napierian logarithm base (dimensionless).

F radiant flux (W); Helmholtz free energy (J mole^{-1}); view factor (decimal fraction); duration of oscillating temperature cycle (s).

f force (N); fraction of saturation for air exhaled from the lungs (decimal fraction); a temperature function $f(T)$.

G heat flow by conduction (W m^{-2}); Gibbs free energy (J mole^{-1}); storage term (W m^{-2}).

$_h G$ total daily shortwave radiation incident on a horizontal surface (J m^{-2} day^{-1}).

$_h G'$ total daily shortwave radiation incident on a horizontal surface on a cloudy day (J m^{-2} day^{-1}).

$G(T_K)$ temperature function for photosynthesis.

g acceleration by gravity (m s^{-2}).

$_h g, \; _h \bar{g}$ global and mean hourly global radiation incident on a horizontal surface at the ground (J m^{-2} hr^{-1}).

$_s g$ global shortwave radiation incident on a slope (W m^{-2}).

H convection coefficient (W m^{-2} °C^{-1}); enthalpy (J mole^{-1}); elevation of a site above mean sea level (m).

$_h \bar{H}_0$ average hourly extraterrestrial direct solar radiation incident on a horizontal surface (J m^{-2} hr^{-1}).

h height (m); length of a cylinder (m); relative humidity (decimal fraction); Planck's constant (J s); hour angle (rad).

h_c convection coefficient (W m^{-2} °C^{-1}).

h_s hour angle at sunrise or sunset, or half-day length (rad).

I radiant intensity (W m^{-2}) (subscripts: o, along direction of a normal to the source surface; λ, monochromatic intensity); thermal insulation (m^2 °C W^{-1}) (superscript a, apparent value); moment of inertia of a molecule (kg m^2).

i angle of incidence (degrees).

J upward flux of radiation in a leaf (W m^{-2}); proportionality constant in respiration relationship (mmole m^{-2} s^{-1}); rotational quantum number of a molecule.

J_j flux density of molecular species j (water vapor, oxygen, or carbon dioxide) (kg m^{-2} s^{-1}).

K thermal conductance (W m^{-2} °C^{-1}) (subscripts f, fur; s, skin; sf, skin and fur; o, overall heat conductance for an animal; c, surface boundary layer for convection); an affinity coef-

	ficient for carboxylation equal to the chloroplast concentration of CO_2 at which $P = P_M/2$ (mmole m^{-3}).
K_L	value of L at which $P_m(L) = P_{ML}/2$ (mmole m^{-3}, E m^{-2} s^{-1}, or W m^{-2}).
K_{rl}	value of L such that $r_s = 2(_{min}r_s)$ (E m^{-2} s^{-1} or W m^{-2}).
K_O	an affinity coefficient for oxygenation equal to the oxygen concentration (vol/vol) which will double K or at which $W_p = W_M/2$ with $C_c = 0$.
k	thermal conductivity (W m^{-2} °C^{-1}) (subscript f, fat); Coulomb force constant, 8.99×10^9 N m^2 C^{-2}; von Karman constant (dimensionless); Napierian absorption coefficient (decimal fraction).
k_1, k_2	proportionality constants.
k_T	velocity constant for chemical reaction.
k_d	ratio of skylight to direct sunlight incident at the earth's surface (decimal fraction).
$_\lambda k$	monochromatic attenuation coefficient (m^2 kg^{-1}); monochromatic "Napierian" absorption coefficient (m^{-1}) (subscripts: s, scattering; a, Mie scattering; r, Rayleigh scattering).
L	photosynthetically active radiant flux, between 400 and 700 nm (E m^{-2} s^{-1} or W m^{-2}).
L_H	a constant such that when $L = L_H$, exp $[(-\ell n\ 2)(L/L_H)] = 1/2$ (E m^{-2} s^{-1}).
ℓ	length of a cylinder (m); distance (m).
l_{cr}	critical length for buckling (m).
M	metabolic rate (W m^{-2}) (superscript *, dry metabolic heat production); molecular weight of a gas (kg mole^{-1}) (subscripts: a, dry air; w, water vapor); ratio of resistance R_2/R_4.
$\bar{M}$	average number of days in a month with 1 in or more of snow cover.
m	mass (kg); atomic mass (kg or g); air mass (dimensionless).
m_w	molecular weight (kg mole).
N	number of stomates; number of elements in a leaf (number m^{-2}).
N_0	Avogadro's number (6.023×10^{23}).
N_r	molecular density (molecules m^{-3}).
NR	net radiation (W m^{-2}).
n	index of refraction (fractional number) (subscript λ, at a given wavelength); number of quanta; number of kilogram-moles of

a gas; number of absorbing and scattering elements in a leaf (number m^{-2}).

$[O_2]$ oxygen concentration (vol/vol).

P photosynthesis rate (mmole m^{-2} s^{-1}) (subscripts: M, at saturating CO_2; ML, at saturating light and saturating CO_2; MLT, at optimum temperature, saturating light, and saturating CO_2 1, between external air and intercellular air space; 2, intercellular air space and chloroplasts; 3, intercellular air space and respiration sites; 4, respiration and fixation sites; 5, at the chloroplasts); power (W) (subscript max, maximum for muscle contraction); pressure (N m^{-2}); slope of line giving dry metabolic heat production versus temperature (W m^{-2} $°C^{-1}$).

p atmospheric pressure (N m^{-2} or mbar) (subscript 0, at sea level).

Q rate of heat flow (W m^{-2}); irradiance (W m^{-2}) (subscripts: c, on a cylinder; s, on a sphere; h, on a horizontal flat plate; superscripts: u, upward facing; d, downward facing).

Q_a amount of radiation absorbed (W m^{-2}); subscripts: c, by a cylinder; s, by a sphere; h, by a horizontal flat plate; superscripts: u, upward facing; d, downward facing).

$_hQ_o$ daily amount of radiation incident on a horizontal surface just outside the earth's atmosphere (J m^{-2} day^{-1}).

$_h\overline{Q}_o$ monthly average extraterrestrial daily insolation incident on a horizontal surface (J m^{-2} day^{-1}).

Q_r thermal radiation emitted by an animal surface (W m^{-2}).

Q_{10} factor by which the rate of activity (such as oxygen consumption) increases with each 10°C rise in temperature, (dimensionless number).

q quantity of heat (J) (subscript c for local amount); specific humidity (dimensionless).

R resistance to CO_2 diffusion (s m^{-1}) (subscripts: m, mesophyll; s, stomatal; min, minimum stomatal; 1, between external air and intercellular air space; 2, intercellular air space and chloroplasts; 3, intercellular air space and respiration sites; 4, respiration sites and chloroplasts); gas constant (8.31 J $mole^{-1}$ $°K^{-1}$); ratio of photosynthetic to transpiration rate (molecules of H_2O/molecules of CO_2).

R_λ spectral emittance.

$\mathcal{R}$ radiant emittance (W m^{-2}) (subscripts: a, from atmosphere; g, from ground surface; i, a particular source; λ, monochromatic); linearized radiation coefficient $4\epsilon\sigma\overline{T}_r^3$ (W m^{-2}).

r reflectance (decimal fraction) (subscripts: g, of ground or lower leaf surface; S, to direct sunlight; s, to skylight); diffusion resistance to water vapor (s m^{-1}) (subscripts: a, boundary layer; c, cuticular; i, intercellular; ℓ, leaf internal; m, mesophyll; s, stomatal); resistance to heat flow (s m^{-1}) (subscripts: h, ha, boundary layer; j, diffusion resistance to molecular species j; radius (m) (subscripts: b, of inside body wall of animal; r, of outer surface of animal).

r_d ratio of average hourly diffuse to the average daily diffuse radiation received on a horizontal surface (decimal fraction).

S amount of sunlight incident upon the earth's surface (W m^{-2}) (subscripts: c, on a cylinder; cv, on a vertical cylinder; ch, on a horizontal cylinder; h, on a horizontal surface; λ, monochromatic); entropy (J mole^{-1} °C^{-1}); equivalent network resistances (s m^{-1}); (subscript λ, monochromatic).

S_o, $\bar{S}_o$ instantaneous and average amount of sunlight incident at a point just outside the earth's atmosphere (W m^{-2}) (subscript h, on a horizontal surface).

$_{ch}^{o}S$, $_{ch}^{90}S$ irradiance of a horizontal cylinder with $\alpha - \alpha_c = 0°$ and $90°$, respectively (W m^{-2}).

s intensity of scattered skylight (W m^{-2}); radius of a stomate (m); scattering coefficient (decimal fraction).

sd   saturation deficit (N m^{-2} or mbar).

T temperature (°C) (subscripts: a, air; b, body; c, animal casting or cold side of thermal barrier; e, operative environmental; g, ground surface; h, hot side of thermal barrier; K, leaf; ℓ, leaf; r, animal surface; S, sky; v, wall; v, vegetation; o, initial or of surroundings; ∞, final); transmitted flux through leaf (W m^{-2}).

T_b body temperature (°C) (superscripts: c, lower setpoint; h, upper set point; superscripts: lc, lower equilibrium; 1h, upper equilibrium); equivalent blackbody temperature (°C).

T_c critical body temperature (°C) (subscripts: ℓ, lower; u, upper).

T_d daily average transmittance of the atmosphere to diffuse radiation (decimal fraction).

$\bar{T}_d$ monthly average daily transmittance of the atmosphere to diffuse radiation (decimal fraction).

T_e operative environmental temperature (°C) (superscripts: a, apparent; lc, lower; 1h, upper).

T_p	temperature at which the dry metabolic heat production has a linear increase with body temperature (°C).
T_Δ	physiological offset temperature (°C) (superscripts: a, apparent; lc, lower when shuttling; lh, upper when shuttling).
T_t	daily average transmittance of the atmosphere to total shortwave radiation (decimal fraction).
$\overline{T}_t$	monthly average daily transmittance of the atmosphere to total shortwave radiation (decimal fraction).
TV	tidal volume for breathing (m³ or cm³).
t	transmittance (decimal fraction); time (s).
U	internal energy content of a system (J mole⁻¹).
u	optical thickness (kg m⁻²).
V	velocity (m s⁻¹); volume (m³) (subscript w, partial molar); photopic efficiency (decimal fraction).
v	vibrational quantum number of a molecule.
W	leaf dimension perpendicular to wind (m); weight (kg); work (J); respiration rate (mmole m⁻² s⁻¹) (subscripts: m, dark; p, light).
W_M	maximum rate of photorespiration at saturating oxygen concentration (mmole m⁻² s⁻¹).
W_{ML}	value of W_M at zero light intensity and optimum temperature (mmole m⁻² s⁻¹).
w	mixing ratio for air (dimensionless) (subscript s, at saturation).
X	heat in storage (W m⁻²).
x	distance (m); thickness (m); ratio of length to radius for a cylinder (decimal fraction).
y	distance (m).
z	distance, or height above a surface (m); zenith angle (degrees).

Dimensionless Groups

Sc	Schmidt number
Sh	Sherwood number
Nu	Nusselt number
Re	Reynolds number
Pr	Prandtl number
Gr	Grashof number
Le	Lewis number

Greek Symbols

α	initial slope of photosynthesis versus light intensity curve (mmoles E^{-1}); angle (rad or degrees); azimuth angle of the sun (degrees).
α_c	azimuth of the axis of a cylinder (degrees).
α_s	azimuth angle of a normal to a slope (degrees).
β	slope angle, direction of normal from horizontal (degrees); coefficient of volumetric thermal expansion ($°C^{-1}$); competitive inhibition factor (dimensionless); proportionality coefficient for Mie scattering; Bowen's ratio, $C/\lambda E$; psychrometer constant (m $°C^{-1}$); parameter (exponent) for Mie scattering.
Δ	change of saturation vapor pressure with temperature (N m^{-2} $°C^{-1}$ or mbar $°C^{-1}$).
Δe	concentration difference of water vapor (kg m^{-3}).
ΔH_1	activation energy for denaturation equilibrium of the photosynthetic reaction (J $mole^{-1}$).
$\Delta H^{\neq}$	activation energy for enzyme-catalyzed photosynthetic reaction (J $mole^{-1}$).
ΔS	entropy of the denaturation equilibrium (entropy units).
δ	declination of the sun (degrees); boundary-layer thickness (m).
ϵ	emissivity (decimal fraction) (subscript a, of atmosphere). angle (rad $degree^{-1}$); angle between axis of a cylinder and direction of the sun's rays; equivalent temperature ($°C$).
λ	latent heat of vaporization (J kg^{-1}); wavelength of radiation (m, cm, μm, nm, Å).
μ	viscosity coefficient (N s m^{-2}). chemical potential of water (J $mole^{-1}$).
ν	frequency of radiation (s^{-1}); kinematic viscosity coefficient (m^2 s^{-1}).
$\tilde{\nu}$	wavenumber of radiation (m^{-1} or cm^{-1}).
π	osmotic pressure (bars or N m^{-2}).
ρ	density (kg m^{-3}) (subscripts: a, of dry air; j, of molecular species; o, of standard air; w, of water vapor); radius of a cylinder (m); reflectivity (decimal fraction).
$_s\rho_a\,(T_a)$	water vapor density in the air at saturation as a function of air temperature T_a (kg m^{-3}).
$_s\rho_\ell\,(T_\ell)$	water vapor density in the leaf at saturation as a function of the leaf temperature T_ℓ (kg m^{-3}).

Σ number of hours of sunshine.

σ Stefan-Boltzmann constant, 5.673×10^{-8} W m^{-2} °K^{-4}; compressive stress on a cylinder (N m^{-2}) (subscript max for maximum).

τ transmittance (decimal fraction) (subscripts: d, to diffuse radiation; t, to total radiation); thermal time constant (s or hr) (superscript a, apparent).

$_\lambda\tau$ monochromatic transmittance (decimal fraction) (subscripts: a, Mie scattering; m, minor constituents; o, ozone; r, Rayleigh scattering).

τ_t average hourly transmittance of the atmosphere to shortwave radiation (decimal fraction).

τ_c length of time an animal can spend in a cold microclimate when shuttling (s).

τ_h length of time an animal can spend in a warm microclimate when shuttling (s).

ϕ latitude (degrees); angle between scattered and incident light directions (degrees); profile function.

Ψ water potential (bars or N m^{-2}).

ω angular velocity of the earth (rad day^{-1}); solid angle (sr).

Chemical Reaction Symbols

[A]   concentration of acception molecule for photosynthesis.

[A$_o$]   concentration of total carbohydrate pool.

[ACO$_2$]   concentration of Calvin–Benson pathway intermediates or carboxylated acceptor.

[AO$_2$]   concentration of oxygenation product of the photorespiration reaction.

[B]   concentration of intermediates formed in the regeneration of RuDP in photosynthesis.

[H]   concentration of reducing power (NADPH).

K rate constants (subscripts 1, 2, 3, 4, and 5 for various reactions).

[O]   concentration of oxygen.

[X]   concentration of a regulating enzyme.

[π]   concentration of the carboxylation product.

Chapter 1

Introduction

Ecology is the study of the relationship of plants and animals to their environment and to one another and of the influence of man on ecosystems. The word ecology is derived from the Greek words *oikos,* meaning house or place to live, and *logos,* meaning science or study. The German zoologist Ernest Haeckel was an early user of the word *Ökologie* in 1866 and described it as a separate field of scientific knowledge, "the relation of the animal to its organic as well as its inorganic environment, particularly its friendly or hostile relations to those animals or plants with which it comes in contact." A book with the word *Ökologie* in the title was published by Hans Reiter in 1885, but it is difficult to say precisely when the science of ecology began to take form as a discipline since it has always been inextricably interwoven with natural history. In America, the field of ecology became active about the turn of the century. In 1899, Henry Cowles, of the University of Chicago, published his classic ecological study of the sand dunes of Lake Michigan. Soon after that, ecology was recognized as a distinct professional discipline. In 1907, Victor Shelford, of the University of Illinois, reported on succession among communities of tiger beetles in direct association with plant succession. An excellent summary of the history of ecology is given by Kendeigh (1974) and of plant ecology by McIntosh (1974).

The definition of ecology makes it clear that it is a science which necessitates understanding of the physical environment, involving the fields of physics, meteorology, geology, chemistry, and so forth, combined with an understanding of biology, including systematics, community dynamics, anatomy, physiology, genetics, and other subjects. The science of ecology, by its very nature, is among the most complex of all the sciences and, because of this inherent complexity, must draw upon knowledge

from the other sciences. Ecology is done poorly if either the biotic or
abiotic aspects of the subject are not treated in a fully correct and rigorous
scientific manner. Each ecological process or event must be studied in its
full complement of physical and biological components. This requires that
the physical principles of ecology be dealt with by the ecologist as
thoroughly and correctly as the physicist deals with physics and the
chemist with chemistry. At the same time, the ecologist must have a com-
petent understanding of physiology, genetics, systematics, and other
branches of biological science. This is a difficult order, yet a necessary
one. Mathematical skills are also needed. Ecology, to be done well, must
involve all the techniques of modern science. Fortunately, the modern
computer is a very sophisticated instrument, capable of enormous data
storage and complex mathematical manipulations.

All of life involves energy flow and material flow. Not a single animal
or plant lives or breathes without the transformation of energy. The most
microscopic change within an organism involves utilization of energy. Energy is involved whether it is the coursing of blood through the veins and
arteries of an animal, the transfer of electrons in the photosynthetic
process of plants, the division or expansion of cells, the beating of a heart,
the flying of a bird, or simply the bending of a branch in the wind. Fundamental to the study of ecology is an understanding of energy flow and of
energy transfer from one form to another within the biological and physical systems. Also fundamental to ecology is an understanding of mass
transport within the environment. Life is not a static process within the
organism; every cell, tissue, and organ is at all times chemically and physically active. An ecologist cannot remove him- or herself from understanding these factors, for they are often important in determining how an
organism will respond to the forces and factors of the environment. Biophysical ecology is basically, therefore, an approach to ecology founded
upon a thorough understanding of the sciences of energy and fluid flow,
gas exchange, chemical kinetics, and other processes. This understanding
is enhanced by using mathematical formulations of physical processes
and relating them to the unique properties of organisms.

If we look about the world we live in, it is obvious that there are fairly
distinct communities of organisms, such as those comprising a forest,
prairie, pond, or stream. Not only are there a variety of communities in
the world, but each community has a distinct set of edaphic environmental features. The term *ecosystem* is a convenient concept first proposed by Professor A. G. Tansley in 1935 to describe the collective sum of
biotic and abiotic components of a segment of the landscape. The definition of the term used here is that proposed by J. W. Marr (1961): "An ecosystem is an ecological unit, a subdivision of the landscape, a geographic
area that is relatively homogeneous and reasonably distinct from adjacent
areas. It is made up of three groups of components—organisms, environmental factors and ecological processes." The term ecosystem may be

applied to a meadow, forest, lake, sand dune, or another readily recognized unit of the landscape. The ecosystem includes interactions between the plants and animals in an area with the climate and physiography of the region. In order to understand the response of a particular organism to its environment, however, knowledge of the climate and physiography of a region is not sufficient. The microclimate and physiography in the immediate vicinity of the organism must be known as well. Traditionally, ecologists have preferred to study ecosystems from a macroscopic standpoint by attempting to describe the community structure, identify the species present, and understand the distribution and association of plants and animals, population dynamics, and the general interaction of climate with the plant and animal community. Other ecologists have been concerned with understanding the trophic levels within ecosystems and the flow of energy and nutrients among the various trophic levels. Each of these approaches is very necessary and worthwhile in its contribution to our understanding of ecosystems. Another approach is also required, however, in order that a better understanding of the detailed processes underlying the major events occurring within ecosystems is achieved. This is a reductionist approach to ecology; it involves understanding the detailed processes going on within ecosystems at the level of the individual organism, as well as organism-to-organism interactions. Much of this falls within the subdiscipline known as physiological ecology. Very often in physiological ecology, however, the physical aspects of ecology are not as rigorously treated as they might be. For this reason, and in order to give additional emphasis to the physics and biophysics of the subject matter, I have used the term biophysical ecology to describe the subject of this book. The term biophysical ecology is necessarily redundant because the study of ecology, by definition, includes the physics of the environment and the biophysics of physiological processes. Nevertheless, this is the term that best describes the subject matter to which this book is devoted.

A Reductionist Approach

The events occurring within an organism, and between an organism and the immediate environment, are fundamental to our understanding of ecological processes. The reductionist approach is predicated on the fact that the ecosystem is the sum and product of its parts, when all interactions are taken into account. Whether a particular approach to the study of ecology is considered reductionistic or holistic does not really matter. The important thing is to study ecology with full use of the skills, tools, analyses, and insights of modern science.

A detailed approach to ecosystem study naturally involves knowledge gained from cellular biology, molecular biology, physiology, biochemistry, anatomy, systematics, physics, geology, meteorology, climatology, and so on. The ecologist is the synthesizer of all this information as applied to description and understanding of the ecosystem. This does not mean the ecologist must be a physiologist, cellular biologist, or biochemist. In fact, the ecologist must be quite careful to avoid being diverted into another specialty since every student of ecology will find one or more of these other disciplines especially intriguing. In order to operate effectively as an ecologist, one must learn these other disciplines well yet avoid being diverted by them, all the while applying the principles of these disciplines to the problems of ecology. Since the ecologist is a synthesizer, and because understanding an ecosystem requires understanding biological as well as physical events, it is necessary that the ecologist also be well grounded in the subject matter of physics, chemistry, and mathematics. As a general rule, it is easier and more useful if the student is trained in these hard-core, analytical subjects early and then increases the amount of coursework in biology. The reverse procedure is generally less satisfactory, but this is a matter of personal choice.

Because of the complexity of the subject matter, our ability to understand cause and effect within ecosystems is severely limited and requires the application of various tools of modern science. The single most important new tool available to the ecologist today is the computer. By this means, we extend our capacity, not for logic, but for data storage and processing. Just as sensory perceptions can be extended through the use of electronic detectors, photographic film, recorders, and other means, so also can we extend our capacity for storage and handling of large amounts of data by the use of computers.

In order to achieve an understanding of ecosystems, the ecologist asks a variety of questions. If all life on earth is supported by primary productivity in the form of photosynthesis, how do plants of various sizes, shapes, and structures capture sunlight and grow in a variety of environments? Why are plants and animals distributed the way they are, and what regulates their distribution? If an ecosystem undergoes great disturbance from wind, fire, insects, or pathogens, there follows a series of successional stages in recovery toward the original state. What are the forces and factors that affect this plant and animal succession? How do plants and animals compete for essential materials and factors? For example, how does an aspen compete with a pine for sunshine, water, carbon dioxide, or nutrients? What regulates and determines which species occupy a given habitat at a specific time? What climatic and edaphic factors influence the behavior of an animal, and how do they do so? What determines the niche or habitat that a given plant or animal may occupy? These are each extremely complex questions and lead to even more specific and detailed problems.

It is possible to approach these questions in a purely phenomenological manner. It is also possible to take a much more reductionistic approach, considering as self-evident that any organism is coupled to its environment through an exchange of energy and matter and, furthermore, that such exchanges must obey the basic laws of physics and chemistry as expounded by modern science. Biophysical ecology is based on this premise and the conviction that many of the questions posed above can only be answered in detail by a reductionist approach. Other methods do, of course, add much useful information to the body of ecological knowledge, and for many aspects of ecology, descriptive and phenomenological approaches continue to be very useful.

Physical Factors

This book deals with the physical factors that characterize the environments of plants and animals and the way in which these physical factors interact with the large variety of organisms in the world. It does not, however, concern itself with every kind of biophysical interaction, only with some of the primary factors that characterize the world in which we live. Some of the obvious physical factors in our environment include gravitational, electric, and magnetic fields, electromagnetic radiation, fluids and solids, chemical elements and compounds, sonic fields, temperature, and fluid motion. Clearly, to deal with each of these in detail is too great an undertaking for one volume. I have, therefore, elected to treat energy and gas exchange as subjects of primary importance to the response of plants and animals in their habitats.

Some physical factors, such as the gravitational, electric, or magnetic fields, are omitted from this treatise not because they are insignificant, but because their effects are not of first-order magnitude with respect to the energy status or gas-exchange rates of organisms. We live in a gravitational field which varies extremely slowly with time or changing position on the earth. The gravitational field enters into some considerations but is not a dominant factor. Gravitation becomes an important environmental factor if an animal falls off a precipice and, in addition, affects the direction of growth for the seed hypocotyl and apical meristem, but these are not subjects with which we will be dealing. The ground surface and atmosphere are filled with massive electric fields and enormous surges of electric currents, but again, these phenomena are not of primary concern here. The sounds and noises that fill our environment, from the random noises of Brownian motion to the massive thunderclaps of lightning bolts is another topic that will not be discussed. Nor will topics purely internal to an organism be discussed unless they are a direct part of the process of energy and gas exchange between the organism and the environment.

A Multiplicity of Variables

The process of photosynthesis is fundamental to the growth and response of plants to environmental conditions. Many questions about adaptation, competition, succession, productivity, and other activities concerning plants are directly related to the process of photosynthesis. Plant photosynthesis and respiration respond directly to energy and gas exchange, which are in turn affected by certain environmental variables. Likewise, the metabolic activity of animals responds directly to various environmental factors. Radiation, air temperature, substrate temperature, wind speed, and humidity are all environmental factors that affect the exchange of energy. Carbon dioxide concentration, oxygen concentration, and humidity affect the exchange of gases. A plant or animal has many properties that allow it to respond to environmental factors with sensitivity or insensitivity. For example, the absorptance of a leaf or animal surface to solar radiation determines the degree to which it is warmed by sunshine and, for plants, the extent to which photosynthesis is carried on. The size of an organism directly influences the rate of exchange of energy and gases through the depth of the boundary layer of air adhering to the organism's surface. The size, shape, and orientation of an organism determine the degree to which the wind affects the temperature of the organism. The rates at which gases (water vapor, carbon dioxide, and oxygen) are exchanged depend upon the permeability or resistance of the plant or animal surface. Without going into more detail, it is easy to see that the ecology of a plant or animal may be directly involved with eight or more independent environmental variables and a half-dozen or more organism parameters. Important dependent variables are leaf temperature and transpiration photosynthetic, and respiration rates for plants and body temperature and metabolic rate for animals.

The problem faced by the ecologist in understanding the interaction or organisms with their environment is the problem of too many variables. For this reason, it is crucial to recognize which variables are of primary importance and not include more than are necessary. It is also because of too many variables that it is essential for ecology to become as analytical as possible. This is one of the most difficult points to understand. The large number of variables leads many people to believe that it is hopeless to attempt physical and mathematical analysis of such complex problems. But actually, the contrary is true: analysis must be used. There are two important reasons for this. The first is the fact that, during the last 15 years, analytical techniques have been successful in yielding answers to many previously unanswered problems. The second reason is that modern science can contribute very much more to the study of ecology than it has to date. Although systems ecology has given us some very significant advances, there is still a serious lack of understanding concerning mechanisms by which plants and animals interact with their environments. Relatively few persons have really addressed the problems of ecology using techniques commensurate with those of modern physics, chemistry, engineering, mathematics, and other analytical disciplines. The limitation is not one of technique or equipment but is due rather to a lack of understanding of what must be done.

Organization of the Book

This book is organized to give the reader an overall view of biophysical ecology, together with examples of its application. In order to show the necessity for the more-detailed treatment given in subsequent chapters, the introductory chapters contain simplified examples of the application of biophysical theory. Following the introductory chapters is a chapter concerning the fundamental laws of energy and radiation, including a thorough discussion of units. Radiation is the most ubiquitous environmental factor of the world in which we live, and all terrestrial organisms are subjected to radiation fields throughout their lives. (Aquatic organisms, by contrast, live in greatly attenuated radiation fields and often at night are free of radiation entirely.) Because of the enormous complexity of the topic of radiation, several chapters are required for its complete discussion. One chapter is devoted to solar radiation, one to thermal and total radiation, and one to the spectral properties of radiation and organisms. Much of the material presented in these chapters is not found in any other book, and an attempt is made to be as complete as possible.

Aquatic organisms have their energy status strongly coupled to the ambient temperature of their environment by means of the thermal conductivity of water. Terrestrial organisms are loosely coupled to air temperature by conduction and convection because of the very poor thermal conductivity of air. The subject of convective heat transfer has been thoroughly worked out by engineers, but application of these principles to biological problems is relatively recent. Those aspects of the subject important to problems of air or water flow around plants and animals are described in Chapter 9. All terrestrial organisms lose water vapor to some degree, and for many organisms, this is an important part of the energy budget. The diffusion of water vapor from an organism to the air is described in Chapter 10. The topics of radiation, convection, and evaporation are treated in the next two chapters, which give a thorough description of energy budgets for plants and animals. Chapter 13 deals with the time-dependent energetics of animals, and Chapter 14 treats the subject of photosynthesis.

The energy status of an organism determines its temperature, and whether or not an organism is too warm or too cold is important to its viability. The ability of organisms to undergo cold or heat resistance determines survival during temperature extremes. For these reasons, and to give the reader an overall view of the subject, a large chapter on the temperature tolerances of organisms (Chapter 15) is included.

Photosynthesis

One objective of this book is to understand as thoroughly as possible how a plant grows in response to environmental factors. Since plant growth is largely related to photosynthesis and respiration, it is important to understand how photosynthesis and respiration respond to environmental factors. A great deal is known about the biochemistry of photosynthesis and respiration, the structure of higher plants, the location of chloroplasts, peroxisomes, and mitochondria within the cell, and the behavior of stomates. Until recently, however, relatively little has been done to make an analytical model describing how the entire system works. A plant leaf exchanges energy and, as a result, has a certain temperature and loses a certain amount of water. While water vapor diffuses outward, carbon dioxide diffuses into the leaf. At the chloroplast, the carbon dioxide is oxidized and assimilated. Oxygen gas is released during photosynthesis and diffuses out from the leaf. The rate of photosynthesis is dependent on temperature, light, carbon dioxide concentration, oxygen concentration, and other factors such as the water status of the leaf. Leaf temperature is determined by the energy budget, and light intensity, which drives the photochemistry, is a part of the radiation term in the energy budget of the leaf. The rate of diffusion of water vapor outward is directly related to the diffusion of carbon dioxide inward, which in turn is commensurate with the rate of consumption in photosynthesis and the rate of release through respiration. All of these events and processes can be described by analytical techniques involving plant parameters such as leaf size, shape, orientation, diffusion resistances, and absorptance and emissivity, as well as the various reaction-rate coefficients that characterize the particular plant species. These ideas are brought together in the chapter concerning photosynthesis.

A plant is a complex biological unit, and the total array of mechanisms by which it works is complicated. Nevertheless, if we are to understand how a plant leaf functions in its habitat, we must formulate the complete analysis as one unit. By the application of mathematics to the description of the physical processes of energy exchange, gas exchange, and chemical kinetics, a complete analytical expression for leaf productivity can be achieved. The steps in this process are presented in Chapter 14. The analytical model may not be complete or fully refined as yet; certainly, it has not been thoroughly tested. The methodology which must be used is clear, however. The purpose of this book is to lead the way to the frontiers of knowledge in biophysical ecology.

Animal Energetics

The energetics of animals is fundamental to their behavior, adaptation, growth, reproduction, and distribution. Ecologists and physiologists have long been interested in the energy relations of animals. However, because of the complexity of natural environments and the diverse shapes, sizes, surfaces, and other features of animals, it has been difficult to produce a complete understanding of energy exchange between an animal and its environment. By making certain simplifying assumptions concerning the geometry of animals, however, it is possible to approach the problem of animal energy budgets in a systematic manner. Recently, many important advances have been made. These new analytical models give us much better procedures for collecting field data and applying these data to animal energetics. The enormous space and time variability of radiation, temperature, wind, and humidity in microenvironments in the vicinity of animals necessitates unique averaging techniques and improved measurement methods. One means of doing this involves the use of inanimate animal models of the same size, shape, and surface characteristics as the living animal. An "environmental" temperature is obtained from the model and represents the sum of radiation and convection effects on the animal. The living animal's true temperature is then equal to the "environmental" temperature plus a physiological offset temperature produced by metabolism, evaporative cooling, and other physiological processes. Using this technique, one can work from the animal outward to the environment or from the environment inward to its effect on the living animal. It has been possible to predict the climates within which a specific animal lives from analysis of its energy budget. It is also possible to predict for an animal with certain characteristics just where it is likely to be in its habitat as a function of time of day or time of year. In addition, it is possible to predict some predator–prey strategies as a result of a thorough analysis of the energy status of various animals. Once again, because of the large number of variables and parameters involved in the relationship of an animal to its environment, the necessity of using an analytical method based on sound physical and biological principles is demonstrated. These are difficult problems, and many advances are yet to be made. However, our purpose here is to review the present state of our knowledge—knowledge upon which others can build a better understanding of animal ecology.

In the process of describing animal energy budgets, many pieces of physiological information are used. It is not intended here to review the vast amount of physiological information available or to critique its validity, accuracy, or usefulness. However, lack of appropriate analytical models has allowed the accumulation, in scientific literature, of a considerable amount of incorrect, incoherent, and generally useless data. It is hoped that, as better analytical models are produced, improved perspective will be gained as to the type of physiological and ecological information required, whether from laboratory or field measurements. Theory must develop hand-in-glove with experiment and observation. Without theory, laboratory measurements and field observations run the risk of being adrift in a morass of variables without a "compass" to give proper direction. It is the purpose of biophysical ecology to develop procedures for placing theory, experiment, and observation into a coherent framework. If we fall short of this goal, it is because the science of biophysical ecology is young and many significant advances are yet to be made.

Microclimates

The environments in which plants and animals live are diverse and involve a broad continuum of microclimates. It is the climate in the immediate vicinity of a plant or animal that is of primary importance to the organism. Environmental and climatic conditions often change enormously near the surface of the ground, with changes in topography and in radiation, temperature, wind, and precipitation. The manner in which wind, temperature, and humidity change with height in the ground boundary layer is of importance to organisms. Both plants and animals respond with considerable sensitivity to the height profiles of these factors. The exchange of energy and gases between the vegetated surface and the atmosphere above it is a complex process that requires special analytical treatment by aerodynamic methods. These analytical methods have been described by Monteith (1975, 1976) and Rosenberg (1974). An excellent description of microclimates has been provided by Geiger (1965).

Knowledge of Ecosystems

The study of ecosystems and their physical and biological components is a challenging subject which requires all the tools and disciplines of modern science. Ecology needs all the methodology that the ingenuity of the human mind can direct to it. I have always been impressed with the enormous amount of ecological information that ecologists store in their minds. Their depth and breadth of understanding of ecosystem structures and processes is truly amazing. Yet, despite many advances in recent years, lack of analysis is of great concern when thinking of what has been done and what is potentially possible. The International Biological Program (IBP) was a great stimulus and brought many advances to the field of ecology, including the extremely useful methodology of systems ecology. However, the advances that might have been made were limited by lack of adequate descriptions of processes and their mechanisms at the physiological level. On the other hand, the IBP gave us an indication of the need for the kind of analysis provided by the field of biophysical ecology.

Chapter 2

Energy and Energy Budgets

Introduction

Organisms are comprised of cells, which are highly organized groupings of complex molecules. To maintain the degree of organization making up life, energy of the right magnitude is necessary. One learns from thermodynamics that the natural trend of the universe is to go toward increased entropy, i.e., from a degree of order toward more disorder. Life on earth runs contrary to this general inexorable trend, but it does so at the expense of the sun.

The sun sends us the high-energy quanta of radiation which are essential for the formation of molecules. The entire process of primary productivity, by which green plants and some bacteria fix carbon dioxide into carbohydrates and form high-energy phosphate bonds, is driven by high-energy photons streaming down upon the earth as sunlight. In this way, more order is created from less order to form life. However, since this is an open system involving the sun as a source and the cosmic cold of space as a sink, it does not violate the principle of increasing entropy. The sun itself is evolving toward a cooler, less organized state. Of all the energy the sun emits into interplanetary space, the earth intercepts only a very small fraction. Of this fraction, less than a few percent is captured to sustain life.

The sun emits gamma radiation, x-rays, ultraviolet radiation, visible light, infrared radiation, radio waves, and, in fact, the entire electromagnetic spectrum of radiation. High-frequency, extremely energetic radiation such as gamma rays, x-rays, and ultraviolet light is destructive to life. Conversely, low-frequency, low-energy quanta of the infrared and radio spectrums are not sufficiently energizing to build molecular bonds. Only a narrow spectrum of visible light, from the ultraviolet to the infrared, has quanta sufficiently energetic to generate life and not so overly energetic as to destroy it. All solar radiation absorbed at the earth's surface that is not utilized in a photochemical manner is converted to heat. This is the heat which bathes the earth in warmth and makes for an ameliorated, hospitable climate, a climate in which the temperature is neither too warm nor too cold for life to thrive. All radiation absorbed by the planet is eventually returned to space as heat.

Most organisms have a wide range of temperature tolerance. The temperature of an organism is a manifestation of its energy status. An organism exchanges energy with its environment through processes involving radiation, convection, conduction, and evaporation, as well as by mass exchange of nutrients, carbon dioxide, and other materials.

Our first task is to understand the energy budget of organisms and the processes by which energy is exchanged between an organism and its environment.

Organism Temperatures

All of life is immersed in a medium in which thermal energy is constantly ebbing and flowing. Energy flows from regions of higher temperature to regions of lower temperature. For this reason, energy is always escaping from the warm earth to the cold of cosmic space. All chemical reactions are to a greater or lesser extent temperature dependent, and all physiological processes are a function of temperature. Usually, when temperatures are low, physiological processes and chemical events are slowed and when temperatures are high, they are speeded up. When the temperature is above a certain level, the bonds forming molecules and cells begin to break down. All organisms have temperature tolerances that are more or less limited. Active physiological functions occur only when temperatures are between $-2°C$ and $100°C$, largely because of the ubiquitous role that water plays in metabolic processes. For many organisms, temperatures must be between $0°C$ and about $50°C$ for life processes to function. Some organisms, such as a few bacteria, can survive temperatures down to absolute zero or up to as high as $100°C$. Protein stability, fluidity of fats, and permeability of membranes are strongly temperature dependent; the temperature optima, minima, and maxima for active life are determined, in part, by these important properties. The general temperature norm in the world is between $15°$ and $20°C$. Departures from this temperature range cause increasing biological stress as they become more extreme.

Figure 2.1 shows the body temperatures normally tolerated by several different kinds of organisms. Only in a few cases do these ranges apply to a specific animal species, in most instances, the range represents the

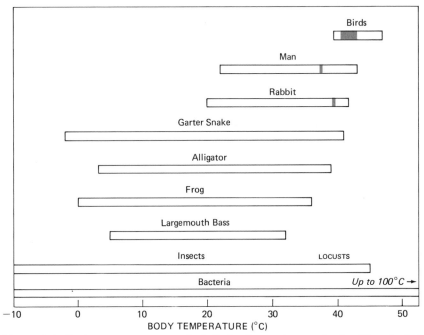

Figure 2.1. Body-temperature survival limits for various animals. The shaded areas represent the normal range of body temperature.

limits which that animal group can tolerate and remain viable. Very small organisms will always have body temperatures very close to the temperature of the air or water in which they are located. This is because the energy content of small organisms, as well as all small inert objects, is strongly coupled by convection to the temperature of the fluid media. But for larger organisms, body temperature may be very different from the environmental temperature. Cold-blooded animals, or *poikilotherms,* whose metabolic energy is not very great, have body temperatures quite close to the temperature of the air or water in which they are immersed, unless they are in a strong radiation field such as sunshine. In strong sunshine, poikilotherms on land may have body temperatures 10°C or more higher than air temperature. Warm-blooded animals, or *homeotherms,* have body temperatures that are regulated to remain within fairly narrow limits through metabolic and behavioral control. The air or water temperatures nearby these animals may be very much lower than the animal's temperature or very much higher. Animals that receive most of their body warmth as energy directly from the environment are called *ectotherms* and are usually poikilothermic. Those animals which generate most of their body warmth from metabolic heat are called *endotherms* and are usually homeothermic. Ectotherms, such as insects and reptiles, generally have very little body insulation. Endotherms, such as birds and mammals, have fur or feathers as insulation.

Plants generally have temperatures close to the air temperature, except in bright sunshine, when leaf temperatures may run 10° to 20°C above air temperature. The temperatures of large succulents such as cacti and euphorbia often depart markedly from air temperature. This is also true of the interior of tree trunks. The larger the trunk, the greater the departure from environmental temperatures owing to a time delay in heat flow to or from the external environment. Often, the departure of a plant or animal temperature from the environmental temperature is extremely significant physiologically. Just as most animals can control their body temperatures by means of behavioral maneuvers, so also do many plants control their leaf, flower, or sometimes even trunk temperature by orientation of plant parts. Generally, in plants, metabolic rates have little or no influence on the temperature of the organism. There are exceptions, however, such as the spathes of arums, wherein high metabolic activity for short periods may generate considerable internal heat.

Microclimate

Climate may be defined as characteristic weather conditions for a given place or region averaged over an extended period of time. Often, when we speak of climatic conditions, we are really speaking of the weather, for we are dealing with the events occurring moment by moment. In any event, it is clear that a plant or animal is affected most directly by environmental conditions close by it. This climate near the organism is defined as the *microclimate*. Rudolph Geiger (1965), the famous German climatologist, defined microclimate as ''the climate near the ground.'' The climate of a region, of course, has a relationship to the microclimate of each and every habitat within the region. Yet we must always keep in mind that, as far as extreme values are concerned, the microclimate of a position in a valley may differ dramatically from the microclimate of a position nearby on a ridge even though the regional climate is supposedly representative of both sites. Aggregates of vegetation taken as a whole, as for example a deciduous hardwood forest, may appear to respond to the regional climate. The productivity and ecology of the forest, however, when looked at in any detail whatsoever, is responding, clearly, to the microclimate in the immediate vicinity of each and every part of the forest.

The Energy Environment

Energy is always flowing within the physical environment. Radiant energy comes to the earth from the sun and escapes from the earth to the cosmic cold of interplanetary and galactic space. This flow of energy from the solar source to the cosmic sink is the dynamic process that drives the ecosystem earth. Organisms are immersed in this ubiquitous flow of energy. Within a given physical environment, there may be a zero net flow of energy; nevertheless, flow takes place. Flow of energy may occur by radiation, convection, conduction, chemical reaction, or the actual mechanical transfer of matter. Different environments have vastly different characteristics in terms of the flow of energy within them.

An organism buried in the soil exists in an environment characterized by flow of energy through conduction and chemical transformations. An organism immersed at depths in water where there is no light undergoes energy transferral through conduction, convection, and chemical mechanisms. An organism near the surface of a lake, where it gets light, exchanges energy through radiation, as well as by conduction, convection, and chemical mechanisms. An organism in a cave or room will have energy exchanged primarily through radiation and to a lesser extent by convection or evaporation, as well as by chemical means. Radiation emitted by the walls of the cave or room is received by the organism within, and the organism, in turn, radiates energy from its surface to the walls. An organism out of doors may receive direct sunlight, skylight, reflected light, and radiant heat from the surrounding surfaces; it will emit radiation and exchange energy by convection, conduction, and chemical reactions.

The climate impinging upon a plant is shown in Fig. 2.2. Similar streams of energy also exist for an animal in the same environment. A plant receives direct sunlight coming from the sun, scattered skylight coming from the sky, sunlight reflecting from clouds, the ground surface, and other objects, thermal radiation streaming downward from the atmosphere hemisphere, and thermal radiation streaming upward from the surface of the ground encompassing the lower hemisphere. Of the energy absorbed by the plant, some of it will be radiated from its surface. Energy is transferred to or from the plant by convection if the plant surface temperature of the plant is different from air temperature. If the air is cooler than the surface of the plant, energy from the plant will be lost to the air by convection. If the air is warmer than the surface of the plant, energy will be delivered to the plant by convection. Energy is lost from the plant by the evaporation of moisture. Plants have the ability to transpire, just as many animals sweat when the heat load on their surface becomes too high. Transpiration and sweating are evaporative cooling processes which provide an effective means for an organism to keep its temperature from rising for limited periods of time. In addition to the streams of energy shown in Fig. 2.2, additional energy is available through metabolism in the form of chemical conversion. If an organism is resting on the surface of the soil or on rock or another substrate, it will have some energy transferred to or from its body by conduction. If the underlying substrate surface is cooler than the organism, heat will be conducted to the substrate. If the substrate is warmer than the organism, heat will be conducted to the organism.

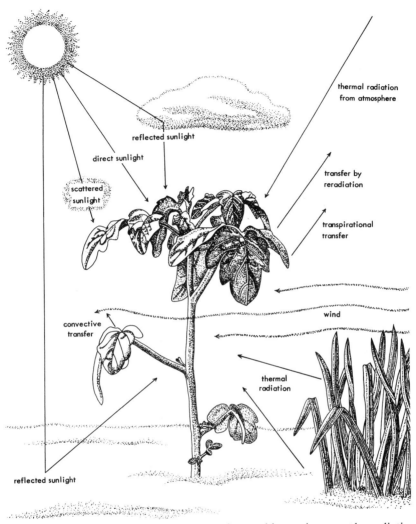

Figure 2.2. Exchange of energy between a plant and its environment by radiation, convection, and transpiration.

Plants are rooted to their environment and cannot escape the searing sun of midday or the cold of night. The environment about the plant may change from moment to moment, often subjecting the plant to enormous extremes of climate and other variations in conditions. Animals can move about to seek a suitable habitat and escape temperature extremes. In either case, plant or animal, the question as to whether the climate changes about the plant or the animal moves within the climate is a relative one. It is our purpose here to understand precisely the manner by which an organism is coupled to the climate around it.

Coupling between Organism and Environment

It is primarily through the flow of energy that climate affects an organism. If the organism is strongly coupled to a given climatic factor, then the organism temperature is strongly affected by this factor. If the organism is weakly coupled to the environmental factor, then the organism temperature and energy content is little affected by this factor.

A plant or animal is coupled to incident streams of radiation by means of the absorptivity of the organism's surface. If the organism has an absorptivity of zero, in other words, if it reflects 100% of the incident radiation, it will be completely decoupled from the incident radiation, and its temperature will not be affected in the least by radiation. If, on the other hand, the organism is black and absorbs 100% of the incident radiation, it will be strongly coupled to the incident flux. In this case, the organism's temperature is intimately affected by the strength of the incident radiation. The absorptance of the organism's surface is often a complicated function of the wavelength, or frequency, of the incident radiation and the geometry of the organism. The variation in surfaces absorptance among various organisms is enormous. Almost every species of animal or plant has a different absorptance as a function of the wavelength. Different organisms receiving the same flux of incident radiation will absorb different amounts of radiation and have, as a result, different surface temperatures.

Energy flows between an organism and the air in which it is immersed by the process of convection, and the rate of this flow is proportional to the temperature difference between the organism and the air. The proportionality constant is the convection coefficient. It depends on the size, shape, and orientation of the organism, as well as the fluid properties of the air. The coupling of plant or animal temperature to the air temperature is expressed by the convection coefficient. A small object has a large convection coefficient, and its temperature is tightly coupled to the temperature of the air or water surrounding it. A large organism possesses a thick boundary layer of adhering fluid, and its temperature tends to be decoupled from the temperature of the surrounding medium.

An organism's temperature is coupled to ground temperature when the organism is in contact with the ground. If the substance in contact with the ground is of high conductivity, then the temperature of the organism will be strongly coupled to the temperature of the contact media. If, however, the material between the body of the organism and the underlying surface is of poor conductivity, then the organism's temperature will be decoupled from the temperature of the underlying media.

The temperature of an organism is also affected by the amount of moisture evaporating from it. Through evaporative loss of moisture, an organism's temperature is coupled to the vapor pressure or relative humidity of the moisture in the air around it. If the organism's skin is impervious to moisture loss, then the organism's temperature is completely decoupled from the vapor pressure.

In addition to these major factors determining the flow of energy between the organism and the environment, there is also the matter of gas exchange, which is significant from a physiological standpoint. One can speak of the degree of coupling between the organism and the environment in terms of gas exchange; this will depend upon the permeability of the organism's surface to oxygen, carbon dioxide, nitrogen, or other gases.

Climate Factors

Climatic factors significant to an organism are radiation, air temperature, wind, relative humidity, and vapor pressure. These climatic factors always act simultaneously and vary with time. The influence of any one climatic factor on an organism can only be understood in the context of all the other, simultaneously acting factors. If the organism's temperature is considered as the dependent variable and the climatic factors are considered as independent variables, it is clear that we are dealing with what, in its most simple form, is a five-dimensional space. It is the simultaneous action of all the climatic factors that makes the study of an organism's interactions with the environment so difficult. Unfortunately, ecologists have by tradition tended to ignore the simultaneity of the factors and attempted to search for cause and effect using only one or two variables at a time. This has been the reason that results have at times been inconclusive and conflicting. It does not make sense to say that an organism will get warm in sunshine unless one specifies, at the same time, the air temperature, wind speed, and humidity of the air. It is not possible to say that the air temperature is too high to maintain acceptable body temperature unless one also specifies, conditions of radiation, wind, and humidity. The character of these simultaneous climatic factors and the way in which any one influences an organism's temperature, as understood in the context of all the others, will be demonstrated in later chapters.

Energy Budgets

The temperature of an animal or plant must be more or less stable with time. Many organisms can raise or lower their temperatures for limited periods of time. It is clear that an organism cannot have its temperature rise indefinitely or it will become too warm and perish. On the other hand, an animal cannot cool indefinitely or it will become too cold and perish. The temperature of an organism adjusts under given environmental conditions until the energy input is equal to the energy output; that is,

$$M + Q_a = \mathcal{R} + C + \lambda E + G + X, \tag{2.1}$$

where M is rate at which metabolic energy is generated, Q_a is amount of radiation absorbed by the surface of the organism, $\mathcal{R}$ is radiation emitted by the surface of the organism, C is energy transferred by convection, λE is energy exchanged by evaporation or condensation of moisture, G is energy exchanged by conduction (through direct physical contact of the organism with soil, water, or substrate), and X is energy put into or taken out of storage within the organism.

   Equation (2.1), which is written in general terms representing all forms of energy transfer available to an organism, may now be written in specific terms involving the significant dependent and independent variables of the situation. Since the surface of the organism is the transducing surface across which all energy flow takes place, we will, for the time being, describe the energy flow with surface temperature as the dependent variable.

Units

Since this is the first time we have encountered the system of units to be used in this book, it requires a brief explanation. Generally, the International System of Units (ISU) will be used, and energy budgets will be given in units of watts per square meter. By tradition, meteorologists and others who have worked in the science of radiation, particularly solar radiation, have used calories per square centimeter per minute. One calorie per square centimeter is one langley, a unit named in honor of the great physicist Samuel P. Langley. Flux density of radiation is given in langleys per minute. In much of the scientific literature, the majority of graphs and tables involving radiation use units of calories per square centimeter per minute. The conversion from one system of units to another is easy, and these factors are given in Appendix 1. Most of non-U.S. scientists have adopted the ISU. We will modify the system for certain dimensions where centimeters are more convenient than meters.

Radiation

All surfaces radiate heat in proportion to the fourth power of the absolute temperature T_s of the surface and according to the emissivity ϵ of the surface. The radiation emitted is given by

$$\mathcal{R} = \epsilon\sigma[T_s + 273]^4, \tag{2.2}$$

where σ is the Stefan-Boltzmann radiation constant, whose value is

5.673×10^{-8} W m^{-2}°K^{-4}. If T_s is in degrees Centigrade, then $T_s + 273$ represents the absolute temperature in degrees Kelvin, the units that must be used for calculation of "blackbody" or "graybody" radiation.

Convection

The rate of heat transfer by convection between an organism and the air or water around it is proportional to the temperature difference between the organism's surface and the fluid. This proportionality constant is called the convection coefficient h_c. The rate of heat transfer by convection is

$$C = h_c [T_s - T_a],\qquad(2.3)$$

where T_s is surface temperature and T_a is air temperature.

The convection coefficient is a complex function of wind speed or rate of fluid flow, properties of the fluid, and characteristics of the surface over which the flow occurs. In order to keep things reasonably simple in this introductory analysis, we shall express the rate of heat transfer by convection as proportional to some power n of the wind speed V and inversely proportional to some power m of the characteristic dimension D of the organism surface. The larger the diameter or width of the organism in the direction of the fluid flow, the thicker the adhering boundary layer of fluid on the surface across which heat transfer occurs. The greater the thickness of the boundary layer, the slower the rate of heat transfer. Therefore, the rate of heat transfer is inversely proportional to the characteristic dimension in the direction of air flow, which is approximately the diameter of the animal or the width of the leaf. A detailed discussion of boundary layers and fluid flow is given in Chapter 9. The rate of heat transfer by convection now can be written as

$$C = k_1 \frac{V^n}{D^m} [T_s - T_a],\qquad(2.4)$$

where k_1 is the proportionality constant. Specific values for k_1 will be given when examples of fluid flow around certain organisms are discussed.

Evaporation

The evaporation of water from an organism, whether by respiration, sweating, or transpiration, results in the loss of energy from the organism of approximately 2.430×10^6 J kg^{-1} at 30°C. This is the latent heat of vaporization of water, designated by the symbol λ, a quantity which is a function of the temperature at which vaporization takes place. The rate at which water is lost from a plant or animal by evaporation is directly proportional to the difference between vapor pressure of water vapor within the organism and that in the free air beyond the boundary layer. The vapor pressure of water vapor in the air is a function of air temperature and relative humidity h. Water-vapor pressure in an organism is a function of the temperature at the site of vaporization, as well as solute concentration and water tension. However, for simplicity, we shall write the vapor pressure in the organism as a function of temperature only. Therefore, the rate of heat transfer by vaporization is a function of the organism's temperature, air temperature, and relative humidity. The heat required to convert liquid water to vapor, the latent heat of vaporization λ, is a function of the surface temperature T_s. Furthermore, the rate at which water vapor diffuses from a plant or animal surface or is expelled by breathing is inversely proportional to the resistance r offered by the pathway. Hence we can write

$$\lambda E = \lambda(T_s)E\ (T_s,\ T_a,\ h,\ r^{-1}), \tag{2.5}$$

where E is the amount of water lost per unit surface area per unit time. For the time being, we will not write this relationship in a more precise functional form, but only note that the rate of vaporization of water depends upon these variables.

Detailed Energy Budget

The total heat budget for an organism is written in the following generalized form:

$$M + Q_a = \epsilon\sigma[T_s + 273]^4 + k_1 \frac{V^n}{D^m} [T_s - T_a]$$
$$+ \lambda(T_s)E\ (T_s,\ T_a,\ h,\ r^{-1}) + G + X, \tag{2.6}$$

with all quantities expressed in watts per square meter.

For any set of environmental factors (i.e., for any given set of values of Q_a, T_a, V, and G and for a given set of values of M, E, and X), the organism has a surface temperature T_s as determined by the energy budget. The mechanisms of what happens are a bit complicated but will become clearer when specific applications to plant leaves or animals are discussed. The proportionality constant in the convection term depends upon the size, shape, orientation, and roughness of the surface of the organism. The conduction term depends upon whether the organism is in contact with the ground surface or other material and the conductance of the material, such as fur or fat, in contact with the ground. Loss of moisture depends upon the availability of moisture and the ability of the organism to transpire or sweat. Some moisture may be lost through respiration or breathing. Many organisms do not have the ability to lose a significant amount of moisture, whereas others can vary the amount of moisture loss and thus regulate evaporative cooling. Some organisms can control their body temperature by changing their color, e.g., by adapting their surface absorptance to the incident radiation, thereby changing the value of Q_a. Some organisms can change their size by spreading out or unfolding portions of their body, in this way presenting a different area to incident radiation or a different body size to convective air flow. Some organisms can change their metabolic rate and adjust their energy budget in order to keep their body temperature within bounds. Birds and mammals can modify the amount of insulation of fur or feathers by means of pilar erection, as well as by seasonal changes in growth of down or fur.

Energy Budget Examples

Figure 2.3 shows the approximate energy loss in percentage for various plant leaves and animals. It is seen that, in all cases, radiant energy loss is the most significant factor. Evaporative cooling plays an inconsequential role in the case of the cockroach and the lizard and a small role in the case of the sheep. Evaporative cooling plays a very significant role for a salamander and a large leaf. Desert plants have small leaves with high resistances to water loss. Evaporative cooling is a relatively insignificant factor here. The role of convection and transpiration in the energy budget of a leaf is always considerably smaller than the role of radiation; losses due to convection and transpiration may be equal or one may exceed the other. The relative amounts of convective, evaporative, and radiant heat exchange between an organism and its environment depend very much on the amount of incident radiation. The relative amounts of energy exchange shown in Fig. 2.3 should not be considered as absolute, but only as roughly illustrative of what may occur in nature. The temperature of an organism is determined by the total energy input and how the organism dispenses it. If energy input is high, temperature of the organism is high. If energy input is low, temperature of the organism is low.

If an organism finds itself in an environment into which energy flow is high and its temperature exceeds the tolerance limit, then the organism will perish unless it moves to another environment. Most animals can migrate when they need to seek a more comfortable, compatible habitat.

In contrast to animals, plants are rooted to their environment and must use means other than moving out of the sun to stay cool. Plants can wilt, thereby changing the angle of the leaves presented to the sun. Many plant species can have their leaves stand vertical in sunshine, so that the leaves absorb less sunshine and remain cooler than they would if horizontal. Except for small adjustments, such as those resulting from wilting, a plant must conform to the temperatures established by energy flow.

By knowing the coefficients of coupling of an organism and its environment, it is possible to predict with precision the organism's temperature. Not only are temperature limits important to an organism, but most physiological and metabolic processes within organisms are temperature dependent. Usually, physiological processes are slowed down at low tem-

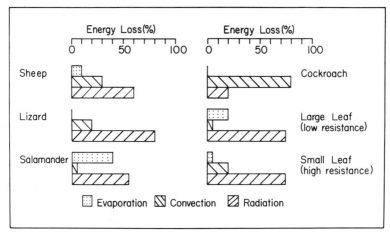

Figure 2.3. Approximate partitioning of energy loss from various organisms by evaporation, convection, and radiation.

peratures, reach a maximum rate at intermediate temperatures, and break down at high temperatures. With plants, for example, the metabolic process of photosynthesis depends upon carbon dioxide and oxygen concentrations, light intensity, and plant temperature. (Within a community of plants, the vital ecological strategies of adaptation, competition, and succession are decided by the enzyme-mediated biochemical reactions within the plant cells.) However, energy must be available to a plant in the proper amount and form, exchange of carbon dioxide, oxygen, and water vapor must occur, nutrients must flow in the stream of sap from roots to leaves, and all the necessary chemical reactions must take place. The entire system must function in a self-consistent, self-compatible manner for the plant to succeed in its habitat.

It is hoped that the methods developed here will advance our understanding of primary productivity and ecological functions. By means of energy budgets, prediction is made of the climates animals are constrained to live in because of their inherent thermodynamic properties. One can also understand the daily and seasonal behavior of animals from an analysis of their energy budgets and, for certain species, such things as altitudinal zonation. Analysis of energy exchange for animals gives us clues to the physiological advantages of warm-blooded animals and the limitations imposed by body size on all kinds of animals.

The purpose of these first four chapters is to give a quick overview of the subject without becoming enmeshed in the detail of the specific topics taken up in subsequent chapters. Chapter 3 describes the processes by which plants are coupled to their environment and the mechanisms utilized for primary productivity, and the energy budgets of animals and how the way in which animals interact with the environment through physical mechanisms and physiological processes are discussed in Chapter 4.

Chapter 3

Application to Plants

Introduction

One must understand, in detail, the functioning of a leaf in order to understand many ecological phenomena concerning plants. Once the mechanisms by which a leaf carries out its vital functions are recognized, one can put together a complete analysis or model relating the properties of the environment to the vital functions of a leaf. Since photosynthesis and respiration are of primary importance to a plant, we shall first look at how these processes take place.

Photosynthesis requires light, carbon dioxide, and a suitable temperature. The temperature of a leaf is the result of energy exchange between the leaf and the environment. The energy exchange involves radiation, and a component of the incoming radiation is the light which drives the photochemical reactions. Exchange of energy is affected by the diffusion of water vapor from the leaf mesophyll out through the stomates to the free air beyond the leaf's adhering boundary layer. Carbon dioxide is supplied to the chloroplasts within the leaf mesophyll by diffusion inward through the stomates, cell walls, and cytoplasm. Once carbon dioxide arrives at the chloroplasts, it must undergo a chemical reaction in the process of photosynthesis compatible with its rate of diffusion to the chloroplasts. These chemical reactions are regulated by light, temperature, and the concentration of carbon dioxide, as well as by many other factors.

Temperature is important to a leaf not only because of its influence on the rates of photosynthesis, respiration, and transpiration, but also because of its effect on the cytological state of plant cells. Protoplasmic streaming and the stability of plant proteins are temperature dependent. Plants must have the ability to withstand the high summer temperatures that may occur during the growing season and to withstand limited amounts of cold at the same time. In one form or another, either as seeds, roots, bulbs, or as a whole, plants must survive winter temperatures that may be very low. Depending upon where the plants are growing, whether in boreal, temperate, desert, or tropical habitats, extreme temperatures may be high or low, or both. Of course, the matter of high or low temperatures is relative. What is low for one plant may be high for another. Evergreens in temperate and boreal regions must be adapted to intense periods of cold, yet must also be able to become active and photosynthetically productive when daytime conditions of sunshine and air temperature warm the plant to temperatures above −10°C. It has been established that photosynthesis occurs at temperatures above 0°C, but there is considerable evidence that photosynthesis will proceed even at lower temperatures. Furthermore, in the spring of the year, most plants generate new leaves; these leaves must withstand brief periods of cold, respond quickly to the presence of warmth and sunshine, and may also have to survive becoming overheated. It is evident that the temperature of a leaf is vital to its survival, its well-being, and its ability to function physiologically. Basic to the process of photosynthesis in a whole leaf is energy exchange. Our first task is to show how energy exchange regulates leaf temperatures, which are in turn extremely important to the vital processes within the leaf.

Energy Budget of a Leaf

The energy budget of a leaf can be written in a somewhat simpler form than the energy budget of an organism. This is because metabolism in plants consumes so little energy that it is negligible when evaluating the temperature of a leaf. Put another way, it is not possible by means of readily available measurements to show that photosynthesis or respiration has a detectable effect on leaf temperature. There are, however, a few exceptions; one example is the substantial increase in temperature, resulting from high respiration rates, exhibited for relatively short periods by the spathes of *Arum*.

A broad, deciduous type of plant leaf in air can be represented by a thin, flat plate which has a certain absorptance to radiation, a characteristic dimension, and a specific diffusion resistance to water vapor. Detailed discussions of these properties may be found in the chapters on radiation, convection, and transpiration. A plant leaf, which is attached to the plant by the petiole, has no significant amount of energy transferred to or from it by conduction. For steady-state situations, the energy storage term is zero. A plant leaf absorbs a certain fraction of the incident radiation and partitions this energy into three outgoing streams: reradiation, convective heat exchange with the air, and evaporation of water, or transpiration. The radiation emitted by the leaf is proportional to the fourth power of the absolute temperature of the leaf. The emissivity ϵ of a leaf is approximately 1.0, or, to be more exact, is usually about 0.96 and may be as low as 0.92. The rate at which a broad leaf (considered as a flat plate) exchanges heat with the air is proportional to the square root of the wind speed V and the temperature difference between the leaf and the air $(T_\ell - T_a)$ and is inversely proportional to the square root of the width D of a leaf in the direction of the air flow, (approximately the characteristic dimension of the leaf). If the leaf absorbs an amount of radiation Q_a, expressed in watts per square meter, and dissipates this energy by radiation, convection, and transpiration, one can write, for steady-state conditions,

$$Q_a = \epsilon\sigma(T_\ell + 273)^4 + k_1 \frac{V^{1/2}}{D^{1/2}} (T_\ell - T_a) + \lambda E, \tag{3.1}$$

where $\sigma = 5.67 \times 10^{-8}$ W m^{-2} °K^{-4}, $k_1 = 9.14$ J m^{-2} s$^{-1/2}$ °C^{-1}, V is in meters per second, D is in meters, E is the rate of transpiration in kilograms per square meter, and λ is the latent heat of vaporization, a quantity which is a function of the temperature of the liquid water. The value of λ is approximately 2.430×10^6 J kg^{-1} at 30°C and 2.501×10^6 J kg^{-1} at 0°C. The transpiration rate is determined not only by the amount of energy available, but also by the vapor gradient between the leaf mesophyll and the free air beyond the boundary layer adhering to the leaf surface. Water vapor must be vaporized at the mesophyll cell walls in the substomatal cavity and then pass through the stomates and across the boundary layer of air. As with a fluid passing through any tube or pipe, there is internal leaf resistance r_ℓ, expressed in seconds per meter, to the fluid flow. In addition, there is a surface boundary-layer resistance r_a. Let $_sd_\ell(T_\ell)$ represent the saturation density (in kilograms per cubic meter) of water vapor in the leaf intercellular air spaces as a function of leaf temperature. Let $_sd_a(T_a)$ represent the saturation density of water vapor in the air as a function of air temperature. Then, if the relative humidity of the air is h, the water vapor density of the air is $h \,_sd_\ell(T_\ell)$. Relative humidity is the ratio of the vapor pressure or vapor density of the air to the corresponding value for saturated air at the same temperature and varies from 0 to 1.0 (this quantity is sometimes also expressed as percent). The rate at which water vapor escapes from a leaf is

$$E = \frac{_sd_\ell(T_\ell) - h \,_sd_a(T_a)}{r_\ell + r_a}, \tag{3.2}$$

where E is in units of kilograms per square meter per second.

The energy-budget relationships for a plant leaf can now be written in the following complete form:

$$Q_a = \epsilon\sigma(T_\ell + 273)^4 + k_1 \frac{V^{1/2}}{D^{1/2}} (T_\ell - T_a) + \lambda(T_\ell) \frac{_sd(T_\ell) - h \,_sd_a(T_a)}{r_\ell + r_a}.$$

$$\tag{3.3}$$

Sample Calculations of Leaf Temperature

For any given set of independent variables [i.e., radiation flux (which determines Q_a), air temperature, wind speed, and relative humidity], a leaf of given properties will have a leaf temperature T_ℓ and transpiration rate E determined by the energy coming in and the energy going out. One can see here, for the first time, the way in which the two dependent variables and Q_a, T_a, V, and h interact simultaneously. From careful study of Eq. (3.3), one can understand the relative importance of each variable and leaf parameter in its influence on leaf temperature and transpiration rate. The simplest way to see the relative importance of each mechanism affecting the transfer of energy between a leaf and its environment is to take them one at a time, adding them up as one goes along. Table 3.1 illustrates the results of this procedure.

Radiation Influence on Leaf Temperature

Initially, assume that an inert, nontranspiring leaf is in a vacuum wherein energy is lost by the leaf only through radiation. If the leaf absorbs a quantity of radiation (shown at the left of Table 3.1), then by the mechanism of radiative equilibrium alone, it will have the temperatures given in the column labeled "reradiation only." These values are simply the blackbody relation between radiation and temperature calculated by Eq. (2.2) and plotted in Fig. 3.1 as the line labeled $\epsilon = 1.0$. The detailed explanation of a blackbody is given in Chapter 7. If the amount of radiation absorbed by a leaf is very large (say 977 W m^{-2}, approximately the maximum amount possible in the natural world), then the leaf will equilibrate at a temperature of 89.2°C. This is very hot indeed, and no living leaf could ever survive such a high temperature. If the radiation load on the leaf were such

Table 3.1. Temperatures (in Degrees Centigrade) of a Leaf under Various Conditions[a]

	Reradiation only	Reradiation and convection			Reradiation, convection, and transpiration		
		$V =$ 0.1 m s^{-1}	$V =$ 1.0 m s^{-1}	$V =$ 5.0 m s^{-1}	$V =$ 0.1 m s^{-1}	$V =$ 1.0 m s^{-1}	$V =$ 5.0 m s^{-1}
Radiation absorbed							
419 W m^{-2}	20.0	27.0	28.7	29.4	23.7	26.0	27.8
698 W m^{-2}	60.0	41.0	34.6	32.3	34.1	31.3	30.7
977 W m^{-2}	89.2	55.0	40.5	35.1	44.5	36.8	34.4

[a] The leaf has a characteristic dimension of 0.05 m and an internal resistance to water loss of 200 m s^{-1}. Air temperature is 30°C, and relative humidity is 50%.

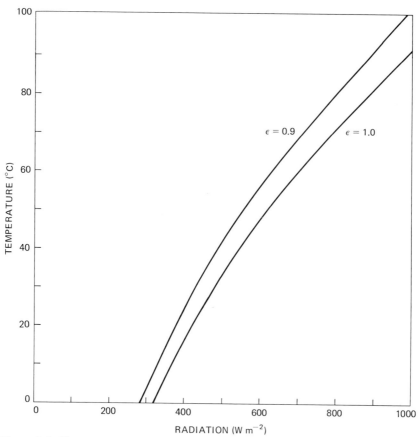

Figure 3.1. Temperature versus amount of energy radiation emitted from surfaces with emissivities of 1.0 (a blackbody) and 0.9 (a graybody).

that it absorbed 698 W m^{-2}, the leaf temperature would be 60°C. If the leaf absorbed 419 W m^{-2}, the temperature would be 20°C. An amount of absorbed radiation of 698 W m^{-2} is very typical of normal midday conditions, and with reradiation as the only mechanism for dumping energy, the leaf becomes very warm. It is only at low amounts of absorbed radiation that leaf temperatures are at all tolerable for a living leaf if the only mechanism for exchanging energy is radiation. It is important to realize that, for this inert, nontranspiring leaf without convection, the only quantities affecting its temperature are amount of incident radiation flux absorbed by the leaf and, of course, emissivity of the leaf in reradiating this energy. The calculation shown in Table 3.1 is based on a leaf emissivity of 1.0. If instead of 1.0, the emissivity is 0.90, then the three temperatures for the condition of reradiation only would be 28°, 69°, and 99°C.

Convection Influence on Leaf Temperature

In this case, instead of being in a vacuum, the leaf is in air. The same exercise may now be repeated with respect to energy exchange. However, we must agree on additional properties for the leaf and some conditions for the air owing to the presence of the convective heat transfer. Assume that the leaf has a characteristic dimension of 0.05 m, which is its width in the direction of the wind flow. Let the air temperature be 30°C and assume that the wind speed takes any one of three values: 0.1, 1.0, and 5.0 m s⁻¹. A rate of 0.1 m s⁻¹ is the amount of air flow one would expect in still air. In fact, use of 0.1 m s⁻¹ for V in the forced-convection term gives a convection coefficient that agrees reasonably well with the convection coefficient for free, or natural, convection.

The simplest way to do this somewhat complex calculation is to take the energy-budget expression for a leaf without transpiration in the following form:

$$Q_a = \epsilon\sigma(T_\ell + 273)^4 + 9.14 \frac{V^{1/2}}{D^{1/2}} (T_\ell - T_a). \tag{3.4}$$

Then, with the air temperature at 30°C, one can calculate the convection term for $D = 0.05$ m, $V = 0.1$, 1.0, or 5.0 m s⁻¹, and various values of T_ℓ. It is really the ratio V/D that is important here; the values of V/D are 2,

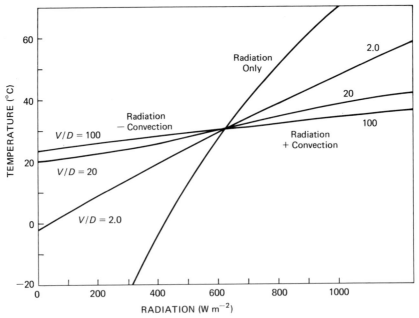

Figure 3.2. Temperature versus energy exchanged by radiation only (emissivity of 1.0) and by radiation and convection combined, at an air temperature of 30°C and ratio V/D of 2, 20, and 100.

20, and 100. Now add the convection term to the reradiation term and plot as shown in Fig. 3.2. Then, for any given value of Q_a, it is possible to find the value of T_ℓ that will satisfy Eq. (3.4).

A glance at Table 3.1 reveals the temperatures the leaf will assume in air at 30°C for three levels of absorbed radiation and wind speed. It is evident that convective heat exchange, either cooling or warming, can have a significant influence on leaf temperature. Let us suppose that, at the maximum amount of radiation absorbed, a leaf's temperature, without the influence of convection, is 89.2°C. Simply placing the leaf in still air ($V = 0.1$ m s^{-1}) at 30°C reduces the leaf temperature to 55.0°C. In winds of 1.0 and 5.0 m s^{-1}, the leaf has a temperature of 40.5° and 35.1°C, respectively. A leaf with a low amount absorbed radiation (419 W m^{-2}) and without convective heat exchange has a temperature of 20°C. With convective heat exchange caused by wind speeds of 0.1, 1.0 and 5.0 m s^{-1}, leaf temperatures become 27.0°, 28.7°, and 29.4°C, respectively. The warm air at 30°C actually warms the leaf by convection for this condition of low radiation. This is typical of a leaf exposed to a clear, relatively cold, night sky on a warm summer night. For moderate amounts of absorbed radiation (698 W m^{-2}), convection produces a 19.0° to 27.7°C drop in temperature of the leaf. When large amounts of radiation are absorbed by a leaf (977 W m^{-2}), leaf temperature is diminished through convection alone by 34.2° to 54.1°C. The role of convection in cooling a leaf is truly impressive when expressed in this manner. A further increase in wind speed, above 5.0 m s^{-1}, produces little further decrease in leaf temperature.

The example just given, it is true, is a theoretical calculation but represents an analytical treatment of the problem of heat transfer between a leaf and its environment based on fundamental physical principles. These principles, elaborated in later chapters, have been established by many years of research in physics and engineering laboratories. The validity of the particular application of these principles to the problem of heat transfer to an from a leaf has been demonstrated by laboratory experiments involving leaves mounted in low-velocity wind tunnels and by field measurements. Clearly, the situation is not quite as simple as expressed here, for leaves are usually of irregular shape and somewhat variable in properties. Nevertheless, one can represent many broad types of leaves, especially deciduous leaves, by flat plates for purposes of heat-transfer analysis.

Transpiration Effect on Leaf Temperature

Living leaves transpire moisture. The temperature of a leaf is influenced by cooling produced by the evaporation of liquid water to water vapor in the mesophyll; this water vapor then escapes by diffusion through the stomates and across the boundary layer to the free air beyond the leaf. In order to estimate the effect of transpirational cooling on leaf temperature, we must make a further decision concerning the properties of the leaf and the air. The air is considered to have a relative humidity of 0.5 (50%), which, at a temperature of 30°C, gives it a water vapor pressure of 21.22 mbar and a density of 15.2×10^{-3} kg m^{-3}. The leaf has a well-defined property for the escape of water vapor from the substomatal cavity through the stomatal opening and across the boundary layer to the free air, i.e., the total diffusion resistance of the pathway. For the example given here, the leaf is assumed to have a relatively low internal diffusion resistance (200 s m^{-1}). Some leaves will have less resistance to the loss of moisture, but many will have larger resistances. When stomates of a leaf are fully open, as they usually are in sunshine when a plant is well supplied with water, minimum internal diffusion resistance to the escape of water will vary from less than 100 to as much as 2000 s m^{-1} or more. The value used here, 200 s m^{-1}, is found frequently in nature. The boundary-layer resistance varies inversely with wind speed and directly with dimensions of the leaf. The larger the leaf, the thicker the boundary layer and the greater the resistance offered to the transfer of water vapor. The greater the rate of air flow across a leaf, the thinner the boundary layer and the lower the resistance. One might guess that, since the rate of convective heat transfer varies with the square root of wind speed and characteristic dimension, the square root might be a factor in these parameters for boundary-layer diffusion resistance. This is essentially the case, but a slight elaboration is needed at this stage. Rather than consider a leaf of dimension D in the direction of air flow, let the leaf have a dimension W at right angles to D. Without further elaboration (the details are described in the chapter on evaporation), we can assert that the boundary-layer resistance

$$r_a = k_2 \frac{W^{0.2} D^{0.3}}{V^{0.5}}, \tag{3.5}$$

where $k_2 = 200$ s$^{1/2}$ m^{-1}.

In order to find the temperature of a leaf undergoing heat transfer by radiation, convection, and transpiration, it is necessary to engage in much more difficult computation than the case in which heat is transferred by radiation and convection only. Equation (3.4) is used for this calculation. Complications arise because saturation water vapor density inside the leaf is a function of the leaf temperature, as is the latent heat of water. The latent heat of water, however, is not a strong function of temperature over the temperature range normally involved. Calculation of leaf temperature when radiation, convection, and transpiration are involved must be done in an iterative fashion. It is not appropriate to explain this entire procedure here since it is done in detail in a later chapter. However, the reader may attempt a solution by guessing a leaf temperature, calculating the three terms on the right side of Eq. (3.3), and adding them up to see if they equal the selected value of Q_a on the left-hand side. Different values of T_ℓ are tried until a balance is achieved. One can see in Table 3.1 the leaf temperatures that result when all three mechanisms of heat transfer (radiation, convection, and transpiration) occur for three absorbed radiation levels and air-flow rates. In a strong radiation field, in which 977 W m^{-2} is absorbed, the leaf's temperatures are 44.5°, 36.8°, and 34.4°C for wind speeds of 0.1, 1.0, and 5.0 m s^{-1}, respectively. These leaf temperatures may be compared with those occurring when the only mechanisms for heat exchange are radiation and convection (55.0°, 40.5°, and 35.1°C, respectively). Hence still-air transpirational cooling reduces the leaf temperature 10.5°C when the amount of radiation absorbed is 977 W m^{-2}. At the more moderate amount of absorbed radiation (698 W m^{-2}), leaf temperature differences are 6.4°, 3.3°, and 1.6°C at wind speeds of 0.1, 1.0, and 5.0 m s^{-1}, respectively. It is clear that when there is some wind on a leaf absorbing moderate amounts of radiation, transpirational cooling may be relatively small, say from 2° to 8°C. In this case, the amount of cooling may not be very significant. However, when radiation loads are large and the leaf is in still air, transpirational cooling can lower the leaf temperature 12°C or more. If the leaf is larger and the rate of heat transfer by convection is reduced, transpirational cooling becomes even more significant. On the other hand, if the leaf is small, convective heat transfer becomes more important, and transpirational cooling is reduced considerably. This will be seen when the results are presented graphically.

A leaf with a low amount of radiation absorbed (419 W m^{-2}) and with the other conditions as given with Table 3.1 is cooled by transpiration. Whereas convection causes an increase in leaf temperature under these conditions, loss of water from a leaf by evaporation will always use energy and cool a leaf.

In Fig. 3.3, one finds transpiration rate and leaf temperature, the two dependent variables, given as the ordinate and abscissa, respectively. The independent variable, air temperature, is shown as a family of lines marked every 5°C. Also shown is the leaf parameter of total diffusion resistance to water vapor. All other independent variables and leaf parameters are kept constant for this figure; their values are listed in the figure caption. A leaf of dimensions 0.05 × 0.05 m and internal diffusion resistance of 200 s m^{-1} is in still air of temperature 20°C at 0.5 relative humidity. The leaf absorbs 698 W m^{-2} of radiation. This leaf will transpire about 5.83 × 10^{-5} kg m^{-2} s^{-1} of water and have a temperature of 27.5°C. If the stomates close somewhat and the diffusion resistance increases to 1000 s m^{-1}, the transpiration rate will drop to 2.17 × 10^{-5} kg m^{-2} s^{-1}, and the leaf temperature will rise to 32°C. If the stomates close entirely and the resistance becomes infinite, the transpiration rate goes to zero, and the leaf temperature rises to 33.5°C, providing air temperature remains at 20°C, relative humidity at 0.5, wind speed at 0.1 m s^{-1}, amount of radiation absorbed at 698 W m^{-2}. If the air temperature changes, one can find the

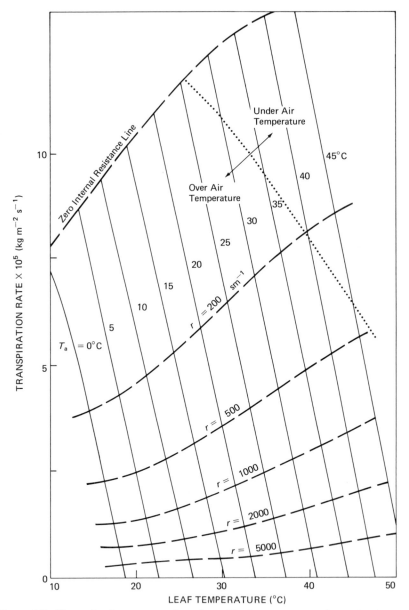

Figure 3.3. Transpiration rate as a function of leaf temperature, for various values of air temperature and total diffusion resistance, with conditions as follows: Q_a, 698 Wm^{-2}; V, 0.1 m s^{-1} h, 0.5; D, 0.05 m; W, 0.05 m.

transpiration rate and leaf temperature bymeans of the lines of constant resistance.

Leaf Dimension Influence

The relationships between the dependent variables (leaf temperature and transpiration rate), independent variables (radiation absorbed, air temperature, wind speed, and relative humidity), and leaf parameters [size and diffusion resistance, as represented by Eq. (3.3)] are conceptually part of a multidimensional space. Figure 3.3 is the first example of a graphical cross section through this multidimensional space. Another cross-section through this space is given in Fig. 3.4. Here, instead of air temperature, leaf dimension is given as a variable parameter, along with the total diffusion resistance. The air temperature is fixed at 40°C, relative humidity at 0.2, wind speed at 1.0 m s^{-1}, and amount of radiation absorbed by the leaf at 837 W m^{-2}. These conditions are typical of a hot, dry, slightly windy desert day with clear sky and full sun. The first thing to observe here is that a small leaf, such as one of dimensions 0.01 × 0.01 m, has a leaf temperature within about 3°C of the air temperature for any value of internal diffusion resistance. A large leaf, however, such as one 0.1 × 0.1 m or 0.2 or 0.2 m, may have a temperature 12° to 15°C above the air temperature when resistance is large or when little water is available. Normally, when air temperatures are moderate, say 20° or 30°C, this would not seem to matter, but when the air temperature is high, such as 40°C, leaf temperatures above 50°C can be catastrophic. Of course, if there is plenty of water available and if the plant leaf has a low resistance to water diffusion, evaporative cooling may bring the leaf temperature close to the air temperature even for a large leaf. But if available water is in short supply or diffusion resistance is large, then, for the conditions existing here (namely, hot, dry, windy, desertlike situations), leaf temperature will be very much above the air temperature, and the proteins in the leaf will denature irreversibly.

Gates et al. (1968) worked with desert plants (excluding the large succulents) with small leaves, and showed that leaf temperatures are always within 2.5°C of the air temperature for these plants. This included leaves of plants such as paloverde, creosote bush, cat's-claw, and sagebrush. All small organisms have a large convection coefficient, which, in terms of the end result, always tends to couple the organism's temperature very strongly to the air temperature. Hence the observations in the field confirmed precisely the predictions of leaf temperature for hot, dry desert conditions as recorded in Fig. 3.4.

Wind Influence

The influence of wind on leaf temperature is an interesting phenomena. The question is often asked whether an increase of wind around a plant will increase or decrease the rate of water loss from the plant. This is a

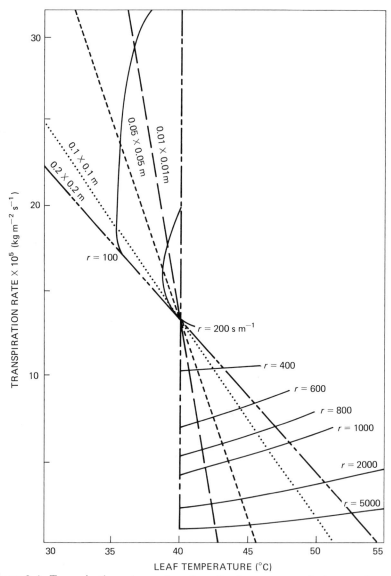

Figure 3.4. Transpiration rate as a function of leaf temperature for leaves of various sizes and total diffusion resistances, with conditions as follows: Q_a, 837 Wm^{-1}; T_a, 40°C; h, 0.2; V, 1.0 m s^{-1}.

question that is not easily answered, because the answer depends upon conditions. Since the various factors affecting energy exchange for a leaf can vary a great deal, and the relative influence of radiation, convection, and transpiration can vary enormously, one can predict an exact relationship between transpiration rate and the wind speed only from the full ana-

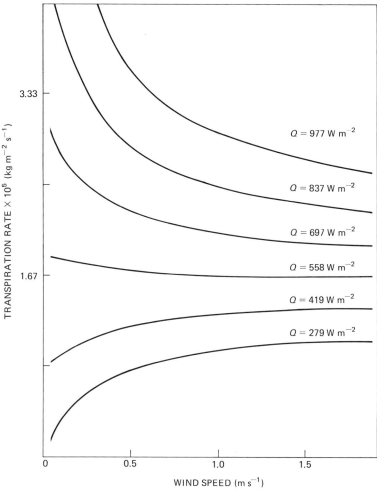

Figure 3.5. Transpiration rate as a function of wind speed for various amounts of absorbed radiation with conditions as follows: T_a, 30°C; h, 0.5; D, 0.05 m; W, 0.05 m; r, 1000 s m^{-1}.

lytical expression for the energy budget of a leaf. Figure 3.5 shows the transpiration rate for a leaf of dimensions 0.05 × 0.05 m and diffusion resistance 1000 s m^{-1} in air of temperature 30°C and relative humidity 0.5. Here one sees that at high radiation levels, an increase of wind speed results in a decrease of the rate of transpiration, but that for low amounts of radiation absorbed, the transpiration rate increases with an increase in wind speed. If the amount of radiation absorbed is fairly intermediate, then, for the conditions of Fig. 3.5, the transpiration rate is relatively uninfluenced by wind speed.

Field Observations

The more one looks at the matter of leaf energies, the more one becomes convinced that many leaves of plants undergo some sort of thermal regulation. There is far more leaf movement in nature in response to radiation intensity and other factors than one imagines from casual observation. A dozen different plants all side by side in the same environment will respond differently to the environmental factors and have different leaf temperatures and transpiration rates; they will also, of course, have various photosynthetic, respiration, and growth rates. These relationships may be observed in any garden, and leaf temperature differences may be felt by comparing sensations of coolness or warmth from plant to plant.

Various strategies are used by plants to adjust the energy exchange rate with the environment. One of the most striking examples ever noted by the author was the behavior of the leaves of *Erythrina indica* and certain other plants which were observed near Sydney, Australia, at noon on a warm summer day. Air temperature was 32°C, radiant temperature of the sky (a term explained in Chapter 7) was 16°C, and there was no wind. Overhead was a thin overcast of cirrus clouds. The *Erythrina* leaves were approximately 8 × 10 cm in dimension, and most were standing in a nearly vertical position. It was an impressive sight to see nearly all the leaves of a good-sized tree standing vertically in the midday sun. The sunlit, vertical leaves all had temperatures between 36° and 38°C. There were a few leaves which, for some reason, were not vertical but horizontal; their temperatures were from 42° to 44°C. The *Erythrina* leaves in the shade were all at 33°C. A few meters away was a cottonwood, *Populus deltoides,* an exotic from North America, with leaves approximately the same size as the *Erythrina* leaves but primarily in a horizontal position. Those cottonwood leaves, which were fully exposed to direct sunlight, had temperatures of 32 to 34°C. At first, this was a great surprise, but I quickly realized that these leaves might be transpiring strongly and keeping their temperatures from rising very much by evaporative cooling. When I measured the temperatures of the shaded leaves of the cottonwood, I found them to be at 30°C. Since the shaded leaves of the cottonwood were cooler than the shaded leaves of the *Erythrina,* I knew that the cottonwood leaves were transpiring much more. Near these trees was a jacaranda tree, *Jacaranda acutifolia,* with small leaves exposed to the full sun, usually in nearly a horizontal position, but occasionally vertical. These leaves had temperatures within 3°C of air temperature. Here we had, with three different plants, a demonstration of three entirely different mechanisms for keeping down the heat stress during a warm summer day. *Erythrina* kept its temperature down by having vertical leaves and not absorbing very much radiation. *Populus* used transpirational cooling effectively, whereas *Jacaranda* kept its leaves tightly coupled to the air temperature by their small size and large convection coefficient. Other plants nearby, which seemed to utilize none of these mechanisms, be-

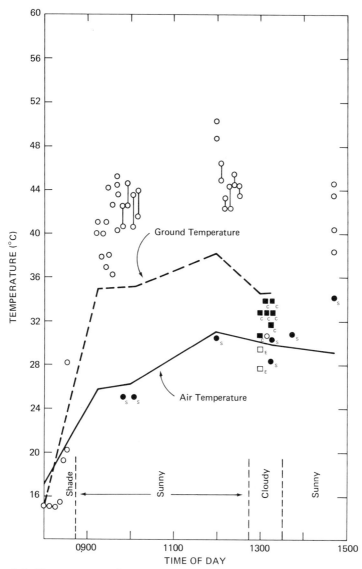

Figure 3.6. Temperatures of *Quercus macrocarpa* leaves as a function of time of day: (○) full sun on leaf; (●) shaded leaf; (□) exposed leaf; (■) cloudy. Connected open circles represent measurements done on upper and lower leaf surfaces. Also shown are (– –) ground and (—) air temperature.

came very warm. Some wilted, and some showed signs of denaturation and drying.

When, on 16 September 1961, I measured the temperature of the leaves of a bur oak, *Quercus macrocarpa*, growing in Boulder, Colo., I obtained the results shown in Fig. 3.6 (see Gates, 1963). In full sunshine, leaf temperatures often exceeded the air temperature by 10° to nearly 20°C. Note that when the air temperature was about 31°C, there was a leaf temperature of almost 50°C. We know that sustained temperatures of 46°C or greater are potentially damaging to the leaves of many plants.

The classical research of Clausen et al. (1940) concerning the experimental taxonomy and genetics of *Mimulus,* as studied in the transplant gardens of the Carnegie Institution, is well known. The transplant gardens were located at Stanford University, in the foothills of the Sierra Mountains, and also in the High Sierras near Tioga Pass, just beyond the boundary of Yosemite National Park. In order to study the nature of the habitat and the temperature response of *Mimulus,* I was asked by Dr. William Hiesey to meet with him and two of his associates, Mr. Harold Milner and Dr. Malcomb Nobs, at Yosemite National Park in August 1963. I came prepared with a relatively new kind of temperature-measuring device, an infrared radiometer, which measures leaf temperatures without contacting the leaves directly. The infrared radiometer is responsive to infrared radiation emitted by the surface of any object, including leaves, according to the fourth power of the absolute temperature of the surface. A simple calibration procedure converts the amount of energy received by the radiometer to the temperature of the surface according to the blackbody radiation law. The instrument used was a Stoll-Hardy radiometer (Gates, 1962).

After some preliminary observations, we decided to observe, during as close a time interval as possible, the leaf temperatures of *Mimulus* plants growing from near timberline in the Sierras above Tioga Pass to the lower foothills at the edge of the hot interior lowlands west of the mountains. *Mimulus,* known as monkey flower, is a member of the Scrophulariaceae. It is a plant which normally requires considerable amounts of water and is usually found growing along stream banks, drainage ditches, or in wet meadows. Dr. Hiesey and his associates had studied *Mimulus* in the growth chamber and knew the manner in which its photosynthetic rate depend upon temperature. They had not yet measured leaf temperatures in the field, or for that matter in the laboratory. Earlier work I had done with the measurement of leaf temperatures of plants growing in Boulder, Colo., had shown me that leaf temperatures could at times exceed the air temperature by as much as 15° to 20°C. So it was with considerable excitement that we set forth by jeep to visit clones of *Mimulus* located near timberline at an elevation of about 3200 m above sea level and along a transect down to Priest's Grade at an elevation of about 400 m. We were to observe plants growing in undisturbed natural sites, as well as those transplanted to gardens located at the Timberline and Mather stations. In the undisturbed habitats at and above the Smoky Jack site (2300 m), we were working with *Mimulus lewisii* and with a hybrid *Mimulus lewisii* × *cardinalis* in the transplant gardens. Most of the observations were made during one day, 14 August 1963, with the exception of the timberline measurements, which were made during several days, between 12 and 16 August. The weather was ideal for the observations since the skies were remarkably clear and very deep blue in appearance. This meant that the amount of solar radiation reaching the ground surface would not change drastically during the time between 1000 and 1500. The elevation of the various sites were as follows: the Timberline talus slope, 3200 m; the Timberline station and slope garden, 3000 m; the Smoky Jack site, 2300 m; Carlin, 1520; Mather, 1400 m; and Priest's Grade, 400 m.

The results of our measurements of leaf temperatures are shown in Fig. 3.7. The subscript C indicates thin clouds interfering with the sun at the time of the measurements. Although the day during which most observations were taken was absolutely clear, some of the measurements shown in Fig. 3.7 were taken during a day when there was intermittent cloudiness. The subscript S indicates leaves in the shade, and the subscript W indicates some wind. Unless indicated as windy or cloudy, conditions at the time of the measurements were still air and clear sky. Air temperatures are indicated by solid lines. The striking thing to notice in Fig. 3.7 is the relatively isothermal character of the temperatures of sunlit leaves of *Mimulus*. Temperatures ranged from 23° to 35°C, with a few exceptions, whereas the air temperature varied from 17°C at Timberline to 37°C at Priest's Grade. This was in stark contrast to what all of my earlier experience with the measurement of leaf temperatures had led me to expect. Until we made these measurements, I would have expected the sunlit leaves of *Mimulus* to be about 23° to 30°C when the air temperature was 17° or 18°C and to be 40° to 50°C when the air temperature was 37°C. Inspection of Fig. 3.7 shows that the air temperature increased rather rapidly as we reached the lower elevations but that leaf temperatures of *Mimulus*, which were well above the air temperature when the air was cool, dropped decidedly below the air temperature when the air was warm. In fact, the leaves of *Mimulus* were remarkably close to being isothermal throughout the transect. At each of the sites, the *Mimulus* plants were growing in wet soil or in water.

The following spring, we decided to try to simulate the field conditions with *Mimulus* plants in a growth chamber. Air temperature in the chamber was similar to that in the field, as was radiation intensity. The chamber observations essentially confirmed the field observations. Below an air temperature of approximately 30°C, leaf temperatures were higher than air temperature. Above 35°C or so, the leaves were cooler than the air. We were able to establish that this phenomena was caused by the influence of transpiration on the energy budget of the *Mimulus* leaf. This is described in more detail in Chapter 10. One can then imagine my surprise when, on 14 August 1963 I first observed the temperatures of fully sunlit leaves of *Mimulus* to be below air temperature. We thought at the time that we had made quite a startling discovery. But soon thereafter, I read a publication of Dr. O. L. Lange (1959) and learned that he had ear-

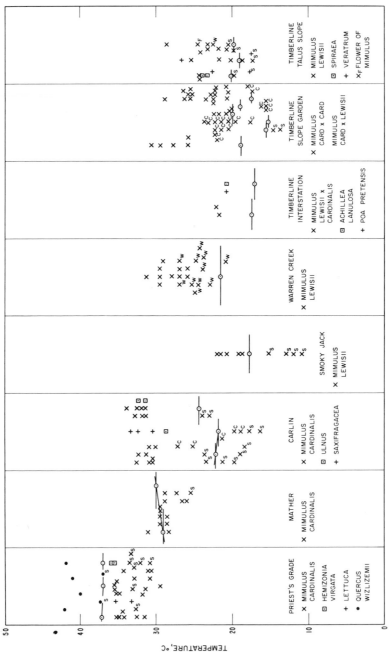

Figure 3.7. Leaf temperatures of various plant species growing at different sites along a transect on the west side of the Sierra Nevada Mountains. Sites and elevations: Priest's Grade, 400 m; Mather, 1400 m; Carlin, 1520 m; Smoky Jack, 2300 m; Timberline interstation and slope garden, 3000 m; Timberline talus slope, 3200 m. An east-side slope at Warren creek, 2700 m, is included. See text for explanation.

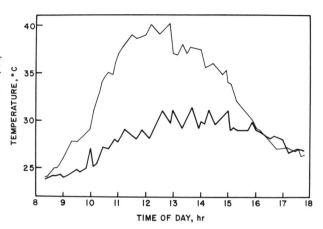

Figure 3.8. Leaf temperature of *Zygophyllum cf. fantanesii* as a function of time of day: (—) leaf temperature; (—) air temperature. (Redrawn from Lange, 1959.)

lier observed such a phenomena with desert and savannah plants of Mauretania.

In Fig. 3.8, leaf temperatures of *Zygophyllum cf. fontanesii,* an "overtemperature"-type plant, and air temperature as a function of the time of day are shown. This is a plant whose leaf temperatures are always above the air temperature when the leaf is in sunshine. Notice that the difference between leaf and air temperature sometimes exceeds 10°C. Figure 3.9 shows leaf temperature of *Citrullus colocynthis,* an "undertemperature"-type plant, and air temperature as a function of the time of day. The temperature difference between leaf and air for this plant often exceeds negative 10°C.

From the leaf energy-budget equation, it is easy to understand the phe-

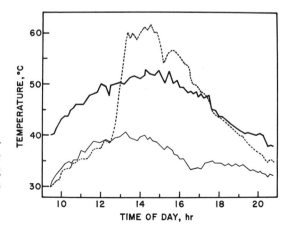

Figure 3.9. Leaf temperature of *Citrullus colocynthis* as a function of time of day; (—) leaf a; (– –) leaf b; (—) air temperature. (Redrawn from Lange, 1959.)

nomena of "overtemperature"- and "undertemperature"-type leaves
when in full sunshine. Let the amount of radiation absorbed by a leaf be
Q_a as before and let R be the amount of thermal radiation emitted by a leaf
$(R = \epsilon\sigma T_\ell^4)$. The net radiation NR absorbed by the leaf is Q_a minus R.
Neglecting photosynthesis, which is a small quantity in the total heat
budget, and writing the energy budget in its simplest form gives

$$NR = \lambda E + h_c \, (T_\ell - T_a) \tag{3.6}$$

$$(T_\ell - T_a) = (NR - \lambda E)/h_c, \tag{3.7}$$

where h_c is the convection coefficient.

It is evident that if the net radiation exchange by a leaf is positive and is
greater than the energy consumed in transpiration, the temperature difference between leaf and air is positive. If, however, the amount of energy consumed in transpiration exceeds the net radiation exchanged by
the leaf, then the temperature difference between leaf and air is negative,
this often is the case when the amount of net radiation exchanged is small.
The interesting fact is that some leaves can overcompensate for rather
large amounts of net radiation by evaporative cooling so that the leaves
are always cooler than the air. The opportunity for a plant leaf to function
in that manner depends, however, on the availability of water to the leaf
and the ability of the leaf to let water escape. This means that the leaf
must have a small diffusive resistance to water vapor. Water loss from a
leaf will be greater when the air is very dry than when the air is extremely
humid. This is the reason that Dr. Lange found these "undertemperature"-type plants performing so well in Mauretania, where the air is very
dry. It is also the reason that we observed the condition for *Mimulus*
under very warm and dry conditions of the lower elevations in the
Sierra Mountains in August during midday. It is also obvious that an
"undertemperature"-type plant leaf will not always perform at a temperature below the air temperature, but will often be at or above air
temperature. It is also clear that an "overtemperature"-type plant leaf
will rarely function with its leaf temperature below the air temperature.
Referring to Fig. 3.3, we see that this is indeed the case. Here the amount
of radiation absorbed by the leaf is 698 W m^{-2}, which is a normal midday
amount for a summer sun in a clear sky. For the leaf temperature to be
lower than the air temperature in still air, the leaf must have a diffusive resistance of 200 s m^{-1} or less, and the air temperature must be above 40°C.
If there is wind movement of 2.0 m s^{-1} and all other conditions are the
same as for Fig. 3.3, leaf temperature will be lower than air temperature
for leaf resistance less than 500 s m^{-1} and air temperatures above 35°C;
this is clearly seen in Fig. 3.10. It is also apparent from these figures that it
is extremely unusual to have leaf-to-air temperature differences greater
than minus 5°C for 698 W m^{-2} of radiation absorbed, even when the internal diffusive resistance is zero. If the leaf size is greater than

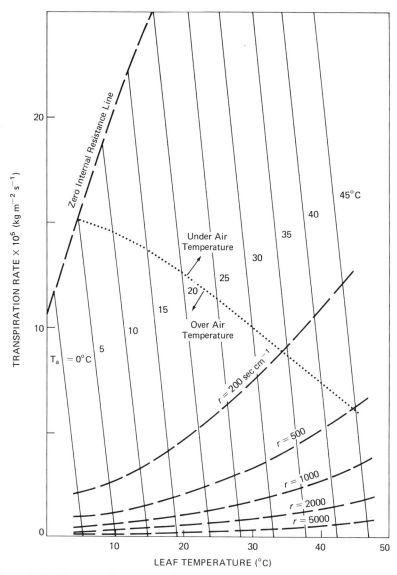

Figure 3.10. Transpiration rate as a function of leaf temperature for different values of air temperature and total diffusion resistance, with conditions as follows: Q_a, 698 Wm^{-2}; V, 2.0 m s^{-1}; h, 0.5; D, 0.05 m; W, 0.05 m.

0.05 × 0.05 m, leaf temperature will depart more from air temperature. If the leaf is smaller than 0.05 × 0.05 m, the temperature will be more tightly coupled to that of the air. At night, when the amount of radiation absorbed is small, it is relatively easy for leaf temperatures to be lower than the air temperature. This is a common observation and is illustrated in Fig. 3.11.

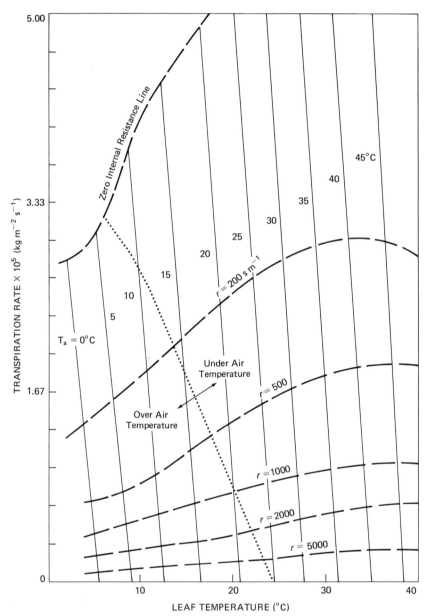

Figure 3.11. Transpiration rate as a function of leaf temperature for different values of air temperature and total diffusion resistance, with conditions as follows: Q_a, 419 Wm^{-2}; V, 0.1 m s^{-1}; h, 0.5; D, 0.05 m; W, 0.05 m.

Photosynthesis by Leaves

The purpose of this chapter is to lay the foundation for the remainder of the book and make clear the objectives of this analytical approach to ecological problems. To understand plant adaptation to habitat, plant competition within a community, plant succession, and the rate of primary productivity, one must understand, among other things, the way a plant leaf functions within its environment. One must understand the mechanisms by which energy is gained or lost by a leaf, gases are exchanged, and nutrients enter and leave the plant and the basic biochemical activities of metabolism and other physiological processes. One must know which properties of the plant leaf are significant parameters and which act as coupling factors between the leaf and the air, soil, and water. Leaf absorptance couples leaf temperature and photochemistry of chloroplasts to the radiation field. Leaf size affects the thickness of the boundary layer of adhering air and acts as the coupling or decoupling factor between air temperature and leaf temperature, as well as affecting the exchange of gases. Leaf resistance of the stomates and cytoplasm affects the flow of gases between the leaf and the air. Rate coefficients for biochemical reactions determine how a leaf of a certain species responds to a given set of environmental conditions. All these factors, must be understood and dealt with when trying to understand plants in relation to habitat and environmental conditions. This is a reductionist approach to ecology, but it is necessary for a detailed understanding of ecosystem functions. This does not in any way deny that other methods are of great value; the aerodynamic approach to the study of gas and energy exchange between a plant stand and the atmosphere is effective and important, and many general systems models can improve our understanding of ecosystems. Nevertheless, despite the great value gained from more general, larger scale approaches to ecosystem studies, very detailed, very specific analyses are necessary at the same time in order to gain insight into the basic physiological–ecological mechanisms determining the arrangement and activity of biota within ecosystems.

The Concept

The primary process by which a plant leaf lives is photosynthesis. Certainly, in terms of the vital functions of a plant, the process of photosynthesis is of utmost importance. With this premise, we shall outline the steps necessary to place the photosynthetic activity of a leaf in analytical form. The basic idea is that a leaf takes up carbon dioxide from the air by means of gas diffusion. By means of photosynthesis within the chloroplasts, the leaf synthesizes carbohydrates and also manufactures amino acids, proteins, and ATP (adenosine triphosphate). The chemical reaction of photosynthesis depends upon the availability of light and carbon dioxide, and the rate of the reaction is temperature dependent. Leaf temperature is determined by the energy budget of the leaf, which relates such quantities as radiation, air temperature, wind speed, and relative or absolute humidity to the energy state of the leaf. Incident radiation, the quantity of which in part determines the leaf temperature, contains a light component that drives the photochemical events of photosynthesis. Diffusion of water vapor out of the leaf determines the transpiration rate, which affects the energy budget and leaf temperature. The rate at which carbon dioxide diffuses into the leaf is closely related to the rate at which water vapor diffuses out since the same pathway through the stomates is involved—except for additional resistance offered to the flow of carbon dioxide by the mesophyll. The rate at which carbon dioxide diffuses into the leaf mesophyll to arrive at the chloroplasts is commensurate with the rate of photosynthesis, which in turn depends upon amount of light, leaf temperature, and carbon dioxide concentration at the chloroplasts. In other words, all of these processes, namely, energy exchange, gas diffusion, and chemical kinetics, proceed simultaneously. In order to formulate these events in an analytical way, one needs to place them in systematic order and determine their relationships with one another. The energy exchange for a leaf has been described already, so now one may write a mathematical description of the gas-diffusion relationship for carbon dioxide and for the chemical kinetics within the chloroplasts. (What is given here is simply a brief introduction to the subject; additional details and a much expanded treatment involving many more variables is given in Chapter 14.)

The Analysis

The analysis of photosynthesis by a whole leaf is given here. In order to keep the algebra involved simple, the analysis is given for a nonrespiring leaf. Diffusion of carbon dioxide from the air into a leaf is given by Fick's law,

$$P = \frac{C_a - C_c}{R}, \tag{3.8}$$

where P is the photosynthetic rate, in millimoles per square meter per second; C_a and C_c are the CO_2 concentrations in the air and in the chloroplasts, in millimoles per cubic meter; and R is the resistance to CO_2 diffusion from the air through the leaf boundary layer, stomata, intercellular air spaces, cell walls, and cytoplasm into the chloroplasts, in seconds per meter.

The chemical process of CO_2 fixation within the chloroplasts depends upon the concentration of carbon dioxide in a manner typical of many chemical reactions. Dependence of the photosynthetic rate on CO_2 concentration is illustrated in Fig. 3.12. This functional dependence is easily

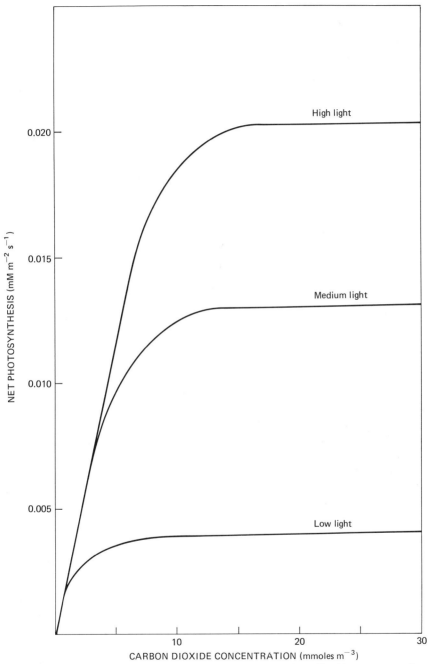

Figure 3.12. Net photosynthesis as a function of the carbon dioxide concentration at low, medium, and high light intensity.

described mathematically in the form of the classical Michaelis–Menten equation for the rate of an enzymatic reaction (photosynthesis is not, however, a true Michaelis–Menten reaction, since a complex series of reactions is involved):

$$P = \frac{P_M}{1 + \dfrac{K}{C_c}}, \tag{3.9}$$

where P_M is the maximum rate of photosynthesis (in millimoles per square meter per second) at saturating C_c, sometimes called "photosynthetic capacity," and K is a constant equal to the chloroplast concentration of CO_2 at which $P = P_M/2$.

Since the concentration of CO_2 at the chloroplasts is not known but its concentration in the air beyond the leaf's boundary layer is known, it is best to eliminate C_c from Eqs. (3.8) and (3.9). Solving Eq. (3.8) for C_c in terms of P, R, and C_a and substituting C_c in Eq. (3.9) gives the following quadratic equation:

$$RP^2 - (C_a + K + RP_M)\,P + C_a P_M = 0 \tag{3.10}$$

Solving this equation for P gives

$$P = \frac{(C_a + K + RP_M) - [(C_a + K + RP_M)^2 - 4C_a RP_M]^{1/2}}{2R}. \tag{3.11}$$

The above equation describes the photosynthetic rate for a nonrespiring leaf in which the chemical reaction rate depends only on the CO_2 concentration at the chloroplast (although the final expression is in terms of the concentration in the air). One knows, however, that photosynthesis is light and also temperature dependent. These functional dependencies can be included without complicating matters too seriously.

Dependence of the photosynthetic reaction on light intensity is precisely of the same form as the dependence on CO_2 concentration shown in Fig. 3.12. Hence one can use the Michaelis–Menten equation to describe this relation:

$$P_M(L) = \frac{P_{ML}}{1 + \dfrac{K_L}{L}}, \tag{3.12}$$

where P_{ML} is the value of P_M at light and at carbon dioxide saturation (in millimoles per square meter per second); L is the light intensity between 400 and 700 nm (in watts per square meter per second); and K_L is the light intensity at which $P_M(L) = P_{ML}/2$.

What we are doing here is to transform P_M, taking into account the functional dependence on light and temperature in the reaction-rate expression. What started out as P_M, a constant, becomes $P_M(L)$, a function of light intensity, and then $P_M(L, T)$, a function of light intensity and tem-

Figure 3.13. Generalized curve representing the dependence of the reaction
rate of photosynthesis on temperature.

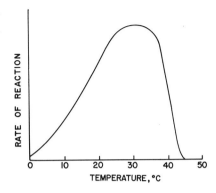

perature. Photosynthesis has a temperature dependence similar to that of
other chemical reactions. A typical temperature-dependence curve is
shown in Fig. 3.13. This curve is described simply as $G(T)$ and can be
solved either numerically or in algebraic form. Dependence of photosynthesis on light intensity and temperature takes the following form:

$$P_M(L, T) = \frac{P_{MLT}\, G(T)}{1 + \dfrac{K_L}{L}}, \tag{3.13}$$

where P_{MLT} is the value of P_M at light and carbon dioxide saturation and
optimum temperature. In other words, this equation represents the maximum photosynthetic capacity of the leaf. The procedure now is to substitute $P_M(L, T)$ directly into Eq. (3.11) in place of P_M in order to obtain P as
a function of the following variables and coefficients: C_a, K, R, K_L, L,
P_{MLT} and T. One can assign values to these quantities and calculate the resulting value of P. It may be useful at this stage to list the normal range of
values of these quantities. Eventually, one expects to use the computer to
do all the arithmetic and in that way save countless hours of tedious computation. It is useful, however, to have a few values available which can
be introduced into Eq. (3.11) in order to calculate values for P.

Normal Values

The normal concentration of C_a in air is 12.5 mmole m^{-3}, and values may
range from about 9 to 15 mmole m^{-3}. A typical value for K is 10 mmole
m^{-3}, but it may range from 0.01 to 100 mmole m^{-3}. The value for K_L is normally about 100 W m^{-2}, whereas the light intensity within the range of
photosynthetically active wavelengths (400–700 nm) is about 400 W m^{-2}
for a leaf in full sunlight. Gross photosynthesis at saturating carbon
dioxide, saturating light, and optimum temperature (P_{MLT}), is typically
about 0.05 mmole m^{-2}. The total resistance R to carbon dioxide diffusion
can vary from 100 to infinity; a typical value is 200 s m^{-1}.

Figure 3.14 gives net photosynthesis P as a function of leaf temperature

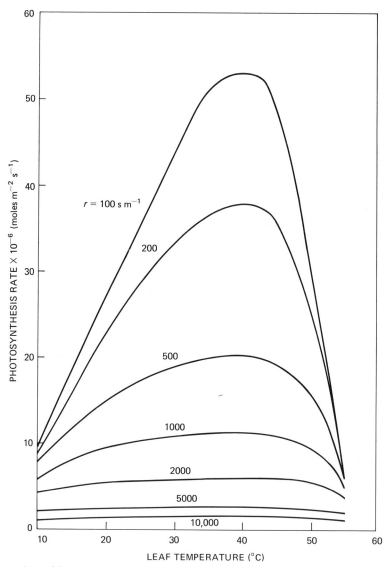

Figure 3.14. Net photosynthesis as a function of the leaf temperature for various values of total diffusion resistance, with conditions as follows: C_a, 12.5 mmoles m^{-3}, P_M, 50 mmole m^{-2} s^{-1}; K, 60 mmole m^{-3}.

and diffusion resistance as calculated using Eq. (3.11) and the typical values given above. These results are for a nonrespiring leaf. It is evident that when resistance is high, photosynthesis depends almost entirely on rate of diffusion and is nearly independent of temperature. In fact, one can show mathematically that as resistance approaches infinity, Eq. (3.11) reduces to

$$\lim_{R \to \infty} P = \frac{C_a}{R}. \tag{3.14}$$

This equation states that at high values of resistance, every CO_2 molecule reaching the chloroplasts is fixed so quickly that C_c in our original diffusion expression, Eq. (3.8), is essentially zero. Hence the rate of CO_2 diffusion strictly determines photosynthetic rate for large resistances.

For the limiting case of low resistance, one can show mathematically that the net photosynthetic rate reduces to

$$\lim_{R \to 0} P = \frac{P_M}{1 + \dfrac{K}{C_a}}. \tag{3.15}$$

This equation shows that at very low resistance, photosynthetic rate is determined by the rate of the biochemical processes of CO_2 fixation and, of course, is not limited by the resistance of the diffusion pathway.

Examples

It is always advisable to try a few numerical calculations so that the mathematics does not seem too formidable. We wish to calculate P for the following conditions: $L = 400$ W m^{-2}; $G(T) = 1.0$; $P_{MLT} = 0.05$ mmole m^{-2} s^{-1}; $R = 200$ s m^{-1}; $C_a = 12.5$ mmole m^{-3}; $K = 10$ mmole m^{-3}; and $K_L = 100$ W m^{-2}.

Equation (3.13) gives $P_M(L, T) = 0.04$ mmole m^{-2} s^{-1}. Then, from Eq. (3.11), $P = 0.0185$ mmole m^{-2} s^{-1}.

In order to test the sensitivity of photosynthesis to various factors, one can simply vary a given quantity and recalculate P. When R increased by a factor of two, from its previous value of 200 s m^{-1} to 100 s m^{-1}, we expect P to increase, but not by a factor of two. Just how much P will increase can be determined only by calculation. The value for P_M is the same as above, and Eq. (3.11) gives a value for P of 0.022 mmole m^{-2} s^{-1}. We see that reducing R by 50% increases the value of P by 18.9%.

Imagine that the amount of light is reduced from 400 to 100 W m^{-2}, a reduction of 75%, and calculate the effect on P if $R = 200$ s m^{-1} and the other conditions are as specified above.

In this case, $P_M(L, T) = 0.025$ mmole m^{-2} s^{-1}, and $P = 0.0125$ mmole m^{-2} s^{-1}.

We see that a reduction of 75% in the amount of light incident on the plant reduces the rate of photosynthesis by 32.4%, providing all other conditions remain constant. This is not what usually happens, however. A reduction of the light level reduces the amount of radiation absorbed by the leaf, which in turn reduces leaf temperature. This changes the value of $G(T)$ from its optimum value of 1.0 to something slightly less. If the leaf temperature is on the low-temperature side of $G(T)$, the effect of a change in leaf temperature on photosynthesis will be significant, but certainly not as dramatic as if it were on the high-temperature side. In the chapter on photosynthesis, a much more thorough analysis is given of the the energy budget of a leaf and the gas diffusion–chemical kinetic events of photosynthesis.

Problems

1. Make a graph with temperature (°C) as the ordinate (y-axis) and radiation intensity (W m^{-2}) as the abscissa (x-axis) and plot the blackbody radiation law,

$$R = \sigma(T_\ell + 273)^4.$$

The temperature scale should go from 0° to 60°C and the radiation scale from 300 to 700 W m^{-2}.

2. The rate at which energy C is transferred by convection to or from a flat plate (such as a broad leaf) and the surrounding air is given by the equation

$$C = h_c[T_\ell - T_a] = k_1 \frac{V^{1/2}}{D^{1/2}} (T_\ell - T_a),$$

where h_c is the convection coefficient, given by

$$h_c = k_1 \frac{V^{1/2}}{D^{1/2}}.$$

Compute a table of convection coefficients (in watts per square meter per degree Centigrade) for the following values of V and D:

	V		
D	0.1 m s^{-1}	1.0 m s^{-1}	5.0 m s^{-1}
0.01 m			
0.05 m			
0.10 m			
0.30 m			

The amount of energy going into convection can be estimated by multiplying these values by a temperature differential between the leaf and air. Assume that this difference is 5°C and determine the amount of energy consumed by convection if $D = 0.05$ m and $V = 1.0$ m s^{-1}.

3. The rate at which water vapor diffuses from a leaf is given by

$$E = \frac{{}_sd_\ell(T_\ell) - h \, {}_sd_a(T_a)}{r_\ell + r_a},$$

and the energy consumed by evaporation is λE, where λ is a function of the leaf temperature. Values of $_sd_\ell(T_\ell)$, $_sd_a(T_a)$, and λ are found in the Smithsonian Meteorological Tables, edited by List.

Draw a graph with leaf temperature as the ordinate (y-axis) and λE as the abscissa (x-axis) for air temperature of 30°C and relative humidities of 0.1(10%), 0.5(50%), and 0.9(90%) for a leaf with $r_\ell + r_a = 200$ m s^{-1}.

Compare the amount of energy consumed by transpiration with that consumed by convection for various assumed conditions.

4. Draw a graph with T_ℓ as the ordinate (y-axis) and $R + C + \lambda E$ as the abscissa (x-axis) for the following conditions: $D = 0.05$ m, $r_\ell + r_a = 200$ m s^{-1}, $V = 1.0$ m s^{-1}, $T_a = 30$°C, and $h = 0.5$ (50%).

Using this graph, determine the leaf temperature when the amount of radiation absorbed by the leaf is $Q_a = 700$ W m^{-2}. What percentage of the total radiation absorbed is reradiated (R), lost to convection (C), and goes into the evaporation of water (λE)?

What is the rate of water loss in kilograms per square meter per second?

5. For an amount of absorbed radiation of 500 W m^{-2}, determine the leaf temperature. Again, what percentage of the total radiation absorbed is reradiated (R), lost to convection (C) and goes into the evaporation of water (λE)?

What is the rate of water loss in kilograms per square meter per second?

6. For a relative humidity of 0.1 (10%) and all other quantities the same as in Problem 4, determine the leaf temperature when $Q_a = 700$ W m^{-2}.

How are the percentages of R, C, and λE changed?

What is the rate of water loss in kilograms per square meter per second?

7. If $W = m D$, substitution into Eq. (3.5) gives

$$r_a = k_2\, m^{0.2}\, \left(\frac{D}{V}\right)^{1/2}.$$

Determine values of the boundary layer resistance in seconds per meter for the following values of m and V/D:

		m	
V/D	0.1 s m^{-1}	1.0 s m^{-1}	10.0 s m^{-1}
1			
10			
100			

8. Determine the leaf temperature and rate of water loss for a leaf with dimensions W and $D = 0.05$ m and internal resistance $r_\ell = 100$ s m^{-1} if air temperature is 30°C, wind speed is 0.1 m s^{-1}, relative humidity is 0.5 (50%), and amount of radiation absorbed is 700 W m^{-2}.

9. In the first example in the chapter, a plant leaf had the following properties: $P_{MLT} = 0.05$ mmole m^{-2} s^{-1}, $R = 200$ s m^{-1}, $K = 10$ mmole m^{-3}, and $K_L = 100$ W m^{-2}. The environmental conditions were $L = 400$ W m^{-2}, $C_a = 12.5$ mmole m^{-3}, and temperature such that $G(T) = 1.0$. Calculation showed that $P = 0.0185$ mmole m^{-2} s^{-1}.

What is the photosynthetic rate if $K = 5$ mmole m^{-3}? (All other parameters and conditions are as specified initially.)

If $K = 20$ mmole m^{-3}?

If $P_{MLT} = 0.025$ mmole m^{-2} s^{-1} and all other parameters and conditions are as specified initially?

If $P_{MLT} = 0.10$ mmole m^{-2} s^{-1}?

If $R = 400$ s m^{-1} and all other parameters and conditions are as specified initially?

10. From the calculations made in the chapter and those in the above problem, rank the following plant parameters with regard to their effect upon photosynthesis, from the most to least important: R, K, and P_{MLT}.

11. Test the photosynthetic sensitivity of the plant to changes in environmental conditions by letting $L = 200$ and then 800 W m^{-2}, with all other conditions constant. Do the same for $C_a = 6.25$ and 25 mmole m^{-3}. Test the sensitivity to $G(T)$ by letting it equal 0.5. From these calculations, rank the response of the plant to light, carbon dioxide concentration, and temperature in order of greatest to least sensitivity.

Chapter 4

Application to Animals

Introduction

An animal functions within its habitat in response to a variety of factors that include the exchange of energy, gases, and nutrients between the animal and the environment. If one is to understand certain aspects of animal behavior, range limitation, and physiological adaptation to habitat conditions, then one must work with a model of the interaction of an animal with its environment that is as complete as possible. Once again, one is confronted with the difficulty of understanding the relationships between several dependent and many independent variables. When challenged with a multidimensional problem, it is essential to place the problem into a good analytical framework.

Energy is a primary requirement of animals just as it is of plants. An animal can function in a viable manner only if it exists in a suitable energy regime. Animals have fairly specific tolerances to temperature. One group of animals are *homeothermic,* or "warm-blooded." Their body temperatures are regulated by physiological mechanisms to remain within rather narrow limits. Homeothermic animals are classed as *endothermic* since their body heat is derived primarily from internal metabolic energy. At the other extreme are the *poikilothermic,* or "cold-blooded," animals, which possess little or no ability to thermoregulate other than the ability to locate themselves in a compatible part of the habitat. The poikilothermic animals are classed as *ectothermic* since their body heat is derived primarily from external energy from the environment, principally solar radiation. In actual fact, the animal kingdom exhibits the full spectrum of temperature regulation. Many warm-blooded animals, such as bears, woodchucks, and chipmunks, hibernate during the winter and allow their body temperatures to approach the cave or soil temperature. The female Indian python can regulate its body temperature during the brooding period in a manner analogous to that of endotherms. Often, an animal's preferred temperature will change with its own temperature history according to whether it has been cool or warm for an extended period of time. Certainly, the environmental temperatures comfortable to a human change according to the temperatures to which one has been conditioned.

Many animals, particularly some invertebrates such as insects, undergo physiological changes which make it possible for them to endure extreme temperatures. Some fish can withstand supercooling and do not freeze at temperatures several degrees below 0°C. (It is known that very pure water in the form of fine droplets can be supercooled to as low as −40°C without freezing.) Many insects generate antifreeze in their body plasma and have low freezing points because of high solute concentrations. The most common antifreeze in insects is glycerol. Many insects that freeze and die when subjected to low temperatures during the spring and summer months are able to survive the same temperatures during the autumn and winter.

Maximum body temperatures tolerated by animals can be as low as 28°C and as high as 50°C. The temperature at or above which an animal fails to survive is known as the Critical Thermal Maximum (CTM). Sometimes, muscle spasm or paralysis occurs near this temperature. Generally, amphibians have lower CTMs than reptiles, and insects have a wide range of tolerance, depending upon the species and degree of conditioning. Death by heat is caused by a combination of factors, including coagulation and destruction of proteins, breakdown of membranes, asphyxiation, and problems connected with accumulation of waste products.

Energy Exchange

The rate at which energy is exchanged between an animal and its environment determines the temperature of the animal in a fairly complex manner. The mechanisms by which energy is exchanged are metabolism, evaporation of water, radiation, convection, and conduction. If an animal is of considerable size, then there is capacity for the storage of energy, and this influences the rate of energy exchange with the environment. A camel, for example, can use its large bulk and enormous heat capacity to store huge quantities of heat during a hot summer day by letting its body temperature rise slowly. Camels have been observed to allow their body temperatures to rise as much as 9°C during the day, returning the stored heat to the environment during the night. For practical purposes, and in order to keep the analysis simple, energy exchange for animals is here given in terms of a steady-state energy budget. It is obvious, however, that much of the time animals are in transient states of energy exchange because of changing environmental conditions. Either the animal is moving about in its habitat or the climatic conditions nearby the animal are changing. The time-dependent analysis of energy exchange for animals will be given in a later chapter.

An animal exchanges energy with its environment across its surface, and this surface has a complex geometry. Metabolic heat is generated by all cells of the body. Moisture is evaporated, with corresponding heat loss, from the lungs or, for some animals, from various parts of the skin surface. Despite all these complexities of form, shape, and process, much can be understood concerning animal energetics by treating an animal as a simple geometrical form with single sources and sinks of energy. Such an analytical model is referred to as a "lumped parameter" model. A "distributed parameter" model requires a great amount of mathematical effort and may not give a great deal of insight to the basic mechanisms of heat transfer for an animal; in addition, for most examples, this kind of analysis requires physical and physiological information about the animal which is not available. Therefore, for our purposes here, an animal is treated as a simple cylindrical form with a single source of total metabolic heat production M expressed in watts per square meter and an evaporative water loss E expressed kilograms per second.

Evaporative water loss is primarily respiratory but some water may be lost by sweating at the skin. For this first treatment of the problem, no distinction is made between the sources of water loss; the distinction will be made in a later chapter. To convert evaporative water loss from kilograms of moisture to equivalent units of heat in joules, it is necessary to multiply E by the latent heat of vaporization λ, which is dependent on the temperature of the evaporating surface. At 30°C, $\lambda = 2.43 \times 10^6$ J kg^{-1}. In the lumped-parameter model, all values of energy are given in energy units per unit surface area of the animal per unit time.

The Analysis

The general energy budget equation for an animal in the steady state is

$$M - \lambda E + Q_a - \epsilon\sigma(T_r + 273)^4 - k_1 \frac{V^{0.5}}{D^{0.5}}(T_r - T_a) - G = 0. \quad (4.1)$$

The quantity $M - \lambda E$ is the net heat, or dry heat production, inside the animal, and Q_a is the amount of radiation absorbed by the surface of the animal. Energy is lost from the animal surface by radiation, convection, and conduction. Energy lost by radiation is represented by the term $\epsilon\sigma(T_r + 273)^4$, where ϵ is the emissivity, σ is the Stefan–Boltzmann constant, and T_r is the temperature of the animal surface. The next term is energy lost by convection, in which k_1 is a proportionality constant, V is the wind speed, D is the "characteristic" dimension of the animal (usually taken as the diameter), and T_a is the air temperature. Finally, G is the quantity of heat lost or gained by conduction to a cold or warm substrate.

Work

Work done by an animal may be included in Eq. (4.1) when it represents energy transferred directly to another object. An example would be a man shoveling snow. Energy is transferred mechanically from the man to the snow. While doing work, an animal will increase its rate of water loss and may even increase its body temperature. A change in potential energy of an animal can be accounted for in Eq. (4.1) in particular instances. Extra metabolic heat generated by work will result in a higher surface temperature, and the extra heat produced will be dissipated by radiation and convection.

Convection

Air moving across the surface of an animal transfers heat to or from the surface depending upon whether the air is warmer or cooler than the surface. The quantities $V^{0.5}$ and $D^{0.5}$ in the convection term represent the convective heat exchange for an object of cylindrical shape. For a cylinder, $k_1 = 3.89$ W m^{-2} s$^{1/2}$°C^{-1}, and the whole convection term is in watts per square meter if V is in meters per second and D is in meters. Detailed discussion of the convection term is reserved for Chapter 9. For the purpose of this introduction, it is assumed that this form of the convection term is reasonably correct for an approximate exchange of energy by an animal.

Body Temperature

Although an animal's surface temperature responds to the flow of energy between the animal and the environment, it is usually internal body temperature that is critical to the animal and is usually measured by an observer. Therefore, it is useful to change surface temperature to body temperature in Eq. (4.1). This is done by considering in simplest form the transfer of the net heat production $M - \lambda E$ from inside the animal through a layer, which may be fat, fur, or feathers, having an insulation I. The net heat production is the metabolic heat produced less the energy transferred in the evaporation of water expelled from the body cavity by breathing. The net heat production must pass to the external environment by flowing across the insulation. The rate of heat flow is inversely proportional to the quality of the insulation and directly proportional to the temperature difference $T_b - T_r$, where T_b is the body or core temperature and T_r is the surface temperature. Hence

$$M - \lambda E = \frac{1}{I}(T_b - T_r). \qquad (4.2)$$

Solving for T_r gives

$$T_r = T_b - I(M - \lambda E). \qquad (4.3)$$

Substituting this value into Eq. (4.1) gives the following expression for the energy budget of an animal in terms of its body temperature:

$$M - \lambda E + Q_a - \epsilon\sigma[T_b + 273 - I(M - \lambda E)]^4$$
$$- k_1 \frac{V^{0.5}}{D^{0.5}}[T_b - T_a - I(M - \lambda E)] - G = 0. \qquad (4.4)$$

Physiological Properties

A particular animal species has a set of physiological and physical properties which are reasonably fixed in value. Body temperature may vary with environmental conditions. This variation may be considerable for a poikilotherm, but even for homeotherms there exists a maximum value, under heat stress, and a minimum value, under cold stress. The metabolic rate, and the rate of water loss generally, change with the environmental temperature or chamber temperature, as shown in Fig. 4.1 for a homeotherm. In the case of a poikilotherm, both the metabolic rate and the water-loss rate increase monotonically with temperature. The metabolic and water-loss rates are affected by the degree of heat or cold stress to which the animal is subjected, as well as by the amount of muscular activity. Nevertheless, despite this range of variation, every species of animal has a characteristic set of values, including the maximum and minimum values for its metabolic and water-loss rates. A particular species has a specific body size and characteristic coloration and quantity of insulation at any stage of development. Body size, in terms of a characteristic dimension for convection, can be varied through pilar erection of feathers, change in thickness of fur with season, by the animal's modifying its shape by hunching or curling, or, of course, by growth. Coloration changes modify the absorptance to radiation of the animal's surface. The quantity of insulation can be varied a great deal from summer to winter, both in fat thickness or amount of fur or feathers. For many animals, the pilar erection of fur will increase insulation at times of cold stress. Nevertheless, despite all the possible variation in parameters, one can specify reasonably well the set of values which characterize a particular species. The parameter values for a chipmunk are not the same as those for a rabbit, sheep, or cardinal, but in each case they can be specified.

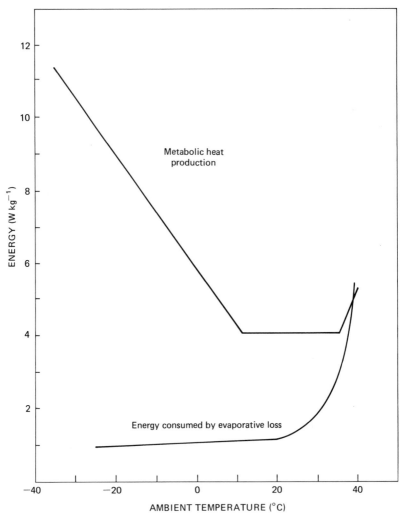

Figure 4.1. Metabolic heat production and energy consumed by evaporative water loss as a function of ambient temperature for a homeotherm.

Every animal accounts for the energy available to it in the manner expressed by Eq. (4.4). This, in fact, is the simplest possible expression of the energy exchange for an animal, the lumped-parameter model, because it assumes that the animal is of simple cylindrical shape.

Heat input to an animal includes metabolic heat production and radiation. Convective heat input may also be a factor if air temperature is greater than surface temperature. Usually, air temperature is less than surface temperature, and convective cooling exists. Energy is also lost by radiation and by evaporation. The conduction term may be either positive or negative, or zero. Most aspects of an animal's energy exchange can be understood using this expression. Radiation is often a very dominant environmental factor in the real world, a phenomena which is only appreciated after working with the numbers involved.

If one knows the values of the parameters of an animal that are important to its energy exchange and can specify the body temperature under conditions of maximum and minimum heat stress, one can determine the set of values of the environmental variables which must exist in order for the animal to survive. In other words, if one specifies the animal's absorptance and emissivity to radiation, metabolic and water-loss rates, insulation, body diameter, and body temperature, one can predict the set of values of radiation, air temperature, and wind speed which must exist for the animal to balance its energy budget. In doing so, analysis is simplified by assuming that no energy goes into storage and that the conduction term is zero. Conduction may be small, depending upon body position relative to a substrate. Actually, a correction for conduction can easily be made, but at this stage of the discussion it is not worthwhile to do so. For large animals, the capacity to store heat over considerable periods of time affects the thermal limits at which survival is possible, but for small animals this is not a significant factor.

Sample Calculations

A series of simple calculations are now shown so that the reader may realize how the theory given above is applied. More elaborate calculations are given in the chapter concerned with energy budgets of animals (Chapter 12).

Radiation Exchange Only

For the sake of illustration, it is assumed that the convection and conduction terms are zero. The energy budget for an animal without convection is then

$$M - \lambda E + Q_a = \epsilon \sigma [T_b + 273 - I(M - \lambda E)]^4. \qquad (4.5)$$

In order to do a numerical example, one needs to work with an animal for which there is good physiological information. Because of the work by Dawson (1958) with the cardinal, it is possible to obtain the values of the pertinent parameters for that animal. The values given by Dawson have been converted to the appropriate quantities necessary for this calculation (the units used there were calories per square centimeter per minute); details are given in the paper by Porter and Gates (1969).

For the cardinal, $M = 75.8$ W m^{-2}, $\lambda E = 75.8$ W m^{-2}, $I = 0.201$ m^2°C W^{-1}, and $T_b = 42.5$°C at thermal maximum. Substitution into Eq. (4.5) gives $Q_a = 562$ W m^{-2}. This simply says that a cardinal without con-

vective exchange cannot survive in a radiation regime in which the
average amount of radiation absorbed by its surface exceeds 562 W m⁻².
The blackbody temperature equivalent to this amount of radiation is
42.5°C, which is exactly the body temperature since $M - \lambda E = 0$ at
thermal maximum.

At thermal minimum, $M = 106.8$ W m⁻², $\lambda E = 3.5$ W m⁻², $I = 0.573$
m²°C W⁻¹, and $T_b = 38.5$°C. Hence $M - \lambda E = 103.3$ W m⁻². Therefore,
substitution into Eq. (4.5) gives $Q_a = 431$ W m⁻².

For these very simplified conditions, under which a cardinal's energy
exchange is not influenced by the air temperature, we find that a cardinal
cannot survive in a radiation regime with less than 431 W m⁻². The blackbody temperature equivalent to this amount of radiation is −34°C.

Radiation and Convection in Still Air

Having completed this simple calculation, one can now solve the
energy-budget equation that includes the influence of air temperature and
wind speed. The easiest way to do this is to select a value for the wind
speed and then find the pairs of values of T_a and Q_a that will balance the
equation. The calculation is set up precisely as before but now includes
the convection term. Assume first that a cardinal is in still air. For this
case, the wind speed is represented not by zero, but by 0.1 m s⁻¹; the
reason is that in still air there is always a little air movement caused by
free or natural convection resulting from natural temperature differences
within the environment. With convection but without storage or conduction, the energy budget of an animal is

$$M - \lambda E + Q_a = \epsilon\sigma[T_b + 273 - I(M - \lambda E)]^4$$
$$+ k_1 \frac{V^{0.5}}{D^{0.5}} [T_b - T_a - I(M - \lambda E)] \quad (4.6)$$

Now one must estimate the size of the animal, in this case the cardinal,
in order to determine the role played by convection in the energy
exchange. Here D is the diameter of the cardinal, which is being approximated by a cylinder. The energy budget is calculated as follows.
At the thermal maximum, $M = 76.8$ W m⁻², $\lambda E = 76.8$ W m⁻², $I =$
0.201 m²°C W⁻¹, $D = 0.05$ m, $T_b = 42.5$°C, and $V = 0.10$ m s⁻¹. Hence
$M - \lambda E = 0$.   Therefore, $Q_a = 5.67 \times 10^{-8}$ $(273 + 42.5)^4 + 3.89$
$[0.10/0.05]^{0.5} [42.5 - T_a]$. This reduces to the following after finding
$[0.10]^{0.5} = 0.316$ and $[0.05]^{0.5} = 0.224$. Therefore, $Q_a = 562 + 5.50$
$[42.5 - T_a] = 796 - 5.50 T_a$. This equation is the locus of all values of
Q_a and T_a that define the thermal maximum environmental conditions
for the cardinal in still air. Any combination of values of absorbed radiation
and air temperature that satisfy this equation are allowed. Characteristic
values are as follows: $T_a = 30$°C, $Q_a = 631$ W m⁻²; $T_a = 40$°C, $Q_a = 576$ W

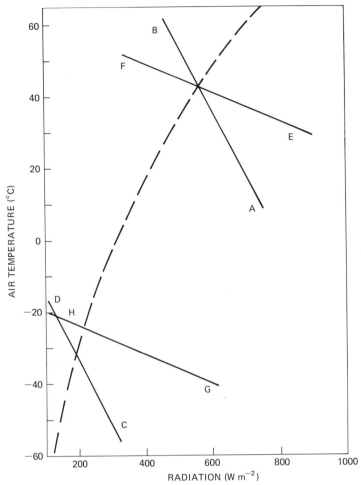

Figure 4.2. Upper and lower thermal limits for a cardinal in climate-space. Lines AB and CD are for a wind speed of 0.1 m s⁻¹ (essentially, still air), EF and GH are for a wind speed of 2.0 m s⁻¹.

m⁻²; T_a = 50°C, Q_a = 521 W m⁻²; and T_a = 60°C, Q_a = 466 W m⁻². These values are plotted in Fig. 4.2 as line AB, in a graph which has air temperature T_a as the ordinate and amount of radiation absorbed Q_a as the abscissa. A graph of this kind is referred to as a "climate space" diagram. At the moment, it contains only two elements of climate, i.e., air temperature and radiation. Presently, wind speed will be included as a variable. Later in the book, when this analysis becomes more sophisticated, the water vapor pressure will also be included. At this point, a very important relationship, the blackbody curve, is added. Any room, cave, burrow, or other enclosure has a well-defined amount of thermal radiation, known as blackbody radiation. The blackbody radiation within such a cavity is a function of the fourth power of the absolute temperature of the walls and also the air within the cavity. The radiation emitted by a blackbody cavity is represented by the broken line in Fig. 4.2. The reason for adding this curve will become more apparent later on; for the time being, it can serve as an excellent reference level for the amount of energy available at any particular temperature.

At the thermal minimum, $M = 106.8$ W m^{-2}, $\lambda E = 3.5$ W m^{-2}, $I = 0.573$ m^2°C W^{-1}, $D = 0.05$ m, $T_b = 38.5$°C, and $V = 0.10$ m s^{-1}. Therefore, $103.3 + Q_a = 5.67 \times 10^{-8} [273 + 38.5 - 59.2]^4 + [38.5 - T_a - 400] + 5.48 [38.5 - T_a - 59.2]$.

$$Q_a = 13.7 - 5.48 \, T_a.$$

This equation now represents the locus of all values of Q_a and T_a that define the thermal minimum environmental conditions for the cardinal in still air. Characteristic values are as follows: $T_a = -20$°C, $Q_a = 123.3$ W m^{-2}; $T_a = -30$°C, $Q_a = 178$ W m^{-2}; $T_a = -40$°C, $Q_a = 233$ W m^{-2}; and $T_a = -50$°C, $Q_a = 288$ W m^{-2}. These values are plotted in Fig. 4.2 as line CD, which represents the lower boundary in climate-space for a cardinal in still air. The actual value of thermal insulation used by Porter and Gates (1969) at the thermal minimum for the cardinal was slightly different than the value used for the calculation given above. The lower bounds in the climate-space diagrams differ slightly in these two calculations. Also, Porter and Gates used $V^{0.33}/D^{0.66}$ rather than $V^{0.5}/D^{0.5}$. The latter is found to be more suitable at the wind speeds normally encountered.

Radiation and Convection in Wind

Any wind speed may be used to calculate the upper and lower limits in climate-space. As an example, select a wind speed $V = 2$ m s^{-1}. Now $V^{0.5} = 1.414$, $D^{0.5} = 0.224$, and $k_1 V^{0.5}/D^{0.5} = 24.6$. The basic energy budget equation for thermal maximum is now

$$Q_a = 1608 - 24.6 \, T_a.$$

The set of values which balance this equation are $T_a = 30$°C, $Q_a = 870$ W m^{-2}; $T_a = 40$°C, $Q_a = 618$ W m^{-2}; and $T_a = 50$°C, $Q_a = 378$ W m^{-2}. These values are plotted as line EF in the climate-space diagram presented in Fig. 4.2. The reader should notice several things concerning this line. The first is that the line EF is rotated closer to the horizontal than line AB. In other words, energy content of the cardinal is more tightly coupled to air temperature for this wind speed. The second feature of interest is that the lines AB and EF intersect precisely at the blackbody line. This occurs only when $M - \lambda E = 0$. Clearly, the crossover point of lines AB and EF occurs when the rate of convective heat transfer is zero, which for $M - \lambda E = 0$ is when $T_a = T_b$. For this situation, the total energy budget reduces to $Q_a = \epsilon\sigma[T_b + 273]^4$, the blackbody line in Fig. 4.2. Hence the intersection of lines AB and EF occurs at the blackbody line.

At the thermal minimum, for the same values as previously, the basic energy budget becomes

$$Q_a = -383 - 24.6\ T_a.$$

The quantities that satisfy this equation are $T_a = -20°C$, $Q_a = 109$ W m^{-2}; $T_a = -30°C$, $Q_a = 355$ W m^{-2}; and $T_a = -40°C$, $Q_a = 601$ W m^{-2}. These values are plotted as line GH in Fig. 4.2. Note that once again, as a result of increased wind speed, the line has rotated to a position closer to horizontal. The increased convective heat transfer has coupled the energy content of the cardinal more closely to the air temperature. This time, however, the two lines CD and GH do not intersect on the blackbody line. The reason they do not is that the net heat, or dry heat production, $M - \lambda E$, is not zero as at the thermal maximum. The cardinal survives the winter in Minnesota and New England, where temperatures occasionally reach $-30°C$. Under these circumstances, a cardinal would be careful to avoid wind. Although a cardinal can maximize its insulation by means of the pilar erection of feathers, wind would tend to compress the feathers and reduce insulation somewhat.

Climate-Space

The climate-space graph is not complete, since the lines are apparently unbounded at their upper and lower ends. The cardinal can withstand any combination of T_a and Q_a along AB and CD in still air and along EF and GH for wind of 2.0 m s^{-1}; however, a little knowledge concerning the combinations of air temperature and radiation fluxes which occur in the real world will show very distinct limits to the lengths of the lines. An indication of a precise relationship between temperature and radiation is the blackbody line, which, as stated earlier, is the amount of radiation within any box, room, cave, burrow, or other cavity as a function of air and wall temperature. At equilibrium, the air and wall temperatures are the same for a blackbody cavity.

Consider the situation at night on the surface of the earth. The ground, comprised of soil, rock, water, or vegetation, will radiate approximately as a blackbody. The emissivity of the ground at ordinary temperature, $300° \pm 50°K$, is nearly 1.0. Consider an organism on the ground surface, sandwiched between the ground surface and the cosmic cold of space, during the night. If the clear sky emitted no radiation, an organism on the surface would receive, on the average, one-half of the blackbody flux

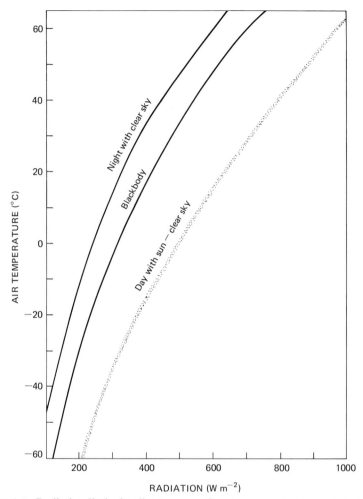

Figure 4.3. Radiation limits in climate-space for clear sky at night and during the day. The radiation intensity for a blackbody is also shown. Solar absorptance is 0.8.

emitted by the ground. A clear sky, however, emits "graybody" radiation from the water vapor and carbon dioxide gases of the atmosphere. Most of this radiation originates in the lower troposphere, where the density of these gases is greatest. This means that the amount of atmospheric radiation emitted downward toward the ground is a function of the air temperature near the ground. An organism on the ground surface under a clear night sky receives a weak flux of downward radiation incident on its upper surface and a flux of blackbody radiation incident on its under surface. The net result of this is that the average flux of radiation received by an organism at night is substantially greater than half the blackbody values, as represented in Fig. 4.3 by the line marked "night with clear sky." This line represents the minimum amount of radiation flux incident upon an organism in a natural environment at the surface of the earth. It is true that there can be some variation of this line because of the variable conditions of humidity and air density that occur on earth, but variation is remarkably small.

If the sky is overcast rather than clear, an organism receives blackbody infrared radiation from the base of the clouds according to the base temperature, as well as blackbody infrared radiation from the ground. The average radiation flux incident on an organism located between the ground and clouds is greater than the line marked "night with clear sky" and less than the line marked "blackbody." Therefore, all atmospheric conditions—clear sky, overcast, or partially cloudy—are approximately represented by a position on this graph corresponding to the air temperature at the particular site.

The daytime situation is more complex than that at night because of the presence of sunshine in addition to the ubiquitous infrared thermal-radiation regime of the environment. Early or late in the day, the sun is near the horizon, and the sun's rays traverse a long, slanted path through the atmosphere. The amount of solar radiation incident upon a surface normal to the sun's rays is relatively weak early or late in the day. At these times of day, air temperature is less than at midday. If the receiving surface is turned at an angle to the direct rays, the amount of solar radiation received is reduced by the cosine of the angle of incidence. At noon, the sun is high in the sky, and the sun's rays are more intense than early or late in the day. At noon, however, or, more precisely, in early afternoon, the air temperatures are greater than at other times of the day as the result of solar heating. Also, in the winter, when air temperatures are low, the sun is lower in the sky during midday than during the summer, and the solar intensities are reduced. In other words, there is a relatively good general correlation between the intensity of solar radiation and the air temperature at the site. Admittedly, this correlation between solar radiation intensity and air temperature is not precise, but is, generally speaking, valid. The justification for using this relationship is even greater when one considers the absorptance factor of most organisms and the amount of radiation actually absorbed by the organism. Since, for most animals, absorptance to solar radiation is between 0.6 and 0.9 and that to infrared thermal radiation is 0.95 or greater, solar radiation is not as strongly coupled to the animal's temperature as thermal radiation. The fact that the sun is a point source rather than an extended source also reduces the contribution of solar radiation to the total energy budget because of the angle between the sun's rays and the normal to the animal's surface.

The result of all these considerations is that an upper boundary for the contribution of solar radiation to the climate-space of an animal may be drawn (see Fig. 4.3). An example is shown for an absorptance of 0.8, but

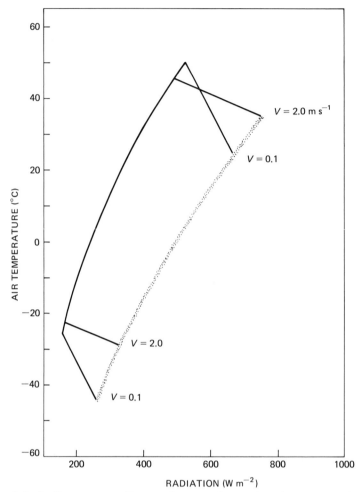

Figure 4.4. A climate-space diagram for a cardinal, with wind speeds of 0.1 and 2.0 m s⁻¹.

with the sun's rays normal to the animal's surface. The line is stippled because its position is less exact than the positions of the other lines in the figure. Any object whose absorptance to solar radiation is 0.8 will not, in general, absorb more radiation (solar plus thermal) than that shown by the stippled line. It is a valuable exercise for any student interested in this subject to consider the various streams of radiation in the natural daytime environment and see if he or she can arrive at an amount of absorbed radiation more or less the same as the right-hand line of Fig. 4.3. An organism that absorbs less than 0.8 of the solar radiation will have a right-hand limit in the climate-space diagram somewhere to the left of the stippled line. Of course, if absorptivity were zero, then, even in the daytime, the position of this limit would be essentially that of the left-hand line, marked "night with clear sky."

A combination of Figs. 4.2 and 4.3 gives the climate-space diagram for a cardinal shown in Fig. 4.4. The upper and lower bounds of the climate-space are basically physiological limits, the left- and right-hand bounds are essentially environmental limits. It is clear, however, that the upper and lower bounds are influenced by convective heat transfer, which involves the air temperature and wind speed. There is a very slight physiological influence on the left- and right-hand boundaries by the animal's absorptance. Therefore, none of the climate-space limits is completely either physiological or environmental. There are small differences in the calculated upper and lower bounds of the climate-space diagrams here and in some of the published literature owing to the use of slightly different convection coefficients. The convection coefficient is one of the more difficult quantities with which to be precise, and various values have been used. Generally, values do not vary by more than about 20%.

The climate-space illustrated in Fig. 4.4 was derived by using physical and physiological information representative of the cardinal. Similar climate-spaces can be derived for any other animal, either warm or cold-blooded. The climate-space for a cold-blooded desert iguana is shown in Fig. 4.5. Calculation of this climate-space is left as an exercise at the end of this chapter. It is clear from Fig. 4.5 that the climatic conditions within which a desert iguana is constrained to live are much more restrictive than those for a cardinal. The iguana can withstand substantially warmer air temperatures while in the sun than the cardinal. The reverse is true for cold-temperature tolerance. Whereas the cardinal can withstand extremely low air temperatures, even with some air movement, the desert iguana is unable to tolerate air temperatures very much below the freezing point for water, except in still air and full sunshine.

The climate-space diagram is intended to show the broad limits that an animal can tolerate in steady-state conditions as the result of certain inherent thermodynamic properties, both physical and physiological, of the animal. An animal in its natural habitat will, whenever possible, avoid exceeding its climate-space limits for long periods of time. Animals will, however, often exceed the boundaries of their climate-space when in transient rather than steady-state situations, providing average conditions do not exceed the limits of their climate-space. Environmental conditions in the vicinity of an animal are constantly changing or else an animal is moving about from one set of conditions to another. An animal may emerge from its nest, den, or burrow, bask in the sun until conditions become too warm, and then move into the shade. An animal basking in the sun will encounter constantly changing conditions such as the change from still air to wind and from direct sunlight to diffuse sunlight, as well as variable air temperatures and humidity. Whether the animal is moving about within the environment of its habitat or environmental conditions

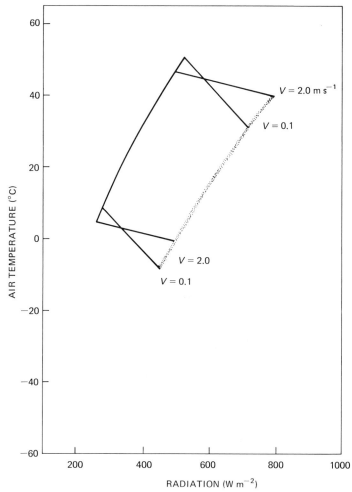

Figure 4.5. A climate-space diagram for a desert iguana, with wind speeds of 0.1 and 2.0 m s⁻¹.

are changing with respect to the animal does not really matter. The significant fact is that, one way or the other, environmental conditions are changing relative to the animal. Every animal has a time constant related to its rate of change of temperature for a given amount of heat gain or loss. This time constant is characteristic of the animal's heat capacity, which in turn is proportional to the animal's mass, or weight. A large animal, like an elephant or hippopotamus, will have a time constant of many hours, perhaps even a day or more, whereas small insects may respond to heat input or output in a second or less. Many animals of moderate size will have time constants of a few minutes. The length of time an animal can remain in an environmental situation which exceeds the boundaries of its climate-space depends upon the animal's time constant and the particular conditions to which it has adjusted or equilibrated before exceeding the limits of its climate-space. Time-dependent energy budgets and climate-spaces for animals will be given in a later chapter.

Rather than predicting the climate-space limits for an animal of known metabolic rate, water-loss rate, thermal insulation, body size, and so forth, one can assume certain climatic conditions, as well as some animal characteristics, and predict the metabolic rate which the animal must have. From the metabolic rate as a function of climatic conditions, one can predict the quantity of food of specific energy content necessary to the animal. One can then consider the distribution of potential food supply and estimate the number of animals within a certain area. It is possible to estimate the competition for food among various species if the energy demand of each species is understood.

Our understanding of predator–prey relationships is advanced knowledge of the energy tolerances and climatic limits, particularly in time-dependent climate-space, of predator and prey. One can predict from energy-budget considerations where a prey must locate itself within a habitat as a function of the time of day. There will be some positions within the habitat from which an animal is excluded from an energy standpoint. There are other regions within the habitat where the animal cannot spend too much time, as for example in a burrow, if it is going to gather food. One can therefore map out the horizontal and vertical positions that an animal can occupy within a habitat as a function of the time of day. If this is done for the predator as well as the prey, one can predict the amount of overlap in time and space of the positions of the two animals. From this, one can show when and where the predator has a chance to find the prey.

The geographical distribution of animals in the world can be predicted if one understands the detailed energy requirements of each animal and its climate-space. It is also possible to make an analysis of the energy required by an animal during migration. The energy required for incubating eggs and feeding young can be determined if one understands the flow of energy between the environment and the nest or den. A good deal concerning animal behavior is learned by understanding the energy restrictions imposed on an animal either by its own physical and physiological characteristics or by habitat conditions.

Problems

1. Determine the climate-space for a desert iguana, *Dipsosaurus dorsalis*, with the characteristics $\epsilon = 1.0$ and body diameter $D = 0.015$ m. At thermal maximum, $M = 10.5$ W m^{-2}, $I = 0.005$ m^2°C W^{-1}, $\lambda E = 6.3$ W

m^{-2}, and $T_b = 45°C$. At thermal minimum, $M = 0.1$ W m^{-2}, $I = 0.005$ $m^2°C$ W^{-1}, $\lambda E = 0.1$ W m^{-2}, and $B_b = 3°C$.

For air flow speeds of $V = 0.1$ m s^{-1} and 2.0 m s^{-1}, determine the climate-space limits at thermal maximum and thermal minimum in the form

$$Q_a = A + B\, T_a.$$

Plot these lines on a graph with the same scales as Figs. 4.2 and 4.3. Now place on the graph the radiation limits as shown in Fig. 4.3. These lines now form an approximate climate-space for the desert iguana.

2. Determine the climate-space for a masked shrew, *Sorex cinereus*, with the characteristics $\epsilon = 1.0$ and body diameter $D = 0.018$ m. At thermal maximum, $M = 140$ W m^{-2}, $I = 0.085$ $m^2°C$ W^{-1}, $\lambda E = 28$ W m^{-2}, and $T_b = 41°C$. At thermal minimum, $M = 349$ W m^{-2}, $I = 0.125$ $m^2°C$ W^{-1}, $\lambda E = 3.5$ W m^{-2}, and $T_b = 37.5°C$.

For an air flow speed of $V = 0.1$ m s^{-1}, determine the climate-space limits at thermal maximum and thermal minimum in the form

$$Q_a = A + B\, T_a$$

Plot these lines on a graph with the same scales as Figs. 4.2 and 4.3. Now place on the graph the radiation limits as shown in Fig. 4.3. These lines now form an approximate climate-space for the shrew.

(Show by cross-hatching the part of the graph in which the shrew actually lives.)

3. If the shrew with the same characteristics described in Problem 4.2 is without any convective heat exchange, what will its climate-space limits be? What do these calculations suggest to you in terms of the habitat in which the shrew actually lives?

4. You are holding a shrew in the palm of your hand. Consider your hand to be a blackbody cavity at 32°C. Will the shrew survive a steady-state condition in this situation? (Use the climate-space diagram determined in Problem 4.2.)

5. Determine the climate-space for a domestic sheep, *Ovis ovis*, with the characteristics $\epsilon = 1.0$ and body diameter $D = 0.25$ m. At thermal maximum, $M = 91$ W m^{-2}, $I = 5.1$ $m^2°C$ W^{-1}, $\lambda E = 95$ W m^{-2}, and $T_b = 41.7°C$. At thermal minimum, $M = 70$ W m^{-2}, $I = 4.0$ $m^2°C$ W^{-1}, $\lambda E = 13$ W m^{-2}, and $T_b = 39.5°C$.

Chapter 5

Radiation Laws, Units, and Definitions

Introduction

Electromagnetic radiation is ubiquitous in the world in which we live. We know this radiation as visible light, infrared radiation (radiant heat), microwaves (radio waves), ultraviolet radiation, X-rays, gamma rays, etc. Regardless of the names we apply to various wavelength bands of radiation, they are all part of the continuum of electromagnetic radiation that exists in the universe.

Solar radiation heats the planet earth, and infrared thermal radiation emitted by the earth's surface to space cools the planet and keeps it in energy balance with the sun. The mean temperature of the earth is determined by this radiation balance. The extremes of temperature with season and time of day are directly related to the radiative exchange of any position on earth with the sun and outer space. Too much or too little radiation would not have permitted life to evolve in the form we know today on earth. Other forms of life may form under circumstances different from those under which life on earth has evolved, but life is not likely to have come about under conditions of intense or weak radiation. All primary productivity in the earth ecosystem depends on the energy from sunlight, but these processes will not function if the temperature is either too high nor too low. Generally, life on earth exists in areas in which the mean annual temperatures is between 0° and 40°C. Temperatures above 100°C are fatal to most forms of life. Reproduction does not occur at extremely low temperatures, and although some organisms can survive to nearly absolute zero, temperatures must approach 0°C for reproduction to occur. Temperature is a measure of the energy status of matter, but the energy status of ecosystems on earth is a consequence of the energy balance of the planet earth, existing in the environment of the sun, an environment which is primarily one of radiative transfer. To understand the functioning of the earth's ecosystem, one must first understand the process of radiation.

Radiation Laws

Radiation is the energy emitted from the surface of all bodies in the form of electromagnetic waves. Radiation travels at the "speed of light" through a vacuum and is also transmitted through air and other semitransparent media. All surfaces emit radiation proportional to the fourth power of their absolute temperature. The quantity and character of radiant energy emitted by a surface per unit time and unit area of surface depends on the nature of the surface as well as its temperature. A surface at a low temperature emits radiation of infrared wavelengths only, but a surface at a high temperature, such as an incandescent tungsten filament in a light bulb, emits radiation of shorter wavelengths, i.e., in the visible part of the spectrum, as well as infrared wavelengths. The sun, radiating at a very high temperature, emits a broad spectrum of radiation, including X-rays, ionized particles, ultraviolet radiation, visible light, infrared radiation, and radio waves; however, only a portion of this spectrum penetrates the earth's atmosphere. Most sources emit a broad spectrum of radiation. Some substances, however, particularly metallic vapors and many gases, either emit monochromatically or exhibit a spectrum characteristic of the quantum states of the atoms or molecules comprising the substance.

The laws of radiation physics are well understood and form the foundation of much of classical and quantum physics. The electromagnetic theory, established by James Clerk Maxwell (1831–79) about 1870, explained the propagation of radiation and showed that radiation travels with the speed of light in a vacuum and at a slower speed in any other medium. An attempt by Max Planck to explain the observations of radiation physics led to the development of quantum theory and modern atomic physics.

The Stefan–Boltzmann Law

It was first suggested by Stefan, in 1879, that the total radiation emitted by a heated body is proportional to the fourth power of its absolute temperature. In 1884, Boltzmann showed that the fourth-power law could be derived from thermodynamics. Boltzmann's reasoning was based on the concept of radiation contained within a blackbody cavity being in equilibrium with the temperature of the walls. A blackbody is, by definition, a surface or an object that absorbs completely any radiation incident upon it. Most surfaces reflect some radiation and are not, therefore, "black." A small hole or aperture in a box or cavity is considered a truly "black" surface. Any radiation entering the aperture is trapped within and has only a remote chance of ever getting back out. The aperture appears perfectly black to any observer on the outside. The enclosure itself is known as a blackbody cavity. When the enclosure is maintained at a constant temperature, a certain amount of radiation per second passes out through the aperture. This radiation is known as "blackbody" radiation and has a unique energy spectrum, which exhibits a special relationship to the temperature of the enclosure. Many situations in nature approximate blackbody conditions. The interior of a forest with a closed canopy approximates a blackbody, and a room in a house or a cave or burrow exhibits the properties of a blackbody if the walls are at a uniform temperature.

The Stefan–Boltzmann radiation law states that the rate of emission of energy from a blackbody is proportional to the fourth power of the absolute temperature of the blackbody. Its mathematical form is as follows:

$$\mathcal{R} = \sigma[T + 273]^4, \tag{5.1}$$

where σ is a proportionality constant known as the Stefan–Boltzmann constant, T is the surface temperature in degrees Centigrade, and $\mathcal{R}$ is the radiant emittance of the source in watts per square meter. The value of σ has been determined with considerable precision and is

$$\sigma = 5.673 \times 10^{-8} \text{ W m}^{-2}{}^{\circ}\text{K}^{-4}.$$

A plot of blackbody radiation as a function of surface temperature is given in Fig. 7.1 in Chapter 7. Because of the uses to which it will be put later, the graph gives the temperature scale in degrees Centigrade. In the calculation given above, however, it is necessary to use the absolute temperature, i.e., degrees Kelvin, or T (in degrees Centigrade) + 273. Tables of blackbody radiation are given in the *Smithsonian Meteorological Tables*, edited by List (1963).

Most objects are not perfect blackbody radiators but have a radiant emittance that is a fraction of blackbody radiant emittance. All such objects are known as gray bodies, and their radiant emittance is given by

$$\mathcal{R} = \epsilon\sigma[T + 273]^4, \tag{5.2}$$

where ϵ is the emissivity of the surface and has a value between 0 and 1.0. The emissivity of the surface is always equal to the absorptivity of the surface or, stated another way, a good absorber is a good emitter and a poor absorber is a poor emitter.

Linearized Approximation of Radiative Exchange

An object exchanging energy with its surroundings by radiation emits a quantity of energy proportional to $(T + 273)^4$, where T is the surface temperature of the object, and receives a quantity of energy proportional to $(T_0 + 273)^4$, where T_0 is the uniform temperature of the surroundings. If the object and the surroundings each have an emissivity ϵ, then the net amount of energy dE gained or lost by an object of area A per unit surface area and per unit time is

$$\frac{1}{A}\frac{dE}{dt} = \epsilon\sigma[(T + 273)^4 - (T_0 + 273)^4]. \qquad (5.3)$$

It is convenient, in this case, to linearize the fourth-power blackbody law by approximating the difference between T and T_0 by using the following mathematical expansion. Replacing T by $T_0 + \Delta T$ gives

$$\frac{1}{A}\frac{dE}{dt} = \epsilon\sigma[(T_0 + 273 + \Delta T)^4 - (T_0 + 273)^4].$$

Then, by expansion,

$$\begin{aligned}\frac{1}{A}\frac{dE}{dt} &= \epsilon\sigma[4(T_0 + 273)^3 + 6(T_0 + 273)^2\,\Delta T \\ &\qquad + 4(T_0 + 273)\Delta T^2 + \Delta T^3]\,\Delta T \\ &\simeq 4\epsilon\sigma(T_0 + 273)^3\,\Delta T, \qquad (5.4)\end{aligned}$$

or

$$\frac{1}{A}\frac{dE}{dt} = K(T - T_0),$$

where $K[= 4\epsilon\sigma(T_0 + 273)^3]$ is a constant.

This is now a convenient linearized approximation to radiative heat transfer. Actually, it is a good approximation only for small temperature differences, and otherwise is considered a poor approximation. If $T_0 + 273 = 300°K$, for example, then heat flow as calculated by the linearized approximation is too low by 22% if $\Delta T = 50°K$, 9% if $\Delta T = 20°K$, and 5% if $\Delta T = 10°K$. The linearized approximation may be used with reasonable accuracy only when temperature differences are less than $10°K$ and the emissivities are nearly equal. If the emissivities differ, then the total error depends upon whether the error due to emissivity differences adds to or cancels that due to temperature differences.

Some authors refer to the linearized law of radiative exchange as Newton's law of cooling, and there has been much confusion concerning this matter in the scientific literature. The problem of nomenclature is clarified by the paper by Bakken and Gates (1974) and the papers referred to therein. More will be said concerning linearized heat transfer in the chapters concerning convection and its application to heat loss or gain from animals.

Planck's Distribution Law

One of the most significant properties of the radiation emitted by a blackbody is its wavelength, or frequency, distribution. Early observations showed that the radiation emitted by a blackbody has a broad spectral dis-

BLACKBODY RADIATION

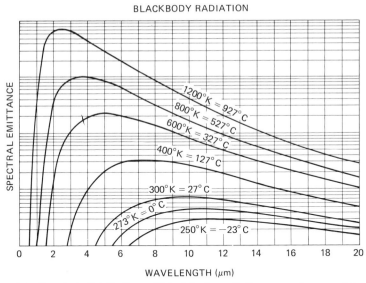

Figure 5.1. Spectral distribution of blackbody radiation at various emitting temperatures.

tribution with a maximum amount of energy radiated at a particular wavelength, the position of which shifts with the temperature of the source.

The spectral-distribution curves for blackbodies at various temperatures are shown in Fig. 5.1. Radiant flux per unit increment of wavelength, or spectral emittance, is plotted against wavelength to show the distribution of energy with wavelength. It can be seen that the wavelength corresponding to maximum radiant flux decreases with increasing temperature. The spectral properties of radiators are commonly experienced. Anyone sitting before a roaring campfire and watching it burn out and grow cold sees the embers turn from 'white hot' to bright orange, to red, to a dull red and then fade from view, while continuing to give off heat. What is being experienced is the continuous shift of the radiation spectrum as fire dies out and the embers cool. Similarly, if the voltage to an incandescent bulb is slowly reduced, one sees the lamp, which had been glowing white-hot giving off a brilliant light fade from orange to deep red, until, finally, it is not visible, though still giving off infrared warmth.

Max Planck (1858–1947), the great German physicist, worked hard and long to achieve a theoretical derivation of the spectral distribution of the radiation emitted by a blackbody. Unable to derive a consistent and complete theoretical derivation purely from classical physics, Planck invoked a radical new concept, i.e., that the interaction between matter and radiation is not a continuous process, but that matter acts as a collection of linear oscillators from which radiation is emitted or absorbed in multiples of some small energy unit. This was the birth of quantum theory and brought with it a complete theoretical derivation of the spectral distribution law for blackbody radiation. The law which Planck derived for the spectral emittance of a blackbody is expressed as follows:

$$\frac{d\mathcal{R}}{d\lambda} = c_1\lambda^{-5}\,(e^{c_2/\lambda T} - 1)^{-1},$$ (5.5)

where λ is the wavelength in meters, e is the base of the natural logarithm, T is the absolute temperature in degrees Kelvin, c_1 is a constant equal to 3.740×10^{-8} W m^{-2}, and c_2 is a constant equal to 143.85 m °K. This gives $d\mathcal{R}/d\lambda$ in watts per square meter per meter of wavelength or watts per cubic meter. If Planck's spectral-distribution law is integrated over wavelength, the Stefan–Boltzmann radiation law [Eq. (5.1)] is obtained. Tables listing the spectral emittance of blockbody radiation are given in the *Smithsonian Meteorological Tables*, edited by List (1963).

Wien's Displacement Law

By differentiating Planck's law and equating the derivative $\partial^2\mathcal{R}\lambda/\partial\lambda^2$ to zero, i.e., evaluating the equation for zero slope or the position of the maximum in the wavelength distribution, Wien's displacement law is obtained. This law states that the product of the wavelength for the maximum intensity multiplied by the absolute temperature is a constant and is written

$$\lambda_{\max}\,T = 289.7 \times 10^{-5} \text{ m°K}.$$ (5.6)

It is clear that if $T = 289.7$°K, $\lambda_{\max} = 10^{-5}$ m (10 μm), if $T = 579.4$, $\lambda_{\max} = 0.55 \times 10^{-5}$ m (5 μm), and if $T = 5794$°K—approximately the apparent "blackbody" temperature of the sun—$\lambda_{\max} = 0.05 \times 10^{-5}$ m (0.5 μm). One micrometer (abbreviated μm) is equal to 10^{-6} m, and this unit is conventionally used for the measurement of wavelength, particularly for infrared radiation. The plots shown in Fig. 5.1 indicate the shift in spectral distribution of radiant energy with change in temperature of the radiating body. We will leave out the units in Fig. 5.1 of only show relative values.

Wave Nature of Radiation

The electromagnetic theory of Maxwell indicated that radiation, including light has a wave nature, but final experimental verification was accomplished by Heinrich Hertz (1857–94) in 1887, when he produced electromagnetic waves by means of oscillating circuits. The fundamental wave equation deduced by Hertz, which was derived earlier by Maxwell as a part of the electromagnetic theory, states that radiation travels with a speed c equal to the product of the frequency ν wavelength λ of the radiation. Hence

$$c = \nu\lambda. \tag{5.7}$$

In a vacuum or in free space, $c = 3 \times 10^8$ m s^{-1}. The unit of frequency has been named for Hertz, and one hertz (abbreviated Hz) equals a frequency of one cycle per second.

An electromagnetic wave, is a transverse wave with electric and magnetic fields oscillating at right angles to one another and energy being propagated in a third direction, at right angles to the plane containing the electric and magnetic field vectors. Radiation obeys the well-known relationships of geometrical and physical optics, including reflection, transmission, absorption, refraction, scattering, diffraction, interference, and polarization.

Wavelength is measured in various units, including meters (m), centimeters (cm), micrometers (μm) and angstroms (Å). More recently, the nanometer (nm) has come into use as a standard unit of wavelength. These various units are related as follows:

$$1 \text{ m} = 10^2 \text{ cm} = 10^6 \text{ } \mu\text{m} = 10^9 \text{ nm} = 10^{10} \text{ Å}.$$

By using the relationship $c = \nu\lambda$, it is easy to determine the frequency associated with a given wavelength. Frequency is measured in cycles per second, or hertz. Spectroscopists often use a frequency unit called the wavenumber $\tilde{\nu}$, which expresses the number of oscillations the wave takes in traveling a distance of 1 cm. The wavenumber is simply the reciprocal of the wavelength and has the dimensions of number of oscillations per centimeter. Hence

$$\tilde{\nu}\lambda = 1. \tag{5.8}$$

In expressing the frequency distribution of radiation, it is often necessary to describe the monochromatic radiant emittance in power per unit of area per unit of frequency. In order to do this, it is necessary to determine the amount of energy per increment of wavenumber, or frequency. The relationships between the different factors are determined by differentiation of Eqs. (5.7) and (5.8) and are as follows:

$$d\lambda = \frac{d\nu}{\nu^2} \; d\lambda = \frac{c}{\nu^2} \, d\nu \tag{5.9}$$

$$d\nu = \frac{d\lambda}{\lambda^2} \quad d\nu = \frac{c}{\lambda^2} \, d\lambda. \tag{5.10}$$

It is evident that the translation from an increment of wavelength to an increment of frequency or wavenumber, or vice versa, is always nonlinear and increases according to the square of the quantity.

Some of the much-used categories of radiation, along with their wavelengths, frequencies, and wavenumbers, are listed in Table 5.1.

Système International

Every student of science should have a clear understanding of the basic units of physical measurement. The student of biology is no more removed from the need for these basic concepts than the student of physics. In this book, we adopt *Système International* (SI) units, which are basically those of the meter–kilogram–second (MKS) system. For practical reasons, it is occasionally necessary to depart from this practice. Often, convention dictates the use of certain units. For example, the entire chapter on photosynthesis uses non-SI units and some graphs in the book will be expressed in units of calories per square centimeter per minute. We also use units of the centimer–gram–second (cgs) system so that the student can realize their relationship to the MKS system.

Force

A force f is a quantity which imparts to a mass m an acceleration a. This relationship is defined by the equation

$$f = ma, \qquad (5.11)$$

where f is in newtons, m is in kilograms, and a is in meters per square second. The weight of an object is the force acting on the object owing to the attraction or pull of gravity. The acceleration of an object by gravity is 9.80 m s^{-2}. Hence a 1-kg mass has a force exerted on it by gravity of 9.80 N. The 1-kg mass is said to weigh 9.80 N or have a weight of 9.8-N force. Mass is a scalar quantity, whereas weight and force are vectors. A vector has magnitude and direction; a scalar has magnitude only. Weight is the force acting toward the center of the earth owing to gravity.

Table 5.1. Relationship between Wavelength, Frequency, and Wavenumber of Radiation. The Interval by Type of Rad Expressed in W, F, and W.

	Wavelength (nm)	Frequency (s^{-1})	Wavenumber (cm^{-1})
Type of radiation			
Ultraviolet	10–400	3.00×10^{16}–7.50×10^{14}	1×10^6–2.50×10^4
Visible	400–700	7.50×10^{14}–4.28×10^{14}	2.50×10^4–1.43×10^4
Blue	480	6.25×10^{14}	2.08×10^4
Green	555	5.40×10^{14}	1.80×10^4
Yellow	600	5.00×10^{14}	1.67×10^4
Red	660	4.54×10^{14}	1.52×10^4
Infrared	700–10^6	4.28×10^{14}–3.00×10^{11}	1.43×10^4–10

Energy

Energy is defined as the ability for doing work. Work is done when a force f acts on a body to move it through a certain distance x. The amount of work done is given by

$$W = fx. \tag{5.12}$$

If f is in newtons and x is in meters, work is in units of newton-meters, or joules. In calorimetry, the branch of physics dealing with heat, it has been the custom to use a unit known as the calorie as a measure of energy. One calorie is defined as that quantity of heat (or energy) which will raise the temperature of 1 g of water from 14.5° to 15.5°C. Often, the "large calorie," or kilocalorie, is used for expressing the energy value of food. One kilocarlorie equals 1000 calories. Since the calorie is fundamentally an energy unit, it can be related to the basic energy units defined by mechanics. By means of experiments, the "mechanical equivalent of heat" has been determined to be 4.185 J cal^{-1}.

Power is defined as the rate at which work is done or energy is expended:

$$P = \frac{dW}{dt}. \tag{5.13}$$

In SI units, power is expressed in joules per second, or watts.

Quanta

Radiation is represented as electromagnetic waves in the "wave theory" of light and as bundles or packets of energy (known as quanta) in the "particle theory" of light. For many biological phenomena, the number of quanta of a given frequency associated with a particular radiant flux is important. Individual quanta invoke specific photochemical responses in living systems. According to Planck's quantum law, a single quantum of radiation contains energy E proportional to the frequency according to the relationship

$$E = h\nu, \tag{5.14}$$

where h is Plank's constant and is equal to 6.62×10^{-34} J s. Then, from the basic relationship between frequency, wavelength, and the speed of light given by Eq. (5.7),

$$E = hc/\lambda. \tag{5.15}$$

Substituting the values $h = 6.62 \times 10^{-34}$ J s and $c = 3 \times 10^8$ m s^{-1} gives

$$E = 19.86 \times 10^{-26} \text{ J m}/\lambda.$$

For one quantum per second, the equivalent energy rate, or power, unit, in watts (joules per second), is

$$dE/dt = 19.86 \times 10^{-26} \text{ W m}/\lambda,$$

when wavelength is given in meters.

Later on in this chapter, it is shown that at the peak sensitivity of the human eye, i.e., at a wavelength of 555 nm, 1 lm (lumen), the unit of luminous flux, is equivalent to 0.00147 W. Since 1 quantum s^{-1} of energy at a wavelength of 555 nm represents a power of 3.59×10^{-19} W, then 1 lm of luminous flux at a wavelength of 555 nm is $n/dt = 4.08 \times 10^{15}$ quanta s^{-1}. Similarly, 1.47×10^{-3} W of green light of wavelength 555 nm is equivalent to 4.08×10^{15} quanta s^{-1}.

The number of quanta equivalent to the same amount of energy at other wavelengths is greater owing to decreased sensitivity of the human eye at wavelengths other than 555 nm. At a wavelength of 680 nm, 1 lm is equivalent to 242×10^{15} quanta s^{-1}, whereas at 450 nm, 1 lm is equivalent to 108×10^{15} quanta s^{-1}.

Einstein

Often, the use of non-SI units is established by convention. An example is the einstein, which is used to measure light intensity, or photon flux, for photosynthetically active wavelengths. This unit will be used in the discussion of photosynthesis (Chapter 14). One einstein (abbreviated E) is defined as a number of quanta equal to Avogadro's number $(6.023 - 10^{23})$, i.e., the number of quanta per molar volume. Hence 1 E $= 6.023 \times 10^{23}$ quanta. If $\lambda = 500$ nm $(5 \times 10^{-7}$ m), then 1 E $= 2.39 \times 10^{5}$ J. The radiant flux density of monochromatic light of wavelength 500 nm is 100 W m^{-2}, which is equivalent to 4.18×10^{-4} E $m^{-2} s^{-1}$.

Electron Volt

In electromagnetic theory and in atomic physics, a unit of energy known as the electron volt is often used. This unit arose from study of the motion of charged particles in electric fields. When physicists developed the atomic theory, they described an atom as a nucleus with positive charges surrounded by one or more electrons of negative charge moving in the electric field of the charged nucleus. In order to describe the energy status of the electrons surrounding the atomic nucleus, the electron volt (abbreviated eV) was defined. One electron volt is defined as the work done on, or the energy gained by, an electron when it is moved through a potential difference of 1 v. A potential difference of 1 V exists in an electric field if 1 J of work is done 1 C of charge in moving it from infinity, where the electric field is zero, to the particular place in the electric field. The student may pursue this topic in any standard physics text.

An electron was found to have a charge of 1.602×10^{-19} C. Therefore, the electron volt is related to the joule as follows: 1 eV equals the product of the charge on the electron (1.602×10^{-19} C) times the potential difference of 1 V, or 1.602×10^{-19} J.

An electron volt is a very small quantity of energy. In many fields, as for example chemistry, it is conventional to use larger units of measurement. Among these units is the mole, which is that quantity of a substance whose mass in grams equals the atomic or molecular weight. The gram-molecular weight of hydrogen is 2.0156, and that of water is 18.0156. A mole of a given substance always contains the same number of atoms or molecules (6.023×10^{23}, Avogadro's number). Hence a mole of electrons accelerated by an electric potential of 1 V has an energy gain given by the number of quanta (6.023×10^{23}) times the energy gained per quantum (1.602×10^{-19} J), or 96,490 J^{-1}. Many chemical reactions involve energies of approximately this amount. The units of calories or kilocalories are frequently used to measure energies in chemistry; electron volts or joules are usually used in physics.

Chemical Bond Energy

Often, one needs to evaluate the energy content of a quantum of radiation of specific frequency in order to determine its effectiveness in producing a chemical reaction. One mole of radiation contains a number of quanta equal to 6.023×10^{23} (indicated by the symbol (N_o). The energy content of a mole of quanta is thus $N_o\, h\nu$, or 3.98×10^{-10} J s times the frequency ν. If $\nu = 10^{15}$ s^{-1} (a wavelength of 300 nm), $N_o h\nu = 3.98 \times 10^5$ J. This amount of energy is equivalent to 95,056 cal, 95.06 kcal, or 2.48 eV. Similarly, if $\nu = 4.3 \times 10^{14}$ s^{-1} (a wavelength of 700 nm), $N_o h\nu = 1.72 \times 10^5$ J.

The absorptive, transmissive, and emissive properties of the atmosphere and the photoreactions in materials and organic substances vary according to the frequency of the radiation. Most reactions involving chemical bonds require energies of 30 kcal $mole^{-1}$ or more, as is seen in Table 5.2. A bond energy of 30 kcal $mole^{-1}$ is equivalent to the energy content of a quantum of light of wavelength 950 nm. Bond energies of 60 and 120 kcal $mole^{-1}$ correspond to radiation of wavelengths 475 and 237.5 nm, respectively. Table 5.2 shows that nearly all chemical bonds have energies corresponding to the energy of radiation in the visible or ultraviolet range, and only rarely in the very near infrared. Radiation of wavelength greater than 1000 nm simply does not have sufficient energy per quanta to produce any effects on chemical bonds and, therefore, does not produce photochemical reactions.

Most of the electronic transitions of molecules from the ground state (lowest energy state) involve the absorption of radiation in the far-ultraviolet range. Absorption bands observed in the near-ultraviolet and visible range are caused by transitions of the outermost, more loosely

Table 5.2. Bond Energies[a,b]

C—C	80.5		O—O	34
C=C	145		O—H	109.4
C≡C	198		H—H	103.2
C—H	98.2		N—N	37
C—Cl	78		N—H	92.2
C—O	79		H—Cl	102.1
C=O	173		H—Br	86.7
C—Br	54		Cl—Cl	57.1
			Br—Br	46

[a] K. S. Pitzer, *Quantum Chemistry*, Prentice-Hall, Englewood Cliffs, N.J., 1953, p. 170.
[b] All values are in kilocalories per mole.

bound electrons of a molecule. Single-ring aromatic compounds absorb in the vicinity of 250 nm, naphthalenes near 300 nm, and anthracenes and phenanthrenes near 360 nm. Saturated hydrocarbons absorb radiation of wavelength less than 180 nm. Chromophores are atomic groups, such as C—C, C=O, —N=N—, and —N=O, that cause absorptions at wavelengths greater than 180 nm. The double bond C=C usually absorbs at wavelengths of 180–190 nm, whereas ketone and aldehyde groups, involving the double bond C=O, have absorption maxima at 270 to 290 nm. The location of these various absorptive maxima indicates the energies required for photochemical activity.

Radiometric Terminology

Radiant Flux

Radiant energy E is measured in joules. A summary of radiometric terminology is given in Table 5.3. The radiant flux F is energy flow per unit time, either being emitted from a source or incident upon a particular surface. Radiant flux is equivalent to power and is measured in watts. The defining relationship is

$$F = \frac{dE}{dt}. \tag{5.16}$$

Solid Angle

A geometrical concept important for our purposes here is that of the solid angle. Just as a plane angle $d\alpha$, measured in radians, is a segment of arc $d\ell$ divided by the radius r of the circle, a solid angle $d\omega$, measured in steradians, is an area dA on the surface of a sphere divided by the square

Table 5.3. SI Units of Luminous- and Radiant-Energy Measurement

Quantity	Symbol	Units	Abbreviation	Defining equation
Radiant energy	E	Joules	J	
Flux				
Radiant flux	F	Watts	W	(5.16)
Luminous flux		Lumens	lm	
Intensity				
Radiant intensity	I	Watts per steradian	W sr^{-1}	(5.19)
Luminous intensity		Lumens per steradian, candelas	lm sr^{-1}, cd	
Flux density at emitting surface				
Radiant emittance	$\mathscr{R}$	Watts per sq. meter	W m^{-2}	(5.20)
Luminous emittance		Lumens per sq. meter	lm m^{-2}	
Flux density at receiving surface				
Irradiance	Q	Watts per sq. meter	W m^{-2}	(5.23), (5.24)
Illuminance		Lumens per sq. meter, lux	lm m^{-2}, lx	
Radiance	$\mathscr{B}$	Watts per sterdian per sq. meter	W sr^{-1}m^{-2}	(5.21), (5.22)
Luminance		Candelas per sq. meter	cd m^{-2}	
Luminous efficacy		Lumens per watt	lm W^{-1}	
Luminous efficiency	V	Decimal fraction		

of the radius r of the sphere. These cases are illustrated in Fig. 5.2. Hence

$$d\alpha = \frac{d\ell}{r} \tag{5.17}$$

$$d\omega = \frac{dA}{r^2}. \tag{5.18}$$

A circle contains 2π radians (abbreviated rad) and a sphere contains 4π steradians (abbreviated sr). (A hemisphere contains 2π sr.)

Radiant Intensity

The radiant intensity I of a source is given by:

$$I = \frac{dF}{d\omega}, \tag{5.19}$$

where dF represents the radiant flux and $d\omega$ is the solid angle whose vertex represents the source. The units of radiant intensity are watts per steradian. For a uniform 100-W spherical source 10^{-2} m in diameter, the radiant intensity will be 100 W/4π sr, or 7.96 W/sr. If the source does not

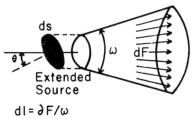

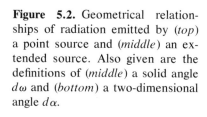

Figure 5.2. Geometrical relationships of radiation emitted by (*top*) a point source and (*middle*) an extended source. Also given are the definitions of (*middle*) a solid angle $d\omega$ and (*bottom*) a two-dimensional angle $d\alpha$.

$$dl = \partial F/\omega$$
$$R = \partial F/\omega \, ds \cos \theta$$

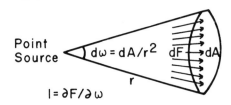

$$I = \partial F/\partial \omega$$

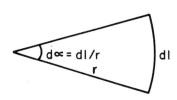

emit uniformly in all directions, the radiant intensity varies with direction; in other words, for nonuniformly radiating sources, the intensity is a function of the angle θ between the normal to the surface and the direction of viewing.

Radiant Emittance

The radiant flux density at the emitting surface, or radiant emittance, is the amount of energy emitted per unit time per unit area of source surface of the radiant emittance, or

$$\mathcal{R} = \frac{dF}{dA}, \tag{5.20}$$

where $\mathcal{R}$ is in watts per square meter.

If our source, which emits a total power, or radiant flux, of 100 W and has a surface area of $10^{-4}\,\pi$ m^2, (the surface area of a sphere 10^{-2} m in diameter), then the radiant emittance of the source is 100 W/$10^{-4}\,\pi$ m^2, or 3.18×10^5 W m^{-2}.

Radiance

The radiance B of a source with surface area A is defined as the power emitted per unit solid angle per unit area of spherical surface. It is expressed as follows:

$$\mathcal{B} = \frac{d^2F}{dA_p \, \partial\omega} = \frac{dI}{dA_p}, \tag{5.21}$$

where ∂A_p is the projected area of the source on a surface perpendicular to the direction of viewing. Radiance is measured in units of watts per steradian per square meter. In the case of a spherical source, the surface of the sphere is the projected surface area.

If the surface of the source is not spherical but of irregular shape, as is generally the case, then the projected surface area ∂A_p is the actual surface area projected on the surface of a sphere, or $\partial A \cos \theta$, where θ is the angle between the normal to the actual area and the radius vector of the sphere. Then, the radiance is given by

$$\mathcal{B} = \frac{\partial I}{\partial A_p} = \frac{\partial I}{\partial A \cos \theta} = \frac{\partial^2 F}{\partial A \cos \theta \partial\omega} = \frac{\partial \mathcal{R}}{\partial\omega \cos \theta}. \tag{5.22}$$

Radiance and radiant emittance are not dependent on the size of the source, since they are expressed per unit area of source surface, but radiant intensity is a function of source size. A small tungsten filament at a given temperature has the same radiance for a specified angle of emission and the same radiant emittance as a large filament at the same temperature. Most sources of radiation are not uniform and yield different radiances in different directions.

The radiance of the 100-W source is either 7.69 W sr^{-1}/$(\pi/4)^{-1} \times 10^{-4}$ m^{-2} or 31.84×10^4 m^{-2}/π sr, which gives 10.13×10^4 W sr^{-1} m^{-2}. It should be noted that, although there are 2π sr in a hemisphere, the radiance of a source, which is the emittance per unit solid angle, is the emittance divided by π sr. The average value of the projected area of a plane surface viewed from all possible directions on one side is one-half of its actual area, hence the division by π sr rather than by 2π sr.

Irradiance

The irradiance Q of a surface describes the rate incident radiant energy received per unit area:

$$Q = \frac{dF}{dA}. \tag{5.23}$$

Irradiance is measured in watts per square meter. Photon irradiance is measured in quanta per square meter per second or einsteins per square meter per second.

The irradiance of an element of surface area dA at a distance ℓ from a source of radiant intensity I when the angle between the normal to the receiving surface and the direction to the source is θ is given by

$$Q = \frac{dF}{dA} = \frac{I \, dA \cos \theta}{dA \, \ell^2}$$

or
$$Q = \frac{I \cos \theta}{\ell^2}, \qquad (5.24)$$

where $dA \cos \theta / \ell^2$ is the solid angle subtended by the receiving area at a distance ℓ from the source.

A surface at a distance of 2.0 m from the 100-W source has, when the angle between the normal to the receiving area and the direction of the source is 20°, is an irradiance $Q = 1.87$ W sr^{-1} m^{-2}.

Lambert's Cosine Law

Perfectly diffusing nonglossy surfaces obey Lambert's cosine law, which states that radiant (or luminous) intensity is proportional to the cosine of the angle between the direction of emission (or direction from which the surface is viewed) and the normal to the surface. A blackbody obeys Lambert's cosine law perfectly, but glossy or semiglossy surfaces do not. Lambert's cosine law states that the ratio of the radiant intensity in the direction θ to the radiant intensity of the source along the direction of the normal to its surface is given by

$$\frac{I(\theta)}{I_0} = \cos \theta. \qquad (5.25)$$

Radiance as a function of the direction from the normal to the emitting surface is written

$$\mathscr{B} = \frac{dI}{dA \cos \theta} = \frac{dI_0 \cos \theta}{dA \cos \theta} = \frac{dI_0}{dA} = \mathscr{B}_0 \qquad (5.26)$$

or
$$\frac{\mathscr{B}}{\mathscr{B}_0} = 1. \qquad (5.27)$$

Hence the radiance of the source is a constant irrespective of the angle of viewing. A small aperture in a blackbody cavity always appears of the same radiance irrespective of the angle at which it is viewed. If a perfectly diffusing surface is irradiated uniformly, its radiance is independent of the angle of view. Unfortunately, most surfaces do not obey Lambert's cosine law, and a more complicated analysis is necessary.

Photometric Terminology

Very often, the amount of light incident upon a plant community is described in units of illumination rather than irradiation. It was natural that, for many years, scientists evaluated light intensity of environments in terms of illumination units since these were the traditional units of light measurement used by photographers, architects, engineers, and others and most light-measuring equipment was calibrated in illumination units. Yet, fundamentally, illumination units have reference to the spectral response of the human eye and, strictly speaking, are not appropriate for the specification of light in general. Nevertheless, since the use of photometric (or illumination) units is widespread, it is necessary that they be included here. The book by Middleton (1952) is a useful for further information.

Luminous Flux

When radiant flux is evaluated with respect to its capacity to invoke response by the human eye, it is called luminous flux. The human eye has a sensitivity to wavelengths between about 390 and 780 nm. The light-adapted eye and the dark-adapted eye have different light responses because of the different sensors involved, i.e., the cones for photopic vision and the rods for scotopic vision. The visual spectrum is usually referred to as being between 400 and 700 nm. The average response of the human eye is defined by the standard relative-luminosity curve. The human eye has no visual sensitivity to ultraviolet or infrared radiation, low sensitivity to blue or red light, and maximum sensitivity to light of wavelength 555 nm, which is yellow-green.

Luminous flux corresponds to radiant flux and has the dimensions of power, or energy flow per unit time. The unit of luminous flux is the lumen (abbreviated lm). A radiant flux of 1 W monochromatic light of wavelength 555 nm corresponds to a luminous flux of 680 lm. Another way of stating this is that if the luminous flux in a sample of wavelength of 555 nm is 1 lm, the radiant flux is 1/680, or 0.00147, W. The lumen does not correspond to a definite number of watts except at a specific wavelength, since it was originally defined in terms of the response of the human eye. For example, at a wavelength of 600 nm, the relative luminosity is 0.62, and 1 W of monochromatic light of wavelength 600 nm generates a luminous flux of 422 lm. At a wavelength of 400 nm, the relative luminosity is 0, and 1 W of monochromatic light of this wavelength produces a luminous flux of 0.

Luminous Intensity

The unit of luminous intensity corresponding to radiant intensity, the candela, has had a long and varied history and originally was the light intensity provided by an actual standard candle of specified construction. A candela (abbreviated cd) is now defined as one-sixtieth of the luminous intensity of 1 cm² of the surface of a blackbody at the temperature of freezing (or melting) platinum (2042°K). A luminous intensity of 1 cd produces 1 lm of luminous flux issuing from one-sixtieth of a square centimeter of opening of a standard source, melting platinum in a blackbody within a solid angle of one steradian. A source of 1 cd emits in all directions at a total flux of 4π lm. The candela is equivalent to a lumen per steradian.

Luminous Emittance

The luminous emittance of an emitting surface is the flux emitted per unit emissive area as projected on a plane normal to the direction of viewing. Luminous emittance is measured in lumens per square meter or lumens per square centimeter (lamberts, abbreviated L). The luminous emittance of a blackbody radiator at 2042°K is 6×10^5 cd m^{-3}, and the luminous intensity of the source is 60 cd.

Because most lamps are not point sources and emit from somewhat irregular filaments, the outputs are given as total luminous flux in lumens. The total luminous flux is 4π times the mean spherical intensity (usually measured by a diffusely reflecting integrating sphere). Thus a tungsten lamp with an output of 300 lm has a mean spherical intensity of 23.9 cd.

Illuminance

When luminous flux strikes a surface, we say that the surface is illuminated. The illuminance of the surface is the luminous flux per unit area, or the flux density. Illuminance may be measured as lumens per square centimeter, meter, or foot. The quantity 1 lm cm^{-2} is a phot 1 lm m^{-2} is a lux, and 1 lm ft^{-2} is a footcandle (1 lm cm^{-2} = 929 fc). European scientists have commonly used the lux and American scientists the footcandle. There are 10.764 lx in 1 fc. Typical values of illuminance are as follows: full sunlight plus skylight at the ground on a clear day, about 100,000 lux and on an overcast day, about 10,000 lux; good indoor artificial illumination, 100 lux; full moonlight outdoors, 0.2 lux; and starlight, 0.0003 lux.

Luminance

Luminance is the photometric term corresponding to radiance. It has the dimensions of power per unit area of source (projected in the direction of viewing) per unit solid angle. The units are lumens per square meter per steradian, or lux per steradian, or candelas per square meter. The luminance of a source does not vary with the distance of the observer from the source, since the solid angle of a given source surface as viewed by the observer varies inversely with the square of the distance of the observer.

Luminous Efficacy

The luminous efficacy of any source is defined as the ratio of luminous to radiant flux emitted by the source. The luminous efficacy of radiation of wavelength 555 nm is 680 lm W^{-1}, and that of radiation of wavelength 600 nm is 410 lm W^{-1}. Because of the broad spectral distribution of radiation from any lamp or radiating surface, luminous efficacy is generally low. In other words, there is a lot of heat for little light. The maximum luminous efficacy for a blackbody occurs at a blackbody temperature of 6300°K and is 93 lm W^{-1}. This is remarkably close to the color temperature of the sun, although the sun cannot be considered to radiate precisely as a blackbody. The luminous efficiency of radiation of wavelength less than 400 nm or greater than 700 nm is zero since the human eye lacks visual sensitivity at these wavelengths.

A summary of photometric and radiant-energy units is given in Table 5.3. Thus it may be seen that the photopic sensitivity of the human eye is intimately involved in the specification of the photometric units of intensity and illumination. It is preferable to measure luminous flux in lumens, luminous intensity in candelas per (candelas), luminous emittance in lumens per square meter, illuminance in lumens cm per square meter (or lux), and luminance in lumens per square meter per steradian^{-1} (candelas per square meter). Historically, however, illuminance has been measured in lumens per square foot, a unit known as the footcandle. Today, most photometric measurements are made photoelectrically, rather than visually as they were once done. The relative spectral response of a filter–phototube combination is adjusted to correspond that of a "standard eye."

Brightness

The term luminance was once called brightness. Now the term photometric brightness is the same as luminance, and brightness refers to the visual sensation produced by a source. The luminous efficiency of the human eye is the ratio of the luminous efficacy at any particular wavelength to the maximum luminous efficacy and is given as a decimal fraction $V(\lambda)$. The brightness of a source is the total response it produces on the eye by all wavelengths between 400 and 700 nm. Therefore,

$$\mathcal{B} = \int_{400}^{700} I(\lambda) \, V(\lambda) \, d\lambda. \tag{5.28}$$

If the integral of Eq. (5.29) is evaluated for a standard source and a "standard eye," the brightness is found to equal 0.2772 W. A standard source emits a total luminous flux of 60π lumens. Therefore, one gets for the "least mechanical equivalent of light" a value of 0.00147 W ℓm^{-1}. Thus 1 W of monochromatic radiant power emitted at a wavelength of 555 nm, the peak sensitivity of the human eye, a response equivalent to 680 ℓm.

Sources of very different power distribution, as expressed in watts per square meter, may produce the same amount of illumination on a surface, expressed in lumens per square meter. It is absolutely essential to know the spectral distribution of the power emitted by a source if an adequate evaluation is to be made of radiant and luminous intensities, as well as the irradiation and illumination of a surface.

Light measurements for photosynthesis have often been given in foot-candles. For example, the compensation point for many plants is estimated at about 100 fc (or 1076 lx), and full sunlight on a clear summer day is about 10,000 fc, (107,600 lx).

Solar Illumination and Irradiation

It is of enormous value to be able to convert from units of radiant energy to units of luminous energy especially when dealing with sunlight and sky-light. For many stations in the world, solar radiation measurements are made only in units of radiant energy. Radiant-energy density is usually measured with a radiometer, pyrheliometer, or actinometer; luminous-energy density is usually measured with a suitably filtered selenium pho-tocell. It is not easy to make conversions between the two systems, and very careful calibration procedures must be followed. If a photocell is calibrated by using a standard incandescent laboratory source, for example, a calibration correction is required when sunlight or skylight is measured. This is not always done, and sometimes data are suspect for this reason.

Many light measurements are given in foot candles, and radiation measurements are given in Watts per square meter or calories per square centimeter per minute. The relationship between irradiation and illumination units depends upon the spectral composition of the light emitted by the source.

An excellent paper reviewing the state-of-the-art concerning the luminous efficacy of daylight is that of Drummond (1958). Drummond states (p. 150), "the precision of this conversion, e.g., from calories cm^{-2} hr^{-1} or milliwatts cm^{-2} into kilolux-hours or foot-candle hours, will depend, to some extent, upon such factors as latitude, synoptic air mass (with which is associated the dust and water vapor concentrations of the lower atmosphere), and optical air mass (or relative path length of the solar beam through the terrestrial atmosphere layer)."

Drummond's measurements were made at Pretoria, South Africa, where the summer sky is generally very clear and the winter sky somewhat dusty. The average luminous efficacy of daylight at Pretoria was 104 ℓm W^{-1} for cloudless conditions and for skylight only 132 ℓm W^{-1}. For average daylight conditions throughout the year, luminous efficacy varied from 99 to 108 ℓm W^{-1} and the skylight from 127 to 137 ℓm W^{-1}. The luminous efficacy of the sun outside the earth's atmosphere is about 105 ℓm W^{-1}. Larsen (1966) reports values for Bergen, Norway and Kimball (1924) for Washington, D.C. Kimball reported luminous efficacy for overcast days of 119 ℓm W^{-1} and for clear days of about 108 ℓm W^{-1}.

Problems

1. A surface is emitting 459.5 W m^{-2}. Determine the temperature of the surface for a surface emissivity (a) 1.0, (b) 0.9, and (c) 0.8.

2. Using Eq. (5.3), calculate the radiant energy exchange between two objects with emissivities of 0.9 and surface temperature are 20° and 50°C. What is the approximate net radiant exchange between the two calculated according to the linearized approximation [Eq. (5.4)]? What is the percent error by this method?

3. For a blackbody surface radiating at a temperature of 1000°K, calculate the spectra distribution of the energy radiated using six wavelengths over the range 2×10^{-6} m to 12×10^{-6} m. Plot the results for energy versus wavelength.

4. For a blackbody surface radiating at a temperature of 1000°K, determine the wavelength corresponding to maximum monochromatic intensity. Does this agree with your answer in Problem 3? If the temperature is 500°K, where is the maximum intensity? If the temperature is 300°K?

5. At a wavelength of 1.0 μm, how much energy is there per quantum of radiation? At a wavelength of 555 nm? Calculate the luminous flux for these two values.

6. Calculate the energy content of a mole of light quanta of wavelength 555 nm. Is this sufficient energy to break a carbon–carbon single bond or an oxygen–oxygen single bond?

7. If ultraviolet light of wavelength 280 nm were to reach the earth's surface, would it be destructive to molecular bonds? Calculate the energy content per mole of quanta of this wavelength.

Chapter 6

Solar Radiation

Introduction

The earth is in orbit around the sun at a distance which keeps the mean temperature of the earth at about 286°K. If the earth were closer to the sun, its mean temperature would be higher; if it were farther from the sun, its mean temperature would be lower. High-temperature nuclear reactions in the sun produce energetic particles, electrons, protons, and ionized atoms, which stream outward into space at very high speeds. This flux of particles and ions is known as the solar wind. The time necessary for the solar wind to reach the earth is approximately 4 to 5 days.

Every now and then, there are great magnetic storms on the sun from which unusually energetic particles are ejected into interplanetary space. These great eruptions represent huge pulses of particles, which pass beyond the orbit of the earth about 3 days after the solar storm. If the earth happens to be in the path of these particle pulses and intercepts them, they bombard the earth's atmosphere. Normally, the earth's magnetic field screens it from the solar wind by deflecting the ions into trajectories around the earth, and relatively few of the particles penetrate very far into the earth's atmosphere. When the more energetic particles from a solar storm do reach the earth, they distort and penetrate the magnetic shield. The result is an intense bombardment of the upper atmosphere. We see this in the form of the aurora borealis (or northern lights) in the north-polar regions and the aurora australis (or southern lights) in the south-polar regions. The auroras are red and green rays of light emitted by the excitation of oxygen and nitrogen atoms high in the earth's atmosphere when bombarded by the solar-storm particles. An aurora occurs about 3 days after a magnetic storm on the sun.

The planet earth, continuously swept over by the solar wind and bombarded by occasional bursts of solar storm particles, is irradiated by a broad spectrum of electromagnetic radiation. It is this electromagnetic radiation which we see as sunlight and feel as heat. This electromagnetic radiation is emitted by the photosphere of the sun, a region of much lower temperatures (approximately 6000°K) than the interior, where the nuclear reactions occur. Even this broad-band electromagnetic radiation contains a vast array of frequencies characterized as ultraviolet radiation, visible light, infrared radiation, microwaves, and radio waves. This is the radiation which warms the earth and maintains it at temperatures suitable for life. The fragile atmosphere of the earth has evolved synergistically with the life of the planet. The atmosphere screens out much of the actinic (biologically reactive) ultraviolet radiation from the sun and shields the earth's surface from excessive heat loss to space. Radiation from the sun and the radiative balance of the earth is the key to evolution of life on this planet, as well as the character of ecosystems.

Sun and Earth Geometry

The amount of sunlight incident upon the earth's surface at any point in space and time is a function of many variables. Time of year determines the position of the earth in its orbit around the sun and, therefore, the distance between sun and earth. Latitude and longitude, as well as time of day, determine the amount of sunlight received at any time of year. And finally the turbidity or cloudiness of the atmosphere determines the amount of solar radiation that penetrates to the earth's surface and the amount of terrestrial radiation that escapes to space. Because of the complexity introduced by the atmosphere, the basic geometrical relationship between sun and earth and the amount of solar radiation received at the outer extremity of the atmosphere will be considered first.

The sun, as the result of nuclear and atomic reactions, emits a total power of about 3.84×10^{26} W. The sun radiates an amount of energy equivalent to that which a blackbody at a temperature of 5760°K would radiate. The radiant emittance of the solar surface is $\sigma[T + 273]^4$, or 6.244×10^7 W m^{-2}. The radiance of the solar surface is the radiant emittance per unit solid angle. Since any element of surface radiates into a solid angle of π, the radiance is $6.244 \times 10^7/\pi$, or 1.988×10^7 W m^{-2} sr^{-1}.

The Solar Constant

The earth is in orbit around the sun at a mean distance $\overline{d}$ of 1.497×10^{11} m. The total flux of radiation passing out into space through the surface of the sun is $4\pi R^2 \sigma[T + 273]^4$, where R is the radius of the sun

(6.965×10^8 m). A flux of radiation also passes across the surface of the earth. This radiant flux is termed irradiance and at the distance of the earth is $\overline{S}_0$, then this flux is $4\pi \overline{d}^2 \overline{S}_0$. Hence

$$\overline{S}_0 = \frac{4\pi R^2}{4\pi \overline{d}^2} \sigma[T + 273]^4 \qquad (6.1)$$

This is equivalent to 1360 W m^{-2}, or 1.95 cal cm^{-2} min^{-1}.

The irradiance of an area perpendicular to the sun's rays just outside the earth's atmosphere and at the mean distance of the earth from the sun is defined as the solar constant. The value of the solar constant has been one of the most difficult fundamental geophysical parameters to determine. Measurement is difficult because of interference by the atmosphere for observations made from the earth's surface and the fact that measurements from satellites or space platforms are difficult to make with accuracy. A very careful observation of solar irradiance has been made by Laue and Drummond (1968) from aircraft flying at altitudes up to 83 km, and a value of 1360 W m^{-2} is given as the most probable value of the solar constant. Johnson (1954) made an extremely careful analysis of a vast number of earlier observations, both spectral and total, and concluded that the most likely value of the solar constant is 1395 W m^{-2}. An earlier value, estimated by staff members of the Smithsonian Institution from many years of measurement at the earth's surface by means of pyrheliometers, was 1353 W m^{-2}, and many of the tabulations and graphs in the literature are based on this value. The difference between the values 1353 and 1360 is only 0.5%, well within the error of observation.

Total Energy Intercepted by the Earth

The total amount of energy intercepted by the earth, considered as a disk of radius $a = 6.37 \times 10^6$ m, is

$$\begin{aligned} \pi a^2 \overline{S}_0 &= \pi (6.37 \times 10^6)^2 \, 1360 \\ &= 17.34 \times 10^{16} \text{ W} \end{aligned} \qquad (6.2)$$

Average Solar Energy on Horizontal Surface Outside the Atmosphere

If the total energy intercepted by the earth were distributed uniformly over the earth's surface, the average amount of energy received per unit of horizontal surface area per unit time at the outer edge of the atmosphere would be

$$_h\overline{S}_0 = \frac{\pi a^2 \overline{S}_0}{4\pi a^1} = \frac{\overline{S}_0}{4}, \qquad (6.3)$$

which gives a value of 340 W m^{-2}.

This figure is useful for comparative purposes only since the actual distribution of energy over the earth's surface is not uniform. The annual amount of energy received at the equator is about 2.4 times that received at the poles. The average amount of solar energy incident on the sunlit hemisphere of the earth is 680 W m^{-2}.

The Geometrical Relationships

The earth travels in an elliptical orbit, tracing out a plane, called the plane of the ecliptic, around the sun, while spinning about its polar axis. It takes 365 and 1/4 days for the earth to move around the sun in a complete ellipse. The irradiation of any segment of the earth by the sun is directly related to the arrangement of these two bodies with respect to each other (see Fig. 6.1). The amount of solar radiation incident at a point just above the earth's surface depends on time of year, time of day, and latitude. The distance d of the earth from the sun varies with the time of year and has a mean value $\overline{d}$. Since the density of the radiation passing from the sun out into space varies inversely with the square of the distance, the actual amount of solar radiation received at any point in the earth's orbit also varies inversely with the square of the distance of the earth from the sun. Hence the instantaneous amount of solar radiation received on a surface perpendicular to the sun's rays at any time of the year is $S_o = \overline{S}_o\,(\overline{d}/d)^2$. The sun's rays make a zenith angle z with the radius vector of the earth, or the local zenith. The amount of sunlight incident on a horizontal surface

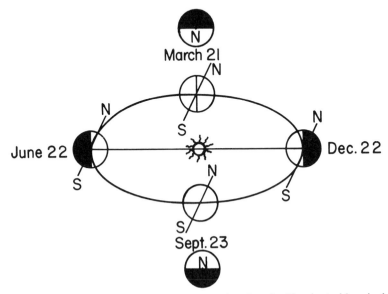

Figure 6.1. The earth's orbit around the sun, showing the illuminated hemisphere at times of solstice and equinox.

just outside the earth's atmosphere $(_hS_o)$ is given by

$$_hS_o = \overline{S}_o \left(\frac{\overline{d}}{d}\right)^2 \cos z, \qquad (6.4)$$

where $\overline{S}_o$ is the solar constant. The factor $(\overline{d}/d)^2$ varies throughout the year, from 1.0344 on 3 January to 0.9674 on 5 July, but is never more than 3.5% away from 1.0. The earth is actually closer to the sun, and the irradiation of the earth is correspondingly greater, during the Norther-Hemisphere winter than during the Northern-Hemisphere summer.

The earth's axis of rotation is inclined at an angle of 66.5° to the plane of the ecliptic, and the axis is always directed toward the same point in the sky, i.e., the pole-star. Hence as the earth moves in its orbit around the sun, the earth's axis moves parallel to itself. The plane of the earth's orbit around the sun is termed the plane of the ecliptic, and the plane through the equator of the earth is called the equatorial plane. The angle between the equatorial plane and the ecliptic is 23.5°. The angle which the line of the sun's rays at the earth makes with the equatorial plane is called the declination δ of the sun (see Fig. 6.2). In other words, declination is the angular distance of the sun north (positive) or south (negative) of the equatorial plane. As the earth orbits around the sun, the declination varies. For the Northern Hemisphere, maximum declination varies from $-23.5°$ at the winter solstice on 22 December to $+23.5°$ at the summer solstice on 22 June. Declination is independent of the calendar year and the latitude of the observer and is a function only of day of the year. For an observer in the Southern Hemisphere, the declination angle simply changes sign. For example, the Southern-Hemisphere summer solstice is on 22 December, when the declination is $+23.5°$. During summer for the Northern Hemisphere, the North Pole is tilted toward the sun, and the sun's rays irradiate the northern hemisphere more strongly than the Southern Hemisphere. During winter for the Northern Hemisphere, the North Pole is tilted away from the sun, and the Southern Hemisphere is irradiated more strongly. The Northern and Southern hemispheres are irradiated equally at the time of the equinox.

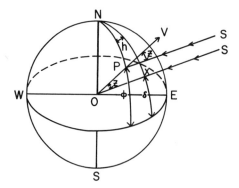

Figure 6.2. Geometrical relations of solar zenith angle z to latitude ϕ, solar declination δ, and hour angle h. The observer is at P.

The amount of solar radiation incident at a point just outside the atmosphere is a function of time of year, time of day, and latitude. In Fig. 6.2, P is the point irradiated, and the line OV is the vertical, or zenith–nadir, direction for an observer at point P. The angle VOS or VPS is the zenith angle z of the sun, which is determined from other fundamental angles, as defined in astronomy and related to one another by means of spherical trigonometry.

Let X be a point on the sphere directly under the sun, or on a radial line connecting the center of the earth with the sun. The arc NX is equal to $90° − δ$. The arc NP is equal to $90° − φ$, where $φ$ is the latitude of P on the earth. The hour angle h is the angle through which the earth must turn to bring the meridian of P directly under the sun and is a function of time of day. The altitude a of the sun is the elevation of the sun above the horizon for an observer at point P; it is the complement of the zenith angle (sin $a = \cos z$). The azimuth $α$ of the sun is the angle between the sun and true north as measured in the horizontal plane of the observer; it is the angle between the observer's meridian plane and the plane containing the observer, the sun, and the center of the earth. Spherical trigonometry gives the following relationships among the angles for the solar altitude a and azimuth $∝$ as seen from the position P:

$$\cos z = \sin φ \sin δ + \cos φ \cos δ \cos h \qquad (6.5)$$

$$\sin a = \sin φ \sin δ + \cos φ \cos δ \cos h \qquad (6.6)$$

$$\sin ∝ = − \cos δ \sin h/\cos a, \qquad (6.7)$$

where a is the altitude of sun (angular elevation above the horizon), $φ$ is the latitude of the observer, $δ$ is the declination of the sun, h is the hour angle of sun (angular distance from meridian of observer), and $∝$ is the azimuth of sun (measured eastward from north).

The altitude of the sun is dependent upon latitude $φ$ of the observer, declination $δ$ of the sun, and hour angle h of the sun, i.e., is dependent upon latitude, time of year, and time of day. The azimuth of the sun does not depend upon the latitude of the observer, only on the time of year and time of day.

It is interesting to note that, at the poles, $φ = 90°$, $\cos φ = 0$, and $\sin φ = 1$, and Eqs. (6.5) and (6.6) give $\cos z = \sin δ = \sin a$, or $a = δ = 90° − z$. Hence at the poles, the elevation angle of the sun always equals the declination angle. During the 6 months of daylight, an observer at the pole sees the sun circle around the horizon, never more than 23.5° above the horizon. At the time of the equinoxes, an observer at the poles sees the sun on the horizon throughout a 24-hr period. Then, for 6 months, the pole receives no direct rays of the sun but only twilight, until, in the depth of winter, total darkness ensues.

At solar noon at any latitude, $\cos h = 1$ (since $h = 0$), and $z = φ − δ$. The zenith angle of the sun is the latitude of the observer less the declination. At latitude 40°N on 21 June, $\delta = 23.5°$ and $z = 16.5°$. Only at a latitude of 23.5° will the sun be directly overhead at the time of the summer solstice. At sunrise or sunset at any latitude (except the poles), $\cos z = 0$ (since $z = 90°$), $a = 0°$, and the hour angle h is half the length of the daytime, referred to as the half-day length h_s. Although one thinks of the day length in hours, it can be expressed in degrees of rotation at the rotation rate of the earth. Solving Eq. (6.5) for $\cos h$ gives

$$\cos h_s = -\tan \phi \tan \delta. \tag{6.8}$$

If either $\tan \phi = 0$ (which is always true at the equator) or $\tan \delta = 0$ (which is true during the equinoxes at all latitudes except the poles), then $h_s = \pm 90°$, the half-day length is 6 hr, and, of course, the full day from sunrise to sunset is 12 hr. The latitude north of which there are no direct rays of the sun is the boundary of the polar night, at which latitude $h_s = 0$. From Eq. (6.8), $\tan \phi = -\cot \delta$ or $\phi = 90 - |\delta|$ in the winter hemisphere. On 22 December (the winter solstice) in the Northern Hemisphere, the polar night occurs north of latitude 66.5°N.

At any time of the year, the sun rises closer to true east for an observer on the equator than an observer at any other latitude. At sunrise or sunset, $z = 90°$, $a = 0$, and $\sin a = 0$, $\cos a = 1$, and $h = h_s$. From Eq. (6.7), combined with Eq. (6.8), one gets, for an observer at any latitude, at sunrise or sunset

$$\sin \propto = \cos \delta \sin h_s = \cos \delta(1 - \tan^2 \phi \tan^2 \delta)^{1/2}. \tag{6.9}$$

At the equator, $\phi = 0$ and $\sin \propto_s = \cos \delta$. For any other latitude, $\phi \neq 0$ and $\sin \propto_s < \cos \delta$. Thus the angle $\propto_s$ departs more from true east for an observer at latitude $\phi \neq 0$ than $\phi = 0$. Inspection of Figs. 3 through 8 in the Appendix clearly shows the position of the rising and setting sun relative to due east.

As already stated, solar declination is a function only of day of year and is independent of location. Values for solar declination are listed in *The Nautical Almanac*, published by the U.S. Government Printing Office, Washington, D.C. The hour angle is zero when the sun is directly north or south of the point of observation and increases by 15° for every hour before or after solar noon. True solar noon occurs at 1200 only for a position on the standard meridian for the standard time zone within which the observer is located. To arrive at true solar time, one must determine first what is called the local mean solar time and add to this the equation of time obtained from the ''Ephemeris of the Sun'' given in *The Nautical Almanac*. The equation of time and the solar declination are listed in Table 169 of the *Smithsonian Meteorological Tables*, edited by List (1963). To determine local mean solar time, one must add to local standard time 4' for each degree of longitude east of the standard meridian or subtract 4' for each degree west of the standard meridian. This quantity may now be used in the equation of time to determine true solar time. List (1963) gives the procedure for the determination of solar altitude and azimuth. It is evident by now that the geometry of the earth and sun is not simple; fortunately, there are charts available which permit one to read off the altitude a and azimuth $\propto$ angle, and it is also possible to program Eqs. (6.6) and (6.7) on a computer. Typical charts of solar altitude and azimuth are given in Appendix 4.

Direct Solar Irradiation without Atmosphere

It is a straightforward procedure to combine the geometrical relationships with the value of the solar constant to obtain the instantaneous amount of direct sunlight incident on a horizontal surface outside the earth's atmosphere. Attenuation by the earth's atmosphere may then be taken into account.

Instantaneous Direct Solar Irradiation

The instantaneous amount of solar radiation $_hS_0$ incident on a horizontal surface just outside the earth's atmosphere is arrived at by Eqs. combining (6.4) and (6.5) to give

$$_hS_0 = \overline{S}_0 \left(\frac{\overline{d}}{d}\right)^2 (\sin \phi \sin \delta + \cos \phi \cos \delta \cos h). \tag{6.10}$$

Daily Total Direct Solar Irradiation

For many purposes, but particularly for considerations of primary productivity, it is useful to be able to determine the total amount of solar radiation irradiating any element of the earth's surface per day. Before considering the amount of irradiation at the ground, it is necessary to obtain an exact expression for positions outside the earth's atmosphere. If Eq. (6.10) is integrated over the period of a day at each position above the earth, one gets for the daily amount of undepleted solar radiation incident on a horizontal surface

$$\begin{aligned}
_hQ_0 &= \int_{-t_s}^{t_s} {}_hS_0 \, dt \\
&= \overline{S}_0(\overline{d}/d)^2 \int_{-t_s}^{t_s} (\sin \phi \sin \delta + \cos \phi \cos \delta \cos h) \, dt.
\end{aligned} \tag{6.11}$$

But dh/dt is equal to ω, the angular velocity of the earth (2π rad day^{-1}). Hence $dt = dh/\omega$, and one can express the time to sunrise and sunset in terms of hour angles. Hence

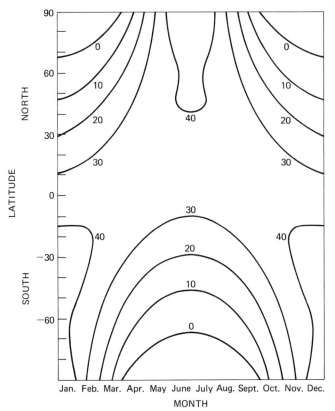

Figure 6.3. Daily totals of solar irradiation per unit horizontal area of the earth's surface, without atmospheric attenuation, as a function of latitude and time of year. Numbers give irradiation in megajoules per square meter per day.

$$_hQ_0 = \overline{S}_0 \left(\frac{\overline{d}}{d}\right)^2 \int_{-h_s}^{h_s} (\sin\phi\,\sin\delta + \cos\phi\,\cos\delta\,\cos h)\,\frac{dh}{\omega}. \quad (6.12)$$

Since the solar constant $\overline{S}_0$ is in watts per square meter and we wish to get total irradiation during a day, we must multiply by 86,400 s day^{-1}. We get, after integration, the following:

$$_hQ_0 = \frac{86,400}{\pi}\,\overline{S}_0 \left(\frac{\overline{d}}{d}\right)^2 (h_s\,\sin\phi\,\sin\delta + \cos\phi\,\cos\delta\,\sin h_s). \quad (6.13)$$

Another expression for the same quantity is derived by multiplying both sides of Eq. (6.8) by $\tan h_s$ and then using the fact that $\cos\phi\,\cos\delta\,\sin h_s = -\sin\phi\,\sin\delta\,\tan h_s$. The equation arrived at is

$$_hQ_0 = \frac{86,400}{\pi}\,\overline{S}_0 \left(\frac{\overline{d}}{d}\right)^2 \sin\phi\,\sin\delta\,(h_s - \tan h_s), \quad (6.14)$$

where h_s is expressed in radians. Since h_s is a function of latitude and time of year and both d and δ depend only on time of year, we find that $_h Q_o$ is a function of only two variables, latitude and time of year. The numerical values of Eq. (6.13) or (6.14) are plotted in Fig. 6.3. It is interesting to note that the peak value of nearly 50.0 MJ m^{-2} day^{-1} occurs at the South Pole on 22 December.

Attenuation by the Earth's Atmosphere

The amount of sunlight incident at any position on the earth's surface depends on atmospheric conditions, in addition to the geometrical relations already described. The atmospheric and meteorological factors involved are the extinction coefficients of the clear sky, water and dust content of the atmosphere ozone content, cloudiness, and surface albedo. In addition, amount of insolation received at any position on the earth's surface depends on slope aspect of the surface, nature of the horizon, and, of course, elevation above sea level. It is a tall order to describe all of these effects on insolation in an analytical fashion. We first describe the total flux of radiation incident at the ground surface for various conditions and then discuss the details of these fluxes, including their spectral character, geometry, and time dependence.

Atmospheric Geometry

First, we shall consider the quantity of solar radiation penetrating the earth's atmosphere, without attempting to understand the mechanisms affecting intensity and spectral composition. It is clear that for the situation in which the direct rays of the sun traverse a stratified atmosphere above a plane surface, as shown in Fig. 6.4a, the ratio m of the path length in the direction of the sun at zenith angle z to the path length in the vertical direction is given by

$$m = \sec z = 1/\cos z. \tag{6.15}$$

The quantity m is known as the air mass. Since $\cos z = \sin a$, one can use Eq. (6.6) to derive m. When the zenith angle z is less than about 60°, the air mass is indeed accurately given by the secant of z, but when the zenith angle is greater than 60°, the secant gives values which are increasingly too great because atmospheric refraction and the curvature of the earth have not been taken into account. Inspection of Fig. 6.4b shows that the earth's curvature tends to reduce the length of the sun's slant rays compared with the depth of the atmosphere in the zenith direction. A detailed discussion of this matter is given in Robinson (1966). A few values of the corrected air mass are as follows: $z = 0°$, $m = 1.00$; $z = 30°$, $m = 1.15$; $z = 40°$, $m = 1.30$; $z = 50°$, $m = 1.55$; $z = 60°$, $m = 2.00$; $z = 70°$, $m = 2.90$; $z = 80°$, $m = 5.60$; and $z = 85°$, $m = 10.39$. Additional values

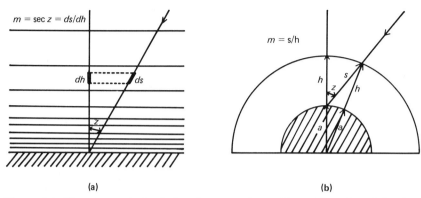

Figure 6.4. The geometrical relations between the zenith path length and the slant path length through an atmosphere above (a) a plane surface and (b) a curved earth.

are given in Table 137 of the *Smithsonian Meteorological Tables,* edited by List (1963). These values are only useful for stations near sea level; when working at higher altitudes, the value for air mass must be corrected for reduction in atmospheric pressure by the factor p/p_0, where p is the pressure at the station and p_0 is the standard sea-level pressure of 1013.25 mbar, or 29.921 in of Hg. (The reader is reminded that 1 mbar = 10^3 dyne cm^{-2}.) The actual atmospheric pressure at any given station depends not only upon altitude above sea level but also upon temperature of all parts of the column of air overhead. Some typical values of atmospheric pressure are 835 mbar for Denver, Colo., which is at an altitude of 1609 m; 644 mbar for the alpine tundra at an elevation of 3658 m; and 310 mbar for the top of Mt. Everest, at 8930 m. It is evident that considerably more solar radiation is incident on surfaces at the tops of mountains than near sea level. This is a phenomenon that becomes very apparent to hikers and skiers at high altitudes.

Transmittance of the Atmosphere

In order to obtain the amount of direct solar radiation that traverses the earth's atmosphere and is incident at the ground, we need to know the transmission properties of the atmosphere. The transmittance τ to solar radiation of the atmosphere is that fraction of the radiation incident at the top of the atmosphere which reaches the ground along the vertical (or zenith) path, which is the shortest path length between outer space and the surface. If the slant path is m times the zenith path, then the transmittance along the slant path will be τ^m. If, for example, the transmittance for the zenith path is τ and the paths being considered are two or three times this in length, then the respective transmittances are $\tau \cdot \tau = \tau^2$ or $\tau \cdot \tau \cdot \tau = \tau^3$ respectively. The amount of direct solar radiation on a surface perpendicular to the solar rays is

$$S = S_0 \bar{\tau}^m, \tag{6.16}$$

where $\bar{\tau}^m$ is the value of the atmospheric transmittance averaged over all wavelengths.

We can also define the monochromatic transmittance $_\lambda \tau$, such that:

$$_\lambda S = {}_\lambda S_0 \, _\lambda \tau^m, \tag{6.17}$$

where $_\lambda S$ and $_\lambda S_0$ are the monochromatic flux of solar radiation at the earth's surface and outside the atmosphere, respectively.

By integrating over the full span of wavelengths from 0 to ∞, one obtains the total flux of solar radiation received at the earth's surface per unit area per unit time:

$$S = \int_0^\infty {}_\lambda S \, d\lambda = \int_0^\infty {}_\lambda S_0 \, _\lambda \tau^m \, d\lambda. \tag{6.18}$$

The mean transmittance $\bar{\tau}^m$ is the mean value of transmittance with respect to wavelength when averaged across the solar spectrum.

$$\bar{\tau}^m = \int_0^\infty {}_\lambda S_0 \, _\lambda \tau^m \, d\lambda \Big/ \int_0^\infty {}_\lambda S_0 \, d\lambda. \tag{6.19}$$

From here on, we shall let $\tau^m = \bar{\tau}^m$, where τ is the mth root of the mean and is assumed to be independent of m.

Measured Values of Transmittance

The values of direct solar transmittance τ vary considerably with location. Liu and Jordan (1960) measured the solar transmittance at Hump Mountain, N.C. (1463 m above sea level), Minneapolis, Minn. (272 m), and Blue Hill, Mass. (192 m). The highest values were $\tau = 0.745$ and 0.743, recorded at Blue Hill and Hump Mountain, respectively. The maximum at Minneapolis was $\tau = 0.706$. The lowest values were $\tau = 0.425$ and 0.450, at Hump Mountain and Minneapolis, respectively. A large number of values at Blue Hill were between 0.60 and 0.72. Hump Mountain had many values between 0.48 and 0.68, and Minneapolis had many between 0.58 and 0.68. This indicates that values of $\tau = 0.60$ or 0.70 are quite realistic for clear-day situations.

Mountain locations at high elevations, such as the Sierra Nevada and Rocky Mountains, with extremely clear, dry skies, should have transmittances up to 0.80. This is confirmed by spectral observations by Stair and Ellis (1968) taken from an elevation of 3400 m on Mauna Loa, Hawaii, and by Stair et al. (1954) at 2805 m on Sacramento Peak, N. Mex. Minimum values of atmospheric transmittance at low-elevation sites for apparently clear sky might go as low as 0.40, depending on the turbidity.

Beer's Law

There is another form in which the transmission of the atmosphere may be written, known as Beer's law:

$$_\lambda S = {}_\lambda S_o \, e^{-_\lambda k \, um}, \tag{6.20}$$

where u is the optical thickness of the absorbing material, or the amount of absorbing material per unit area (in kilograms per square meter), $_\lambda k$ is the monochromatic attenuation coefficient of the absorbing material for a wavelength interval between λ and $\lambda + d\lambda$, and m is the air mass. If u is expressed in kilograms per square meter, $_\lambda k$ is in units of square meters per kilogram.

Beer's law derives from the fact that the incremental absorption $d_\lambda S$ produced by an incremental amount of absorbing material mdu is proportional to the intensity of the radiation $_\lambda S$. Hence

$$d_\lambda S = -_\lambda k \, {}_\lambda S \, mdu. \tag{6.21}$$

Integration of this equation gives Eq. (6.20). Beer's law applies to any wavelength interval over which k is a constant. Often, it is used when k is not constant, but it is not very accurate in such cases (see the section dealing with the spectral transmission of the earth's atmosphere).

From Eqs. (6.17) and (6.20), it is apparent that the monochromatic transmittance $_\lambda \tau$ is given by

$$_\lambda \tau = e^{-_\lambda k \, u}. \tag{6.22}$$

If attenuation of the solar beam is purely the result of absorption, then one can write the relationship for the amount of monochromatic, direct-beam solar radiation at the ground surface as

$$_\lambda S = {}_\lambda S_o \, {}_\lambda \tau_a^m. \tag{6.23}$$

If, however, the process of attenuation is purely a result of scattering, one can write

$$_\lambda S = {}_\lambda S_o \, {}_\lambda \tau_s^m. \tag{6.24}$$

When both absorption and scattering play a role in the attenuation of the solar beam in its passage through the earth's atmosphere, then

$$_\lambda S = {}_\lambda S_o \, {}_\lambda \tau_a^m \, {}_\lambda \tau_s^m = {}_\lambda S_o \, {}_\lambda \tau^m. \tag{6.25}$$

Hence

$$\lambda\tau = \lambda\tau_a \; \lambda\tau_s. \tag{6.26}$$

Both absorption and scattering are important processes in the earth's atmosphere and each is strongly wavelength-dependent.

Direct Solar Irradiation with Atmosphere

It is an easy step to combine the solar constant and the geometrical relationship between sun and earth with the transmission coefficient to obtain the instantaneous and daily amount of direct solar irradiation of a horizontal surface at the ground.

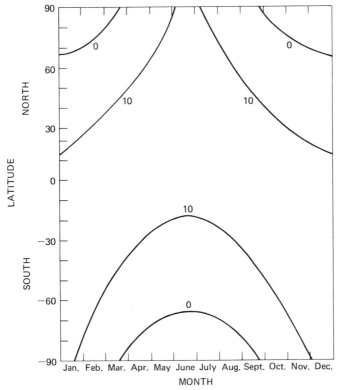

Figure 6.5. Daily totals of solar irradiation per unit horizontal area of the earth's surface with atmospheric transmittance $\tau = 0.6$. Numbers give irradiation in magajoules per square meter per day.

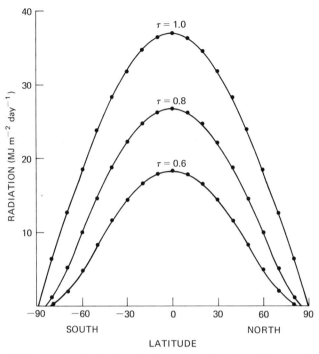

Figure 6.6. Daily totals of solar irradiation per unit horizontal area of the earth's surface as a function of latitude at the time of the equinox for various values of atmospheric transmittance.

Instantaneous Direct Solar Radiation at the Ground

Instantaneous amount of solar radiation received on a horizontal surface at the ground is obtained by combining Eqs. (6.10) and (6.16) to give

$$_hS = \overline{S}_o \left(\frac{\overline{d}}{d}\right)^2 (\sin \phi \sin \delta + \cos \phi \cos \delta \cos h)\, \tau^m, \qquad (6.27)$$

where m is equal to $1/\cos z$ [or $1/(\sin \phi \cos \delta + \cos \phi \cos \delta \cos h)$] plus a refraction correction for long slant paths through the atmosphere. Equation (6.27) applies for direct sunlight through a sky without clouds. Values of τ are usually 0.4 to 0.7. When the air is extremely clear, particularly at high mountain elevations, τ is about 0.8.

Daily Total Direct Solar Radiation at the Ground

If instantaneous amount of solar radiation received on a horizontal surface at the ground from sunrise to sunset is integrated over time, one obtains the total daily amount. Because of the complexity of the integration, it is necessary to use numerical integration. When this is done for a value of τ of 0.6, one obtains the results shown in Fig. 6.5 as a function of time of year and latitude.

It is interesting to compare Figs. 6.5 and 6.3 The distribution with time of day and latitude of direct solar radiation at the earth's surface on a horizontal surface is very different from the distribution outside the atmosphere. The long slant path of the sun's rays in polar regions greatly attenuates the direct beam of sunlight. Maximum values are about 33.5×10^4 J m^{-2} day^{-1} instead of 82.7×10^4 J m^{-2} day^{-1} and occur at latitude 30°N in June and 20°S in early February. Another manner of plotting the same data in order to show the changes with transmittance and latitude is shown in Fig. 6.6 for the equinox and in Fig. 6.7 for the summer solstice in the Northern Hemisphere.

Diffuse and Global Shortwave Irradiation with a Clear Sky

In addition to direct solar radiation, scattered skylight also reaches the ground. Amount of skylight varies enormously with the turbidity of the sky. Total amount of shortwave radiation from the direct solar beam and the scattered skylight received on a horizontal surface is known as the global radiation and represents the amount which any plant or animal living in the open on the earth's surface will experience. It is necessary first to derive the analytical expressions that give the amount of global shortwave radiation received with clear sky and then to do the same thing for overcast conditions.

Diffuse Skylight Irradiation at the Ground

It is much easier to estimate the amount of diffuse radiation reaching the ground from the hemisphere of the clear sky than to estimate it from a cloudy, partly cloudy, or polluted sky. A clear sky without dust or pollution contains molecules of air and also exhibits extremely tiny air-density fluctuations, both of which scatter light. These air molecules and density fluctuations are small compared to the wavelength of light and scatter light inversely proportional to the fourth power of the wavelength, a process known as Rayleigh scattering. This process gives the sky its blue color since the blue wavelengths of sunlight are scattered much more strongly than the red wavelengths. Dust particles are of approximately the same size as the wavelengths of visible light, with the result that dust in the sky produces an entirely different effect, i.e., a scattering which is inversely proportional to the 1.5 power of the wavelength. This is known as Mie scattering and gives the sky a whitish appearance. A pure molecular sky, without dust, is called a Rayleigh sky. Air molecules, which are

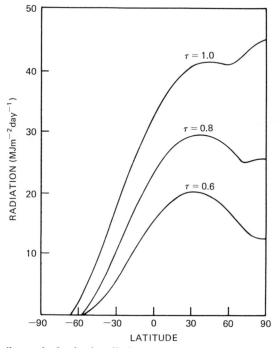

Figure 6.7. Daily total of solar irradiation per unit area of the earth's surface as a function of latitude at the time of the solstice for various values of atmospheric transmittance.

nitrogen, oxygen, ozone, carbon dioxide, water vapor, and other gases, produce some selective absorption of radiation.

The amount of scattering from the solar beam during its traversal of the atmosphere depends upon the path length of the solar beam through the atmosphere, which is a function of the zenith angle z. The amount of scattering is also dependent upon total quantity of atmosphere above the ground, which is, of course, less for station at a high altitude than one at sea level. Since the altitude of the station is defined in terms of its atmospheric pressure p, amount of scattering is a function of atmospheric pressure. If $_hS(z, p)$ is the amount of direct-beam solar radiation incident on a horizontal surface at the ground as a function of solar zenith angle z and atmospheric pressure p of the station, and if $_hd(z, p)$ is the amount of scattered skylight incident upon a horizontal surface, then the ratio of scattered skylight to direct solar radiation received at the ground is

$$k_d = \frac{_hd}{_hS}. \tag{6.28}$$

An analysis by McCullough and Porter (1972) shows that, for a pure Rayleigh sky, k_d is about 0.07 for a surface albedo of 25%, slightly less for an albedo of 10%, and about 0.115 for an albedo of 80%; these calculations were made for solar zenith angles from 0° to 40° and for a sea-level station. For larger zenith angles, the value of k_d increases considerably. Only rarely are we dealing with a pure Rayleigh sky; dust and pollutants are usually present. Remote mountain regions may approximate pure Rayleigh sky conditions occasionally but to a lesser extent year by year as population and industrialization continue to increase throughout the world. One can often judge the "quality" of the sky by its blueness. Having once encountered the deep-blue color characteristic of a pure Rayleigh sky with dry air, one begins to realize what a rare event it has become in many parts of North America; a Rayleigh sky is seen best from an aircraft flying at an altitude above 10,000 m.

Table 6.1 gives observed intensity as a function of solar zenith angle for the direct solar beam on a surface perpendicular to the beam and on a horizontal surface and diffused sunlight on a horizontal surface for clear-sky conditions without significant pollution, a cloudless industrial atmosphere, and an overcast sky with various types of cloud cover. For most purposes, one can assume that the scattered-skylight contribution to the irradiation of a horizontal surface is about 15% of the direct-beam solar radiation when the sun is relatively high in the sky. For ecological purposes, the accuracy of most measurements of $_hS$ does not justify the effort involved in determining a more precise value of $_hd$.

From Table 6.1, we see that when the sun is low in the sky, the diffuse component makes up between 25 and 50% of the total irradiation of a horizontal surface, although diffuse radiation is only about 11% of the direct-beam intensity on a surface perpendicular to the sun's rays. It is noteworthy that a deck of high, cirrostratus clouds lets through more solar irradiation on a horizontal surface than an industrial cloudless atmosphere. Other types of cloud cover produce significantly more attenuation of sunlight. Very often, a different ratio is used, namely, the ratio of scattered diffuse skylight incident on a horizontal surface to total flux incident on the surface.

Another example of observed diffuse shortwave radiation incident upon a horizontal surface at the ground, direct solar radiation, and total global shortwave radiation is illustrated in Fig. 6.8 from data taken by Monteith (1973) at Sutton Bonington, England, on 16 July 1969. This cloudless day was of average turbidity. At very low solar elevations, diffuse radiation makes up 50% or more of the global radiation, whereas at very high solar elevations diffuse radiation is about 20% of the total. Hence the ratio $_hd/_hg$ depends on elevation of the sun and turbidity of the atmosphere. When the atmosphere is very clear, the ratio $_hd/_hg$ may be as low as 10%.

Figure 6.9 shows the daily amounts of global radiation and diffuse radiation as a function of the time of year, as measured by Turner and Tranquillini (1961) for a station in Austria. Under clear-sky conditions, diffuse sky radiation is about 15% of global radiation in the summer, when the

Table 6.1. Values of Direct and Diffuse Solar Flux for Different Atmospheric Conditions.[a]

Solar zenith angle	Air mass	Clear atmosphere				Industrial cloudiness				Overcast			
		Direct-beam normal	Direct-beam horz	Diffuse horz	Total horz	Direct-beam normal	Direct-beam horz	Diffuse horz	Total horz	Ciro-Stra	Alto-Cumu	Strato-Cumu	Fog
0	1.00	928	928	112	1040								
10	1.02	921	900	112	1012								
20	1.06	914	865	112	977	593	558	216	775	816	502	349	154
30	1.15	893	775	105	879	572	509	202	712	740	475	314	140
40	1.30	865	670	105	775	544	419	181	600	642	405	265	126
50	1.55	802	530	98	628	495	321	174	488	516	328	223	112
60	2.00	740	370	91	461	363	216	140	356	377	237	160	84
70	2.90	621	216	70	286	328	112	98	209	223	160	112	63
80	5.60	391	70	42	112	181	35	56	91				
85	10.39	209	21	21	42	105	7	28	35				

[a] Values of solar flux are in units of watts per square meter.

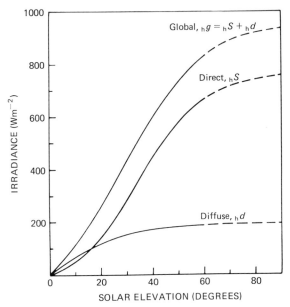

Figure 6.8. Amount of direct, diffuse, and global short-wave radiation on a horizontal surface at Sutton Bonington, England, for 16 July 1969. (Redrawn from Monteith, 1973.)

sun is high in the sky, and about 25% in the winter. Also compare amount of radiation during overcast sky conditions with that received during clear-sky conditions.

Clear-Sky Transmittances

A great deal of data pertaining to the relationship of diffuse skylight to direct sunlight incident upon a horizontal surface have been summarized by Liu and Jordan (1960). Let τ be the instantaneous transmittance of the atmosphere to direct sunlight in the zenith direction, previously defined as

$$\tau^m = S/S_o = {}_h S/{}_h S_o, \tag{6.29}$$

and let τ_d be the instantaneous transmittance to diffuse skylight, defined as

$$\tau_d = {}_h d/{}_h S_o, \tag{6.30}$$

where ${}_h d$ is the intensity of diffuse radiation from the sky incident on a horizontal surface. Liu and Jordan found the following relationship for clear-sky, but not dustfree, atmospheric conditions:

$$\tau_d = 0.271 - 0.294\, \tau^m. \tag{6.31}$$

This relationship shows, of course, that the greater the direct solar beam

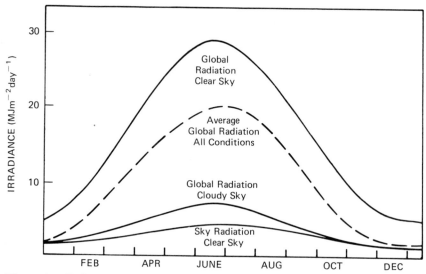

Figure 6.9. Daily amount of global and diffuse radiation as a function of the time of year for a station near Innsbruck, Austria. (Redrawn from Turner and Tranquillini, 1961.)

transmittance, the smaller the transmittance to scattered skylight. Typical values of direct-beam transmittance for a dust-free, clear sky range from 0.400 to 0.800, and the corresponding diffuse skylight transmittance varies from 0.153 to 0.037.

Total instantaneous radiation incident on a horizontal surface is known as global radiation $_hg$ and is the sum of the intensities of direct $_hS$ and diffuse $_hd$ radiation:

$$_hg = {_hd} + {_hS}. \tag{6.32}$$

Total instantaneous transmittance τ_t of the atmosphere is the ratio of total radiation received at the surface to that received outside the atmosphere:

$$\tau_t = {_hg}/{_hS_o} = {_hd}/{_hS_o} + {_hS}/{_hS_o}. \tag{6.33}$$

Then,

$$\tau_t = \tau_d + \tau^m. \tag{6.34}$$

Remember that τ^m is the transmittance of the atmosphere to the beam of direct sunlight only. This transmittance results from attenuation of the direct solar beam by scattering and absorption, as given by Eq. (6.26).

If we solve Eq. (6.34) for τ^m and substitute this value into Eq. (6.31), we obtain another useful relationship:

$$\tau_d = 0.384 - 0.416 \, \tau_t. \tag{6.35}$$

Hence when $\tau^m = 0.400$, $\tau_d = 0.153$ and $\tau_t = 0.555$, and when

$\tau^m = 0.800$, $\tau_d = 0.038$ and $\tau_t = 0.838$. For a range of total transmittance from 0.800 to 0.500, diffuse skylight will contribute from 6.3 to 35.2% of the total radiation incident on a horizontal surface at the ground for clear-sky, dust-free conditions. The empirical results taken from Liu and Jordan (1960) agree reasonably well with the results given in Table 6.1, except at very large solar zenith angles.

Global Radiation with a Clear Sky

Global radiation received at the ground is the sum of the direct solar radiation and the scattered skylight received on a horizontal surface. It is written as

$$_hg = {}_hS + {}_hd = S_o \, \tau^m \cos z + S_o(0.271 - 0.294 \, \tau^m) \cos z. \quad (6.36)$$

This is one of the most useful formulations with which we will work.

Geometrical Distribution of Skylight

Figure 6.10 shows the intensity distribution (in relative units) of sunlight scattered by the atmosphere toward the ground under clear-sky conditions. The sky is brightest near the sun and darkest in a zone approximately 90° from the sun in the plane of the sun and observer. This dark spot in the sky is shown in Fig. 6.10 by the isophote marked 8. The sky just above the horizon is generally brighter than regions higher in the sky,

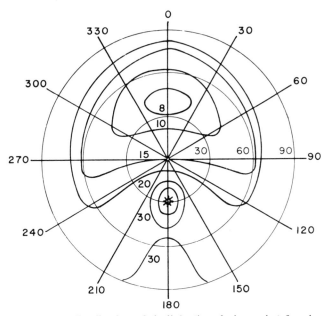

Figure 6.10. Intensity distribution of skylight (in relative units) for clear sky conditions as a function of azimuth. (Redrawn from Robinson, 1966.)

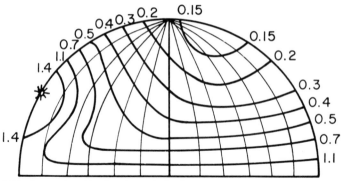

Figure 6.11. Intensity distribution of skylight (in relative units) under clear sky conditions for a vertical plane including the observer and the sun. (Redrawn from Robinson, 1966.)

except those regions near the sun. This is seen more clearly in the cross section of the sky hemisphere, shown in Fig. 6.11, which contains the plane of the sun, the observer, and the zenith. The higher intensity near the horizon is due to the fact that light is always most strongly scattered by molecules or small particles in the forward or backward directions and least scattered at right angles to the direction of the primary beam.

Polarization of Skylight

For many purposes, the sky is considered to scatter sunlight in an isotropic, or uniform, manner. In actuality, however, the sky is extremely nonuniform in intensity. The sky is also very nonuniform with respect to emission of thermal radiation, as we shall see in another chapter.

Although direct sunlight is always unpolarized, the scattered skylight of the blue sky is partly polarized. When sunlight strikes the molecules of the atmosphere, it sets these molecules into vibration at right angles to the direction of propagation of the sunlight. These air molecules act like small electrical antennas and scatter light of varying intensity and polarization in all directions. The result is that skylight is essentially unpolarized near the sun and has a maximum polarization approximately 90° from the sun in the vertical plane. Percent of polarization has never been observed to exceed about 60%, and theoretical calculations involving scattering theory indicate that the degree of polarization should not exceed 70%. Figure 6.12 shows the degree of polarization of skylight at various zenith angles in the vertical plane containing the sun, observer, and zenith for two positions of the sun, at solar zenith angles of 23° and 66°. These curves are drawn from data reported by de Bary (1964) for green light of wavelength 548 nm. Other wavelengths in the visible range exhibit just about the same degree of polarization. Volz and Bullrich (1961) give observational data in the visible and near-infrared and show that there is a small increase in degree of polarization of infrared over visible radiation out to a wavelength of 2000 nm.

Type of ground surface, whether soil, grass, or sand, affects the intensity and polarization of skylight. Coulson (1966) has made extensive studies of these phenomena and given detailed calculations and many curves showing the influence of ground surface on the character of skylight. Degree of polarization of light reflected from the ground surface is given by Coulson.

Hence skylight is polarized to varying degrees, depending upon the position of the sun. At sunrise or sunset, the sun's rays are horizontal, and the light from the zenith sky is strongly polarized. Many people can discern the polarization of skylight with the unaided eye. When looking at polarized light, they see in the center of their field of vision a faint blue-and-yellow figure known as Haidinger's brush. This phenomenon can also be observed by reflecting the most strongly polarized skylight off a glass surface or by using a polarizing filter. Haidinger's brush is caused by the dichroism of the yellow spot of the retina. Not everyone sees Haidinger's brush in the same manner, apparently because of a difference in shape and structure of the yellow spot. One can practice seeing Haidinger's brush and readily discern the polarization of the sky. A splendid discussion of this phenomenon is given by Minnaert (1954) in his famous book, *The Nature of Light and Color in the Open Air.* Anyone interested in atmospheric optical phenomena will find this book a great joy to read.

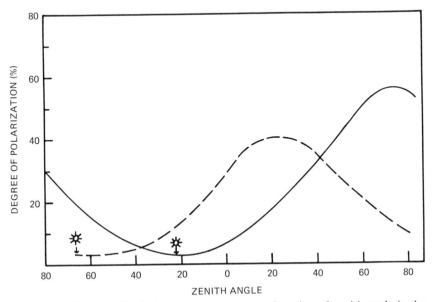

Figure 6.12. Degree of polarization in percent as a function of zenith angle in the vertical plane of the observer and the sun for solar zenith angles of (—) 23° and (--) 66°.

Many animals make use of the polarized skylight for orientation and navigation; excellent discussions of this subject are given by Frisch (1950), Waterman (1955), and Wellington (1974).

Global Shortwave Irradiation with an Overcast Sky

The flux of direct and diffuse solar radiation incident at the earth's surface is highly variable under conditions during which the sky is partially or totally overcast, and specific values at any one instant are nearly impossible to predict. Measurements can be made by means of radiation instruments of short time constant. The only meaningful way in which a relationship can be established between diffuse radiation and total radiation for cloudy days involves statistical averages of observational data collected over long periods of time.

Daily Average Transmittances with an Overcast Sky

Let the daily diffuse radiation on a horizontal surface be $_hD$ and the daily total global radiation falling on a horizontal surface be $_hG$. The following dimensionless parameters for daily transmittance are then defined in terms of the extraterrestrial daily insolation $_hQ_o$ received on a horizontal surface:

$$T_d = {_hD}/{_hQ_o} \tag{6.37}$$

$$T_t = {_hG}/{_hQ_o}. \tag{6.38}$$

Whereas the quantities τ_d and τ_t are the instantaneous transmission coefficients, the quantities T_d and T_t are the daily averages of diffuse and total shortwave transmittance taken over the hours of any single day.

Liu and Jordan (1960), from analysis of a great quantity of data, derived the relationship between T_d and T_t shown in Fig. 6.13. Since daily total amount of solar radiation received at many localities is measured and extraterrestrial amount of daily solar radiation is easy to calculate for any location, one can determine daily total transmittance for global shortwave radiation. Then, from Fig. 6.13, it is easy to get an estimate of daily diffuse transmittance for that location. It is evident that when the sky is very clear, the amount of diffuse radiation or skylight is low. When the sky is heavily overcast, the amount of diffuse radiation is also low. There is an optimum in between, probably when there is a thin deck of cirrus or stratus clouds, which diffuse the sunlight strongly yet have a good transmittance. Figure 6.13 shows that total daily transmittance is about 0.40 for the maximum amount of diffuse radiation, under which conditions the diffuse radiation amounts to about 25% of extraterrestrial insolation. Fig-

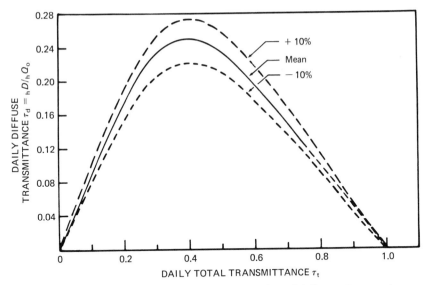

Figure 6.13. Daily diffuse transmittance as a function of daily total transmittance. (Redrawn from Liu and Jordan, 1960.)

ure 6.14 shows the relationship between the ratio of diffuse and global daily radiation $_hD/_hG$ and total transmittance T_t. On completely overcast days, when the ratio of $_hG/_hQ_o$ is small, incident radiation is diffuse only, and the ratio $_hD/_hG$ approaches 1.0. When the sky is cloudless and extremely clear, Rayleigh scattering predominates, and the ratio $_hD/_hG$ approaches a limiting value of about 0.10. This is the horizontal line from values of T_t of about 0.8 to 1.0. Often of interest are monthly averages of the daily transmittances, designated as $\overline{T}_d$ and $\overline{T}_t$. For many general purposes, such as considerations of agricultural productivity or solar heating, monthly averages of atmospheric transmission are adequate. The quantity $\overline{T}_t$ is referred to as the cloudiness index. A locality with a cloudiness index of 0.30 is considered to be extremely cloudy, whereas one with $\overline{T}_t = 0.70$ is very sunny.

Empirical Equations Relating Global Radiation to Cloud Cover

Many investigators have derived empirical equations to express the relationship between amount of global radiation received during a cloudy day to that received during a clear day. Just as the transmittances derived by Liu and Jordan (1960) are generally useful, so also are these empirical equations. Only a few examples are given here.

Fritz and MacDonald (1949) used the following relationship between $_hG'$, amount of global radiation received on a cloudy day, and $_hG$, amount

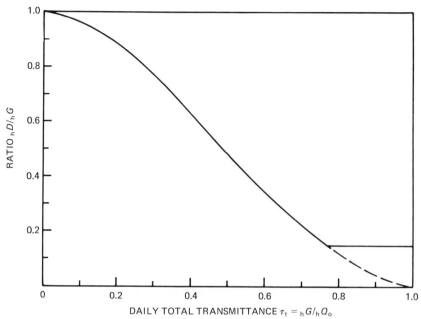

Figure 6.14. Ratio of the daily diffuse radiation and the total daily radiation incident on a horizontal surface $_hD/_hG$ as a function of daily total transmittance T_t.

received on a cloudless day:

$$_hG' = {_hG}(a + b\Sigma), \tag{6.39}$$

where a and b are constants and Σ is the number of hours of sunshine, recorded instrumentally. Evaluation of the constants is usually done with the use of monthly averages of $_hG$ and Σ. Fritz and MacDonald, using data from 11 stations, derived values for a and b of 0.35 and 0.61, respectively. These values clearly do not add up to 1.0, for the reason that the data applied to a limited range of Σ from 0.35 to 0.97 and the fact that days on which $\Sigma = 1.0$ may have had some cloud cover. Kimball (1919, 1927), Angstrom (1924), and Haurwitz (1934) obtained slightly different values for a and b with similar formulations. With earlier meteorological data however, it is not always certain that the definitions of $_hG$ and Σ as used instrumentally are always the same.

Black (1956), using data from 88 stations distributed over the globe, deduced the following empirical formula to express the relationship between monthly mean daily amount of solar radiation $_h\overline{G}$ received at the ground on a horizontal surface (as a function of the monthly mean amount of solar radiation $_h\overline{Q}_o$ incident on a horizontal surface in space) and monthly mean amount of sky covered by clouds $\overline{C}$ (a decimal fraction):

$$_h\overline{G} = {_h\overline{Q}_o}(0.803 - 0.340\,\overline{C} - 0.458\,\overline{C}^2). \tag{6.40}$$

This formula assumes that the maximum amount of sunlight received at the surface on a cloudless day is only 0.803 of the amount received outside the earth's atmosphere. Obviously, this assumption is not completely valid, since atmospheric conditions such as amount of water vapor, dust, and pollution may vary.

Bennett (1965) used monthly means of daily insolation from 41 stations to derive a regression equation relating monthly mean daily global radiation at the earth's surface $_h\overline{G}$ to monthly mean daily insolation on a horizontal surface just outside the atmosphere $_h\overline{Q}_o$:

$$_h\overline{G} = {_h\overline{Q}_o} \, (203 + 5.13\overline{\Sigma}) \times 10^{-3}, \qquad (6.41)$$

where $\overline{\Sigma}$, as before, is the mean monthly number of hours of sunshine, expressed as percent of total possible hours of sunshine. Hence if $\overline{\Sigma} = 100\%$, then $_h\overline{G}/_h\overline{Q}_o = 0.716$; and if $\overline{\Sigma} = 0\%$, $_h\overline{G}/\overline{Q}_o = 0.203$. A snow-covered ground surface produces increased insolation because of multiple reflections between the snow surface and the clouds. With $\overline{M}$ indicating the mean percentage of days during the month with 1 in or more of snow cover Bennett (1965) used the following relationship:

$$_h\overline{G} = {_h\overline{Q}_o} \, (210 + 6.15\overline{\Sigma} + 2.14\overline{M}) \times 10^{-3}. \qquad (6.42)$$

Bennett also found that the coefficients in Eq. (6.42) depended on station elevation H and cosine of the station latitude ϕ. The regression equation is

$$_h\overline{G} = {_h\overline{Q}_o} \, [(201.8 + 0.3658H) + (2.755 - 0.00308H + 3.201 \cos \phi) \, \overline{\Sigma}]. \qquad (6.43)$$

The results Bennett obtained are given in the next section describing actual measured amounts of global radiation.

It is important to be able to determine amount of solar radiation received at the earth's surface when only degree of cloud cover is known. This is frequently the only information known for regions such as the oceans, for example, where cloud cover is estimated from satellites. According to Lumb (1964), mean hourly amount of solar radiation reaching the sea surface $_h\overline{g}$ is a monotonic function of the mean sine of the solar altitude for any given type of cloud cover. He writes the relationship as

$$_h\overline{g} = A \, \overline{\sin a} + B \, \overline{\sin^2 a}, \qquad (6.44)$$

where A and B are constants. In terms of the solar constant S_o, the equation is written

$$_h\overline{g} = S_o \, \overline{\sin a} \, (A' + B' \, \overline{\sin a}). \qquad (6.45)$$

The quantity $_h\overline{g}/S_o \sin a$ is the fraction of the total solar radiation at the outer limits of the earth's atmosphere that penetrates to the sea surface. This is our earlier value of instantaneous total transmittance τ_t averaged over an hourly period. Hence we can write

$$_h\overline{g} = S_o \, \tau_t \, \overline{\sin a}, \qquad (6.46)$$

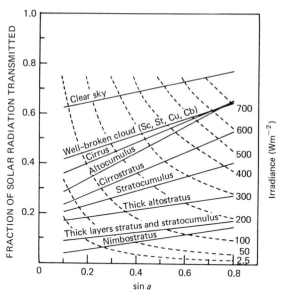

Figure 6.15. Fraction of total solar radiation transmitted through the atmosphere as a function of the sine of the solar altitude for various cloud types. Measurements were taken in the North Atlantic (52°N, 20°W). Abbreviations: Sc, stratocumulus; St, stratus; Cu, cumulus; Cb, cumulonimbus. Broken lines represent irradiance as a function of attitude. (Redrawn from Monteith, 1973.)

where

$$\bar{\tau}_t = A' + B' \overline{\sin a}. \tag{6.47}$$

Lumb evaluates A' and B' by application of a least-squares method to substantial numbers of observations taken by weather ships in the North Atlantic Ocean. The results of these determinations are shown by the straight lines in Fig. 6.15. The dashed lines, calculated by means of Eq. (6.46), represent the hourly totals of solar radiation falling on a horizontal surface.

Hourly Averages of Shortwave Irradiation

Liu and Jordan (1960) have derived the relationship between hourly and daily amounts of diffuse and total radiation as related to the time of day, expressed as number of hours from solar noon. Again, it must be emphasized that these are averages and that variable cloudiness precludes the possibility of obtaining true instantaneous radiation intensity during cloudy days, except from direct observation.

Average daily transmission coefficient of the atmosphere to diffuse radiation T_d is the ratio of average daily diffuse radiation on a horizontal surface at the ground to total daily direct radiation on a horizontal surface outside of the earth's atmosphere. That is,

$$\overline{T}_d = {}_h\overline{D}/{}_h\overline{Q}_o. \tag{6.48}$$

The assumption is made that average daily transmission to diffuse radiation is also equal to average hourly transmission as given by

$$\overline{T}_d = {}_h\overline{d}/{}_h\overline{H}_o, \tag{6.49}$$

where ${}_h\overline{d}$ is the average hourly amount of diffuse radiation at the ground and ${}_h\overline{H}_o$ the average hourly amount of direct solar radiation outside the earth's atmosphere, each received on a horizontal surface.

The ratio of the average hourly diffuse radiation on a horizontal surface to the average daily diffuse radiation is written

$$r_d = {}_h\overline{d}/{}_h\overline{D}. \tag{6.50}$$

Substituting for ${}_h\overline{d}$ from Eq. (6.49) and for ${}_h\overline{D}$ from Eq. (6.48) gives

$$r_d = {}_h\overline{H}_o/{}_h\overline{Q}_o. \tag{6.51}$$

It is now possible to express r_d purely in terms of geometrical considerations. This is true since, from Eq. (6.13), we have ${}_h\overline{Q}_o$ as a function of

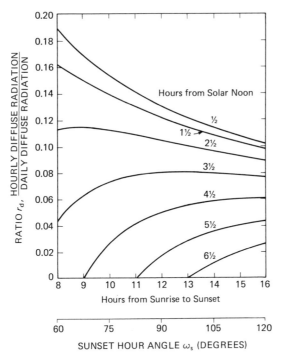

Figure 6.16. Ratio of hourly to daily diffuse radiation as a function of number of hours from sunrise to sunset (length of day) for the time from solar noon. (Redrawn from Liu and Jordan, 1960.)

geometrical factors. We shall take the hourly average value $_hH_o$ to be the same as $_nS_o$ at the midpoint of the hour interval as given by Eq. (6.10). This is an approximation but is sufficiently accurate for our purposes here. Hence

$$r_\text{d} = \frac{\pi}{86,400} \frac{\sin \phi \sin \delta + \cos \phi \cos \delta \cos h}{h_\text{s} \sin \phi \sin \delta + \cos \phi \cos \delta \sin h_\text{s}}. \qquad (6.52)$$

The sunset or sunrise hour angle h_s is given by Eq. (6.8) for any latitude ϕ and time of year (which determines the value of δ). Substituting into Eq. (6.52), one gets

$$r_\text{d} = \frac{\pi}{86,400} \frac{\cos h - \cos h_\text{s}}{\sin h_\text{s} - h_\text{s} \cos h_\text{s}}. \qquad (6.53)$$

This equation is plotted in Fig. 6.16, which does not show the observational values contained in the original plot by Liu and Jordan, although these author's data fit the curves exceedingly well. The data came from observations made at Blue Hill Observatory, Milton, Mass., and Helsingfors, Finland.

A similar argument and analysis applied to the hourly and daily total radiation gives the results seen in Fig. 6.17. Again, the observational data fit the theoretical curves extremely well.

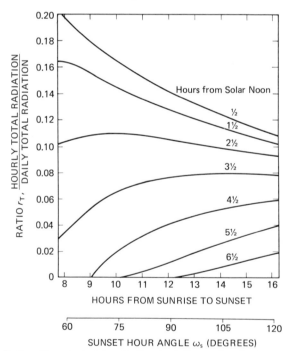

Figure 6.17. Ratio of hourly total radiation to daily total radiation as a function of number of hours from sunrise to sunset (length of day) for the time from solar noon. (Redrawn from Liu and Jordan, 1960.)

The phenomena exhibited in Figs. 6.16 and 6.17 are most interesting. Note that near midday, within 1.5 hr of solar noon, there is a monotonic decrease of r_d or r_t from short days (winter) to long days (summer). Midday hourly diffuse and total radiation constitute a greater proportion of the daily amounts in the winter than in the summer. The opposite is true for hours early or late in the day, when the winter contribution to the total daily amount is much less than the summer. For the hour 2.5 hr from solar noon, the ratio varies little from winter to summer. The curves also show that the midday hours always contribute more to the daily total than hours early or late in the day.

Measured Amounts of Global Radiation

Theory gives a good idea of the maximum amounts of direct solar radiation and approximate estimates of global radiation received at any position on the earth's surface. From the theoretical formulation, one can determine the momentary values and, by integration, the hourly, daily, monthly, and annual totals of direct solar radiation. Atmospheric conditions are constantly changing, however, and clouds affect the amount of shortwave radiation received in a manner difficult to predict from theory alone. The empirical equations for average transmittances with overcast conditions have been developed for these reasons. There is usually a biological application for various types of time averages of incident shortwave radiation, whether it be the instantaneous values or the hourly, daily, monthly, or annual totals. The observed ranges for some of these values follow.

Instantaneous Global Radiation

Instantaneous amount of global radiation incident on a horizontal surface may vary from about 50 to 1250 W m^{-2}, depending on sky conditions. Typical recordings of instantaneous global radiation are shown in Fig. 6.18. The global-radiation record for a perfectly clear day will be a simple bell-shaped curve symmetrical about solar noon. Clouds produce the variations seen in Fig. 6.18. It is evident that heavy cloud cover may reduce global radiation by as much as 82%. The very highest values of instantaneous global radiation occur when there are scattered cumulus clouds in a clear sky; in this case, direct sunlight and scattered skylight reach the ground from the open sky, and the edges of the cumulus clouds near the sun reflect additional sunlight to the ground. In high, mountain locations, when the sky is very clear and dry but contains scattered cumulus clouds, instantaneous global radiation may reach 1500 W m^{-2}.

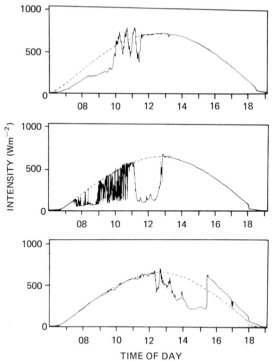

Figure 6.18. Instantaneous amount of global solar radiation incident on a horizontal surface as a function of time of day. Three typical recordings are shown.

Hourly and Daily Total Global Radiation

When values for hourly and daily total global radiation are obtained, a certain amount of averaging occurs for whatever variable sky conditions exist. Hourly total global radiation may range from 150 kJ m^{-2} hr^{-1} to 3.6 MJ m^{-2} hr^{-1}.

Monthly averages of total daily global radiation are particularly useful in estimating agricultural productivity. Typical average daily amounts of global radiation range from 4.0 MJ m^{-2} day^{-1} for an overcast winter day at mid-latitudes to as high as 30 MJ m^{-2} day^{-1} for a very clear day on the summer solstice.

Kondratyev (1969) gives many tables which summarize solar radiation data on a worldwide basis. He makes particular use of the work of Berland and Danilchenko (1961). Other important reference works include Dirmhirn (1964), Schulze (1970), and Coulson (1975).

Maps of Daily Global Insolation

Bennett (1965) presents maps for the mean daily solar insolation for the United States for each month of the year. The maps shown in Figs. 6.19 and 6.20 are examples for the months of January and July. Whereas the

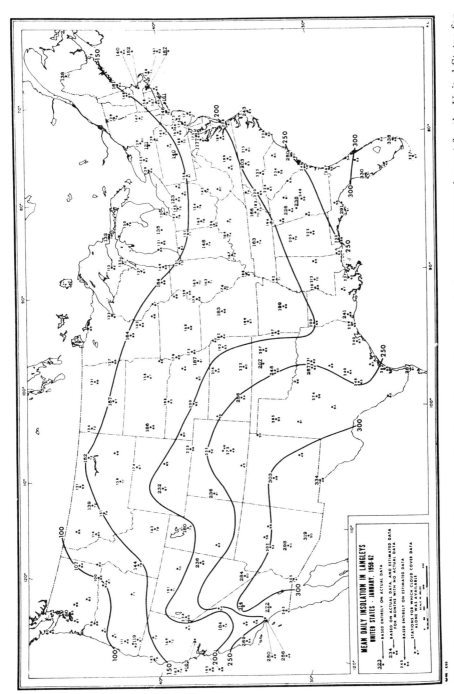

Figure 6.19. Mean daily insolation on a horizontal surface in langleys (calories per square centimeter) for the United States for January, 1958 to 1962. (From Bennett, 1965.)

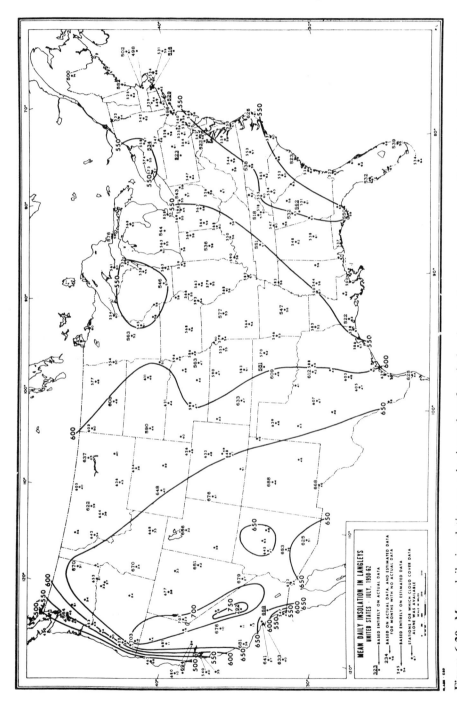

Figure 6.20. Mean daily insolation on a horizontal surface in langleys (calories per square centimeter) for the United States for July, 1958 to 1962. (From Bennett, 1965.)

isolines of solar insolation run primarily east and west across the conterminous United States in January, they exhibit a strong north-south trend in July. During the winter months, there is a strong north-south gradient in insolation. In January, insolation values greater than 12.6 MJ m^{-2} day^{-1} are found in southern Florida and the southern United States, whereas the insolation is about 4.2 MJ m^{-2} day^{-1} along the Canadian border. During July, the gradient of insolation reverses itself, and higher values are found in the area of the Great Lakes than along the Gulf Coast. A similar condition exists in the West, where the monthly means of daily insolation for Spokane, Wash., and Boise, Idaho, are each 28.1 MJ m^{-2} day^{-1}, compared to 27.3 and 26.2 MJ m^{-2} day^{-1} for Phoenix and Tucson, Ariz. A notable feature of most of the monthly maps, especially pronounced in the winter maps, is the southward swing of the isolines over the Mississippi Valley owing to low elevation and the southward push of storm tracks and the northward swing of the isolines over the High Plains and the Basin and Plateau Region owing to high elevation and a general weakening and lower incidence of storms.

The reader is reminded that the average flux received by a horizontal surface located outside the earth's atmosphere is 29.4 MJ m^{-2} day^{-1}. For the world as a whole, however, an average of only 57.8% of the flux outside of the earth's atmosphere reaches the ground (17.0 MJ m^{-2} day^{-1}).

Revfeim (1978) gives a method for obtaining estimates of global daily radiation on a sloping site by means of data from nearby horizontal sites.

Map of Annual Global Insolation

Several investigators have summarized the radiation records of weather stations distributed over the earth's surface to obtain annual average solar irradiation of a horizontal surface at any position. To draw a map of generalized isopleths of global surface radiation clearly requires extrapolation across many regions, particularly the oceans, where weather stations are not located. Recent summaries of this information are given by Landsberg (1961) and Budyko (1963). The map given by Landsberg is shown in Fig. 6.21. It is seen that strong heat sources exist over the continents, the largest being the heat source centered over northern Africa and extending over the northern Indian Ocean. The most northern heat source is the one centered over the southwestern United States and northern Mexico. The difficulty of drawing accurate maps on a worldwide basis is appreciated when it is realized that in 1930 there were only 32 insolation stations in the entire world with data suitable for climatic analysis. By 1945, there were 58 stations; by 1948, 89; and by 1954, only 139. There are perhaps 200 stations in operation today. Unfortunately, for many stations, the records are too short or, for some, too intermittent for good map determinations. In determining the energy budgets of plants and animals, annual isolation budgets may not be as meaningful as daily, hourly, or instantaneous amounts of insolation determined for specific locales.

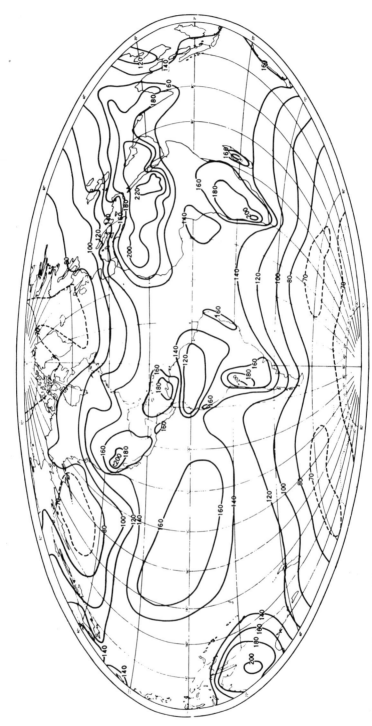

Figure 6.21. Mean annual insolation throughout the world. (From Landsberg, 1961.) Calories per square centimeter per year.

Daylight and Twilight

Length of Daylight and Twilight

A great deal of animal behavior, and some plant responses, takes place during the twilight periods before sunrise and after sunset. It is necessary to have an indication of amount of time each day when twilight occurs and also a measure of amount of illumination during this period. The *daylight* period is defined as the time between sunrise and sunset. *Civil twilight* is defined as the interval of time between sunrise or sunset and the time when the true position of the center of the sun is 6° below the horizon; during this time, stars and planets of the first magnitude are just visible, and twilight glow is seen on the horizon. *Astronomical twilight* is the interval between sunrise or sunset and the time when the sun is 18° below the horizon; during this time, stars of the sixth magnitude are visible near the zenith, and there is no trace of twilight glow on the horizon. Values for the duration of daylight, civil twilight, and astronomical twilight are given in List (1963).

Intensity of Twilight

Kimball (1938) reviewed the information available concerning twilight and found that the observations made at Mount Weather, Va., in 1913 and Salt Lake City, Utah, in 1914 and 1916 define the line shown in Fig. 6.22. Twilight intensity is given in footcandles. When the upper limb of the sun coincides with the horizon, i.e., when the center of the sun is about 50' below the true horizon, zenith illumination is about 33 fc. At the end of civil twilight, illumination is about 0.4 fc. By comparison, the sun at zenith is about 9600 fc, the full moon at zenith 0.02 fc, a quarter moon 0.002 fc, and the starlit sky 0.00008 fc.

At the instant of sunrise or sunset, illumination is about 1650 times as intense as that from the zenith full moon, and at the end of civil twilight, it is about 20 times as intense; when the sun is about 8°40' below the horizon, twilight intensity equals that of the zenith full moon. On a clear, dark night, starlight illumination is about 0.004 as intense as the zenith full moon.

The numbers here refer to average clear-sky conditions. When the atmosphere is very dry, twilight illumination will be more intense than when the atmosphere is moist. In addition, clouds, haze, smoke, or dust will reduce light intensity.

Moonlight

Relatively little research has been done concerning the responses of plants and animals to moonlight. Tansey and Jack (1975) reviewed the literature and found quite a few reports of plant and animal response to

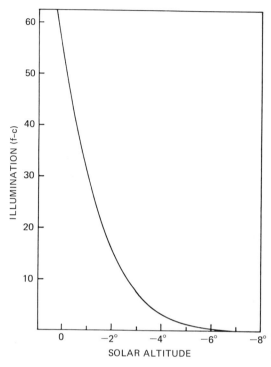

Figure 6.22. Twilight intensity in foot-candles as a function of altitude of the sun below the horizon. (Redrawn from Kimball, 1938.)

moonlight. In particular, a variety of responses has been reported for fungi, including vegetative growth and CO_2 fixation, growth of sporangiophores, elongation of conidiophores, induction or inhibition of sporulation, spore liberation, and stimulation of spore germination. In most cases, however, these responses occur in many species. These authors point out that ''lunar timing of spore release of insect and plant pathogens may allow dispersal to suitable hosts at times when deleterious effects of direct solar radiation are absent, and beneficial effects of high relative humidity and dew are optimal for spore germination and penetration of the hosts (p. 404).''

Moonlight illuminance levels are about 0.68 lx (0.063 fc) for full moon at a lunar altitude of 60° with clear sky. This amount of moonlight is equivalent to about 2.3 mW m^{-2}.

Plant Productivity

It is of interest to insert here a topic of biological importance concerning gross and net primary productivity of the entire earth and an estimate of the global efficiency for photosynthesis. Gross productivity of the entire earth's biosphere is estimated at 6.53×10^{18} kJ yr^{-1}. If we divide this value by the surface area of the earth, taken to be 510×10^{12} m^2, we get an average gross productivity per unit area of 12.8 MJ m^{-2} per year, or

35.0 kJ m⁻² per day; this value is equivalent to 0.405 W m⁻². Average daily amount of solar radiation at the earth's surface is 17.0 MJ m⁻² day⁻¹. Hence the primary productivity of all the ecosystems of the world has an overall efficiency of 0.207%.

Net productivity is considerably less than gross productivity, since respiration consumes, on the average, 3.81×10^{18} kJ of the 6.53×10^{18} kJ of gross productivity per year, giving a value of 2.72×10^{18} kJ for net productivity. When divided by the earth's surface area, this is equivalent to 14.7 kJ m⁻² day⁻¹. The percentage of solar radiation which reaches the ground and shows up in net primary productivity is thus 0.086%.

Crops are designed for high productivity and make a more efficient use of incident sunlight than the natural ecosystems of the world on the average. Crops are maintained by means of a tremendous energy input by man, however. Loomis and Williams (1963) have made very careful estimates of the maximum potential gross primary productivities of crops. They find that when the daily input of solar radiation is 20.9 MJ m⁻² day⁻¹, potential plant productivity is 0.107 kg m⁻² day⁻¹, equivalent to 1.67 MJ m⁻² day⁻¹ if the caloric value of the plant material is 15.66 MJ kg⁻¹. This represents an efficiency, in terms of total available light, of 8%. If one assumes that 44% of total sunlight is contained in the photosynthetically useful wavelengths of the visible spectrum, then the efficiency, in terms of utilization of visible light, is 18%. It is the figure of 8%, however, which should be compared with the efficiency of the global ecosystem (0.207%) Loomis and Williams went on to measure the primary productivity of Sudan grass growing at Davis, Calif., when the average amount of sunlight was 28.9 MJ m⁻² day⁻¹. Gross productivity was 1.75 MJ m⁻² day⁻¹ and net productivity, 0.85 MJ m⁻² day⁻¹, which represents a percent utilization of total sunlight of 6 and 3%, respectively. The average annual crop productivity for the United States has an efficiency of about 1.7%, and 4-month averages for the summer growing season in the central United States are about 2.4%.

Once again, we can compare these efficiencies with those characteristic of specific natural ecosystems: a salt marsh in Georgia, 0.8%; a pine forest in England, 0.5%, a tallgrass prairie in Nebraska, 0.1%; and a desert in Nevada, about 0.04%. A tropical forest has a large gross productivity, utilizing sunlight with an efficiency of about 4.4%, but also has a very high maintenance cost through respiration, so that net productivity is less than 20% of gross productivity.

Solar and Skylight Irradiation of Slopes

The climate conditions for a slope depend to a considerable degree upon the amount of direct solar irradiation. Temperature, windiness, degree of dryness, and amount of light depend very much on solar irradiation. Although describing the amount of direct solar radiation incident on a slope is fairly straightforward, it is difficult to determine accurately the diffuse skylight incident on various slopes at a given hour and time of year at the earth's surface. We first describe the purely geometrical relationships between slope aspect and direct solar irradiation and then add the effect of diffuse irradiation by skylight.

Geometry of Slopes

Our treatment of the geometry of slopes is a direct extension of the description of the relationship between the sun, the earth, and the position of an observer using spherical trigonometry. The easiest way to describe the aspect, or direction, of a slope is to describe the direction which the normal, or perpendicular, to a slope assumes; the normal to the plane can take only a single direction in space. Figure 6.23a shows an inclined plane. The normal to the plane has azimuth angle α_s and altitude angle a_s. The slope angle β from horizontal has a value of $90° - a_s$. The direct rays of the sun are incident on this slope along a line of azimuth angle α and altitude angle a, as shown in Fig. 6.23b. The direct beam of sunlight makes an angle of incidence i to the slope. Angles of incidence are always defined relative to the normal to the surface. If the flux of direct solar radiation is S, then the flux $_sS$ falling at angle i on a slope is given by

$$_sS = S \cos i = S_o \tau^m \cos i. \tag{6.54}$$

In order to assist the reader in visualizing the angles α_s and a_s for various slopes, examples are given in Table 6.2. Keep in mind that azi-

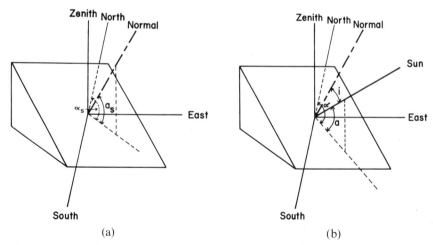

(a)   (b)

Figure 6.23. (a) Geometrical relations defining the normal to an inclined slope in terms of its azimuth angle α_s and altitude angle a_s. (b) Geometrical relations between the angle of incidence i to the normal to a slope for solar rays of azimuth angle α and altitude angle a.

Table 6.2

Slope aspect	a_s	α_s
Horizontal		
Vertical north-facing	0°	0°
Vertical east-facing	0	90
Vertical south-facing	0	180
Vertical west-facing	0	270
45° North-facing	45	0
45° East-facing	45	90
45° South-facing	45	180
45° West-facing	45	270

Solar position	a	α
1. Sun due east on horizon	0°	90°
2. Sun due east at 45° elevation	45	90
3. Sun in the SE at 60° elevation	60	135
4. Sun due south at 30° elevation	30	180

Angles of incidence for different slope aspects	Solar position			
	1	2	3	4
Horizontal	90°	45°	30°	60°
Vertical north-facing	90	90	110.7	150
Vertical east-facing	0	45	69.3	90
Vertical south-facing	90	90	69.3	30
Vertical west-facing	180	135	110.7	90
45° North-facing	90	60	68.7	105
45° East-facing	45	0	30.5	69.3
45° South-facing	90	60	30.4	15
45° West-facing	135	90	68.7	69.3

muth is the angle of the normal from the direction of north. A horizontal surface always has its normal pointing to the zenith, hence an altitude of 90°, whereas a vertical surface, irrespective of its azimuthal orientation, has an altitude of 0°. An upward-facing slope possesses an altitude angle a_s between 0° and +90°; a downward-facing slope, such as the underside of an overhanging rock ledge or the lower side of a leaf or animal, has an altitude angle between 0° and −90°.

The angle of incidence i between the incident solar rays and the normal to the surface is given by

$$\cos i = \sin (90 - a_s) \cos a \cos (\alpha - \alpha_s) + \cos (90 - a_s) \sin a \quad (6.55)$$
$$= \cos a_s \cos a \cos (\alpha - \alpha_s) + \sin a_s \sin a.$$

The reader is reminded that, for azimuth angles in the first quadrant (between 0° and 90°), the sines and cosines are positive; for angles in the second quadrant (between 90° and 180°), the sines are positive and the cosines are negative; for angles in the third quadrant (between 180° and 270°), the sines and cosines are negative; and for angles in the fourth quadrant (between 270° and 360°), the sines are negative and the cosines are positive. For altitude angles between 0° and 90°, sines and cosines are positive, but for altitude angles between 0° and −90°, the sines are negative and the cosines are positive. Angles of incidence to various slopes for several solar positions are given in Table 6.2. When the angle of incidence is greater than 90°, the slope will not receive direct-beam irradiation.

It is evident that if the sun is in the meridian plane with altitude a and azimuth 180°, slopes tilted north or south in the meridian plane have the angle of incidence given by

$$\cos i = \cos a_s \cos a + \sin a_s \sin a = \pm \cos (a_s \mp a) \qquad (6.56)$$

$$i = a_s \mp a. \qquad (6.57)$$

Tilting a surface in the meridian plane north or south from horizontal is the equivalent of going north or south in latitude by the same number of degrees.

Direct Solar Irradiation of Slopes

Equations (6.54) and (6.55) may be used to calculate the amount of direct solar radiation incident upon any slope. Values are given in Table 6.3 for a position at latitude 40°N at the time of the summer solstice for a variety of slope aspects. [See the tabulation by Buffo et al. (1972).]

The direct-beam irradiation of various slopes at 40° north latitude on 22 June by solar radiation unattenuated by the earth's atmosphere is shown in Fig. 6.24. Although some of the significant features of slope aspect and solar irradiation show up here, considerable modification of these curves occurs when atmospheric attenuation and scattering are taken into account. Sunrise and sunset are very abrupt for these vertical slopes, particularly for the vertical north-, east-, and west-facing slopes.

If the same calculation is done for various slopes at the earth's surface by using a value for atmospheric transmittance of $\tau = 0.6$, one obtains the lower set of curves shown in Fig. 6.24. The values are given in Table 6.3. Changes in some of the curves are very striking, particularly during the early and late part of the day, when atmospheric attenuation is very strong. Some slopes have a much-delayed sunrise or premature sunset. One should notice that most slopes have a single diurnal cycle of irradiation beginning at sunrise and ending at sunset. A north-facing vertical slope has a double diurnal cycle of direct solar irradiation. At times between the vernal and autumnal equinoxes, the summer sun rises to the north of east and sets to the north of west. This means that a north-facing slope is illuminated by the direct rays of the sun early in the day

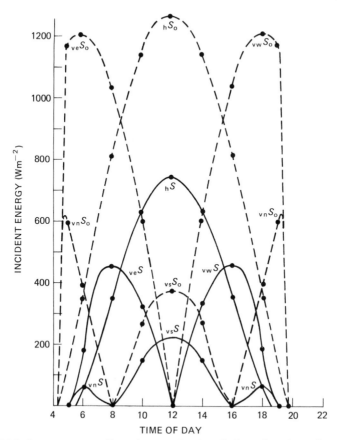

Figure 6.24. Instantaneous direct-beam solar irradiation of various slopes at 40° north latitude on 22 June as a function of time of day (−−) without atmospheric attenuation and (—) with a transmittance $\tau = 0.6$. Slopes: h, horizontal, vn, vertical north-facing; vs, vertical south-facing; ve, vertical east-facing; vw, vertical west-facing.

and again late in the day, spending most of the midday period shaded. At the same time, of course, the south-facing vertical slope does not receive the direct rays of the sun until the north-facing vertical slope is in the shade.

The horizontal slope gets more daily irradiation on 22 June than the other slopes. However, the greatest amount of direct solar irradiation is experienced by a slope normal, or perpendicular, to the solar rays at noon. Since the earth's axis on 22 June is tilted 23.5° north of the equator, for an observer at 40° north latitude, the zenith angle of the sun is $40° − 23.5° = 16.5°$. A slope at 40° north latitude that is tilted southward in the meridian plane by 16.5° receives maximum daily insolation on 22 June. Put another way, a horizontal surface at a latitude of 23.5°N

Table 6.3. Angles, Trigometric Functions, and Incident Direct, Diffuse, and Global Shortwave Radiation for Various Slopes at 40° North Latitude on 22 June for $\tau = 0.6$, $S_0 = 1317$ W m^{-2}, and $r = 0.2$.[a]

Time	a	$\cos a$	α	$\cos \alpha$	z	$m(=\sec z)$	τ^m	S	τ_d	$_hd$	$_vd$	$_{45}d$
0500	5	0.996	63	0.454	85	11.47	0.003	4	0.270	31	16	24
0600	16	0.961	72	0.309	74	3.63	0.157	206	0.225	82	41	62
0800	38	0.788	89	0.018	52	1.62	0.436	574	0.143	116	58	87
1000	60	0.500	114	-0.407	30	1.16	0.554	730	0.108	123	62	92
1200	73.5	0.285	180	-1.000	16.5	1.04	0.587	773	0.098	124	62	93

Horizontal

Time	$\cos z$	$_hS_0$	$_hS$	$_hg$	r_hg
0500	0.087	115	0	31	6
0600	0.276	363	57	139	28
0800	0.616	811	354	470	94
1000	0.866	1141	632	755	151
1200	0.959	1262	741	865	173

Vertical north-facing / Vertical south-facing

Time	$\cos i$	$_{vn}S_0$	$_{vn}S$	$_{vn}g$	$\cos i$	$_{vs}S_0$	$_{vs}S$	$_{vs}g$
0500	0.452	595	2	21	-0.452	0	0	19
0600	0.297	391	61	116	-0.297	0	0	55
0800	0.014	18	8	111	-0.014	0	0	169
1000	-0.203	0	0	138	0.203	268	148	286
1200	-0.285	0	0	149	0.285	375	220	369

Vertical east-facing / 45° south-facing

Time	$\cos i$	$_{ve}S_0$	$_{ve}S$	$_{ve}g$	$\cos i$	$_{45s}S_0$	$_{45s}S$	$_{45s}g$
0500	0.888	1169	3	22	-0.258	0	0	26
0600	0.914	1204	189	244	-0.015	0	0	69
0800	0.788	1038	453	558	0.426	561	244	355
1000	0.457	602	333	471	0.756	996	552	682
1200	0	0	0	149	0.879	1158	680	816

45° north-facing / 45° east-facing

Time	$\cos i$	$_{45n}S_0$	$_{45n}S$	$_{45n}g$	$\cos i$	$_{45e}S_0$	$_{45e}S$	$_{45e}g$
0500	0.382	503	2	28	0.689	907	3	29
0600	0.405	533	83	152	0.841	1108	173	242
0800	0.445	586	255	366	0.992	1306	569	680
1000	0.468	616	342	472	0.935	1231	683	813
1200	0.477	628	369	505	0.678	893	531	667

[a] Angles are in degrees, and radiation is in watts per square meter.

receives a maximum diurnal insolation on 22 June. At the time of the equinox, of course, a horizontal surface at the equator receives a maximum direct-beam daily insolation.

Total Shortwave Irradiation of Slopes

In order to obtain the total shortwave irradiation of a slope, one must add to the direct-beam solar irradiation the amount of diffuse skylight and ground-reflected sunlight and skylight reaching the slope. The sky is considered to be isotropic, or nondirectional, with respect to the flux of scattered light coming from it. Reflected ground radiation will first be treated as isotropic for the diffusely reflected shortwave radiation, and then specular, or mirrorlike, reflection will be taken into consideration. Actually, light from the sky is not isotropic, as we have seen elsewhere, but for analytical purposes, it is much easier, and sufficiently accurate for our purposes here, to consider skylight as uniform in intensity over the sky hemisphere.

The simplest approach to the problem of irradiation of a slope is to consider how much of the sky is seen by a small element of unit area located beneath a large hemisphere representing the sky. As a small element of upward-facing area is tilted from horizontal, it sees less sky and more ground surface, from which it receives reflected light. The ground surface is also considered to be a hemisphere. The sum of irradiation of a slope by direct sunlight, scattered skylight, and reflected global shortwave radiation is given by

$$_s g = S \cos i + {}_h d \left(\frac{90° + a_s}{180°} \right) + r \, {}_h g \left(\frac{90° - a_s}{180°} \right), \qquad (6.58)$$

where a_s is the altitude of the normal to the slope, measured in degrees, and r is the reflectance of the ground surface.

A somewhat different analytical approach to the problem of shortwave irradiation of a slope is sometimes used. The irradiation of a tilted plane by a hemisphere, such as the sky, is shown in Fig. 6.25. If the altitude of the normal to the plane is a_s, then the slope angle θ from the horizontal is

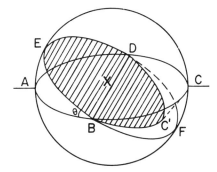

Figure 6.25. Irradiation of a tilted plane by the sky hemisphere for a position X on the plane. See text for explanation.

$90° - a_s$. If a hemisphere of unit radius is projected on the upward-facing surface of a tilted plane or slope, the amount of projected area irradiated by the hemisphere is DEBX + DC'BX = $\pi/2 + (\pi \cos \theta)/2$. The first term $(\pi/2)$ is clearly the area of a semicircle, and the second term $[(\pi \cos \theta)/2]$ is the area of a semiellipse. The sky irradiates a horizontal area π equal to the base area of the hemisphere with a total power of $\pi_h d$, where $_h d$ is the diffuse-skylight flux, expressed as power per unit area. The upward-facing tilted plane receives an amount of diffuse sky-light equal to $(\pi/2) (1 + \cos \theta) _h d$. Using the trigonometric identity $(1 + \cos \theta)/2 \equiv \cos^2 \theta/2$, one gets for the amount of skylight on the tilted plane $\pi(\cos^2 \theta/2)_h d$. With this formulation, total shortwave irradiation of a tilted surface, on a unit-area basis, by direct sunlight, diffuse skylight, and ground-reflected radiation, is given by

$$_s g = S \cos i + _h d \cos^2 (\theta/2) + r_h g \sin^2 (\theta/2). \tag{6.59}$$

Ground-reflected radiation is proportional to the product of ground reflectance r, global radiation $_h g$ incident on the ground, and area illuminated, which, on a unit area basis, is $(1 - \cos^2 \theta/2) = \sin^2 \theta/2$. Equation (6.59) can be rewritten as follows:

$$\begin{aligned} _s g &= S\cos i + _h d \cos^2 (\theta/2) + r(_h S + _h d) \sin^2 (\theta/2) \\ &= S \cos i + r _h S \sin^2 (\theta/2) + _h d[\cos^2(\theta/2) + r \sin^2 (\theta/2)]. \end{aligned} \tag{6.60}$$

For most of the applications with which we are concerned here, there is not a great deal of difference between the use of Eq. (6.58) and Eqs. (6.59) or (6.60). The results are nearly the same, and for the small amounts of flux from the ground and sky, the difference in final results is relatively small.

In Eq. (6.30), a transmittance factor was defined for the amount of diffuse sky radiation as $\tau_d = _h d/_h S_0$. From the observations of Liu and Jordan (1960), a relationship was established between τ_d and τ. Before we apply this relationship, a parenthetical note is in order concerning the use of τ and m. As defined originally, τ represented the transmittance of the atmosphere for the zenith direction. When the solar beam traverses a slant path through the atmosphere, the transmission along the path is τ^m, where m (= sec z) is the air mass. When Liu and Jordan developed the empirical relationship between τ_d and τ, they used τ for the direct-beam transmittance along any path and their τ was equivalent to τ^m. Their definition of τ is not used here.

We can now write the following equation for the diffuse skylight radiation:

$$_h d = _h S_0 \tau_d = S_0 \tau_d \cos z = S_0 \tau_d \sin a. \tag{6.61}$$

Using Eq. (6.31), which relates τ^m and τ_d, one gets for the diffuse skylight the expression

$$_h d = S_0 (0.271 - 0.294 \tau^m) \sin a. \tag{6.62}$$

The amount of global shortwave radiation incident on the horizontal surface is

$$
\begin{aligned}
_h g = {}_h S + {}_h d &= S_0 \, (\tau^m + 0.271 - 0.294 \, \tau^m) \sin a \\
&= S_0 \, (0.271 + 0.706 \, \tau^m) \sin a.
\end{aligned}
\tag{6.63}
$$

The irradiation of any slope at the surface of the earth with an upward- or downward-facing plane surface is given by substituting Eq. (6.62) and (6.63) into Eq. (6.58):

$$
\begin{aligned}
_s g = S_0 \, \tau^m \cos i &+ S_0 \, (0.271 - 0.294 \, \tau^m) \left(\frac{90° + a_s}{180°} \right) \sin a \\
&+ r \, S_0 \sin a \, (0.271 + 0.706 \, \tau^m) \left(\frac{90° - a_s}{180°} \right).
\end{aligned}
\tag{6.64}
$$

An alternative method of writing Eq. (6.64) is as follows:

$$
\begin{aligned}
_s g = S_0 \, \tau^m \cos i &+ S_0 \, (0.271 - 0.294 \, \tau^m) \cos^2(\theta/2) \sin a \\
&+ r \, S_0 \sin a \, (0.271 + 0.706 \, \tau^m) \sin^2(\theta/2),
\end{aligned}
\tag{6.65}
$$

where θ is the slope angle from horizontal and is equal to $90° - a_s$.

If ground reflectance is added to the irradiation of slopes according to Eq. (6.64) and $r = 0.20$, then for $\tau = 0.6$ at the summer solstice at 40° north latitude, one gets the curves shown in Fig. 6.26 and the values given in Table 6.3. The table is sufficiently complete that the reader can follow through a computation of the numbers involved. The character of the curves changes even more as reflected light begins to add a midday peak to the irradiation of surfaces having no direct solar radiation incident at noon. Higher ground reflectance, owing for example to the presence of snow instead of grass, enhances this phenomenon even further.

A north-facing vertical surface receives an impulse of direct sunlight early in the day and again late in the day, but with a considerable amount of light reflected throughout the day, total irradiation becomes substantial. An east-facing vertical surface receives a very strong impulse of direct sunlight during the morning and continues to receive quite a bit of scattered and reflected light throughout the late afternoon (it has the same curve as the vertical north-facing wall during this time of day). Sunrise is relatively late for the south-facing vertical slope, and sunset comes early, but diffuse light continues to irradiate the surface until astronomical sunrise or sunset.

It has been assumed thus far that the sky is isotropic with respect to diffuse radiation. This is a reasonable approximation for our needs here. We know that the sky exhibits considerable anisotropy, as illustrated in Figs. 6.10 and 6.11. The sky has a higher luminance near the sun than at 90° or 180° from the sun. This anisotropy becomes more significant when an object is exposed to some particular small segment of the sky and not to a large portion over which luminance can be integrated. The sky appears to be much brighter near the horizon when there is snow on the

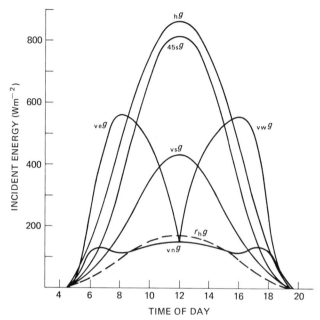

Figure 6.26. Instantaneous global irradiation of various slopes at 40° north latitude on 22 June as a function of time of day with atmospheric transmittance $\tau = 0.6$ and ground reflectance $r = 0.2$. Subscripts referring to slope orientation: h, horizontal; ve, vertical north-facing; vs, vertical south-facing; ve, vertical east-facing; vw, vertical west-facing; 45S, 45° south-facing. The line $r_h g$ is the amount of ground-reflected global radiation.

ground than when the ground is bare. When clouds are present, the whole problem of sky luminance becomes much more difficult. Clouds usually exhibit bright edges and dark centers, but the exact manner in which clouds reflect light depends upon their angular distance from the sun and whether they consist of ice crystals or water droplet.

Calculations are made more difficult by the fact that cloudy days dominate many environments, and perfectly clear days are in the minority. Clouds are highly variable in form, size, density, height, and duration, and it is thus extremely difficult to calculate the irradiation of surfaces during cloudy or partly cloudy days. Very thin, transparent cirrus clouds have little influence on global radiation, whereas thick, dark thunderstorm clouds may reduce radiation to 1% of its normal value.

Climatologists are interested in total global radiation received during a day or over a period of several hours. For such determinations, it is possible to use statistical averages of conditions and utilize the observed duration of sunshine out of a maximum possible value for that date and place. But for most biological purposes, one wishes to know instantaneous values, and calculations are much more difficult with clouds present. Observations have shown that the highest value of diffuse luminance from a partly cloudy sky occurs when about 0.7 of the sky is covered with clouds; lower values are obtained when the cloud cover is less than or greater than 0.70. Empirical relations giving the amount of global radiation with cloudy sky in terms of the global amount with clear sky are given as a function of cloudiness elsewhere in this chapter. Robinson (1966) has calculated irradiation of slopes using these empirical relations for global irradiation under cloudy conditions.

Specular reflection of direct sunlight and skylight may occur from water, ice, rock, or even vegetation. Specular reflectance is highly directional and may reach very high values, as for example the reflectance of the sun at grazing incidence over snow, ice, or water. Amount of reflectance depends on the geometry involved and the character of the reflecting surface. An additional term would be added to the Eqs. (6.63) or (6.64) to account for specular reflection.

Measured Irradiation of Slopes

The amount of observational data concerning irradiation of slopes under different conditions is sparse. Some of the best data are available in the Russian literature (see review by Kondratyev, 1969). Conditions under which observations might be made are so variable that almost any particular situation will differ somewhat from any other, although observations for clear-sky conditions are reasonably reliable and reproducible. There is a complication which has not been allowed for, however. If the albedo of the ground surface is high, then there is much multiple reflection between ground and sky, and both the downward and upward flux of diffuse radiation will be enhanced. The best way to illustrate this is to look at the ratio of diffuse plus reflected shortwave radiation incident on a slope to diffuse radiation incident on a horizontal surface as a function of slope angle and azimuthal aspect. Kondratyev (1969) has reported measurements made at Leningrad and Karadag, Crimea. Smoothed curves of this ratio are shown in Fig. 6.27 for a solar altitude of 26° and surface albedos of 59% (when snow covered the ground) and 20% (during the summer). With low sun, as is the case in this example, the tilted surface generally receives more diffuse radiation, scattered from the sky and reflected from the ground, than the upward-facing horizontal surface. It is the reflected component which really makes the difference because the tilted surface actually receives less skylight than the horizontal surface. This is the reason for the dip in some of the curves below a ratio of 1.0 when ground albedo is low. The situation illustrated in Fig. 6.27 for an albedo of 59% is just about the extreme case encountered in the natural world. Steep slopes, of course, get the greatest irradiation enhancement by reflected radiation. At higher solar altitudes, the enhancement effect is considerably reduced. The results are nearly independent of the azimuthal position of the sun since the ratio plotted involves diffuse quantities only. If specular reflection or

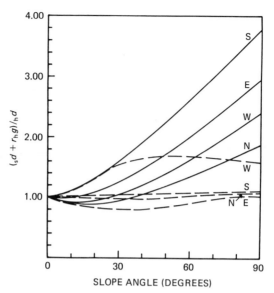

Figure 6.27. Ratio of diffuse plus reflected short-wave radiation incident on a slope to diffuse radiation incident on a horizontal surface as a function of the slope angle for a solar altitude of 26° and for ground reflectance of (--) 0.20 and (—) 0.59. Capital letters indicate slope orientation.

direct solar irradiation were considered the azimuthal angle between slope aspect and solar position would be very important. Kondratyev (1969) gives many curves which illustrate this effect for various slopes and azimuth angles. Many surfaces of interest in biology, such as flat leaves or the undersides of animals, involve downward-facing surfaces located at some height above the ground. For such cases, the amount of ground-reflected radiation becomes quite significant.

When the ratio plotted in Fig. 6.27 is determined for fully overcast conditions, it is seldom greater than 1.0, even for an albedo of 70%, and is usually considerably less than 1.0 for conditions of low surface albedo. In fact, the ratio becomes smaller as the angle of inclination becomes greater because less sky is exposed to the surface and ground-reflected radiation is weak.

Problems

1. Formulate a table similar to Table 6.3 for various slopes at 40° north latitude on 22 December as a function of time for $\tau = 0.6$, $S_o = 1406$ W m^{-2}, and $r = 0.4$. Set up the table in a similar manner but for times of day of 0800, 0900, 1000, 1100, 1200 and the following slopes: horizontal; vertical north-facing; and 45° south-facing.

2. Formulate a table similar to Table 6.3 for various slopes at 40° north latitude on 22 March or 22 September as a function of time for $\tau = 0.6$, $S_0 = 1351$ W m^{-2}, and $r = 0.2$. Set up the table in a similar manner but for times of day of 0700, 0800, 0900, 1000, and 1200, and the following slopes: horizontal; vertical north-facing; and 45° south-facing.

3. Determine the angle of incidence of the direct beam of solar radiation for an east-facing 45° slope at 40° north latitude for the following dates and times of day.

	22 December	22 March	22 June
0800			
1000			
1200			

4. A slope faces due south and has an altitude $a_s = 30°$. Determine the angle of incidence of the solar rays for the following solar altitudes (a) and azimuth angles (α): $a = 30°, \alpha = 180°$; $a = 60°, \alpha = 90°$; $a = 30°, \alpha = 60°$; $a = 0°, \alpha = 120°$. Use the relationship

$$\cos i = \cos a_s \cos a \cos(\alpha - \alpha_s) + \sin a_s \sin a.$$

5. Using Table 6.3, determine the amount of shortwave radiation incident on a horizontal surface (determine $_hS$, $_hg$, and r_hg) if $\tau = 0.4$ and all other conditions are the same as for Table 6.3.

Time	a	$\cos a$	α	$\cos \alpha$	z	$m(= \sec z)$	τ^m	S
0500								
0600								
0800								
1000								
1200								

Time	τ_d	$_hd$	$\cos z$	$_hS_0$	$_hS$	$_hg$	r_hg
0500							
0600							
0800							
1000							
1200							

Chapter 7

Longwave and Total Radiation

Introduction

All objects in the universe emit radiation proportional to the fourth power of their absolute temperature or some fraction thereof, depending upon their spectral emittance. Whether the emitted radiation is transmitted by the medium surrounding the object depends on the properties of the medium. A vacuum transmits all emitted radiation. Air transmits a broad band of visible, ultraviolet, and infrared radiation, with certain spectral regions selectively absorbed. Liquid water transmits only very-near ultraviolet, visible, and very little infrared radiation. The plants and animals of the world live in an environment generally dominated by radiation. Organisms in terrestrial habitats receive some radiation at all times. Organisms living in aquatic habitats may receive radiation only part of the time since liquid water transmits some sunlight, relatively little ultraviolet, and no infrared.

Thermal, or longwave, radiation is considered, for meteorological and bioclimatological purposes, to be radiation emitted by objects whose surface temperature is less than 600°K. According to the blackbody radiation law discussed in Chapter 5, an object whose surface temperature is 600°K or less emits radiation of infrared wavelength, with no measurable emission in the visible range. Excluding volcanoes, hot magmas, and fires, the surface of the earth is generally at a temperature between 185°K, the coldest air temperature on record at the USSR Vostok Station in Antarctica, and about 373°K, the temperature of boiling water in hot springs at sea level. Most objects, such as air, clouds, water, ice, soil, rock, and vegetation, with which we are generally concerned are at temperatures within this range and therefore emit a broad spectrum of infrared radiation, known as longwave, or thermal, radiation. Shortwave radiation is of wavelength less than 2.5 μm and is emitted by the sun, lights, fires, hot magmas (including volcanoes), and other high-temperature sources. These sources emit longwave as well as shortwave radiation. There is not always a neat or complete dichotomy between the temperature ranges over which various sources emit shortwave or longwave radiation. However, high-temperature sources generally emit both shortwave and longwave radiation, whereas low-temperature sources emit only longwave radiation.

Because of their high infrared absorptance, most organisms have their energy level, or temperature, strongly coupled to longwave thermal radiation from the environment. Organism temperatures or energy levels are usually only weakly or moderately influenced by shortwave radiation, because of low absorptance to visible wavelengths.

Terrestrial Longwave Radiation

It is important to know how much longwave radiation may be emitted or received by any terrestrial object or organism. Whereas shortwave radiation in the natural environment originates entirely from the sun, which is a point source, longwave radiation comes from objects with extended radiating surfaces such as clouds, sky, rock, soil, water, and vegetation or animals. A plant or animal living in a terrestrial environment experiences a longwave radiation flux at all times, although this flux varies in intensity. Shortwave radiation flux from the sun, however, is present only part of the time. The longwave radiation of the terrestrial environment is a complex phenomenon because of the variety of geometrical surfaces involved and the gradations of surface temperatures in the natural environment. It is relatively easy, however, to make reasonable estimates of the longwave radiation present in the terrestrial world. In order to do so, the blackbody radiation emitted by objects at normal temperatures is considered first.

Blackbody Radiation Environments

The amount of blackbody longwave radiation emitted at temperatures from $-40°$ to $50°C$ is shown in Fig. 7.1. These fluxes range from 160 W m^{-2} at $-40°C$ to 615 W m^{-2} at $50°C$. These figures alone tell us a great deal about the radiation environment around us. Between the coldest and warmest temperatures normally encountered on earth, there is less than a fourfold difference in amount of longwave thermal radiation flux. The blackbody curve (D) of Fig. 7.1 represents the total amount of radiation incident upon and absorbed by most organisms in a cave, burrow, house, or the interior of a forest with completely closed canopy. Any enclosed

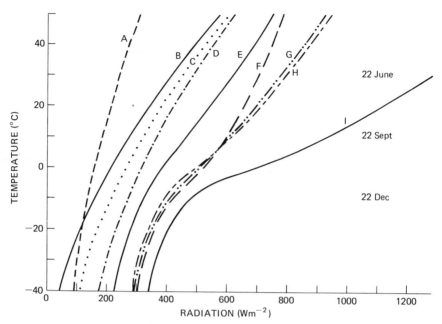

Figure 7.1. Amount of longwave radiation emitted by various natural sources and total radiation (shortwave plus longwave) incident on various objects as a function of air temperature in the outdoor environment: (A) average of ground radiation and nonradiating sky; (B) atmospheric; (C) average ground plus atmosphere; (D) blackbody; (E) downward-facing plate; (F) vertical cylinder; (G) sphere; (H) horizontal cylinder ($\alpha - \alpha_c = 0°$); (I) upward-facing plate.

cavity is automatically a blackbody radiator. If, for example, the air temperature and wall temperature of your house is 20°C, then the amount of radiation from the walls incident upon you is about 418 W m^{-2}.

Radiation Environment between the Ground and an Overcast Sky

The least complex radiation environment in the natural world, next to that of a blackbody, is that of an organism located between the ground surface and a completely overcast sky at night. The lower half of the organism is irradiated by the ground and the upper half by the atmosphere. Both the ground surface and the overcast sky radiate as blackbodies, but generally at different temperatures. It is the temperatures of the cloud base and the ground that determine the amount of radiation emitted. The cloud-base temperature might be 0°C when the ground surface is at 10°C. The stream of longwave radiation flowing upward from the ground surface is then 363 W m^{-2} and that flowing downward is 313 W m^{-2}. The average radiation received by an organism is 338 W m^{-2}. If the absorptance to longwave radiation of an organism's surface is 0.95, the amount of radiation absorbed will be 0.95 × 338 W m⁻², or 321 W m⁻², or 321 W m⁻². With overcast skies, the amount of radiation received and absorbed at night by organisms is a little to the left of the blackbody line in Fig. 7.1.

Longwave Radiation from Clear Skies

If the atmosphere were completely transparent to all wavelengths of radiation, it would not emit any radiation. An organism on or above the surface of the earth would then receive blackbody radiation from the ground and zero radiation from the atmosphere and would radiate heat to the cosmic cold of outer space, which is at about 4°K. In this case, the average amount of energy received by an organism would be one-half of the blackbody radiation emitted by the ground. This amount of radiation is shown in Fig. 7.1 as Line A (average of ground radiation and nonradiating sky).

Water vapor, carbon dioxide, and ozone molecules have strong absorption bands at infrared wavelengths. The atmosphere, in turn, emits infrared radiation according to Kirchoff's law. Since atmospheric temperature is approximately 300°K, the atmosphere emits a broad band of infrared wavelengths with an emission maximum around 100 μm, according to Wien's displacement law. Because of the density distribution of water vapor and carbon dioxide in the atmosphere, much of the radiation emitted originates within the lowest few kilometers of the troposphere.

Although it is possible to give a theoretical analysis of atmospheric radiation, the derivation is extremely complex and difficult mathematically. It is convenient and reasonably accurate to express flux of atmospheric thermal radiation in empirical form. This was done first by Ångstrom (1915), who expressed the atmospheric radiation in terms of air temperature and water vapor pressure near the ground. Subsequently, Brunt (1932) published a somewhat modified formula relating air temperature and water vapor pressure. None of these relationships, however, gives truly accurate predictions when compared with observed values.

Swinbank (1963) reexamined all of the observed data and raised a question concerning the influence of water vapor pressure on radiation flux from clear skies. It is also clear that the Ångstrom and Brunt formulations do not allow for the influence of atmospheric temperature on emissivity. Swinbank has shown circumstances when great variations of atmospheric moisture and tropospheric temperatures occur while emissivity and temperature near the ground remain constant. The meteorological screen or shelter temperature often will not vary when the air temperature of the the troposphere up to the base of the stratosphere (14 to 20 km above the surface of the ground) varies considerably. These situations convinced Swinbank that a different formulation was required. The Swinbank relationship for the downward stream of thermal radiation $\mathcal{R}_a$ from a clear atmosphere as a function of air temperature is

$$\mathcal{R}_a = 1.22 \ \sigma(T_a + 273)^4 - 171, \qquad (7.1)$$

where σ is the Stephan–Boltzmann constant (5.673×10^{-8} W m^{-2} °K^{-4}).

Swinbank argued further that $\mathcal{R}_a$ may not necessarily be a function of $(T_a + 273)^4$ and that some other power might be more justified from theory. Since the 6.0-μm water vapor absorption and emission band is located on the shortwave side of the 10-μm peak of the blackbody spectral distribution at atmospheric temperature and the rotational water vapor band is located on the longwave side of the peak, change in atmospheric temperature may have a more sensitive influence than is indicated by Eq. (7.1). Swinbank obtained the following regression formula as a good fit to observations:

$$\mathcal{R}_a = 53.1 \times 10^{-14}(T_a + 273)^6. \qquad (7.2)$$

Swinbank was able to test the equations both at sea level and at 1400 m and got good agreement with observational values. He suggested that as the atmosphere becomes less dense around high-altitude stations, some departure from the equations might be expected. Whether Eq. (7.1) or (7.2) is used makes a difference of less than 3% over the temperature range from 0° to nearly 40°C. Swinbank's observations were confined to a fairly limited range of temperatures (approximately 8° to 38°C). The first equation probably fits the observed data over a broad range of temperatures slightly better, but the second has somewhat more physical justification. A plot of Eq. (7.1), indicated as "atmospheric" radiation, is given in Fig. 7.1.

If one considers the emissivity of the atmosphere to be

$$\epsilon_a = \mathcal{R}_a / \sigma(T_a + 273)^4, \qquad (7.3)$$

it is clear from Eq. (7.1) that emissivity is about 0.91 at 40°C, 0.81 at 20°C, and 0.69 at 0°C. Colder air contains less water vapor and has a smaller emissivity. The temperature dependence of atmospheric emissivity may also be derived by means of a linearization of the equation for atmospheric emissivity.

Linearized Form of the Equation for Atmospheric Emission

It is often convenient to linearize equations containing power functions. Linearization is often done when working with radiation equations in order to obtain approximations and is a widely used procedure in applied mathematics.

The emissivity ϵ_a increases almost linearly with increasing temperature between $-5°$ and $25°C$. It is reasonable to expand $(T_a + 273)^{-4}$ about $T_a + 273 = 283°K$ as follows:

$$(283 + T_a + 273 - 283)^{-4} = (283)^{-4}\left(1 + \frac{T_a + 273 - 283}{283}\right)^{-4}. \quad (7.4)$$

An algebraic expansion of a polynomial is

$$(1 + a)^n = 1 + na + \frac{n(n - 1)}{2} a^2 + \ldots \quad (7.5)$$

When a is small, one can drop all but the linear term. Recognizing from Eq. (7.4) that $n = -4$ and $a = (T_a + 273 - 283)/283$, one gets

$$\left(1 + \frac{T_a + 273 - 283}{283}\right)^{-4} = 1 - 4\left(\frac{T_a + 273 - 283}{283}\right). \quad (7.6)$$

Substituting this into Eq. (7.4) and then into Eq. (7.3) gives

$$\epsilon_a = 1.22 - \frac{171}{\sigma(283)^4}\left[1 - 4\left(\frac{T_a + 273 - 283}{283}\right)\right] \quad (7.7)$$

or

$$\epsilon_a = 0.674 + 0.007\, T_a. \quad (7.8)$$

This shows that emissivity of the atmosphere is a linear function of air temperature to a first approximation.

Linearized Form of the Equation for Radiation Exchange

In order to linearize Eq. (7.1), one proceeds in the same way. Therefore,

$$\mathcal{R}_a = 1.22\, \sigma(T_a + 273)^4 - 171$$
$$= 1.22\, \sigma(283 + T_a + 273 - 283)^4 - 171$$
$$= 1.22\, \sigma(283)^4\left[1 + 4\left(\frac{T_a + 273 - 283}{283}\right)\right] - 171$$

or

$$\mathcal{R}_a = 6.17\, T_a + 204. \quad (7.9)$$

The linearized approximation is good over the temperature range $-10°$ to $25°C$. In order to be a good approximation over another temperature range, Eq. (7.1) would have to be expanded at a lower or higher temperature than 283°K.

The blackbody law, if expanded in the same manner, is linearized as

$$\mathscr{R} = \sigma[T_b + 273]^4 = 5.05\ T_b + 307. \tag{7.10}$$

These linearized formulas are used to determine approximately the effective radiant sky temperature, which is the temperature T_b a blackbody would have in order to radiate the same amount of radiation as the sky. Equating Eq. (7.9) with Eq. (7.10) gives

$$6.17\ T_a + 204 = 5.05\ T_b + 307 \tag{7.11}$$

$$\begin{aligned} T_b &= 1.22\ T_a - 20.4 \\ &= T_a - 20.4 + 0.22\ T_a. \end{aligned} \tag{7.12}$$

This last equation shows that the radiant temperature of a clear sky is approximately 20°C less than that of the air near the ground. This temperature difference can also be seen by comparing lines B and D in Fig. 7.1. Assume a value on the temperature scale. Drop vertically from its intersection on line B until curve D is intersected at a lower temperature. The displacement on the temperature scale is about 8° at 40°C, 15° at 15°C, and 35° at −10°C. At colder temperatures, the temperature of the radiant sky is proportionately less than that of the air near the ground. This is consistent with the dropping emissivity of the clear atmosphere with lower temperatures and decreased water vapor.

Nighttime Radiation Climate

Now it is possible to complete our estimate of the radiation climate of organisms on the ground surface exposed to a clear sky at night. Using the Swinbank formula, Eq. (7.1), one gets for the amount of radiation from a clear sky line B in Fig. 7.1. Upward radiation from the ground surface is calculated by using the blackbody formula and assuming ground surface temperature to be approximately the same as screen air temperature. This radiation is shown in Fig. 7.1 as line D. An organism receives half its irradiation from the ground and half from the sky. The average of these two radiation fluxes is the total amount of thermal radiation incident on an organism exposed to the sky on a clear night, shown in Fig. 7.1 as line C. This line is the lowest amount of thermal radiation a terrestrial organism is likely to receive. All other radiation climates are to the right of this line. Once again, for comparison, consider the very low amount of radiation incident on an organism when it is between a nonradiating sky and a radiating ground surface. The amount of infrared radiation emitted by water vapor and carbon dioxide in the atmosphere is extremely significant to organisms living on the earth's surface, as demonstrated by the distance between lines A and C in Fig. 7.1. The difference is especially great on warm nights.

Measured Thermal Radiation

Arctic Summer

An organism out of doors between ground and sky is subjected to whatever variations in radiation that occur. Results of radiation studies by Stoll and Hardy (1955) that were concerned with the thermal or cold stress on humans for an arctic summer day at Nome, Alaska, are shown in Fig. 7.2. A thermal radiometer was used for measuring the flux of radiant energy from sky, clouds, or ground, and an Eppley pyrheliometer was used for solar-radiation measurements. Radiation from ground and sky is expressed as equivalent blackbody temperature. Radiation from direct solar radiation and heat load on the human subjects are given in watts per square meter. The air temperature remained nearly constant at about 14°C over a 12-hr period on 25 August 1952. The sky was clear from 0700 to 0830, when clouds began to form. At that time, the amount of direct solar radiation received dropped steeply, the equivalent environmental radiant temperature fell, and the heat load on a person dropped to nearly 85 W m^{-2}. The radiant sky temperature underwent an abrupt increase as clouds began to cover the sky. Net loss of energy by a person exposed to this summer arctic environment would have been even greater if it had not been for the downward flux of thermal radiation from the clouds. As the clouds thinned around noon, solar radiation began to penetrate the clouds and the radiant sky temperature dropped, but the radiant environmental temperature increased, as well as the total heat load on a person. Late in the day, as the sun set, the sky remained nearly clear, the radiant environmental temperature began to drop, and the heat load diminished accordingly. The particularly striking feature of this arctic summer day was the constancy of the air temperature while the radiant environmental temperature and heat load on a person changed dramatically.

Arctic Winter

Measurements made by Stoll and Hardy (1955) during an arctic winter day are shown in Fig. 7.3. Again, this is a day during which air and ground-surface temperature varied relatively little during the day, but radiant temperature of the sky varied greatly. Radiant sky temperature was −73°C when the air temperature was −38°C, but when the sky became overcast and it began to snow, the radiant sky temperature rose by 48°C to a value of −27°C, which was above the air temperature of −35°C at that time.

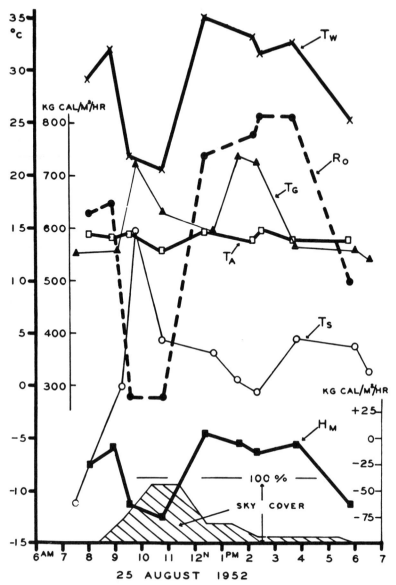

Figure 7.2. Variation in solar radiation and environmental temperature and resultant radiant heat load on a human, 25 August 1952, at Nome, Alaska: T_W, radiant temperature of the enviromment; R_o, direct solar radiation; T_g, radiant temperature of the ground; T_a, air temperature. (From Stoll and Hardy, 1955.)

Yellowstone Park in Winter

Gates (1961) reported a series of measurements of the thermal environment in Yellowstone Park made during February 1961. When air temperature was $-20°C$, radiant temperature of the zenith cloudless sky was

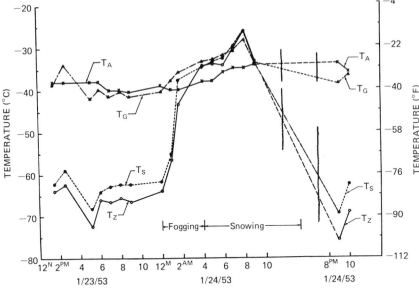

Figure 7.3. Effect of fog and snow on environmental radiant temperatures: T_a, air temperature; T_g, radiant temperature of the ground; T_s, radiant temperature of the sky; T_z, radiant temperature of the sky at the zenith. (From Stoll and Hardy, 1955.)

$-58°C$. An overcast sky produced radiant sky temperatures of $-5°$ to $-10°C$. There is no question that cloud cover has a dramatic influence on the radiant sky temperature. One very clear, cold night, I was measuring radiant sky temperatures in the valley below Old Faithful. Suddenly, the radiant sky temperature jumped from $-50°C$ to $-31.5°C$ and snow flakes began to fall. My colleagues had turned on a silver iodide generator and seeded the cold, supersaturated air overhead, thereby creating a cloud cover.

Desert Environment

Kelly et al. (1957) observed radiant sky temperatures in the Imperial Valley of California during the hot summer months, when cattle undergo the greatest amount of heat stress. For a considerable part of each year, many areas of the world have a climate too hot for economical growth and production of many breeds of European cattle. It is possible that animal shelters, which cut off solar radiation and thereby reduce the radiation heat load on cattle, might be designed with an opening to the cold north sky so that cattle could lose heat by radiation. Kelly et al. found that a point in the north sky 60° above the horizon is usually as cool or cooler than other parts of the sky. This is seen in Figs. 7.4 and 7.5, which show the thermal radiation intensity from the atmosphere and ground surface in a north–south and east–west vertical plane from horizon to zenith to

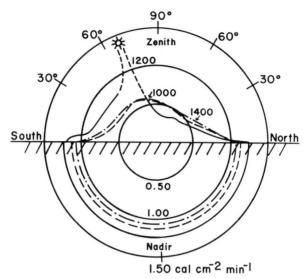

Figure 7.4. Total radiant flux from the sky or ground as measured in a north-south vertical plane in the Imperial Valley, Calif., on 27 April 1949 at three times of day (1000, 1200, and 1400). Units are calories per square centimeter per minute (1.00 cal cm^{-2} min^{-1} = 697.8 W m^{-2}). (From Kelly et al., 1957.)

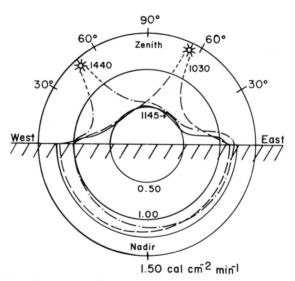

Figure 7.5. Total directional radiant flux from the sky or ground as measured in an east-west vertical plane in the Imperial Valley, Calif., on 27 April 1949 at three times of day (1030, 1145, and 1440). Units are calories per square centimeter per minute. (From Kelly et al., 1957.)

horizon to nadir to horizon for a clear day, 27 April 1949, in the Imperial Valley. It is noticed that the temperature of a clear sky has a minimum at the compass point in the plane of the observer and the sun at approximately 90° from the sun. Clearly, the "cold spot" in the sky moves with the sun, but when designing animal shelters, it is not practical to design a moving opening. Kelly et al. (1957) therefore took the hottest time of day and considered the "cold spot" to be in the north sky at 60° above the horizon for summer months at midlatitudes in North America.

Daytime Radiation Climate

Whereas the nighttime radiation climate contains only thermal-radiation sources, the daytime radiation climate is affected by direct, scattered, and reflected sunlight, as well as thermal radiation. From the discussion of solar radiation and sky radiation given in Chapter 6, it is possible to make good estimates of the flux of solar radiation incident on any surface during a clear day. Intensity of solar radiation varies considerably with atmospheric conditions, particularly atmospheric turbidity. If normal or average turbidity is given, one can calculate the intensity of solar radiation plus diffuse skylight as a function of time of day. For cloudless days, radiation as a function of time of day gives a bell-shaped curve. Solar intensities are high during the summer and low during the winter, as are air temperatures. There is, of course, no precise relationship between solar intensity and air temperature near the ground, but there is a general correlation. In the winter, when air temperatures are cool, the sun is lower in the sky, and the intensity of sunlight is less than in the summer. One can get an estimate of the maximum amount of radiation encountered by organisms by considering conditions at noon on a clear day.

Horizontal Surfaces

In Chapter 6, a method was given for estimating the amount of shortwave radiation, both direct and diffuse, incident on a horizontal surface for any latitude, time of year, and time of day. Since we are interested in estimating the maximum amount of radiation incident on a horizontal surface, we shall calculate these values for noon on 22 June, 22 September, and 22 December at 40° north latitude as an example.

The total downward flux of shortwave radiation incident on an upward-facing horizontal surface is given by Eq. (6.36), which is repeated here:

$$_h g = {}_h S + {}_h d = S_o \tau^m \cos z + S_o (0.271 - 0.294 \tau^m) \cos z, \quad (7.13)$$

where m is sec z, z is the solar zenith angle, and τ is atmospheric transmit-

Table 7.1. Components of Incident Shortwave Radiation for an Upward-Facing Horizontal Surface at Noon at 40° North Latitude for $\tau = 0.6$ and $\bar{S}_0 = 1360 \text{ W m}^{-2}$.

Date	$(\bar{d}/d)^2$	z	$\cos z$	$m(=\sec z)$	τ^m	S_0	S	$_hS$	$_hd$	$_hg$
22 June	0.968	16.5°	0.959	1.043	0.587	1317	773	741	124	865
22 Sept	0.993	40.0	0.776	1.305	0.513	1351	693	531	124	655
22 Dec	1.033	63.5	0.446	2.242	0.318	1406	447	200	111	311

tance. The first term is direct solar radiation, and the second term is diffuse sky radiation incident on a horizontal surface. The quantity S_0 is $\bar{S}_0$ (1360 W m^{-2}) as modified by the distance d of the earth from the sun when multiplied by $(\bar{d}/d)^2$, where $\bar{d}$ is the mean distance.

An upward-facing horizontal surface will receive direct solar radiation, diffuse skylight, and longwave radiation from the atmosphere:

$$_h^uQ = {}_hS + {}_hd + \mathcal{R}_a = {}_hg + \mathcal{R}_a. \tag{7.14}$$

The shortwave components are calculated in Table 7.1. Then, by selecting air temperatures more or less characteristic of the time of year at 40°N, i.e., 30°C on 22 June, 10°C on 22 September, and -10°C on 22 December, longwave radiation from the atmosphere is determined by means of Eq. (7.1). Adding the longwave and shortwave radiation gives the total radiation flux $_h^uQ$ incident on an upward-facing horizontal surface. These values are given in Table 7.2 and form the basis of the curve given in Fig. 7.1.

The radiation fluxes $_h^dQ$ incident on a downward-facing horizontal surface are given by

$$_h^dQ = r_hg + \mathcal{R}_g, \tag{7.15}$$

where r is the reflectance of the ground surface. These values are given in Table 7.2 for a ground reflectance to shortwave radiation of 0.15. The ground-surface temperature is assumed to be equal to their air temperature, and $\mathcal{R}_g$ is calculated from the blackbody radiation law. These calculations form the basis of the curve given in Fig. 7.1. The curves represent

Table 7.2. Components of Total Atmospheric Radiation Incident on and Absorbed by Upward- and Downward-Facing Horizontal Surfaces, for Ground Reflectance $r = 0.15$.[a]

Date	T_a(°C)	$\mathcal{R}_a$	$_h^uQ$	$_h^uQ_a$	r_hg	$\mathcal{R}_g$	$_h^dQ$	$_h^dQ_a$	$_pQ_a$
22 June	30	412	1277	931	130	478	608	543	737
22 Sept	10	273	928	666	98	364	462	413	540
22 Dec	-10	160	471	347	47	271	318	294	320

[a] Shortwave absorptance used are 0.6 for the upper surface and 0.5 for the lower surface. Longwave absorptances are 1.0.

the maximum radiation fluxes likely to be incident on horizontal surfaces at 40°N throughout the year. The curve representing the upward-facing horizontal surface is the least accurate, yet represents a reasonably good approximation. Flat plates of orientation other than horizontal can be treated with the same formulation used for slopes.

Sphere

Another geometrical form which resembles the parts of some organisms is a sphere. Consider a sphere of radius ρ located at a small distance above the ground surface. The surface area of the sphere is $4\pi\rho^2$. At any time of day, the direct rays of the sun strike the sphere along a radius vector. The sphere intercepts an amount of direct sunlight given by $\pi\rho^2 S$, and the average amount of direct sunlight incident per unit surface area of the sphere is $\pi\rho^2 S/4\pi\rho^2 = S/4$. The upward-facing hemisphere receives sky-light and atmospheric longwave radiation over its entire surface, and the downward-facing hemisphere receives the diffuse sunlight and skylight reflected from the ground and ground-emitted longwave radiation. Diffuse sky radiation illuminates the upward-facing hemisphere with a total amount of radiation of $2\pi\rho^2{}_h d$, but the average over the sphere as a whole is $2\pi\rho^2{}_h d/4\pi\rho^2 = {}_h d/2$. The geometry is similar for reflected sunlight and skylight (assumed to be isotropic and diffuse) and longwave atmospheric and ground-emitted radiation. The average amount of radiation incident on a sphere per unit surface area per unit time is

$$_s Q = \frac{S}{4} + \frac{1}{2} (_h d + r_h g + \mathcal{R}_a + \mathcal{R}_g). \tag{7.16}$$

Table 7.3 shows the calculations of the amounts of radiation incident on a sphere above the ground surface at noon at 40°N for an atmospheric transmittance of 0.6 and ground-surface reflectance of 0.15. These results are plotted in Fig. 7.1. A sphere receives considerably less radiation than an upward-facing horizontal surface.

Table 7.3. Shortwave, Longwave, and Total Radiation Incident on a Sphere, Vertical Cylinder, Horizontal Cylinder with Axis Perpendicular to Sun's Rays and in the Plane of the Sun's Rays at Noon at 40° North Latitude for $\tau = 0.6$, $r = 0.15$, $\bar{S}_0 = 1360$ W m^{-2}.[a]

Date	$\sin z$	$_s S$	R	$_s Q$	$_{cv} S$	$_{cv} Q$	$^{90}_{ch} S$	$^{90}_{ch} Q$	$^{0}_{ch} S$	$^{0}_{ch} Q$
22 June	0.284	193	572	765	127	699	197	769	167	739
22 Sept	0.643	173	430	603	167	597	176	606	91	521
22 Dec	0.895	112	294	406	122	416	114	408	11	305

[a] All cylinders with $x = 4.0$. $\mathcal{R} = (_h d + r_h g + \mathcal{R}_a + \mathcal{R}_g)/2$. Values from Tables 7.1 and 7.2.

Cylinder

The parts of many organisms are well-approximated geometrically by a cylinder. The trunks, limbs, and stems of plants and the bodies and appendages of many animals are nearly cylindrical in form.

Cylinder of Any Orientation

It is very easy to obtain the surface area of a cylinder of any orientation projected perpendicular to the direction of the sun's rays. If θ is the angle between the axis of the cylinder and the direction of the sun's rays, the midsectional area of the cylinder of radius ρ and length h is projected as $2\rho h \sin \theta$. The circular ends are projected as ellipses to give a projected area of $\pi\rho^2 \cos \theta$ since the area of an ellipse is π times the product of the semiaxes, which are ρ and $\rho \cos \theta$. Hence the total projected area of the cylinder is

$$_c A_p = 2\rho h \sin \theta + \pi\rho^2 \cos \theta. \qquad (7.17)$$

The total surface area of the cylinder is

$$2\pi\rho h + 2\pi\rho^2. \qquad (7.18)$$

Average irradiation of a cylinder by the direct rays of the sun is

$$_c S = \frac{2\rho h \sin \theta + \pi\rho^2 \cos \theta}{2\pi\rho h + 2\pi\rho^2} S$$

$$= \frac{2h \sin \theta + \pi\rho \cos \theta}{2\pi(h + \rho)} S. \qquad (7.19)$$

By substituting $x = h/\rho$, Eq. (7.19) becomes

$$_c S = \left[\frac{2x \sin \theta + \pi \cos \theta}{2\pi(x + 1)} \right] S. \qquad (7.20)$$

It is relatively easy to get a definition of the angle θ in terms of the altitude a and azimuth α of the sun and the altitude a_c and azimuth α_c of the direction of the cylinder axis. Referring to our discussion concerning the angle of incidence of the sun's rays to a slope whose normal took any direction in space as defined by altitude and azimuth angles, one can write

$$\cos \theta = \cos a_c \cos a \cos (\alpha - \alpha_c) + \sin a_c \sin a. \qquad (7.21)$$

In the case of the vertical cylinder, $a_c = 90°$ and α_c is indeterminate, but Eq. (7.21) reduces to

$$\cos \theta = \sin a = \cos z \qquad (7.22)$$

since the altitude and zenith angles are complementary. Hence, from Eq. (7.20),

$$_{cv} S = \left[\frac{2x \sin z + \pi \cos z}{2\pi(x + 1)} \right] S. \qquad (7.23)$$

When the sun is in the zenith, $z = 0$, $\sin z = 0$, $\cos z = 1$, and the equation for a vertical cylinder is

$$_{cv}S = \frac{S}{2(x + 1)} = \frac{\rho S}{2(h + \rho)} = \frac{\pi \rho^2 S}{2\pi(\rho h + \rho^2)}. \tag{7.24}$$

This quantity is the ratio of the area of one end of the cylinder to the total surface area times the incident flux of solar radiation.

When the sun is at the horizon, $z = 90°$, $\sin z = 1$, $\cos z = 0$, and the equation for a vertical cylinder is

$$_{cv}S = \frac{xS}{\pi(x + 1)} = \frac{hS}{\pi(h + \rho)} = \frac{2\rho hS}{2\pi(\rho h + \rho^2)}. \tag{7.25}$$

This quantity is the ratio of the area of a rectangle (cross section of the cylinder) to the total surface area times the incident flux of solar radiation.

In the case of a horizontal cylinder, $a_c = 0$ and Eq. (7.21) becomes

$$\cos \theta = \cos a \cos (\alpha - \alpha_c), \tag{7.26}$$

and if $\alpha - \alpha_c = 0$, then $\cos \theta = \cos a = \sin z$ and $\sin \theta = \cos z$. Therefore,

$$_{ch}^{0}S = \frac{1}{2(x + 1)} \left(\frac{2x \cos z}{\pi} + \sin z \right) S. \tag{7.27}$$

If $\alpha - \alpha_c = 90°$, $\cos \theta = 0$ and $\sin \theta = 1.0$. In this case, for the horizontal cylinder,

$$_{ch}^{90}S = \frac{xS}{\pi(x + 1)} = \frac{hS}{\pi(h + \rho)} = \frac{2\rho hS}{2\pi(\rho h + \rho^2)}. \tag{7.28}$$

A cylinder, no matter what its orientation, presents approximately the same area to the isotropic radiation of sky, atmosphere, and ground. The total irradiation of a cylinder of any orientation is given by

$$_{c}Q = {}_{c}S + \frac{_{h}d + \mathcal{R}_a + r_h g + \mathcal{R}_g}{2}$$

$$= \left[\frac{2x \sin \theta + \pi \cos \theta}{2\pi(x + 1)} \right] S + \frac{1}{2} (_h d + \mathcal{R}_a + r_h g + \mathcal{R}_g). \tag{7.29}$$

Calculation using the same numbers contained in Tables 7.1 and 7.2 gives the results listed in Table 7.3. These results are shown in Fig. 7.1 for a horizontal cylinder with $\alpha - \alpha_c = 0°$. There is relatively little variation about this line for other orientations of a horizontal cylinder. Because so much of the incident radiation originates with diffuse sources rather than point sources, dependence of total irradiance on azimuthal position of a horizontal cylinder is very slight. If the cylinder rests on the ground rather than being suspended above the ground, there will be a very slight reduction in the amount of incident radiation because of the shadow cast by the cylinder. Since the amount of reflected light is very small, however, a small reduction in this quantity produces an insignificant reduction of the total radiation incident on the cylinder. It is seen that there is substantially less radiation incident on a vertical cylinder at midday during summer months at 40° north latitude than on either a sphere or a horizontal cylinder. One is reminded of the many cacti of the desert that have vertical cylindrical forms. At lower latitudes, where most large cacti grow, this phenomena is more significant over more months of the year than at 40° north latitude.

Absorption of Radiation

An organism, because of its shape, orientation, and absorptance, will absorb some fraction of the total radiation flux incident upon it. Most plants and animals absorb about 95 to 98% of the incident longwave radiation. So at night, the incident amounts of longwave radiation given in Fig. 7.1 are almost exactly the absorbed radiation by the organism. Most plants absorb between 50 and 70% of incident sunlight and skylight. Animals absorb as little as 30% and as much as 85% of incident sunlight and skylight. The surface area of the organism exposed to each radiation source is an important factor in the total amount of radiation absorbed by an organism. In terms of the increments of surface areas A_i of absorptance a_i exposed to each source of radiation $\mathcal{R}_i$, one gets, for the total amount of absorbed radiation Q_a by an organism whose total surface area is A, the following equation:

$$Q_a = \frac{\Sigma a_i A_i \mathcal{R}_i}{A}$$

$$= \frac{\begin{array}{c} a_1 A_1 S + a_2 A_2 {}_h d + a_3 A_3 r({}_h S + {}_h d) \\ + a_4 A_4 \mathcal{R}_a + a_5 A_5 \mathcal{R}_g \end{array}}{A} \qquad (7.30)$$

Specific Absorptances

Eq. (7.30) can be shortened considerably in many cases. The values of a_4 and a_5 are always the same since these are the absorptances of the organism to the two sources of longwave radiation and the spectral distribution of the radiation from each source is nearly the same. In the cases of sunlight, skylight, and reflected light, however, there may be considerable differences because of substantial differences in the wavelength distributions of the emitted or reflected light. Sunlight has a broad spectral range covering the ultraviolet, visible, and near-infrared. Skylight, however, is relatively rich in ultraviolet and the blue portion of the visible but is depleted of red wavelengths and contains essentially no near-infrared. Depending on the spectral reflectance and absorptance of the organism's surface, the average absorptance to sunlight and skylight may be quite different. If direct solar radiation S has a monochromatic intensity distribution S_λ such that $S = \int_0^\infty a_\lambda S_\lambda \, d\lambda$ and an organism's surface has a monochromatic spectral absorptance a_λ, then the average absorptance to sunlight is

$$a_1 = \frac{\int_0^\infty a_\lambda S_\lambda \, d\lambda}{\int_0^\infty S_\lambda \, d\lambda}. \qquad (7.31)$$

Average absorptance to diffuse skylight $_hd_\lambda$ is

$$a_2 = \frac{\int_0^\infty a_\lambda \, _hd_\lambda \, d\lambda}{\int_0^\infty \, _hd_\lambda \, d\lambda}, \qquad (7.32)$$

and absorptance to ground-reflected sunlight and skylight is

$$a_3 = \frac{\int_0^\infty a_\lambda \, r_\lambda(_hS_\lambda + \, _hd_\lambda) \, d\lambda}{\int_0^\infty r_\lambda(_hS_\lambda + \, _hd_\lambda) \, d\lambda}, \qquad (7.33)$$

where r_λ is the monochromatic spectral reflectance of the ground surface. In the case of plant leaves, mean absorptance to skylight is about 0.10 greater than absorptance to direct sunlight. If mean absorptance to direct sunlight is 0.60, absorptance to skylight is about 0.70. Further details are given in Chapter 8.

For many purposes, one can assume that $a_1 = a_2$ and accept the relatively small error this may make in the estimate of total absorbed radiation. The error is usually quite small because the flux of skylight is about 15% or less of the direct sunlight, and the product of this figure and the absorptance of the organism is a small quantity. In the case of ground-reflected radiation, total intensity is fairly low, and it is reasonable to assume that $a_3 = a_1$. Hence, in somewhat simplified form, Eq. (7.30) is written

$$Q_a = \frac{a_1 [A_1 S + A_2 \, _hd + A_3 \, r(_hS + \, _hd)] + a_4(A_4 \mathcal{R}_a + A_5 \mathcal{R}_g)}{A}. \qquad (7.34)$$

Absorbing Areas

The areas A_i which are exposed to the various sources of radiation can sometimes be simplified as well. The only true point source in the system is the sun. All of the outer sources are extended sources. Skylight and longwave radiation from ground and atmosphere come from hemispherical sources. The area of an organism exposed to skylight should be the same as the area exposed to radiation from the atmosphere, so one would expect A_2 to equal A_4. If ground-reflected sunlight and skylight were perfectly diffusely reflected, they could be characterized as a hemispherical source on the downward side from the organism. When there is glint, or specular reflection, a correction may have to be made. Therefore, we shall assume that A_3 equals A_5. Rewriting Eq. (7.34) by using these approximations gives

$$Q_a = \frac{a_1 [A_1 S + A_2 \,_h d + A_3 \, r(_h S + \,_h d)] + a_4 (A_2 \mathcal{R}_a + A_3 \mathcal{R}_g)}{A}. \quad (7.35)$$

Flat Plate Absorption

The geometrical relations between a flat plate and the various sources of radiation are relatively simple. A flat plate as considered here has two surfaces. It is necessary to distinguish the upward-facing surface u from the downward-facing surface d. For a flat plate, $A_1 = A_2 = A_3 = A/2$. In the following formulations, the shortwave absorptances to direct sunlight, scattered skylight, and reflected light are considered to be equal.

Horizontal. The amount of radiation absorbed by the upper surface of a horizontal plate is given by

$$\begin{aligned}
_h^u Q_a &= \,_u a_1 (S \cos i + \,_h d) + a_4 \, \mathcal{R}_a \\
&= \,_u a_1 \,_h g + a_4 \, \mathcal{R}_a.
\end{aligned} \quad (7.36)$$

The amount of radiation absorbed by the lower surface of a horizontal plate is given by

$$_h^d Q_a = \,_d a_1 \, r \,_h g + a_4 \, \mathcal{R}_g. \quad (7.37)$$

Often, $_u a_1 = \,_d a_1$ to a good approximation. Then, the average amount of radiation absorbed by the two surfaces of a horizontal plate is given by

$$_h Q_a = \frac{_h^u Q_a + _h^d Q_a}{2} = \frac{a_1}{2} (1 + r)_h g + \frac{a_4}{2} (\mathcal{R}_a + \mathcal{R}_g). \quad (7.38)$$

Tilted. A flat plate is tilted from the horizontal by an angle θ, with the normal to the plate making an angle θ to the vertical. A fraction of the sky and atmosphere equal to θ/π is lost by the upper surface and gained by the lower surface, whereas the same fraction of the ground emitting hemisphere is gained by the upper surface and lost by the lower surface. When $\theta = \pi/2$, the plate is in a vertical position, and both surfaces are exposed to the same amount of sky and ground, i.e., a half-hemisphere. A hemisphere covers an angular dimension of π; when a sector θ/π is lost from a hemisphere, the amount remaining is $(\pi - \theta)/\pi$.

In the following equations for a flat plate, all shortwave absorptances are considered to be equal.

Amount of radiation absorbed by the upper surface of a tilted plate is given by

$$_uQ_a = {}_ua_1 \left[S \cos {}_ui + \frac{\pi - \theta}{\pi} {}_hd + \frac{\theta}{\pi} r({}_hS + {}_hd) \right]$$

$$+ a_4 \left(\frac{\pi - \theta}{\pi} \mathcal{R}_a + \frac{\theta}{\pi} \mathcal{R}_g \right). \tag{7.39}$$

Amount of radiation absorbed by the lower surface of a tilted plate is given by

$$_dQ_a = {}_da_1 \left[S \cos {}_di + \frac{\pi - \theta}{\pi} r({}_hS + {}_hd) + \frac{\theta}{\pi} {}_hd \right]$$

$$+ a_4 \left(\frac{\pi - \theta}{\pi} \mathcal{R}_g + \frac{\theta}{\pi} \mathcal{R}_a \right). \tag{7.40}$$

It is necessary to distinguish between the angle of incidence of the direct rays of the sun on the upper plate or leaf surface and the angle of incidence on the lower surface (hence the notation $_ui$ and $_di$).

The average amount of radiation absorbed by the flat plate or leaf as a whole is

$$_pQ_a = \frac{{}_uQ_a + {}_dQ_a}{2}. \tag{7.41}$$

The angle of incidence of the direct rays of the sun is easily determined using the same trigonometric formulation as used in Chapter 6 for the irradiation of slopes. The relationship is the same as Eq. (6.55) or (7.21) and is written as follows:

$$\cos i = \cos a_s \cos a \cos (\alpha - \alpha_s) + \sin a_s \sin a, \tag{7.42}$$

where a_s and α_s are altitude and azimuth angles of the normal to the surface of the flat plate or leaf and a and α are altitude and azimuth of the direction of the sun's rays.

Absorption by a Sphere

In general, absorption by a sphere is given by the equation

$$_sQ_a = a_1 \frac{S}{4} + \frac{1}{2} (a_2 {}_hd + a_3 r {}_hg + a_4\mathcal{R}_a + a_5\mathcal{R}_g). \tag{7.43}$$

Absorption by a Cylinder

The absorption of radiation by a cylinder of any orientation is given by

$$_cQ_a = \frac{1}{2} \left\{ a_1 \left[\frac{2x \sin \theta + \pi \cos \theta}{\pi(x + 1)} \right] S + a_2 {}_hd \right.$$

$$\left. + a_3 r {}_hg + a_4 \mathcal{R}_a + a_4 \mathcal{R}_g \right\}. \tag{7.44}$$

Animals usually have different absorptances on their dorsal and ventral sides, and it is necessary to consider the case in which $a_1 \neq a_3$. Usually, however, one can consider $a_1 = a_2$ as an approximation.

Sample Calculations

That calculations using these equations are easy is demonstrated by the following five examples in which the radiation absorbed is determined for a healthy green leaf if the leaf is horizontal, vertical in the east-west plane (normals to the north and south), at a 30° tilt from horizontal with the normal pointing due north (i.e., 30° from the zenith), at a 30° tilt from horizontal with the normal pointing due south (i.e., 30° from the zenith), and at a tilt of 60° with the normal pointing due south. The leaf has a shortwave absorptance of 0.6 on the upper surface and 0.5 on the lower surface. The longwave absorptance is taken as 1.0. All examples are calculated for a clear day at noon on 22 June and a position located at 40° north latitude. The atmospheric transmittance is 0.6 and the solar flux S is 773 W m^{-2}. The values given in Tables 7.1 and 7.2 are used.

Example 1

The upper surface of a horizontal leaf has an altitude angle $a_s = 90°$, and the azimuth angle is indeterminate. The slope angle θ is 0°. The angle of incidence $i(=z)$ is 16.5°.

Using Eq. (7.36), one gets, for the total radiation absorbed,

$$_h^u Q_a = 0.6 \times 865 + 1.0 \times 412 = 931 \text{ W m}^{-2}.$$

The lower surface of the leaf has an altitude angle $a_s = -90°$, and the azimuth angle is indeterminate. The angle of incidence i for the direct rays of the sun is $180° - 16.5° = 163.5°$, which shows that the direct rays do not strike this surface, nor does the skylight illuminate it. Equation (7.37) gives, for the total radiation absorbed,

$$_h^d Q_a = 0.5 \times 130 + 1.0 \times 478 = 543 \text{ W m}^{-2}.$$

The average amount of radiation absorbed by upper and lower surfaces of the leaf is $1/2(_h^u Q_a + _h^d Q_a)$, or 737 W m^{-2}.

Example 2

For a vertical leaf in the east-west plane, $_u a_s = {_d a_s} = 0°$. One surface must be identified as the "upper" and one as the "lower" surface. Let the upper surface face north and the lower face south; then, $_u \alpha_s = 0°$ and $_d \alpha_s = 180°$. The solar altitude a is $90° - 16.5°$, or 73.5°, and the azimuth α

is 180° since the time is noon (assume that measurements are being carried out on the mean meridian for the time zone). The angle of incidence for the upper surface, given by Eq. 7.42, is $\cos {}_u i = -0.284$, and ${}_u i = 106.5°$. The direct rays of the sun will not strike the vertical north-facing surface. The slope angle θ is $\pi/2$, and the upper surface will "see" half a hemisphere of sky and of ground surface. Total radiation absorbed is ${}_u Q_a = 0.6 \times 0.5(124 + 130) + 1.0 \times 0.5(412 + 478) = 521$ W m^{-2}.

The south-facing, "lower" surface has direct sunlight incident upon it such that $\cos {}_d i = 0.284$ and ${}_d i = 73.5°$. Total radiation absorbed is

$$
{}_d Q_a = 0.5 \times 773 \times 0.284 + 0.5 \times 0.5(124 + 130) \\
+ 1.0 \times 0.5(412 + 478) = 619 \text{ W m}^{-2}.
$$

The average radiation absorbed by "upper" and "lower" leaf surfaces is 570 W m^{-2}.

Example 3

The leaf is tilted 30° from the horizontal (normal is 30° from zenith) with the normal to the upper surface pointing due north. The angles are ${}_u a_s = 60°$, ${}_u \alpha_s = 0°$, ${}_d a_s = -60°$, and ${}_d \alpha_s = 180°$. For the solar rays, $a = 73.5°$ and $\alpha = 180°$; $\cos {}_u i = 0.688$ and ${}_u i = 46.5°$. The angle θ is 30°, or $\pi/6$. The upper surface "sees" a fraction of the sky, $(\pi - \theta)/\pi$, which is 5/6, or 0.833, and a complementary fraction of the ground surface, 1/6 or 0.167. Total amount of radiation absorbed is

$$
{}_u Q_a = 0.6 \times 773 \times 0.688 + 0.6 \times 0.833 \times 124 + 0.6 \times 0.167 \times 130 \\
+ 1.0 \times 0.833 \times 412 + 1.0 \times 0.167 \times 478 = 817 \text{ W m}^{-2}.
$$

The lower surface has no direct sunlight on it and "sees" 0.833 of the ground hemisphere and 0.167 of the sky hemisphere. Total amount of radiation absorbed is

$$
{}_d Q_a = 0.5 \times 0.167 \times 124 + 0.5 \times 0.833 \times 130 + 1.0 \times 0.167 \\
\times 412 + 1.0 \times 0.833 \times 478 = 531 \text{ W m}^{-2}
$$

The average radiation absorbed by upper and lower surfaces together is 674 W m^{-2}.

Example 4

The leaf is tilted 30° from the horizontal (normal is 30° from zenith) with the normal to the upper surface pointing due south. The angles are ${}_u a_s = 60°$, ${}_u \alpha_a = 180°$, ${}_d a_s = -60°$, ${}_d \alpha_s = 0°$, ${}_u i = 13.5°$, and ${}_d i = 166.5°$. ($\cos {}_u i = 0.972$ and $\cos {}_d i = -0.972$). Total amount of absorbed radiation is

$$
{}_u Q_a = 0.6 \times 773 \times 0.972 + 0.6 \times 0.833 \times 124 + 0.6 \times 0.167 \times 130 \\
+ 1.0 \times 0.833 \times 412 + 1.0 \times 0.167 \times 478 = 949 \text{ W m}^{-2}
$$

The lower surface "sees" no direct sunlight, and the total radiation is

$$_dQ_a = 0.5 \times 0.167 \times 124 + 0.5 \times 0.833 \times 130 + 1.0 \times 0.167$$
$$\times 412 + 1.0 \times 0.833 \times 478 = 531 \text{ W m}^{-2}.$$

The average radiation absorbed by upper and lower surfaces together is 740 W m^{-2}.

Example 5

A leaf is tilted 60° from the horizontal (normal is 60° from zenith) with the normal to the upper surface pointing due south. The angles are $_ua_s = 30°$, $_u\alpha_s = 180°$, $_da_s = -30°$, $_d\alpha_s = 0°$, $_ui = 43.5°$, and $_d = 136.5°$ (cos $_ui = 0.726$ and cos $_di = -0.726$). The angle θ is 60°, or $\pi/3$. The upper surface "sees" a fraction of the sky $(\pi - \theta)/\pi$, which is 2/3, or 0.666, and a complementary fraction of the ground surface, 1/3, or 0.333. Total radiation absorbed is

$$_uQ_a = 0.6 \times 773 \times 0.726 + 0.6 \times 0.666 \times 124 + 0.6 \times 0.333 \times 130$$
$$+ 1.0 \times 0.666 \times 412 + 1.0 \times 0.333 \times 478 = 846 \text{ W m}^{-2}.$$

The lower surface "sees" no direct sunlight, and total radiation absorbed is

$$_dQ_a = 0.5 \times 0.333 \times 124 + 0.5 \times 0.666 \times 130 + 1.0 \times 0.333$$
$$\times 412 + 1.0 \times 0.666 \times 478 = 519 \text{ W m}^{-2}.$$

The average amount of radiation absorbed by upper and lower surfaces together is 682 W m^{-2}.

It is also possible to take into account details concerning the anisotropy of skylight or the difference in leaf absorptance to direct solar radiation and scattered skylight. Invariably, however, such details change the total amount of absorbed radiation by less than 1%. There is one correction that might have to be made if there is strong specular reflection of direct sunlight from the ground surface. The calculations given above assumed diffuse reflection from the ground surface. If, for example, there is strong specular reflectance, with $r = 0.3$, then an additional beam of light of 232 W m^{-2} will be incident upon the surfaces, depending upon the angle of incidence of this beam. The specularly reflected beam would have angles of altitude and azimuth related to the altitude and azimuth of the sun as follows: $a_r = -a$ and $\alpha_r = \alpha$. From these angles, one can determine the angle of incidence to the leaf surface in the same manner as for direct sunlight.

These calculations are summarized in graphical form in Fig. 7.6. Note the sudden increase in the radiation absorbed by the downward-facing surface as it suddenly becomes exposed to some direct sunlight. If sky conditions are overcast rather than clear, one can make the calculations in

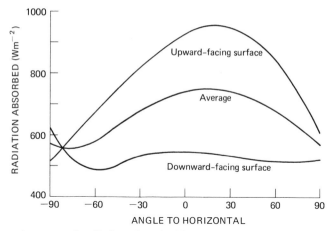

Figure 7.6. Amount of radiation absorbed by a green leaf (assumed to be a flat plate with two surfaces) as a function of tilt from horizontal in the north-south plane at noon at 40° north latitude on 22 June for an atmospheric transmittance of $\tau = 0.6$ and ground reflectance r or 0.15.

the same manner except that there will be no direct sunlight incident on the leaf. One would treat the cloud light as a diffuse hemispherical source with an intensity as discussed in Chapter 6 for overcast conditions.

Radiation Incident on a Sphere and Cylinders

The amount of radiation incident on a sphere, vertical cylinder, horizontal cylinder with axis oriented east and west ($\alpha_c = 90°$) and north and south ($\alpha_c = 0°$) are given in Table 7.3 for noon at 40° north latitude for 22 June, 22 September, and 22 December. The atmospheric transmittance τ is 0.6 and the ground shortwave reflectivity r is 0.15. The cylinders have a length-to-radius ratio x of 4.

The cylinder with its axis in the east-west plane has the same average amount of radiation incident on its total surface as the sphere. The vertical cylinder intercepts decidedly less radiation during the summer than the horizontal cylinders or the sphere. The vertical cylinder intercepts more radiation than the horizontal cylinder at noon during the winter.

Diurnal Radiation Climate

It is easy to continue these calculations to include the variation of the radiation climate with time of day. One must first decide on the course of temperature during the day and the state of the sky, whether overcast, clear, or partly cloudy. One has considerable choice concerning representative conditions for a characteristic day. If a clear day is selected, then the cal-

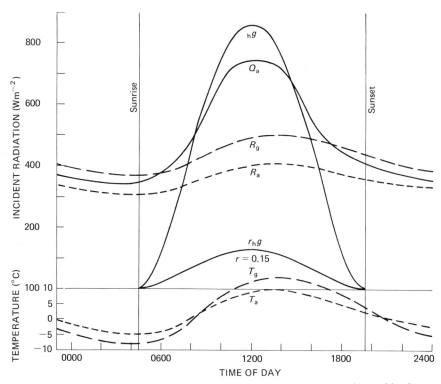

Figure 7.7. The various components of radiation incident on a horizontal leaf as a function of time of day on 22 June at 40° north latitude for an atmospheric transmittance of $\tau = 0.6$. Radiant atmospheric and ground temperatures are T_a and T_g. Ground reflectance is $r = 0.15$. Total amount of radiation absorbed by the leaf is Q_a. Leaf absorptance to shortwave radiation is $a_s = 0.6$ for small angles of incidence and $a_s = 0.4$ for large angles of incidence.

culations represent maximum energy conditions during the daylight hours and minimum energy at night. A completely overcast sky represents minimum energy during the daytime and maximum energy at night. It is certain that whatever conditions are selected for a given date and location, there will be days at that time of year analogous to the selected conditions. Every day is different; some are clear, some are overcast, some are partly cloudy, some are warm, some are cool, and so forth. Nevertheless, for a particular time of the year, there are reasonable limits to the conditions which will occur in nature.

Summer Day

The first characteristic day for purposes of example is a cloud-free summer day on 22 June at 40° north latitude. It is reasonable to assume that the air temperature varies from a minimum of about 14°C just before dawn to a maximum of 30°C in the early afternoon, particularly for a continental climate. The temperature of the ground surface may be about 2.5°C cooler than that of the air during the night and as much as 5°C warmer during the mid-afternoon. Once these temperature relationships are established, $\mathscr{R}_a$ and $\mathscr{R}_g$ can be directly determined. The ground surface radiates as a blackbody. The value for $\mathscr{R}_a$ is given by Swinbank's formula, Eq. (7.1). These values are plotted in Fig. 7.7. Using an atmospheric transmittance $\tau = 0.6$, one can estimate $_hg = (_hS + _hd)$ from Eq. (7.13) for various times of day. In order to do this, the solar zenith angle as a function of the time of day must be known. Ground reflectance is assumed to be 0.15. Global radiation and flux of reflected sunlight and skylight are shown in Fig. 7.7. When the zenith angle of the solar rays is between 17° and 52°, the absorptance of a horizontal leaf is assumed to be 0.6, and when the angle is greater than 52°, absorptivity is assumed to be 0.4; these are fairly arbitrary values. The exact numbers used are not as important as the demonstration of the method used for estimating the total radiation flux incident upon an organism and determining the amount of energy absorbed.

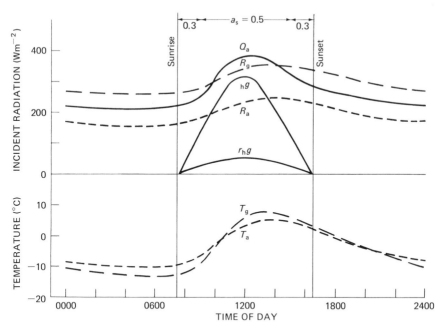

Figure 7.8. The various components of radiation incident on a horizontal leaf as a function of time of day on 22 December at 40° north latitude for an atmospheric transmittance of $\tau = 0.6$. Radiant atmospheric and ground temperatures are T_a and T_g. Ground reflectance is $r = 0.15$. Total amount of radiation absorbed by the leaf is Q_a. Leaf absorptance to shortwave radiation is $a_s = 0.5$ for moderate angles of incidence and $a_g = 0.3$ for large angles of incidence.

Winter Day

The same procedure is used to determine the total radiation flux incident on a horizontal leaf as a function of time of day and the amount of radiation absorbed by the leaf for 22 December at 40° north latitude. Here, the absorptance of the leaf surface is assumed to be 0.5 for solar zenith angles between 63° and 75°. For angles greater than 75°, a value of 0.3 is used. The air temperature during this winter day has a minimum of $-10°C$ and a maximum of 5°C. The ground-surface is assumed to be cooler than the air during the night and slightly warmer than the air during the day. The values for amount of absorbed radiation are very characteristic of a cloudless winter day at 40° north latitude. The fluxes of radiation and the temperatures are plotted in Fig. 7.8. An overcast condition would produce a slightly greater Q_a at night and a considerably reduced Q_a in midday. In fact, it is relatively easy to estimate the value of Q_a under such circumstances. The nighttime flux of longwave radiation is $Q_a = (\mathcal{R}_a + \mathcal{R}_g)/2$. If the air temperature is as shown in Fig. 7.8 for an overcast day, then the nighttime radiation flux absorbed is between the $\mathcal{R}_a$ line of Fig. 7.8 and the $\mathcal{R}_g$ level. In the daytime, one would expect total radiation absorbed on an overcast day to exceed $\mathcal{R}_g$ by a small amount at midday. A leaf is strongly coupled to the longwave infrared radiation field, by absorptivities approximating 1.0, and is only moderately or weakly coupled to the shortwave solar, sky, and reflected radiation, by absorptances of 0.6 or less.

Climate-Space Diagram

Another way of looking at the same results is to plot them on a climate-space diagram with temperature as the ordinate and radiation as the abscissa. The results for a horizontal leaf on 22 June and 22 December are plotted in Fig. 7.9. Amount of energy absorbed by an organism during the nighttime hours is approximately halfway between the sky radiation line and the blackbody line, differing from halfway only because of the difference between the air and ground temperatures in these examples. Climate-space diagrams have been used extensively to describe the steady-state energetics of animals but are useful for plant parts as well. Porter and Gates (1969) originated the idea of the climate-space diagram, and it has had many useful applications since then (see Heller and Gates, 1971; Spotila et al., 1972; Morhardt and Gates, 1974; Morhardt, 1975).

Gates and Papian (1971) presented a climate-space diagram for plant leaves and showed on this diagram an array of climate situations, characterized as follows: alpine winter, temperate winter, desert winter, alpine summer, temperate summer, and desert summer. This diagram is shown in Fig. 7.10 with some modification. It is given for 40° north latitude and an atmospheric transmittance of 0.6 and ground reflectance of 0.15. It is recognized that there are not categorical boundaries between these

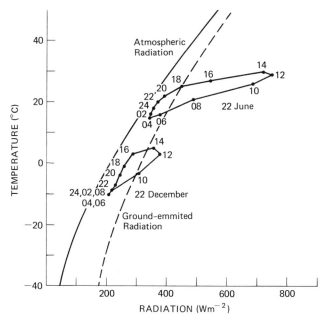

Figure 7.9. A climate-space diagram for a horizontal leaf at 40° north latitude showing the amount of total absorbed radiation as a function of air temperature and time of day for 22 June and 22 December.

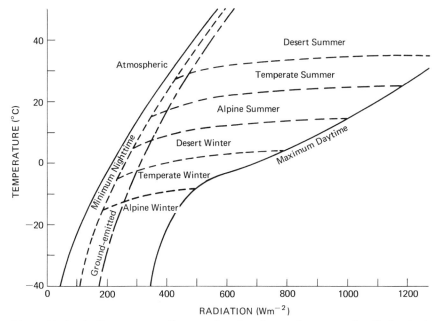

Figure 7.10. A climate-space diagram showing the total amount of radiation incident on a horizontal leaf as a function of air temperature for various locales and times of year at 40° north latitude for an atmospheric transmittance of $\tau = 0.6$ and a ground reflectance to shortwave radiation of $r = 0.15$.

various climates, and in fact, the climate-spaces overlap one another considerably. The purpose of the diagram is to point out that the temperature and radiation climate of the alpine, temperate, or desert regions can be reasonably well described and that the most frequently occurring conditions fall within the regions indicated on the climate-space diagram. Furthermore, the diagram shows that during an annual cycle, the desert could have any or all of the temperature and radiation conditions indicated from summer to winter as a continuum and that these conditions could completely overlap with those of the temperate and alpine summer. Likewise, the transition from summer to winter for the alpine region produces a continuum of conditions overlapping those of the desert and temperate winter. One can be reasonably certain that desert winter conditions will never coincide with alpine winter conditions, nor will alpine summer conditions ever coincide with desert summer conditions. All the other climate conditions could overlap at one time or another with temperate summer or winter conditions.

The line marked "maximum daytime" is the maximum amount of radiation incident on an upward-facing flat horizontal surface at 40° north latitude for an atmospheric transmittance of 0.6. A flat surface perpendicular to the solar rays receives a radiation intensity somewhere to the right of this line. Any cylindrical, spherical, or otherwise-shaped object or organism will receive a radiation intensity to the left of this line. The actual amount of radiation absorbed by an organism will usually be considerably to the left of the line marked "maximum daytime."

From the various calculations presented in this chapter, the reader should be quite familiar with the amount of radiation occurring under various natural conditions above the surface of the earth. It is very simple, from basic principles, to achieve an understanding of the radiation environment of organisms. Just as most people have a sense concerning values of air temperature, humidity, and wind speed, so also one should be familiar with the usual range of values for radiation, a truly ubiquitous phenomenon in the world in which we live.

Geometrical Advantages of Cylinders for Plants or Animals

The amount of radiation absorbed by a sphere or by cylinders of various orientations is readily determined. A few examples will suffice to demonstrate the significance of geometry on the total radiation absorbed.

For a sphere, the rays of direct sunlight and skylight always strike the surface along a radius vector, i.e., are perpendicular to the surface. For these rays, maximum absorptance of the surface will always prevail. In the case of a vertical cylinder, however, the rays of the sun will only strike the main cylindrical part of the surface along a radius vector early or late in the day, that is, when the sun is near the horizon. But at that time of day, the intensity of direct sunlight is much reduced by atmospheric attenuation. For example, from Table 6.3 one finds that intensity of direct sunlight at 0600 and 0800 is 27 and 75%, respectively, of the noon intensity. At 0600, the altitude of the sun is 16°, and the sun's rays will strike the side of a vertical cylinder fairly straight on. At 0800, the solar altitude is 38°, and the sun's rays will already be fairly oblique to the cylinder. As the rays become more oblique, absorptance of the surface to shortwave direct sunlight decreases quite strongly. If absorptance is 0.6 when the rays are nearly perpendicular to the surface, it may be 0.4 when the rays make an angle of about 38° to the radius vector and as low as 0.1 or 0.2 when the angles are about 70° or more.

Shown in Table 7.4 are the amounts of shortwave, longwave, and total radiation absorbed by a sphere, flat plate, vertical cylinder, and a horizontal cylinder with $\alpha_c = 90°$ and $\alpha_c = 0°$. The calculations are done for 22 June as a function of time of day. Atmospheric transmittance is $\tau = 0.6$ and the ground reflectance to shortwave radiation is $r = 0.15$. Here all shortwave absorptance is assumed to be 0.6 for both upper and lower surfaces in order to simplify the calculations somewhat. The values in Table 7.2, however, were calculated using an absorptance of 0.6 for the upper surface and 0.5 for the lower surface. There are thus some small differences between the numbers for the flat plate given in Tables 7.2 and 7.4.

Early in the day, at 0600, the amount of incident direct sunlight is 57 W m^{-2} for the horizontal plate, 56 W m^{-2} for the vertical cylinder and the horizontal cylinder with $\alpha_c = 0°$, 52 W m^{-2} for the sphere, and 40 W m^{-2} for the horizontal cylinder with $\alpha_c = 90$. By 0800, however, there has occurred an enormous increase in the amount of direct sunlight incident on a horizontal flat plate, the value now being 354 W m^{-2}. If this is averaged over the upward- and downward-facing surfaces of the horizontal plate, the amount of radiation still averages 177 W m^{-2}, which is a large amount compared to the values for the other objects. The vertical cylinder receives 151 W m^{-2} of direct sunlight; the horizontal cylinder with $\alpha_c = 0°$, 147 W m^{-2}, the sphere, 144 W m^{-2}; and the horizontal cylinder with $\alpha_c = 90°$, 135 W m^{-2}. Now, at noon, the vertical cylinder has distinctly less direct solar radiation incident upon it than any of the other objects, i.e., 130 W m^{-2}. The horizontal cylinder oriented north and south with $\alpha_c = 0°$ has 167 W m^{-2}; the sphere, 193 W m^{-2}; the east-west oriented horizontal cylinder with $\alpha_c = 90°$, 197 W m^{-2}; and the horizontal plate, 741 W m^{-2} (or, averaged over two surfaces, 370 W m^{-2}). The vertical cylinder receives its maximum direct-sunlight irradiation around 1000.

Not only does the vertical cylinder have less direct solar radiation incident upon it after about 0900 than the other objects, but the solar rays strike its surface at large oblique angles, and the reflectance of the surface increases greatly. The result is that the vertical cylinder absorbs disproportionately less direct solar radiation than any of the other objects. The

Table 7.4. Amount of Shortwave, Longwave, and Total Radiation Absorbed by a Sphere, Flat Plate, Vertical Cylinder, and Horizontal Cylinder with $\alpha_c = 90°$ and $\alpha_c = 0°$ as a Function of Time of Day for $\tau = 0.6$, $r = 0.15$, and $S_0 = 1317$ W m^{-2} on 22 June at 40° North Latitude.[a]

Time	S	$_hd$	$_hg$	r_hg	T_a	$\mathcal{R}_a$	T_g	$\mathcal{R}_g$	$\mathcal{R}'$
0500	4	31	31	5	15.5	310	12.5	370	350
0600	206	82	139	21	16	315	13	375	376
0800	574	116	470	70	20.5	350	19	410	437
1000	730	123	755	113	26	380	28	465	498
1200	773	124	865	130	30	412	32.5	478	521

Time	$_sa_1$	$_sS$	$_sS_a$	$_sQ_a$	$_ha_1$	$_hS$	$_hS_a$	$_pQ_a$
0500	0.6	1	1	351	0.1	0	0	350
0600	0.6	52	31	407	0.2	57	11	382
0800	0.6	144	86	523	0.4	354	142	508
1000	0.6	182	109	607	0.6	632	379	688
1200	0.6	193	116	637	0.6	741	445	743[b]

Time	$_{cv}\theta$	$_{cv}a_1$	$_{cv}S$	$_{cv}S_a$	$_{cv}Q_a$	$_{ch}^{90}\theta$	$_{ch}^{90}a_1$	$_{ch}^{90}S$	$_{ch}^{90}S_a$	$_{ch}^{90}Q_a$
0500	85	0.6	0	0	350	27.5	0.3	1	0	350
0600	74	0.5	56	17	393	23.9	0.3	40	12	388
0800	52	0.4	151	54	491	38	0.4	135	54	491
1000	30	0.3	156	59	557	62.8	0.5	199	100	598
1200	16.5	0.2	130	42	563	90	0.6	197	118	639

Time	$_{ch}^{0}\theta$	$_{ch}^{0}a_1$	$_{ch}^{0}S$	$_{ch}^{0}S_a$	$_{ch}^{0}Q_a$
0500	63	0.5	1	1	351
0600	72.7	0.5	56	28	404
0800	89.2	0.6	147	88	525
1000	101.7	0.6	167	100	598
1200	106.6	0.5	167	83	604

[a] Cylinder with $x = 4.0$; $a_1 = 0.6$ to $_hd$ and r_hg; a_1 to S a function of θ. $\mathcal{R}' = 1/2(a_1\,_hd + a_1r_hg + \mathcal{R}_a + \mathcal{R}_g)$.

[b] This value differs slightly from that given in Table 7.2 since the lower surface absorptance is taken equal to the upper surface absorptance of 0.6 for these calculations.

total integrated amount of direct solar radiation absorbed during the day is very much less for the vertical cylinder than for the other objects.

A horizontal flat plate absorbs very little radiation early or late in the day. A vertical flat plate absorbs little direct solar radiation near midday and generally behaves like a vertical cylinder. A spherical shape is always disadvantageous if absorbing too much radiation is deleterious. Many cacti resemble vertical cylinders, as do all tree trunks. However, cacti are usually spaced apart in the hot, dry desert environment, whereas trees generally grow closely spaced in the forest, where they shade one another. The crowns of trees also shade the trunks. The vertical cylindrical geometry of many cacti, in particular the saguaro, is distinctly an adaptation against overheating by solar radiation. Tree trunks are probably vertical cylinders primarily for structural reasons. Many *Opuntia* cacti have their broad, flat blades aligned vertical in the north-south plane, which probably reduces overheating by solar radiation.

Problems

1. Plot a blackbody curve on a piece of graph paper to the same scale as shown in Fig. 7.1, with temperature as the ordinate and radiation as the abscissa.

2. Plot the amount of radiation from a clear sky according to Swinbank's T^6 formula [Eq. (7.2)] on the same graph used for Problem 1.

It will be interesting to see how latitude affects the radiation climate of organisms on a given date assuming that at each latitude the same atmospheric transmittance exists which is taken as $\tau = 0.6$. The next exercises are designed to elaborate on this question.

3. Determine the solar altitude at solar noon for latitudes $0°$, $20°$, $40°$, $60°$, and $80°N$.

From the solar altitude, determine the air mass m at solar noon for these latitudes.

Using $\tau = 0.6$, determine the intensity of direct solar radiation S, skylight $_hd$, and global radiation on a horizontal surface $_hg$ for solar noon at each latitude. Use $S_0 = 1317$ W m^{-2}.

Assume the ground reflectance r to be 0.10 and determine the flux of reflected sunlight and skylight.

4. Making the assumption that the noon air temperature is $40°$, $35°$, $30°$, $25°$, and $20°C$ at $0°$, $20°$, $40°$, $60°$, and $80°$ north latitude, respectively, and assuming the ground surface temperature to be the same as the air temperature, determine $\mathcal{R}_a$ and $\mathcal{R}_g$ for noon at each latitude.

5. Determine the total radiation flux incident on the following objects at solar noon for the five latitudes given below. Use $x = 6$ for the cylinders.

Latitude	Sphere	Vertical cylinder	Horizontal cylinder ($\alpha - \alpha_c = 45°$)	Horizontal leaf (average upper and lower surface)
0				
20				
40				
60				
80				

6. Determine the diurnal radiation climate and make graphs similar to Figs. 7.7 and 7.8 for a horizontal leaf on 22 June and 22 September at the equator. Assume that the maximum and minimum air temperatures are 40° and 25°C on 22 June and 30° and 20°C on 22 September. Assume that leaf absorptance to shortwave radiation is 0.5 for all solar angles involved, that the longwave absorptance is 1.0, and that atmospheric transmittance is $\tau = 0.6$.

7. For a site at 60° north latitude, determine the diurnal radiation climate and make a graph similar to Fig. 7.9 for a sphere with an absorptance to shortwave radiation of 0.7 and an absorptance to longwave radiation of 1.0. Assume that the maximum and minimum air temperatures are 30° and 15°C on 12 August and 0° and −10°C on 9 February.

Chapter 8

Spectral Characteristics of Radiation and Matter

Introduction

The wave nature of radiation is responsible for many interesting and unique phenomena involving the interaction of radiation and matter. Prisms, gratings, and other devices disperse radiation into many types of spectrums, classified according to the techniques used for their detection or generation. The wavelength regions have historical names related to scientific discovery and development of the particular spectral interval. There are, for example, gamma rays, x-rays, ultraviolet radiation, visible light, and infrared, microwave, and radiowave radiation. Smaller wavelength intervals within these broad spectral regions have also been given names.

The wave nature of radiation and the quantum energy associated with a given frequency or wavelength was described in Chapter 5. Also discussed there was the interaction of radiation and matter and the general concepts of photochemistry, molecular bond strength, and energy content of radiation. In this chapter, the spectral quality of sunlight, skylight, thermal, and ground-reflected radiation is described, as well as the spectral reflectance, transmittance, and absorptance of various substances, including the earth's atmosphere, plant surfaces, animal coats, and soils.

As a general rule, the terms ending in *ance* apply to whole objects and those ending in *ivity* apply to a specific piece of material. For example, one would refer to the *reflectance* of a bird but the *reflectivity* of a particular part of the bird's surface. I have made a choice to use *ance* throughout this book for all uses.

The spectral characteristics of the earth's atmosphere are described before the spectral distribution of sunlight since the atmosphere has a strong effect on the spectral distribution of sunlight received at the earth's surface. The basic theory of molecular spectroscopy is given first in order that the absorptive and radiative properties of the earth's atmosphere can be properly understood.

Molecular Spectra

Atoms have absorption and emission spectra corresponding to discrete changes of the energy states of their electrons. The energy state of an electron bound in orbit around a nucleus is dependent upon the orbital number, angular momentum, and spin of the electron. As an electron moves from an inner to an outer orbit, energy is absorbed. In turn, when an electron jumps from an outer to an inner orbit, radiation is emitted by the atom. Higher energy states characteristic of outer electron orbits converge to smaller energy differences from orbit to orbit until ionization occurs and an energy continuum results. The line emissions or absorptions of such well-known spectral sequences as the Balmer, Lyman, Paschen, Brackett, and Pfund series of atomic hydrogen are manifestations of transitions between various orbital states and changes in spin and angular momentum.

When atoms bind together to form molecules, additional energy states exist because of vibrational and rotational modes within the molecule. A simple diatomic molecule has only one vibrational mode, which is an oscillation of the atoms along the axis of the molecule. This vibrational oscillation modulates the electron energy states, and so the spectra produced are more complex than those of single atoms. The diatomic molecule can tumble or rotate as it is jostled about by collisions in a gas. This rotational motion is quantized and will further modulate the energy states of the electrons and add additional complexity to the absorption and emission spectra. Polyatomic molecules have much more complicated spectra than diatomic molecules. A large portion of the earth's atmospheric absorption spectrum is caused by polyatomic molecules.

Internal Energy of a Molecule

The internal energy of a molecule is the sum of the electronic, vibrational, and rotational energies. Electronic energies states are approximately one hundred times greater than vibrational energies, and these in turn are 10^4 to 10^5 times greater than rotational energies. The internal energy of a molecule, to a first approximation, may be written

$$E = E_e + E_v + E_r, \tag{8.1}$$

where E_e, E_v, and E_r are the electronic, vibrational, and rotational energies, respectively.

Electronic, vibrational, and rotational changes within a molecule occur with the absorption or emission of a single photon of frequency ν as described by the equation

$$h\nu = (E'_e - E''_e) + (E'_v - E''_v) + (E'_r - E''_r), \qquad (8.2)$$

where h is Planck's constant and E' and E'' represent the lower and upper energy states, respectively.

The actual amount of energy involved in each energy mode, electronic, vibrational, and rotational, depends on the mass of atoms or molecules. (It is these great differences in masses which result in the wide separation of the energy states and hence of the frequency of radiation absorbed or emitted.)

Diatomic Molecules

An absorption band of a diatomic molecule results from a large number of individual absorption lines which, taken together, represent all the "allowed" transitions among the various vibrational, rotational, and electronic states. A diatomic molecule has most of its absorption and emission electronic bands at high frequencies, usually in the ultraviolet. The single vibrational mode is of much lower frequency and usually occurs in the infrared portion of the spectrum. For the vibrational mode to be effective in the absorption of radiation, however, the molecule must have a permanent dipole moment. The earth's atmosphere does not contain significant amounts of diatomic gases with permanent dipole moments. The lack of abundance of diatomic molecules, other than N_2 and O_2, in the earth's atmosphere and the relatively low concentrations of polyatomic molecules keep the atmosphere free of absorption throughout most of the visible spectrum.

Vibrational Spectra of Polyatomic Molecules

It is only at infrared wavelengths that the vibrational and rotational absorption and emission spectra of such polyatomic molecules as H_2O, CO_2, and O_3 become effective.

The infrared vibrational spectra of polyatomic molecules are very significant for the exchange of radiation in the earth's atmosphere. The number of fundamental vibrational modes in a molecule is determined from the degrees of freedom which N bodies may have. It takes three coordinates, x, y, and z, to describe a single body. If there are N bodies, it requires $3N$ coordinates to describe them, and we say that the N bodies have $3N$ degrees of freedom. But a molecule of N atoms bound together can be translated in each of three directions, and, therefore, 3 degrees of freedom are assigned to translation. It takes three coordinates to describe the rotation of a molecule, and, therefore, 3 degrees of freedom are consumed in rotation. The number of vibrational modes of a polyatomic molecule is $3N - 6$. In the case of linear molecules, it is $3N - 5$, there being two rotational modes, indistinguishable from one another. Hence a diatomic molecule can have one vibrational mode. A triatomic molecule such as H_2O or O_3 will have three fundamental vibrational absorption bands at infrared wavelengths.

A single transition from one vibrational state to another produces absorption or emission of a single frequency only. The energy content of a single vibrational state of quantum number v is given by

$$E_v = \left(v + \frac{1}{2}\right) h\nu_0, \tag{8.3}$$

where h is Planck's constant, ν_0 is the fundamental vibration frequency, which is related to the atomic mass m and the force constant k as follows:

$$\nu_0 = \frac{\pi}{2}\left(\frac{k}{m}\right)^{1/2}. \tag{8.4}$$

Actually, m is a reduced-mass equivalent of the various atoms involved in the vibrational mode, rather than the mass of a single atom.

Spectroscopists are accustomed to expressing the frequency of a transition by means of the wavenumber $\tilde{\nu}$, which is the reciprocal of the wavelength λ. The frequency of a quantum of energy absorbed or emitted by a molecule is related to the change in vibrational energy by the equation

$$\tilde{\nu} = \frac{1}{\lambda} = \frac{\nu}{c} = \frac{E'_v - E''_v}{hc} = (v' - v'')\frac{\nu_0}{c} = \frac{\pi}{2c}\left(\frac{k}{m}\right)^{1/2}(v' - v''). \tag{8.5}$$

Rotational Spectra

The rotational states of a molecule are quantized, and the energy levels are given by

$$E_r = \frac{h^2 J(J + 1)}{8\pi^2 I}, \tag{8.6}$$

where I is the moment of inertia of the molecule about the axis of rotation and J is the rotational quantum number, an integer. The quantity E_r is given in joules when I is in kilograms per square meter. A pure rotational absorption spectrum can occur only if the molecule has a permanent dipole moment. The only allowed transitions are those in which $\Delta J = \pm 1$. In absorption, $\Delta J = +1$; in emission, $\Delta J = -1$. Hence the frequency in wavenumbers for a transition from a level with quantum number $J - 1$ to the level J is given by

$$\tilde{\nu} = \frac{h}{8\pi^2 I} [J(J + 1) - (J - 1)J]. \qquad (8.7)$$

Most pure rotation absorption or emission spectra are in the far-infrared or microwave region. If a molecule has very light atoms, such as the hydrogen atoms in H_2O, then the moment of inertia is small, and the rotational spectrum is centered at much higher frequencies. Water is a nonlinear molecule; it forms an obtuse triangle, with the hydrogen atoms at an angle of 105° and the oxygen at the apex. The pure rotational spectrum for the water molecule is centered around 50 μm, in the far infrared.

Vibration–Rotation Spectra

A molecule has a particular amount of rotational energy at any moment for a given rate of rotation. A gas or collection of molecules has a broad distribution of rotation rates. This occurs in the same manner as a broad distribution of translational velocities of gas molecules caused by Brownian motion. At a particular temperature, some of the gas molecules will be moving slowly, some will be rotating slowly, and some will be rotating rapidly and moving about rapidly. The higher the temperature, the more molecules in high states of translational and rotational energy. Because the molecules of a gas exist in a variety of rotational states for a given vibrational state, when energy is absorbed, which changes the vibrational state, a transition is made not only to the new vibrational state but to all the rotational states superimposed on the new vibrational state. In this way, the vibrational transition is modulated to contain many rotational absorption lines. A vibration–rotation absorption band has lines of frequency given by

$$\tilde{\nu} = (v' - v'')\frac{\nu_0}{c} + \frac{h}{8\pi^2 I} [J'(J' + 1) - J''(J'' - 1)]. \qquad (8.8)$$

A typical vibrational absorption band for a polyatomic molecule is shown in Fig. 8.1.

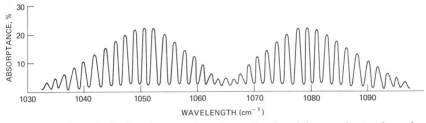

Figure 8.1. A typical vibration-rotation absorption band for a polyatomic molecule. This is a combination band $\nu_3 - 2\nu_2$ for the CO_2 molecule at 9.40 μm.

The heavier the masses of the atoms constituting the molecule, the lower the frequency and the longer the wavelength of radiation absorbed by a fundamental vibrational transition. The water molecule is one of the lightest of the more abundant molecules in nature. The fundamental vibrational absorption bands of H_2O are at as high a frequency as those of almost any molecule. The absorption bands of H_2O, CO_2, and O_3 are those that characterize the absorption and emission spectrum of the earth's atmosphere. These triatomic molecules have three fundamental bands, designated ν_1, ν_2, and ν_3, which will occur in the infrared absorption spectrum if there is a change in dipole moment. In the case of CO_2, the fundamental band ν_1 does not appear in the infrared, but only in the Raman spectrum. Overtone and combination bands are usually of considerably weaker intensity than the fundamental bands and will generally occur at higher frequencies. Difference bands may occur at lower frequencies. The vibrational–rotational absorption bands of the atmosphere dominate the spectrum in the infrared but become very weak as the visible part of the spectrum is approached. The visible "window," between 400 and 700 nm, is relatively free of strong absorptions by atmospheric gases. In the ultraviolet spectrum at wavelengths less than 300 nm, absorption by the electronic bands of atmospheric gases becomes important.

The Atmospheric Absorption Spectrum

The primary constituents of the atmosphere, oxygen and nitrogen, are diatomic molecules and absorb mainly in the ultraviolet, being generally transparent to visible and infrared wavelengths. If the atmosphere were composed only of oxygen and nitrogen, sunlight would reach the ground surface nearly unattenuated, and infrared radiation emitted by the ground would stream out to space without being absorbed by the atmosphere. As a consequence, the sunlit side of the earth would be very much warmer than it is now, and the dark side would be much colder. However, minor atmospheric gaseous constituents such as water vapor, carbon dioxide, and ozone absorb strongly in the infrared, and the temperature of the troposphere is distinctly warmer than it would be without them. These compounds absorb not only sunlight but also the terrestrial radiation emitted by the earth's surface toward space. As the ground-emitted longwave radiation is absorbed by the atmosphere, it is being reradiated in all directions. Approximately half of this radiation returns to the ground, and half streams out to space. This phenomenon makes the atmosphere an effective thermal "blanket" and has been termed a "greenhouse" effect (a misnomer which will be discussed in detail later).

As solar radiation traverses the upper atmosphere, a series of complex photochemical reactions occur which create new compounds such as nitrous oxide and ozone from the simpler nitrogen and oxygen molecules. Although oxygen and nitrogen absorb radiation in the far-ultraviolet, ozone absorbs very strongly throughout the intermediate ultraviolet and forms for life on earth a protective filter against the intense actinic (biologically reactive) rays of direct solar radiation.

Sunlight is also attenuated through scattering caused by atmospheric gases, aerosols, and dust. Molecular, or Rayleigh, scattering occurs when the scattering particles, such as air molecules, are small compared with the wavelength of radiation. This accounts for the blue color of the sky. Large-particle, or Mie, scattering is caused by dust, aerosols, and water droplets of size comparable with the wavelength of the radiation.

The gases of the troposphere and stratosphere are pretty thoroughly mixed by turbulence, but above about 100 km, the process of diffusion begins to dominate. At these heights, the lighter gases tend to concentrate in the upper layers and the heavier gases in the lower regions. Helium and hydrogen are the principle constituents of the very high atmosphere. In the troposphere, particularly near the ground, water vapor is extremely nonuniform because of its special properties, such as the ability to vaporize, condense, freeze, or sublime at ordinary temperatures.

Ultraviolet Atmospheric Absorptions

The ultraviolet region of the spectrum comprises wavelengths less than 400 nm, for which the human eye has no normal visual response. The atomic, molecular, and ionic composition of the upper atmosphere depends rather critically on the character of solar emissions and their interactions with the atmospheric gases. Although occurring high in the atmosphere, these interactions are important to life on the earth's surface. Some of the important molecular absorptions occurring relatively high in the earth's atmosphere are shown in Fig. 8.2. This figure shows the extent to which various atmospheric gases reduce the intensity of incoming sunlight to $1/e$ of its initial intensity.

Nitrogen.   Molecular nitrogen, which makes up 78%, by volume, of the earth's atmosphere, is generally transparent to radiation of more than 100 nm. Some very weak molecular nitrogen absorption occurs between 145 and 100 nm, but below 80 nm, a number of complex and strong absorption bands appear and photoionization of nitrogen occurs. The solar emission lines of the hydrogen Lyman series, including Lyman α, β, etc., begin at 121.6 nm, but Lyman γ, at 97.3 nm, appears to coincide with the peak of the strongest molecular nitrogen absorption band. Molecular nitrogen very rarely dissociates, and relatively little atomic nitrogen exists even at altitudes of several hundred kilometers. The nitrogen molecule produces absorption bands of moderate to weak intensities, with band heads located at wavelengths of 380.5, 357.7, 337.1, 315.9, and 297.7 nm. Nitric oxide ionizes at about 134 nm (9.25 eV), the lowest ionization po-

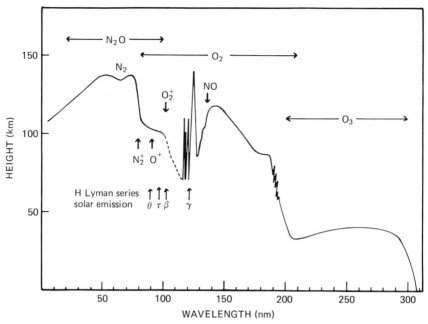

Figure 8.2. Ultraviolet absorption spectrum of the earth's atmosphere. Ordinate is the height in the atmosphere at which solar radiation is reduced to $1/e$ of its initial value.

tential of any of diatomic molecules of the atmosphere. The absorption spectrum of N_2O begins at about 100 nm and plays an important role in the absorption of high-energy ultraviolet radiation.

Oxygen.   Oxygen is uniformly mixed in the lower atmosphere at a concentration of about 21% by volume, but at an altitude of about 90 km, dissociation of molecular oxygen by ultraviolet radiation begins. Concentration of atomic oxygen increases with increasing altitude until, at about 200 km, the amount of atomic oxygen is greater than that of O_2. Molecular oxygen produces the very intense Schumann-Runge bands, found below 210 nm; this broad, intense region of absorption is seen in Fig. 8.2. These bands merge with the intense Schumann-Runge dissociation continuum extending to wavelengths shorter than 175 nm. The absorption spectrum of molecular oxygen is very complex since strong molecular-absorption bands tend to be mixed with ionization continuum from about 105 nm down to less than 20 nm. Very strong atmospheric absorption by O_2 occurs between 240 and 220 nm (the Herzberg bands), caused by the "forbidden" transitions of the electron states of the molecule. Thus ozone and oxygen, to a considerable degree, share the burden of absorbing ultraviolet radiation of wavelengths shorter than 290 nm before it reaches the lower atmosphere. It is this shield against the intense actinic ultraviolet radiation from the sun that is of great concern to scientists, who realize that the exhaust gases from supersonic jet aircraft and the fluorocarbons

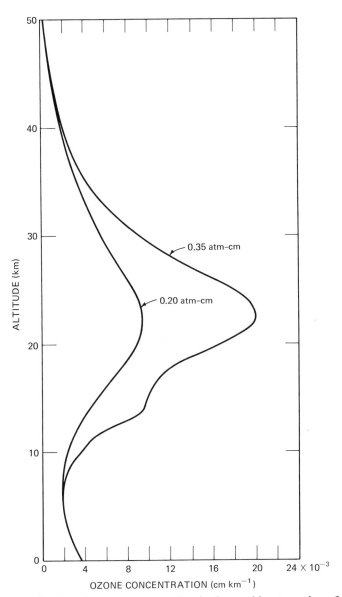

Figure 8.3. Distribution of ozone concentration in the earth's atmosphere for the total amounts 0.20 and 0.35 atm-cm.

or freon released from aerosol spray cans may be destroying stratospheric ozone.

Ozone.   Ozone is a minor constituent of the earth's atmosphere but plays a major role as an atmospheric absorber of radiation. The radiative heat balance and temperature structure of the stratosphere is controlled to a large extent by the distribution of ozone. The height profile of ozone distribution in the earth's atmosphere is shown in Fig. 8.3. Although ozone has its maximum concentration at a height of about 22 km, the warmest region of the stratosphere is near 50 km on the average. The reason maximum heating occurs above maximum ozone concentration is that ultraviolet absorption of sunlight by ozone, even when only present in small amounts, is extremely strong, and the amount of sunlight of these particular wavelengths left over to be absorbed lower in the atmosphere is not very great. Ultraviolet absorption by ozone is in the form of a broad system of bands from 180 to 340 nm, known as the Hartley bands, and the overlapping Higgins bands, from 320 to 360 nm.

In general, ozone absorbs between 1.5 and 3% of the solar radiation incident on the atmosphere, depending on solar elevation and total ozone content along the slant path. The total integrated amount of ozone in the earth's atmosphere may vary between 0.24 and 0.46 atm-cm. (Values for total ozone content are usually given in units of atmosphere-centimeters, where 1 atm-cm would be the length of a column of ozone in a vertical atmospheric column brought to normal surface pressure and temperature.) Total ozone amounts in equatorial regions average about 0.24 atm-cm and vary only slightly throughout the year. The amount of ozone increases toward the higher latitudes and reaches values of 0.40 atm-cm or more at very high latitudes. Outside of the tropical zone, variation with season is approximately sinusoidal, with a distinct maximum in early spring and a distinct minimum in the autumn; the amplitude of variation increases with latitude.

Ultraviolet Solar Spectrum at the Ground

The relative energy distribution of the ultraviolet solar spectrum at the earth's surface is shown in Fig. 8.4, as deduced from observations by Stair (1951) and reconstituted by theory and calculation for the air masses given. The shape of these curves is the result of the irregular character of the solar emission as modified by Fraunhofer absorptions in the solar atmosphere and absorption by ozone in the earth's atmosphere. Scattering by molecules and dust in the atmosphere also has a strong effect on the ultraviolet solar spectrum received at the ground.

Bener (1963) has made extensive measurements of the amount of solar ultraviolet radiation received at the surface in the Alps at Davos-Platz, Switzerland, at an altitude of 1590 m. Figure 8.5 shows the intensity of solar ultraviolet global radiation as a function of solar altitude above the horizon for various wavelengths. It is evident that very little radiation of wavelength shorter than 297.5 nm is received at the ground surface. Only at very high mountain elevations is the ultraviolet cutoff lower than about 295 nm. Ajello et al. (1973) reported on the relative intensity of ultraviolet solar flux at ground level in Los Angeles. They have shown that the cutoff wavelength is 316 nm at a solar elevation of 4° and 295.5 nm at 41°. For higher solar elevations, the cutoff does fall below about 295 nm. At

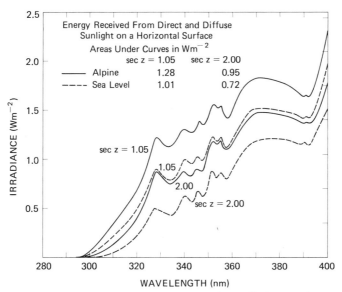

Figure 8.4. Spectral distribution of ultraviolet solar radiation as observed at the earth's surface (−−) at sea level and (—) at about 3650 m elevation each for values of sec z = 1.05 and 2.00.

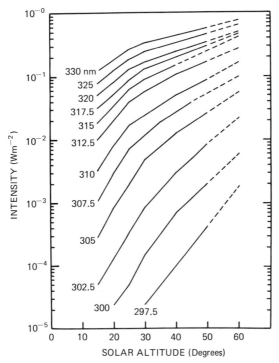

Figure 8.5. Intensity of solar ultraviolet global radiation for specific wavelengths as a function of the solar altitude measured at Davas-Platz, Switzerland, at an elevation of 1590 m. Amount of atmospheric ozone, 0.25 atm-cm.

Climax, Colo., elevation 3,400 m above sea level, Stair (1952) recorded no solar radiation shorter than 295 nm. The ozone amount for which the curves in Fig. 8.5 were determined was 0.25 atm-cm. This amount of ozone is minimal for many stations on a yearly basis. If the ozone amount increases from 0.25 to 0.30 atm-cm, the solar flux of 297.5 nm will decrease from 10^{-4} to 3×10^{-5} W m^{-2} (μm)$^{-1}$ when the solar altitude is 40°. Bener (1963) gives many curves showing the changes of ultraviolet global and sky radiation as a function of wavelength, solar altitude, and ozone amount.

Ultraviolet solar radiation data are given in Urbach (1969). In this book, R. Schulze and K. Grafe give maps of the worldwide distribution of ultraviolet global radiation for cloudless days and for all days on the average as computed from Bener's measurements referred to above. These authors suggest that difference in attenuation of ultraviolet radiation by atmospheric aerosols between urban and rural areas amounts to only 10 to 20%. Later in the same book, J. S. Nader shows that the proportion of intensity of broad-band ultraviolet (300 to 390 nm) radiation incident on Mount Wilson to that incident on downtown Los Angeles was approximately 65% for a moderate-to-heavy smog day, as compared to 87% for a clear day. Cloudiness influences the amount of ultraviolet solar radiation by a factor inversely proportional to the amount of cloud cover. With a cloud cover of 0.3, ultraviolet transmittance is about 83% of that on a clear day, with cloud cover of 0.6, about 66%; and with cloud cover of 1.0, about 43%.

Visible-Region Atmospheric Absorptions

Molecular absorption of sunlight by atmospheric gases throughout the visible region of the spectrum is relatively weak. The visible region is from 400 to 700 nm. Ozone produces a very weak continuum of absorption from 440 to 740 nm, known as the Chappuis bands. Nitrogen molecules have absorption bands of moderate to weak intensities with band heads at 678.7 and 405.9 nm. Except for these absorptions, most of the attenuation of sunlight by the earth's atmosphere is produced by molecular and large-particle scattering.

Infrared Atmospheric Absorptions

The solar spectrum at the earth's surface exhibits a large number of very significant infrared absorption bands caused by polyatomic molecules of gases that are minor constituents in the atmosphere. These absorptions are particularly important for the temperature structure of the atmosphere and especially of the earth's surface.

The absorption spectrum of radiation passing through the vertical atmosphere is shown in Fig. 8.6 (bottom panel). Also shown are absorp-

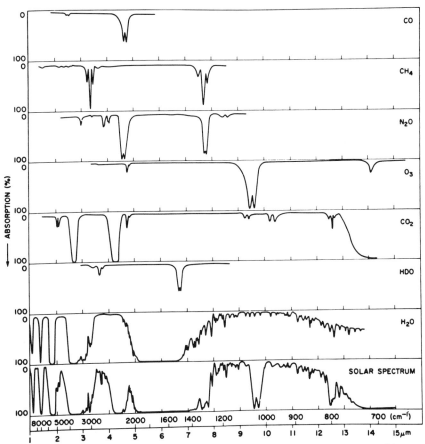

Figure 8.6. (Bottom panel) Atmospheric infrared absorption spectrum and (upper panels) absorption bands of the primary molecular compounds in the atmosphere.

tion spectra for various atmospheric gases as determined in laboratories. These absorption bands consist of many closely spaced absorption lines, which represent rotational transitions in the molecule. Because these regions of absorption have low resolution, individual rotational lines are not observed, and the absorption region appears as a broad band. The principal atmospheric absorption bands in the infrared region are accounted for by water vapor, carbon dioxide, and ozone; minor constituents such as CO, CH_4, N_2O, and HDO contribute some absorption. Since ozone is primarily a constituent of the stratosphere, its absorption will not appear in horizontal paths near the ground but does, of course, appear in the solar spectrum. Molecular oxygen has a strong, narrow absorption band at 759.6 nm and another at 1270 nm.

Water Vapor.   The water molecule is nonlinear, with an angle of 105° between the OH bonds, and has a permanent dipole moment. The molecule has three fundamental bands, ν_1, ν_2, and ν_3, at 2.74, 6.3, and 4.25 μm, respectively, and overtone and combination bands $2\nu_2$ at 3.18 μm, $\nu_2 + \nu_3$ at 1.88 μm, $2\nu_2 + \nu_3$ at 1.46 μm, $\nu_1 + \nu_3$ at 1.38 μm, $\nu_1 + \nu_2 + \nu_3$ at 1.14 μm, and $2\nu_1 + \nu_3$ at 0.94 μm. Lesser bands occur at 0.72 and 0.81 μm. Every one of these bands is significant in absorbing solar radiation except the 6.3-μm band, which is beyond the main solar-energy wavelength distribution. The amount of extraterrestrial solar energy at wavelengths less than 0.71 μm is 50%; at wavelengths less than 1.6 μm, 90%; and at wavelengths less than 3.0 μm, 98%. The water molecule possesses a pure rotational spectrum which consists of many hundreds of single absorption lines beginning at about 12 μm and continuing with increasing intensity toward the band center at 50 μm. The result is that the solar spectrum exhibits some atmospheric absorption in the vicinity of 12 μm, with absorption increasing toward 20 μm and complete absorption occurring beyond 22 μm. Near 14 μm, CO_2 has a very strong fundamental band. The atmosphere is opaque to radiation in this region because of CO_2 absorption superimposed on water-vapor absorption. These absorption bands play a very significant role in limiting the amount of terrestrial radiation that escapes to space.

Water vapor is highly variable in the earth's atmosphere. For extremely cold and dry conditions, the total amount of precipitable water vapor is less than 1 mm, but when conditions are warm and moist, there may be 3 cm or more of moisture in a vertical column of air. By precipitable water is meant the total height of liquid water which would be formed if the moisture in a vertical atmospheric column were condensed upon the surface. Normally, the amount of precipitable water vapor in atmosphere is about 1 cm. Gates (1956) reported a series of spectroscopic measurements of the amount of precipitable water vapor above Denver and Mount Evans, Colo.

Carbon Dioxide. Carbon dioxide, along with water vapor, is an extremely important minor constituent of the atmosphere. The normal concentration of atmospheric CO_2 is 0.033% by volume, or 330 ppm. In contrast to ozone and water vapor, which are distributed nonuniformly in the atmosphere, carbon dioxide is uniformly mixed with nitrogen, oxygen, argon, helium, neon, and other gases. There is considerable evidence, however, that the atmospheric concentration of carbon dioxide has been steadily increasing during much of the past 100 years. Very careful determinations of the carbon dioxide concentration in the air above the trade-wind inversions were begun on Mauna Loa during the International Geophysical Year in 1957 and have continued ever since (see Keeling et al., 1976). These observations, which are summarized in Fig. 8.7, indicate that there is an average annual increase in carbon dioxide concentration of nearly 1.00 ppm per year. At the present rate of increase, the CO_2 concentration of the atmosphere will increase by nearly 10% over the next 30 years. The annual oscillations seen in Fig. 8.7 are the seasonal changes in concentration resulting from assimilation of CO_2 by primary productivity in the Northern Hemisphere.

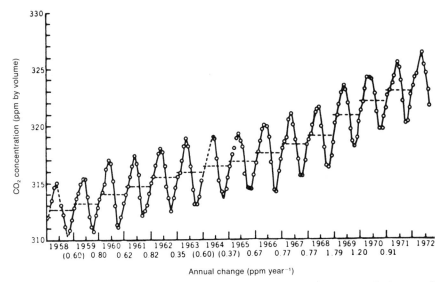

Figure 8.7. Atmospheric concentration of carbon dioxide and annual change of concentration from year to year as measured at Mauna Loa, Hawaii.

Carbon dioxide is a symmetric, linear molecule, with the carbon atom at the center. Carbon dioxide has major vibration–rotation absorption bands at wavelengths of 2.69, 2.77, 4.25, and 15.0 μm and lesser bands at 1.4, 1.6, 2.0, 4.8, 5.2, 5.4, and 10.4 μm.

Atmospheric windows. Transparent regions in the atmosphere are known as atmospheric "windows." These "windows" are evident in the atmospheric transmission spectra shown in Figs. 8.6 and 8.8. It is through these windows that solar radiation streams to the earth's surface at wavelengths shorter than 5 μm and blackbody earth radiation streams out to space at wavelengths greater than 5 μm. In addition to the great transparent span from 300 to about 720 nm and some narrow windows at 760, 820, and 900 nm, there are atmospheric windows in the wavelength regions 1.0–1.1 μm, 1.2–1.3 μm, 1.47–1.8 μm, 2.0–2.5 μm, 3.4–4.0 μm, 7.8–9.3 μm, and 10.0–12.3 μm, and a semitransparent window in the region 17–20 μm.

Thermal Emission Spectra

Emission of Radiation by the Ground Surface

Blackbody terrestrial radiation has an emission peak near 10 μm for ground-surface temperatures of about 300°K and is characterized by a broad spectral distribution which begins at about 5 μm and continues to 20 μm and beyond. For this reason, location of atmospheric windows and absorption bands is particularly significant.

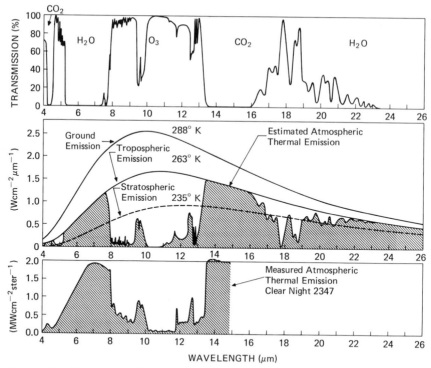

Figure 8.8. (Top) Atmospheric infrared transmission spectrum as a function of the wavelength. (Middle) Calculated emission spectra of ground surface and atmosphere. (Bottom) Downward spectral flux of atmospheric radiation as measured at the ground for a clear night at Columbus, Ohio.

The wavelength distribution of blackbody ground-surface emission is shown in Fig. 8.8 (middle panel) to the same scale as the atmospheric absorption spectrum above it. It is easy to see the importance of the various atmospheric absorption bands on the stream of blackbody ground radiation flowing to outer space. About 70% of the thermal radiation from the ground is intercepted by the atmosphere, where it is reradiated both outward toward space and downward toward the earth's surface. If the atmosphere were totally transparent at infrared wavelengths, ground-emitted thermal radiation would stream uninhibited to the cosmic cold of space. This would result in the earth's surface being very much colder at night than it is at present, and winters would also be much colder. A good example of the extremely low temperatures resulting from radiational cooling through a highly transparent atmosphere is found in Antarctica. The Antarctic continent is a large land mass surrounded by an ocean. The center of the continent is a plateau of ice at 3,050 m above sea level. Because of the cold, there is very little moisture in the atmosphere. In fact, the Antarctic is one of the driest regions of the world. There is much blowing snow, but very little new precipitation. Very dry skies over the interior of the Antarctic plateau and low vapor pressures facilitate strong radiative cooling of the ground surface. The result is the lowest temperatures recorded on earth. The coldest standard air temperature ever recorded was $-88.3°C$ ($-127°F$), measured at the Pole of Cold at the Soviet Vostok Station in Antarctica on 24 August 1960. Elsewhere on earth, winter and nighttime temperatures would be very cold if it were not for the blanket of air containing water vapor and carbon dioxide that absorbs and reradiates back to the ground some of the surface-emitted infrared radiation.

Emission of Radiation by the Atmosphere

If we had infrared-sensitive eyes and looked at the clear sky through a prism, we would see a fascinating spectrum of radiation directed toward the surface of the earth at wavelengths from 5 to 50 μm. It is possible to record the spectrum of infrared radiation emitted by the atmosphere by means of infrared detectors and infrared spectrometers. The result of such observations is the spectrum shown in the lower part of Fig. 8.8. The nature of the infrared atmospheric-emission spectrum is predictable from basic radiation principles. Kirchoff's law of radiation states that, for a given wavelength, a good absorber is a good emitter. For an object to emit radiation, however, it must be at some temperature above absolute zero. A blackbody absorbs 100% at all wavelengths and, therefore, emits with 100% efficiency at all wavelengths. The amount of radiation emitted is proportional to the fourth power of the absolute temperature of the object according to the Stefan–Boltzmann radiation law. The wavelength distribution of the emitted radiation will also depend on the temperature of the object according to Planck's law. An opaque object at 300°K emits radiation with a broad spectral distribution which peaks at a wavelength of about 10 μm, whereas the radiation emitted by an object at 3000°K has a spectral peak at about 1.0 μm. An object that is not completely opaque at all wavelengths but selectively opaque at specific wavelengths radiates at these wavelengths only, but in a manner consistent with the Planck distribution law for a specific temperature.

The atmosphere absorbs very selectively in certain broad bands, as shown in Fig. 8.8 (top). The average temperature of the troposphere (the lower atmosphere), where most of the water vapor and carbon dioxide is concentrated, is about 263°K. Hence, if the atmosphere were completely opaque, it would emit a broad spectrum of infrared radiation from about 5 μm to 50 μm or beyond, with a peak intensity at about 10 μm. Since the atmosphere absorbs in specific bands, it emits in these same bands. The predicted emission spectrum from the lower atmosphere is shown in the center plot of Fig. 8.8. In addition to the emission bands generated by water vapor and carbon dioxide, ozone, emits a band centered at 9.6 μm. The main concentration of ozone in the atmosphere, however, is at an altitude of about 22 km above the surface of the earth, in the stratosphere.

The mean temperature here is about 235°K, and intensity of ozone emission is dependent upon this temperature. The predicted and observed atmospheric emission spectrum are compared in the center and lower plots of Fig. 8.8. A close correspondence between the two is seen. There is a slight discrepancy between the two energy scales because of the instrument used for the measurements. Nevertheless, all of the predicted features of the atmospheric emission spectrum appear in the observed curve, with approximately the same relative strengths.

Water Vapor Influence on Climate

If the atmosphere is very dry and clear, there is a weakening and narrowing of some of the water-vapor emission bands, and less radiant flux is emitted by the atmosphere toward the ground. If the air is very warm and humid, as in tropical regions, the emission bands are broadened and strengthened. (The primary variation occurs in the edges of the bands, where they are semitransparent.) Clear, dry nights tend to be cold nights, partly as the result of radiative cooling of the earth's surface. Clear, humid nights are generally warm because of the lack of radiative cooling. Temperatures of the lower atmosphere and ground surface are dependent upon net flux of radiation. Air and ground temperatures are determined by the difference between upward flux of radiation emitted by the ground surface or air near the ground and downward flux of radiation emitted by the atmosphere. If the sky is overcast, no direct radiation from the ground surface escapes to space, and conditions near the ground are warmer than on clear nights.

Carbon Dioxide Influence on Climate

As seen in Fig. 8.7, there has been a steady increase in the amount of CO_2 in the atmosphere. Concern is growing that this increase of carbon dioxide is producing a change in the temperature of the atmosphere. Carbon dioxide has a massive absorption band centered at about 14 μm which exerts considerable influence on the net radiant flux exchanged between the earth's surface and the atmosphere. The influence of CO_2 is reduced somewhat because of overlap of the CO_2 band with the rotational absorption band of water vapor. Kaplan (1960) calculated the change in net radiative flux expected in the atmosphere with change in the CO_2 concentration, assuming the atmosphere to be uniformly mixed, and found that a 10% change in atmospheric CO_2 content would result in a change of not more than 0.25°C in the surface temperature of the air. Earlier, Plass (1956, 1959) had estimated that a reduction of CO_2 content by 50% would produce a temperature drop of 3.8°C and could trigger an ice age.

In about 1880, the mean temperature of the Northern Hemisphere began to exhibit a definite warming trend. This warming trend continued, despite some short downward trends between 1890 and 1920, until, in

1940, mean annual temperature was nearly 0.6°C higher than in 1880. Callendar (1938) was the first to propose that an increase of atmospheric carbon dioxide produced by industrialization and the burning of fossil fuels might account for the temperature increase. Just when it appeared that the CO_2 increase was the real cause of the warming trend, the mean temperature of the Northern Hemisphere began to decrease. The initial decrease occurred during the 1940s, and the temperature has been dropping ever since. The mean temperature of the Northern Hemisphere is now what it was about 1900. Once the cooling trend was detected, various scientists began to speculate that it was the result of increased atmospheric turbidity caused by air pollutants. There is, however, as yet no proof that this is the case.

According to Broecker (1975), there has been a natural temperature cycle in the atmosphere with a 40-year period, and the warming effect of CO_2 is superimposed on this natural cycle. It appears that during the last 35 years, the cooling phase of the natural cycle has been occurring. Now, as the warming part of the cycle coincides with the warming from increasing CO_2, it is likely that the global climate will become very warm. The onset of a warming trend may begin during the 1980s. A doubling of atmospheric CO_2 concentration may increase the mean global temperature from 1° to 3°C. Schneider (1975) has reviewed the various models of temperature change based on changes in CO_2 concentration. If mankind continues to increase the burning of fossil fuels at the present 4.3% growth rate, atmospheric CO_2 concentration may increase from 3 to 5 times the present concentration by the year 2075. At a 2% rate of growth, which is very minimal, the CO_2 concentration will increase nearly two-fold by the year 2075 according to Baes et al. (1977). A maximum rate of growth would result in a 10°C increase of temperature, coincident with nearly a 5-fold increase in CO_2 concentration, by the year 2075. It is unlikely that this last scenario will occur, but it seems very probable that a more than twofold increase in CO_2 concentration will occur.

Only about half of the CO_2 released by the burning of fossil fuels ends up in the earth's atmosphere. The other half appears to be going into the oceans, where it is tied up in carbonates. If the cutting of forests throughout the world is resulting in increased levels of carbon dioxide in the atmosphere, then sinks other than the oceans need to be found for some of this carbon.

Atmospheric Scattering

Scattering of solar radiation by atmospheric molecules and dust affects the radiation climate at the earth's surface. Without scattering, there would be no skylight or diffuse radiation, and only direct sunlight would illuminate the surface of the earth. Some of the scattered skylight returns to space, but much is scattered toward the ground. In addition, both direct sunlight and scattered skylight are reflected from the ground surface into the atmosphere, and there is repeated reflection and scattering of radiation between ground and sky. The brightness of a scene, the illumination of hillsides, canyons, and walls that are in the shadow of direct solar radiation, and, in general, the optical character of the world in which we live, is determined in large part by multiply scattered and reflected sunlight. The scattering of light by molecules and suspended particles of the atmosphere is a strongly wavelength-dependent phenomenon.

There are several ways to present attenuation of radiation by a medium of thickness x. Monochromatic transmittance $_\lambda\tau$, monochromatic, "decimal" absorption coefficient $_\lambda a$, and monochromatic, "Napierian" absorption coefficient $_\lambda k$ are related in the following manner:

$$_\lambda\tau = {_\lambda I}/{_\lambda I_0} = 10^{-_\lambda a x} = e^{-_\lambda k x} \qquad (8.9)$$

$$_\lambda k = -(\ell n \,_\lambda\tau)/x = {_\lambda a} \,\ell n \,10 = 2.3026 \,_\lambda a. \qquad (8.10)$$

The product $_\lambda k x$ is called the monochromatic extinction optical thickness. A medium such as the atmosphere attenuates a beam of radiation by scattering and selective absorption. One can express a scattering coefficient and an absorption coefficient or a transmittance for each of these mechanisms, independent of the others. Total attenuation is given by the product of the individual transmittances or the sum of individual extinction optical thicknesses.

A monochromatic beam of light of intensity $_\lambda I$ passing through a scattering media of depth dx is reduced by an amount $d_\lambda I$ according to the relationship

$$\frac{d_\lambda I}{_\lambda I} = -_\lambda k_s \,dx, \qquad (8.11)$$

where $_\lambda k_s$ is defined as the scattering coefficient per unit path length. The scattering coefficient measures fractional depletion of a beam of unit area by a unit volume of the scattering medium. There is small-particle scattering, known as Rayleigh scattering, and large-particle scattering, known as Mie scattering. The total scattering coefficient is $_\lambda k_s = {_\lambda k_r} + {_\lambda k_a}$, where $_\lambda k_r$ and $_\lambda k_a$ are the Rayleigh and Mie scattering coefficients, respectively. Small particles such as air molecules, which have a diameter comparable to the wavelength, cause the scattering known as Rayleigh, or molecular, scattering. Larger particles such as dust, whose size is of the order of the wavelength or greater, cause Mie, or large-particle, scattering.

Molecular, or Rayleigh, Scattering

In seeking an explanation for the blue appearance of the sky, Lord Rayleigh showed that air molecules were the cause of light scattering in the atmosphere. Rayleigh derived the following relationship for the scattering coefficient:

$$\lambda k_r = \frac{32 \ \pi^3 (n_\lambda - 1)^2}{\lambda^4 \ N_r} \frac{\rho}{\rho_0}, \qquad (8.12)$$

where N_r is the number of molecules per cubic meter under standard conditions of pressure (1013.3 mbar) and temperature (273°K), n_λ is the index of refraction at wavelength λ for air under these standard conditions; and ρ_0 and ρ are the standard and actual air densities in kilograms per cubic meter.

The striking feature of this relationship is that the scattering coefficient is inversely related to the fourth power of the wavelength, somewhat modified by the fact that n_λ is also a function of the wavelength. This implies that a wavelength of 350 nm is scattered 16 times more than a wavelength of 700 nm. This is the reason that skylight is rich in ultraviolet and blue wavelengths and depleted of red and infrared and thus that the sky appears blue to the eye. It is also the reason that the sun appears red at sunset; because of the long slant rays, the blue wavelengths of the solar spectrum are scattered into the sky at sunset, with only the red rays coming directly to the observer. When larger particles such as dust are present in the atmosphere, Mie scattering occurs, the longer wavelengths are scattered along with the shorter wavelengths, and the sky is whitish in appearance.

Rayleigh found that the directional intensity distribution in the scattered radiation took the form

$$\lambda I(\phi) \propto \lambda I_0 (1 + \cos^2 \phi)(n_\lambda - 1)^2 / \lambda^4, \qquad (8.13)$$

where ϕ is the angle the scattered light makes with the direction of the incident beam. Maximum scattering by small particles or small-scale density fluctuations occurs symmetrically in the forward and backward directions. Rayleigh's theory applies in general when the radius r of the particles is less than 0.1 of the wavelength. When r is very large, 25 times the wavelength or greater, geometrical optics will apply, but for the intermediate situation in which r is greater than 0.1 and less than 25 times the wavelength, it is necessary to use the more complicated scattering theory derived by Mie.

Large-Particle, or Mie, Scattering

Mie scattering in the atmosphere is produced by dust, aerosols, and water droplets. Mie scattering differs from Rayleigh scattering in several respects. The directional dependence of large-particle scattering is very strong in the forward direction and relatively weak toward the sides. In addition, polarization effects are very significant, and both the index of refraction and the absorption coefficient of the scattering particles are important. Scattering phenomena are considerably more complex in the atmosphere than in a medium containing a homogeneous distribution of particles of uniform size and properties. The atmosphere always contains a distribution of particles, whose sizes and indexes of refraction may differ considerably. A great deal of research has been done on atmospheric scattering, and various empirical relationships have been defined describing the amount of scattered light resulting from several distributions of particle size and density with height in the atmosphere. Some of these are given by Kondratjev (1969), Penndorf (1954), and Robinson (1966).

Although it is highly desirable to derive mathematical relationships from fundamental principles, it is often expedient to use an empirical formulation derived from observational data. Furthermore, the theory of Mie scattering is extremely complex. Ångstrom (1951) has shown that the attenuation coefficient for dust in the atmosphere may be written

$$_\lambda k_a = \beta \, \lambda^{-\gamma}, \tag{8.14}$$

where β is proportional to the particle density and γ is a parameter which decreases as the ratio of particle density to wavelength increases. With larger-sized particles, the value of γ diminishes. For average atmospheric conditions, the value of γ is 1.3; in dust storms, it is 0.5. For fog droplets, $\gamma = 0$, a condition, referred to as diffuse reflection, in which scattering is independent of wavelength. This is in striking contrast to the phenomenon of Rayleigh scattering, where attenuation and amount of scattered light vary inversely with the fourth power of the wavelength. A comparison of Rayleigh and Mie scattering is shown in Fig. 8.10 for a horizontal path near the earth's surface.

Atmospheric Attenuation

When considering the infrared transmission of solar radiation by the earth's atmosphere, one must contend with the nonuniform distribution of water vapor, dust, and ozone; the uniform concentrations of carbon dioxide, oxygen, and nitrogen; and the atmospheric pressure, which decreases exponentially with altitude. The total monochromatic transmittance $_\lambda\tau$ of the atmosphere is the product of the monochromatic transmittances due to Rayleigh scattering $_\lambda\tau_r$, Mie scattering $_\lambda\tau_a$, selective ozone absorptions $_\lambda\tau_o$, and selective absorption by water vapor, carbon dioxide, and other minor constituents $_\lambda\tau_m$. Hence

$$_\lambda\tau = {_\lambda\tau_r} \, {_\lambda\tau_a} \, {_\lambda\tau_o} \, {_\lambda\tau_m}. \tag{8.15}$$

It is also possible to add the corresponding "Napierian" absorption coefficients, each multiplied by its equivalent path length integrated over the depth of the atmosphere from the surface to space (monochromatic extinction optical thicknesses) as follows:

$$_\lambda kx = \int_0^\infty {_\lambda k_r} \, dh + \int_0^\infty {_\lambda k_a} \, dh + \int_0^\infty {_\lambda k_o} \, dh + \int_0^\infty {_\lambda k_m} \, dh. \tag{8.16}$$

Figure 8.9. Characteristic distribution of atmospheric dust with height. The two lines represent the minimum and maximum ground-level concentrations, 200 and 800 particles cm^{-3}, respectively.

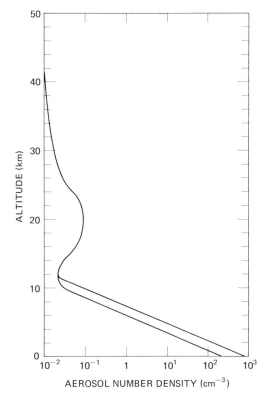

Absorption coefficients are additive, whereas transmittances are multiplicative.

Attenuation of solar radiation by the atmosphere has been described by Gates (1966b). The symbols used in that paper are different from the ones used here. In the Gates paper, the symbol τ is used for the extinction optical thickness, whereas here τ represents atmospheric transmittance. For calculation purposes, change in molecular density N_r with altitude is taken from the "U.S. Standard Atmosphere," published by the Government Printing Office, Washington, D.C., in 1962. For aerosol attenuation number densities, the data of Elterman (1964), based his distributions on observations by Chagnon and Junge (1961), Curcio et al. (1961), and Penndorf (1954) were used. Two concentrations were used for the troposphere: one for a surface concentration of 200 particles cm^{-3} and one for 800 particles cm^{-3}.

Elterman (1964) has determined the aerosol attenuation coefficients for the ultraviolet and visible regions on the basis of the observations of Baum and Dunkelman (1955). A characteristic height distribution of dust in the atmosphere is shown in Fig. 8.9. Two lines are shown in the lower portion of the figure; one is for a surface concentration of 200 particles cm^{-3}, the other for 800 particles cm^{-3}. These are the minimum and maximum dust concentrations normally encountered at the earth's surface. Fig-

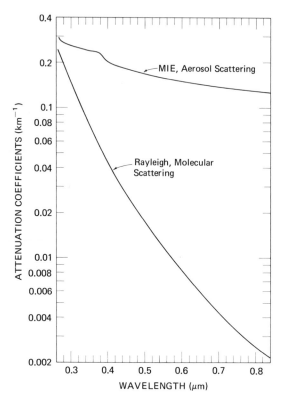

Figure 8.10. Attenuation coefficient of radiation along a horizontal path at the earth's surface by Mie and Rayleigh scattering as a function of wavelength for an aerosol concentration of 200 particles cm^{-3}.

ure 8.10 shows the attenuation per kilometer by molecular and aerosol scattering corresponding to an aerosol concentration of 200 particles cm^{-3} for a horizontal path at the earth's surface. Aerosol scattering predominates over molecular scattering near the ground throughout the visible and infrared regions. At altitudes 2–4 km or more above the surface, molecular scattering dominates at most wavelengths because of the rapid decrease of aerosol concentration with altitude. (The altitudinal distribution of ozone is shown in Fig. 8.3.) Air and aerosol or dust densities, and attenuation of solar radiation by these constituents, decreases with increasing altitude, whereas attenuation produced by ozone dominates shortwave extinction at an altitude of about 22 km and above. In the visible portion of the spectrum, at wavelengths longer than the regions of ozone absorption in the ultraviolet and blue, the atmosphere becomes highly transparent. Only in the far-red and near-infrared region does water-vapor absorption affect atmospheric transmission.

In the near-infrared region, water vapor and carbon dioxide cause absorption bands consisting of hundreds of narrow individual absorption lines. Each line has a narrow peak absorption with broad wings of lesser absorption to either side. The wings of these numerous lines contribute a definite amount to the continuum extinction by the atmosphere. The extinction coefficients for these infrared absorption bands and the windows between the bands have been derived by Gates and Harrop (1963) and Gates (1960) from direct observation of the solar spectrum.

Spectral Distribution of Direct Solar Radiation

The spectral quality of sunlight is of enormous importance biologically because of the enormous number of photochemical reactions occurring within organisms and the thermal effects of radiation upon organisms.

Plot of Intensity versus Wavelength or Wavenumber

The spectral distribution of solar radiation may be presented in two forms: monochromatic intensity versus wavelength or monochromatic intensity versus frequency (or wavenumber). The wavelength plot is the more familiar form, but it has an inherent problem in that the infrared portion of the solar spectrum extends indefinitely toward longer wavelengths and cannot be contained within a finite graph. In the frequency, or wavenumber, plot, the infrared wavelengths rapidly converge toward zero frequency, and the ultraviolet, or high-frequency, end of the scale corresponds to complete absorption of solar radiation by the atmosphere. There is another reason that frequency or wavenumber plots are used by spectroscopists. As shown earlier, all energy transitions in atoms or molecules are proportional to the frequency of radiation absorbed or emitted; the result is that spectral bands or lines are generally symmetrical with frequency.

The solar spectrum is shown in Fig. 8.11 as a function of the wavelength and in Fig. 8.12 as a function of wavenumber. The solar curve peaks at 0.5 μm when plotted as a function of wavelength and at 1.0 μm when plotted as a function of wavenumber. The peak in the spectral distribution of solar radiation is thus a function of the scale against which it is plotted. The reason for this shift is described below. It is important to understand the procedure for converting from a wavelength scale to a wavenumber scale, or vice versa. The unit of wavenumber is the reciprocal of the wavelength and is expressed in centimeters^{-1}. Wavenumber is proportional to frequency, the two being related by the velocity of light. Wavelength λ, frequency ν, and velocity of light c have the relationship $\lambda\nu = c$, and $\lambda = 1/\tilde{\nu}$, where $\tilde{\nu}$ is the wavenumber. Hence $\tilde{\nu} = \nu/c$.

The quantity $_\lambda E$ represents the flux density E per unit increment of wavelength, and $_{\tilde{\nu}} E$ represents the flux density per unit wavenumber in-

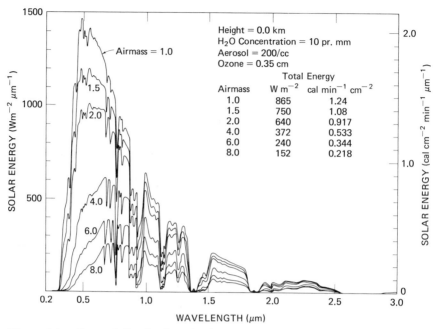

Figure 8.11. Spectral distribution of solar radiation received on a surface perpendicular to the sun's rays at the ground for various air masses as a function of wavelength.

crement. The wavelength increment $\Delta\lambda$ is related to the wavenumber increment $\Delta\tilde{\nu}$ as follows:

$$\Delta\lambda = \frac{\Delta\tilde{\nu}}{\tilde{\nu}^2} \quad \text{or} \quad \Delta\tilde{\nu} = \frac{\Delta\lambda}{\lambda^2}. \tag{8.17}$$

Then, since

$$_\lambda E = \frac{E}{\Delta\lambda} \quad \text{and} \quad _{\tilde{\nu}} E = \frac{E}{\Delta\tilde{\nu}}, \tag{8.18}$$

it follows that

$$_\lambda E = \frac{_{\tilde{\nu}} E}{\lambda^2} = {_{\tilde{\nu}} E}\tilde{\nu}^2 \tag{8.19a}$$

and

$$_{\tilde{\nu}} E = \frac{_\lambda E}{\tilde{\nu}^2} = {_\lambda E}\lambda^2. \tag{8.19b}$$

If $_\lambda E$ is plotted against λ, then to convert to $_{\tilde{\nu}} E$, one must divide $_\lambda E$ by the quantity $\tilde{\nu}^2$ or multiply $_\lambda E$ by λ^2. In either case, the conversion from one scale to the other is nonlinear and produces a change in the shape of the

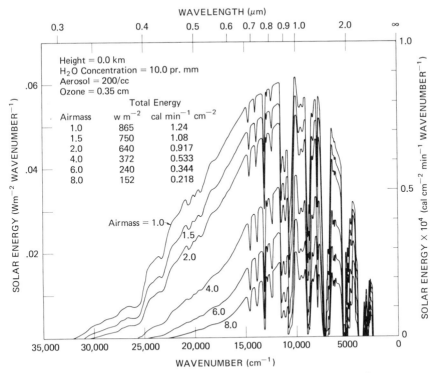

Figure 8.12. Spectral distribution of solar radiation received on a surface perpendicular to the sun's rays at the ground for various air masses as a function of wavenumber.

plot, with the maximum appearing in a different position. The amount of energy contained within a given wavelength interval on the wavelength plot is precisely the same as the amount of energy between the equivalent wavenumber positions on the wavenumber plot.

Spectral Distribution with Air Mass

The spectral distribution of direct solar radiation incident at sea level on a surface perpendicular to the sun's rays is given in Figs. 8.11 and 8.12 as a function of the wavelength or wavenumber for air masses from 1.0 to 8.0. These curves were determined by means of procedures outlined by Gates (1966b). Total amount of precipitable water vapor in the zenith direction is 10.0 mm, the aerosol concentration at the surface is 200 particles cm⁻³, and the total amount of ozone is 0.35 cm. Total amount of direct solar radiation received is given by the area under each curve and is also listed in the columns marked "total energy." These curves give a good idea of the change in spectral composition of direct solar radiation with changing air mass. A strong shift toward the red part of the spectrum and reduction of intensity occurs with increasing air mass.

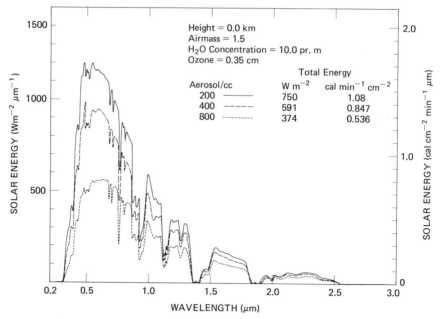

Figure 8.13. Spectral distribution of solar radiation received on a surface perpendicular to the sun's rays at the ground for various aerosol concentrations as a function of wavelength.

Spectral Distribution with Aerosols and Dust

The variation of the spectral distribution of direct solar radiation with aerosol and dust content is shown in Fig. 8.13 for the sun at a zenith angle of 48°11′. Amount of attenuation due to scattering depends very much on the particle size distribution. If the atmosphere is very dusty and filled with large particles, attenuation will be very much greater than for clear sky conditions. More detailed information concerning the determination of these curves is given in Gates (1966b). That paper also gives a set of curves for the spectral solar energy distribution as a function of amount of water vapor in the solar path.

Spectral Distribution with Altitude

The earth's surface ranges in altitude from below sea level to 8848 m above sea level. It is of interest to discover the extent to which the spectral distribution of direct sunlight changes with altitude. With increasing altitude, total amount of atmosphere diminishes, and total number of molecules of air, water vapor, carbon dioxide, or aerosol decreases. The altitudinal distributions of ozone and aerosol concentration are shown in Figs. 8.3 and 8.9, respectively. Molecular density is for standard atmosphere. Since water-vapor concentration in the troposphere is widely variable, a fixed amount of 5 mm of precipitable water was used in the com-

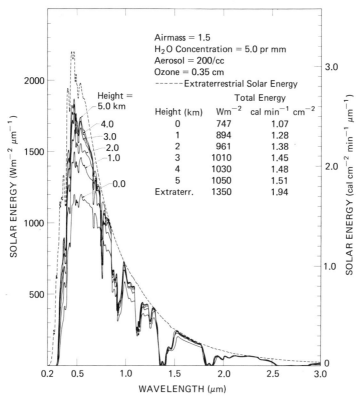

Figure 8.14. Spectral distribution of solar radiation at various altitudes received on a surface perpendicular to the sun's rays from the ground to outer space as a function of wavelength.

putations. If lesser amounts of water vapor exist above any particular altitude, then some correction of the infrared absorptions should be made (Gates, 1966b). Variation of the spectral distribution of direct solar radiation with altitude is shown in Fig. 8.14. As expected, the greatest altitudinal changes in solar radiation reaching any particular level in the atmosphere occur at ultraviolet and visible wavelengths.

Spectral Distribution of Skylight

The intensity of diffuse sky radiation received on a horizontal surface from a perfectly scattering Rayleigh atmosphere is shown in Fig. 8.15. These curves are derived from the calculations and observations reported by Deirmendjian and Sekera (1954). The maximum flux of sky radiation shifts toward longer wavelengths with increasing distance from the solar zenith. It is evident that sky radiation during most daylight hours is relatively rich in ultraviolet and blue wavelengths, with peak values occurring

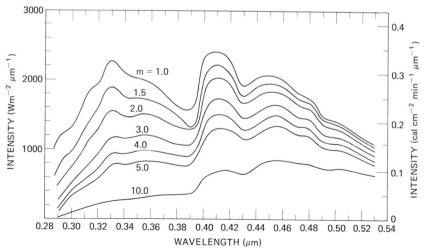

Figure 8.15. Spectral distribution of diffuse sky radiation on a horizontal surface at the ground for various air masses as a function of wavelength.

at 330 and 410 nm. A Rayleigh atmosphere occurs only on the clearest days, and presence of aerosols and dust greatly enhances the scattering in the red. Diermendjian and Sekera (1954) have shown the effect of ground reflection on intensity of sky radiation. A surface reflectivity of 0.25 increases the intensity of scattered skylight by approximately 35% relative to the intensity of skylight above a black surface. Thus the nature of the surface, e.g., vegetation, soil, or sand, strongly influences the downward flux of skylight.

Bennett et al. (1960) have made measurements of the clear-sky radiance for sites at elevations of 305, 1,830, and 4,270 m above sea level. They found that the proportion of forward-scattered sunlight is relatively greater in the region from 600 to 2000 nm than in the visible region. The proportion of near-infrared radiation relative to visible radiation was greater for a hazy atmosphere than for a clear atmosphere. Also, an unexpected secondary maximum of radiance was observed in the near infrared. Increased scattered radiation always occurred during days of high absolute humidity, when much visible haze was usually present.

Spectral Distribution of Global Radiation

The sum of direct-beam solar radiation and scattered skylight received on a horizontal surface is the global radiation. If direct-beam solar radiation incident on a surface perpendicular to the sun's rays is converted to that incident on a horizontal surface and combined with the flux of sky radiation, the curves shown in Fig. 8.16 result. Although direct-beam solar radiation incident on a horizontal surface diminishes rapidly with increasing air mass, flux of scattered skylight adds considerable energy in the

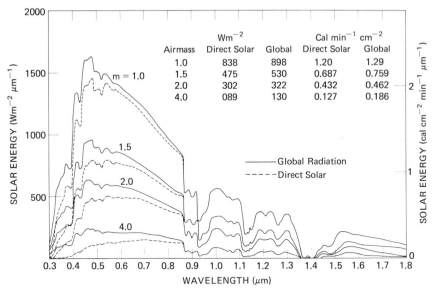

Figure 8.16. Spectral distribution of direct and global solar radiation received on a horizontal surface at the ground for various air masses as a function of wavelength.

ultraviolet and blue. The total energy associated with each curve is given in Fig. 8.16. These values for total energy include small amounts in the longwave infrared not shown in the curves.

Comparative Spectral Distribution of Natural Light

The spectral distributions of sunlight, cloudlight, skylight, and light transmitted through a canopy of vegetation are shown in Fig. 8.17. Direct sunlight is strongly distributed through the visible and infrared. Skylight is rich in ultraviolet and blue wavelengths. Cloudlight is strong in the visible and low in the infrared. Light passing through vegetation has a strong peak in the near infrared. (The spectral properties of leaves will be described later in the chapter.)

Spectral Properties of Liquid Water

Water is an extremely ubiquitous and very important ingredient of the biosphere. The spectral properties of liquid water are highly significant in the interaction of radiation with the earth's surface and life. The degree to which water reflects, absorbs, and emits radiation determines the temperature of bodies of water and, in part, the extent to which radiation pene-

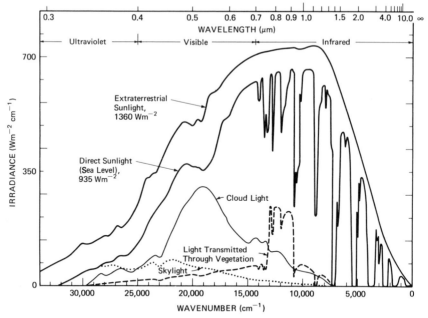

Figure 8.17. Spectral distribution of direct sunlight, cloud light, skylight, and light transmitted through a plant canopy received on a horizontal surface at the ground as a function of wavelength.

trates animal integuments, plant leaves, animal eyes, thermal sensors, and other organic parts.

The spectral reflectance of liquid water is shown in Fig. 8.18. Reflectance of water at any wavelength is a function of the angle of incidence of radiation. Reflectance ρ of a plane water surface to unpolarized light is given by the Fresnel reflection law,

$$\rho = \frac{1}{2}\left[\frac{\sin^2(i-r)}{\sin^2(i+r)} + \frac{\tan^2(i-r)}{\tan^2(i+r)}\right]. \qquad (8.20)$$

The angles of incidence i and refraction r are related through the index of refraction n as follows:

$$n = \sin i/\sin r. \qquad (8.21)$$

The index of refraction of pure water at visible wavelengths is 1.33, and that for sea water of 35% salinity is 1.34, the difference between these two figures being negligible. There is ample observational evidence to indicate that the Fresnel law is a good representation of the reflectance of water surfaces generally. The reflectance of water is low for angles of incidence less than 60°. Reflectance in the visible region is 2.0% at 0° and 10° angle of incidence, 2.1% at 20° and 30°, 2.5% at 40°, 3.4% at 50°, 6.0% at

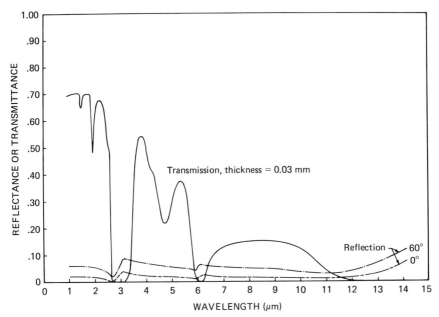

Figure 8.18. Spectral reflectance of liquid water for two angles of incidence (0° and 60°) and spectral transmittance through 0.03 mm of liquid water as a function of wavelength.

60°, and 13.4, 34.8, and 58.4% at 70°, 80°, and 85°, respectively. Most of the solar radiation striking a water surface is absorbed, except when the sun is low in the sky. The human eye is very sensitive at 550 nm, and the solar intensity is very high at this wavelength. The result is that at large angles of incidence to a plane water surface, such as occurs when the sun is near the horizon or for tilted water surfaces such as the slopes of waves, sunlight is seen as a bright glint reflected off the water. This gives us the impression that the reflectance of water is greater than it is in general.

Because of the relative "blackness" of a water surface, sunlight enters the ocean, lake, pond, or stream and penetrates to a depth that depends upon the turbidity of the water. The attenuation coefficients of various natural bodies of water are shown in Fig. 8.19. The reciprocal of the attenuation coefficient gives the distance a ray of light must travel in the water to be reduced to $1/e = 1/2.718$, or 36.8% of its initial intensity. If the attenuation coefficient is 0.5 m^{-1}, intensity of radiation is reduced to 36.8% of its initial value at a depth of 2.0 m. If the attenuation coefficient is 0.1 m^{-1}, the equivalent depth is 10 m. It is evident from Fig. 8.19 that the light intensity in most bodies of water at depths greater than 10 m is low. According to Tyler and Preisendorfer (1962), some lakes are very clear and have attenuation coefficients of less than 0.05 m^{-1} for the blue and green. With this amount of attenuation, 36.8% of the incident light penetrates to a depth of 20 m and 13.5% to 40 m. It is interesting to note

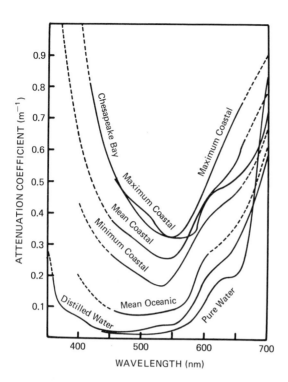

Figure 8.19. Attenuation coefficient for marine waters as a function of wavelength.

that the attenuation of radiation by water increases very rapidly in the ultraviolet and in the red (see Fig. 8.19).

A very thin layer of liquid water (0.03 mm) transmits infrared radiation as shown in Fig. 8.18. As a general rule, however, most bodies of water are totally opaque to infrared radiation. The strong infrared absorption bands, which are distinct regions of absorption, with "windows" between the bands, for water vapor or very small amounts of liquid water, are merged, forming a strong continuum of absorption throughout the infrared.

The light climate for plants in rivers has been discussed by Westlake (1966). The major characteristic of rivers, with respect to light conditions, is the variability due to frequent changes in turbidity resulting from soil erosion and pollution. Furthermore, rivers are often shaded by shrubs and trees growing along the banks, and this results in the same kind of variability of light as encountered in a forest. All of these factors affect the amount of light in a river or stream and thus growth of aquatic plants, behavior of fish, insects, and other animals, and temperature of the water. Westlake (1966) measured the absorption coefficient a for visible radiation for river water in England by means of selenium photocells with an opal glass filter. These values can be converted to the Napierian absorption coefficient k by use of Eq. (8.10). Typical values of a for the River Test at Bossington are 0.26 to 0.52; the River Frome at East Stoke, 0.46 to

2.22; the River Colne at Denham Studios, 0.41 to 0.79; and the Maple Lodge effluent, 0.64 to 0.95. At the outfall of a modern activated sewage treatment plant, however, a value of 15.0 in very turbid water was measured. The River Test is a very clear chalk stream, the Frome is a large chalk river, and the Colne is fed by chalk streams but receives agricultural drainage, canal waters, and other sources of pollution. From a spectral standpoint, when measurements are made using a selenium photocell with a red, green, or blue filter, greatest attenuation is in the blue region and the least in the red and green, with attenuation in the red region slightly greater than that in the green.

The shapes of the extinction curves for rivers are very similar to those for lakes, except that the values are much more variable and are larger. However, details in the curves and comparison of the extinction coefficients for filtered and unfiltered water allow one to separate the effects of dissolved substances and suspended solids. Most of the attenuation in the Colne and Frome rivers is due to suspended solids. Presence of algae in water causes a substantial increase in the extinction coefficients, and selective absorption by algal pigments shows up at specific wavelength intervals. The blue absorption by chlorophylls in blue-green algae is strong; there are absorption peaks at 640 and 680 nm and a peak at 615 nm caused by phytocyanin.

Spectral Properties of Plants

The interaction of radiation with plant leaves is extremely complex. A decade or so ago, the general features of this interaction had been observed, but very few of the spectral features were understood. The spectral characteristics of leaf reflection, transmission, and absorption were reported by Gates et al. (1965). More recently, as the result of careful, detailed experiments concerning the optical properties of plant leaves, an improved understanding of the mechanisms involved has been achieved (see Gausman and Allen, 1973; Allen et al., 1970; and Woolley, 1971). The spectral characteristics of various types of plant leaves will be discussed first, followed by the detailed theory and explanation of the mechanism involved.

Broad Deciduous Leaves

Typical spectral reflectance, transmittance, and absorptance curves for the leaves of cottonwood (*Populus deltoides*) are shown in Fig. 8.20. These curves are typical of all broad green leaves. Some leaves have different values of transmittance or reflectance than the cottonwood leaf, but the shapes of the curves are always the same. The spectral reflectance,

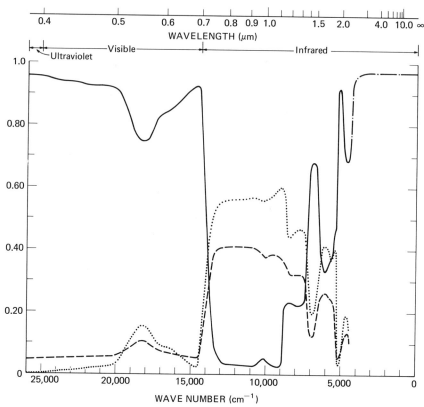

Figure 8.20. Spectral properties of a leaf of cottonwood (*Populus deltoides*) as a function of wavenumber: (—) absorptance; (--) reflectance, upper surface; (· · · · ·) transmittance.

transmittance, and absorptance for a leaf of *Nerium oleander* is shown in Fig. 8.21. Note that reflectance is higher than transmittance in both the green and infrared regions for oleander, whereas for cottonwood, transmittance is higher than reflectance at these wavelengths.

Table 8.1 gives the mean reflectances and absorptances to incident solar radiation of various plant species for a sun low and high in the sky. In all cases, radiation is incident on the upper leaf surface unless otherwise specified. For most leaves, data on which the measurement was made are given, and, where appropriate, there is a statement concerning the size, state of development of the leaf, or type of exposure. Where data is not given, the leaf was mature and in healthy condition. All leaves were measured immediately after removal from the plant.

Absorptance to high and low sun, and overcast. Certain important features concerning the reflectance, absorptance, and transmittance of these leaves are apparent from the values listed in Table 8.1. The mean value of reflectance or absorptance is obtained by multiplying the spectral

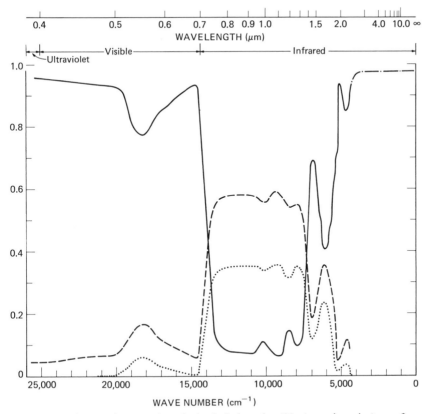

Figure 8.21. Spectral properties of a leaf of oleander (*Nerium oleander*) as a function of wavenumber: (—) absorptance; (––) reflectance, upper surface; (· · · ·) transmittance.

reflectance or absorptance by the spectral intensity of solar radiation and dividing by the total flux of solar radiation. Sunlight from a low sun is relatively richer in red wavelengths and poorer in blue wavelengths than sunlight from a high sun. This accounts for the shift in mean reflectance or absorptance for low and high sun. For most deciduous broad leaves, the mean reflectances for low sun range from 0.26 to 0.32, and mean absorptances from 0.34 to 0.44. For a high sun, most of the mean reflectances range from 0.20 to 0.26, and mean absorptances from 0.48 to 0.56. If one needs to guess the amount of absorption of solar radiation by a leaf, an absorptance of approximately 0.50 is a very good estimate, except very early and very late in the day, when 0.40 would be reasonable. Mean absorptances to high sun generally increases with leaf age by 5 to 9%. For example, the very young *Liriodendron tulipifera* leaf has an absorptance of 0.43 and the mature leaf 0.52; the young *Acer saccharinum* leaf 0.458, and the mature leaf 0.531 to 0.558; and the young *Carya* sp. leaf 0.473 to 0.490, and the mature leaf 0.565. A ''sun'' leaf generally absorbs more

Table 8.1. Mean Solar Reflectances and Absorptances of Plant Leaves for a Sun Low and High in the Sky.[a]

Species	Condition	Date	R_L	R_H	A_L	A_H
Acer saccharinum,	Very small	04/29	0.256	0.206	0.325	0.458
silver maple	Medium	04/29	0.282	0.231	0.374	0.490
	Small	05/07	0.232	0.187	0.396	0.511
	Medium	05/07	0.284	0.225	0.347	0.483
	Medium	05/12	0.286	0.222	0.431	0.558
	Large	05/12	0.305	0.238	0.398	0.531
Aesculus hippocastanum buckeye			0.0295	0.233	0.306	0.453
Ailanthus altissima, tree of heaven			0.277	0.216	0.363	0.508
Andromeda glaucophylla,	Upper shade leaf		0.389	0.368	0.321	0.413
bog rosemary	Lower shade leaf		0.345	0.272	0.373	0.514
	Upper sun leaf		0.357	0.282	0.372	0.504
	Lower sun leaf		0.411	0.400	0.317	0.392
Artemisia sp.,	Young		0.413	0.344	0.454	0.554
wormwood	Older		0.461	0.392	0.462	0.547
Aristida sp., needle grass			0.275	0.220	0.275	0.424
Capsicum annuum, pepper			0.271	0.211	0.397	0.534
Carya sp.,	Very small	04/29	0.282	0.231	0.374	0.490
hickory	Small	04/29	0.303	0.239	0.341	0.473
	Medium	05/07	0.284	0.225	0.427	0.543
	Large	05/07	0.312	0.244	0.414	0.541
	Very large	05/29	0.311	0.245	0.444	0.565
Cercis canadensis redbud	(White)	04/25	0.232	0.189	0.344	0.462
		05/05	0.206	0.163	0.358	0.480
		05/28	0.254	0.203	0.370	0.501
	Sun leaf	06/20	0.304	0.246	0.332	0.465
	Shade leaf	06/20	0.285	0.224	0.365	0.503
Chamaedaphne calyculata,	Upper shade leaf		0.360	0.283	0.335	0.485
leather-leaf	Lower shade leaf		0.325	0.268	0.352	0.487
	Upper sun leaf		0.370	0.294	0.385	0.516
	Lower sun leaf		0.407	0.335	0.344	0.476
Clematis fremontii, clematis		06/20	0.259	0.207	0.488	0.592
Colocasia sp., elephant ear			0.314	0.244	0.417	0.553

Table 8.1. (*continued*)

Species	Condition	Date	R_L	R_H	A_L	A_H
Erythrina sp., coral bean			0.297	0.233	0.342	0.488
Euonymus bungeanus, spindle tree	Very young	03/26	0.268	0.221	0.342	0.465
	Young	04/12	0.304	0.247	0.326	0.460
		05/11	0.304	0.243	0.362	0.498
		06/17	0.309	0.248	0.391	0.524
		06/29	0.331	0.263	0.397	0.531
		08/29	0.331	0.265	0.415	0.541
		10/04	0.345	0.277	0.419	0.542
		11/29	0.356	0.286	0.376	0.502
		12/13	0.347	0.287	0.315	0.441
Fagus grandifolia, american beech		04/25	0.262	0.214	0.355	0.474
		05/05	0.298	0.238	0.400	0.517
		06/09	0.311	0.244	0.375	0.513
Fagus sylvatica sanguinea, european beech	July		0.247	0.196	0.286	0.435
	October		0.263	0.205	0.384	0.525
Ginko biloba, ginko		04/23	0.317	0.252	0.382	0.508
		05/01	0.355	0.286	0.366	0.499
		05/29	0.326	0.258	0.467	0.585
Gossypium hirsutum, cotton			0.277	0.221	0.397	0.524
Hamamelis venalis, witch hazel			0.265	0.207	0.392	0.529
Helianthus annus, sunflower	Second leaf		0.317	0.255	0.398	0.521
	Third leaf		0.269	0.213	0.407	0.530
	Fourth leaf		0.280	0.219	0.443	0.565
	Fifth leaf		0.282	0.220	0.400	0.532
Kalmia polifolia, swamp laurel	Upper surface		0.361	0.285	0.407	0.537
	Lower surface		0.432	0.383	0.322	0.429
Liriodendron tulipifera, tulip tree	Very small	04/04	0.171	0.143	0.312	0.430
	Unfolding	04/16	0.197	0.165	0.312	0.439
		05/18	0.282	0.225	0.358	0.495
		06/08	0.287	0.225	0.393	0.528
		06/29	0.307	0.240	0.383	0.521
		08/29	0.324	0.260	0.388	0.519
		09/27	0.342	0.267	0.387	0.526
	Yellow	10/18	0.378	0.319	0.282	0.401
	Brown	11/01	0.359	0.288	0.295	0.436
Magnolia virginiana, sweet bay	Small	05/06	0.169	0.137	0.393	0.503
	Medium	05/12	0.309	0.245	0.428	0.550
	Large	05/12	0.284	0.229	0.459	0.564

Table 8.1. (*continued*)

Species	Condition	Date	R_L	R_H	A_L	A_H
Musa paradisiaca, banana			0.333	0.262	0.415	0.546
Parmelia sp., lichen			0.378	0.313	0.603	0.672
Parthenium integrifolium, quinine			0.288	0.233	0.379	0.501
Pelargonium sp., geranium			0.276	0.215	0.419	0.548
Platanus occidentalis, american sycamore	Very small	04/17	0.351	0.302	0.455	0.542
	Small	04/23	0.285	0.233	0.358	0.474
	Small	04/29	0.271	0.224	0.307	0.430
	Small	05/05	0.225	0.187	0.326	0.442
	Medium	05/21	0.291	0.227	0.373	0.508
Populus deltoides, cottonwood	Very small	04/27	0.198	0.169	0.296	0.416
		05/07	0.237	0.198	0.324	0.448
		05/21	0.258	0.211	0.379	0.508
		07/01	0.268	0.218	0.375	0.508
		08/16	0.305	0.255	0.360	0.487
		09/13	0.302	0.246	0.373	0.501
		09/20	0.316	0.260	0.366	0.493
	Yellow	10/25	0.362	0.310	0.278	0.389
Prunus persica, peach			0.306	0.246	0.478	0.586
Prunus serotina, black cherry	Small	05/05	0.284	0.229	0.368	0.488
	Medium	05/05	0.312	0.246	0.378	0.509
Quercus alba, white oak		04/24	0.339	0.270	0.431	0.547
		04/30	0.260	0.212	0.440	0.546
		05/02	0.268	0.221	0.311	0.436
		05/22	0.285	0.228	0.389	0.519
		07/16	0.281	0.217	0.476	0.601
Quercus imbricaria, laurel oak		04/25	0.256	0.210	0.433	0.539
		05/29	0.304	0.238	0.441	0.565
Quercus macrocarpa, bur oak		04/24	0.214	0.176	0.395	0.505
		05/05	0.261	0.209	0.283	0.418
		06/09	0.301	0.240	0.377	0.509
Quercus marilandica, black jack oak		04/28	0.308	0.245	0.331	0.467
Quercus velutina, black oak		04/23	0.264	0.216	0.383	0.500
		04/30	0.293	0.241	0.388	0.503

Table 8.1 (*continued*)

Species	Condition	Date	R_L	R_H	A_L	A_H
		05/28	0.300	0.239	0.439	0.559
Reboulia sp., liverwort			0.218	0.169	0.753	0.809
Sassafras albidum, sassafras	Small	04/29	0.246	0.197	0.335	0.463
	Medium	05/21	0.297	0.234	0.347	0.487
	Large	06/09	0.315	0.249	0.346	0.488
Streliztsia sp., bird of paradise			0.291	0.230	0.507	0.617
Verbascum sp., mullein	Young Leaf	7/18/68	0.421	0.363	0.435	0.524
Xanthuim sp., cocklebur	Severed from plant		0.305	0.244	0.352	0.481
	Attached to plant		0.274	0.219	0.410	0.528

[a] R_1, reflectance for low sun; R_H reflectance for high sun; A_L, absorptance for low sun; A_H, absorptance for high sun. All values are for the upper leaf surface unless otherwise specified. All leaves are mature unless indicated as very small, young, or given an early date.

than a "shade" leaf on its upper surface, but less on its lower surface, as shown for *Andromeda glaucophylla* and *Chamaedaphne calyculata*. On an overcast day, the mean absorptance of leaves to light is about 16% greater than that occurring with direct sunlight. If the direct-sunlight absorptance is 0.50, then the cloudy-day absorptance is 0.59. This increase is the result of the spectral distribution of cloudlight on a cloudy day being entirely in the ultraviolet and visible part of the spectrum, where a leaf has its greatest absorption.

Ultraviolet and visible absorptance. A leaf reflects and transmits very little radiation at ultraviolet and visible wavelengths and, as a result, has high absorption at these wavelengths (except in the green). The human eye has its greatest sensitivity to green radiation, and plant leaves appear green to us in both reflection and transmission. High absorption throughout the ultraviolet and most of the visible is caused by the presence of leaf pigments such as chlorophyll, carotene, lutein, anthocyanine, and others, which have electronic absorption bands in this region (Fig. 8.22). Because of multiple scattering of light within a leaf by cell walls, as well as the many particles and structures within cells, the light path length is lengthened, and the absorption bands are broadened and intensified. The result is strong absorption by a leaf throughout the ultraviolet and visible and only a small reduction of absorption in the green gap between the chlorophyll absorption bands.

Infrared absorptance. The abrupt increase of reflection and transmission at a wavelength of 700 nm by a green plant leaf, and the resulting precipitous drop in absorptance, when first encountered, is somewhat star-

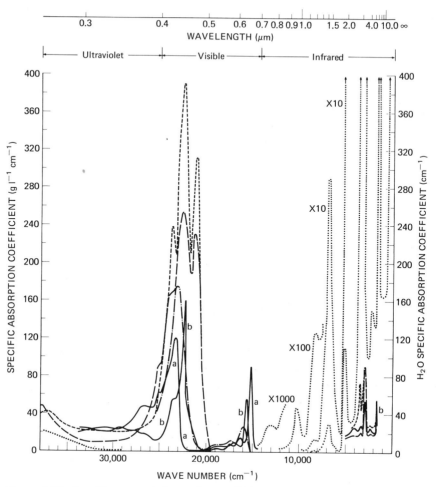

Figure 8.22. Specific spectral absorption coefficient of various plant pigments as a function of wavenumber: (—) chlorophyll; (– · –) protochlorophyll; (——) α-carotene; (– – –) lutein, a xanthophyll; (· · · · ·) liquid water.

tling to the observer. The fact that the spectral properties of a leaf change so suddenly with wavelength is suprising. The high near-infrared reflectance seen in Figs. 8.20 and 8.21 is well-known since infrared photographs of scenes containing vegetation show the leaves as "white" in appearance, in stark contrast to most other objects. The discussion of the origins and nature of molecular absorption bands earlier in this chapter revealed a gap in the absorption spectrum between regions in the ultraviolet and visible, where electronic energy transitions dominate the spectrum, and the intermediate and far infrared, where vibrational and rotational energy transitions dominate. This energy, or wavelength, gap in the absorption spectrum occurs in the near-infrared region, from a wavelength of about

700 nm to 1400 nm. The vibration–rotation absorption bands of liquid water enter the spectrum in the near infrared and, although weak at shorter wavelengths, begin to dominate the spectrum at wavelengths greater than 1400 nm. The result is that reflectance and transmittance of a leaf rapidly diminish at longer wavelengths, and absorptance increases, with nearly complete absorptance occurring at wavelengths greater than 2000 nm. Longwave infrared reflectance for many species of plants has been reported by Gates and Tantraporn (1952). These measurements, which were among the first of their kind, show reflectance in the far infrared to be less than about 0.5 for a 65° angle of incidence and less than 0.3 for 20° angle of incidence. These authors also showed that transmittance in the far infrared is zero, and therefore that absorptance is between 0.95 and 0.97.

Absorptance and energy exchange.   A fundamental law of radiation, Kirchoff's law, states that a good absorber of radiation is a good emitter at the same wavelength. However, an object emits radiation in a broad distribution of wavelengths according to Planck's distribution function [Eq. (5.5)]. The peak of the wavelength distribution, according to Wien's displacement law [Eq. (5.6)], is a function of the absolute temperature. Most leaves are at ambient temperatures in the range 270° to 320°K, or approximately 300°K. Radiation emitted from a blackbody at 300°K has a broad wavelength distribution centered at 10 μm with wavelengths from 4 to 40 μm or beyond. A plant leaf with an absorptance of 0.95 to 0.97 at these wavelengths will have an emittance of 0.95 to 0.97 of blackbody radiation. Hence a plant leaf emits radiation effectively in the far infrared, where it is important to do so in order to dump excess heat. A plant leaf strongly absorbs solar radiation in the visible range, where a small fraction of the absorbed energy is utilized photochemically in photosynthesis and the remainder goes into evaporating water and warming the leaf.

The extraterrestrial solar spectrum has approximately 50% of its energy at wavelengths of 710 nm or above. A plant leaf, by having very low absorptance in the near infrared, is essentially decoupled from this part of the solar spectrum and thereby remains substantially cooler than it would be if it absorbed all incident solar radiation. The product of spectral absorptance and spectral solar intensity gives the spectral distribution of energy absorbed by a leaf, shown for *Populus deltoides* in Fig. 8.23. High absorptance, coupled with the rising amount of energy in the solar curve from the ultraviolet into the visible, gives the curve the shape shown. The sudden decoupling between leaf and radiation at 700 nm and the lack of solar radiation absorbed above this wavelength is evident. One can make a very approximate estimate of the temperature increase for a leaf similar to cottonwood if the amount of radiation absorbed by the leaf was increased about 40%, say from 560 to 780 W m^{-2}. A calculation shows the leaf temperature would increase from about 30° to 38°C in still air at 20°C. Many plant leaves would be substantially warmer and be in danger of de-

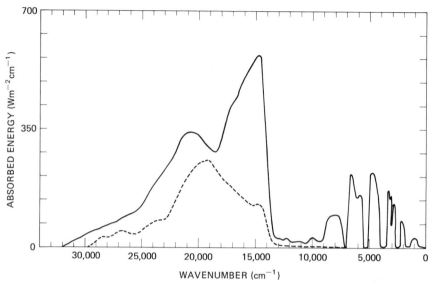

Figure 8.23. Spectral distribution of sunlight absorbed by a leaf of cottonwood (*Populus deltoides*) as a function of wavenumber for (—) clear sky and (– –) overcast conditions.

naturation if they absorbed most of the incident solar radiation. It is hard to know if the lack of strong absorptance in the near infrared by thin, broad plant leaves is a fortuitous result of the absorption-band mechanisms which exist in the physical world or is the result of evolutionary selection. The amount of energy absorbed by a cottonwood leaf on an overcast day is also shown in Fig. 8.23. Because almost no near-infrared solar radiation is transmitted by clouds, no energy is absorbed by a leaf at these wavelengths.

In summary, a thin, broad plant leaf absorbs radiation of wavelength of use photochemically, reflects solar radiation in the near infrared, which would tend to overheat the leaf, and absorbs, thereby also emitting, radiation in the far infrared, and allows it to dump heat to the environment by reradiation.

Desert Plants

In contrast to the spectral properties of the thin, broad deciduous leaves are those of desert succulents such as cactus and agave, shown in Fig. 8.24. Here, only reflectance and absorptance are shown, since these plant parts (trunk and stems) do not transmit radiation. It is important to note, however, the incredible amount of reflectance by these plants of radiation of wavelength 800 nm. The agaves, for example, reflect nearly 80% of incident radiation of this wavelength. This reflectance is so high that it reduces very substantially the amount of near-infrared solar radiation ab-

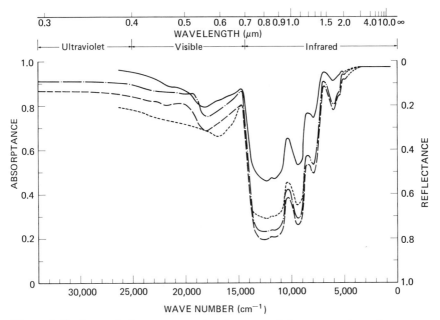

Figure 8.24. Spectral absorptance and reflectance of desert succulent plants as a function of wavenumber: (—) *Cereus giganteus;* (– · –) *Agave americana,* lower surface; (–––) *Agave americana,* upper surface; *Opuntia laevis.*

sorbed by the plant. The leaves of agave and the blades of cacti are often oriented more or less vertically, a position which greatly reduces the amount of solar radiation absorbed during midday. Early and late in the day, however, rays of the sun strike the plant surfaces at near normal incidence but since air temperatures are somewhat reduced at this time, the effects of radiation absorption are not quite so pronounced. Most desert plants show some signs of tissue denaturation during hot days. The energy absorbed by the saguaro cactus, *Cereus giganteus,* is shown in Fig. 8.25. Since the near-infrared reflectance of the saguaro is not as great as that of the agave, the amount of radiation absorbed is substantially greater.

Role of Thorns

Considerable speculation has existed concerning the role of thorns in the heat balance of desert plants. In order to ascertain if there might be an effect, Gates et al. (1965) measured the spectral reflectance of the cactus *Mammillaria lasiacantha,* which is covered with a dense layer of fine thorns, and compared it with that of other cacti such as *Opuntia laevis. Mammillaria* reflects more ultraviolet and visible radiation and less near infrared than the thornless *Opuntia laevis.* On balance, the total amount of

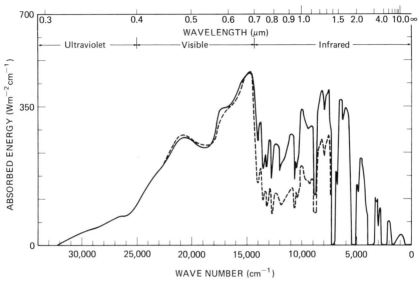

Figure 8.25. Spectral distribution of sunlight absorbed by (—) a cactus (*Cereus giganteus*) and (--) an agave (*Agave lecheguilla*).

direct solar radiation absorbed by *Mammillaria* is less, but the reduction is not significant enough to suggest that this is the reason for the presence of thorns. It is likely, however, that thorns do reduce the radiation load at the cuticular surface since, with thorns present, radiation is absorbed more in the thorns themselves and less in the cuticle and spongy tissue of the plant. The tops of the saguaro, barrel cacti (*Echinocactus*), and cholla cacti are covered with dense mats of thorns, which surely must absorb a high percentage of the incident radiation and thereby screen the underlying tissue from radiation. In addition, if incident radiation is absorbed strongly in the thorn mat, heat will be radiated to the sky and less of it will be transferred by conduction to the underlying cuticle. It is probable that this is an important role for the dense thorn mats, but the function of thorns distributed generally over the surface of a cactus may be quite different. Thorns also have other functions, serving, for example, as a deterrent to predation.

Leaf Pubescence

The role of trichomes, including hairs, thorns, and bristles, on the surface of leaves is not well understood. Suggested functions include reflection of radiation, reradiation, reflection and absorption of radiation to protect plant pigments and cells against too much radiation of certain wavelengths, reflection of radiation onto the mesophyll in order to provide increased light for photosynthesis, shading of the leaf surface, increase of the gas diffusion pathway in order to reduce water loss from a leaf, provi-

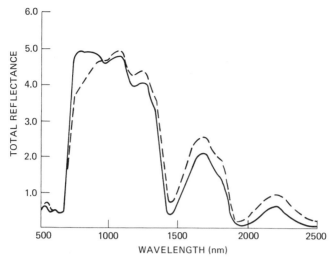

Figure 8.26. Total spectral reflectance of a velvet plant (*Gynura aurantiaca*) (—) with and (− −) without pubescence as a function of wavelength.

sion of moisture traps on a leaf surface to reduce water loss, insulation against heat loss in the manner of fur insulation of an animal, reduction of nutrient leaching, a means to transfer salt from within a leaf to its surface, protection against insect and (in the case of thorns) large-animal predation, protection against pathogens, protection against stomatal clogging by water droplets or dust, and, finally, reduction of the effect of wind on the leaf boundary layer. Each of these hypothesized advantages of leaf pubescence can be tested. The literature concerning this subject is not so often occupied with valid tests as with speculation.

Concerning the effect of thorns on cacti blades, Gates et al. (1965) showed that the presence of thorns could increase visible reflectance and decrease near-infrared reflectance while leaving the total amount of absorbed energy nearly unchanged. However, some of the energy is absorbed by the mass of thorns, and less energy is absorbed in the cactus blade itself. Gausman and Cardenas (1968) report the same phenomenon with the pubescence of leaves of *Gynura aurantiaca*, known as the velvet plant. The reflectance of the velvet plant with and without pubescence is shown in Fig. 8.26. Pubescence significantly increases the total and diffuse reflectance in the 750- to 1000-nm wavelength region but decreases them in the 1000- to 2500-nm region. Pubescence had very little effect on the reflectance of light from 500 to 750 nm. Many years earlier, Coblentz (1913) showed that hairs on *Verbascum thapsus* (mullein) did not increase the reflection of light, a report that was reaffirmed by Shull (1929). Shull also found the same thing for *Abutilon theophrasti*; however, hairiness on the under surfaces of *Populus alba* (silver-leaved poplar) leaves and the bloom of *Magnolia acuminata* did increase reflectance. Billings and

Morris (1951) measured the visible and near-infrared reflectance of *Eurotia lanata* (white sage), whose leaves are densely covered with white, stellate hairs. Visible light was much more reflected by the sage leaves than by the leaves of other plants from the same habitat, but there was little difference at infrared wavelengths.

Pearman (1966) measured the reflectance of visible light from the upper surface of leaves of *Arctotheca nivea,* a sand-dune plant from western Australia. Removal of hairs reduced the average reflectance from 0.317 to 0.15 in the wavelength range from 340 to 620 nm. Pearman concluded that hairs not only increase reflectance for this plant but also decidedly decrease the amount of light entering the leaf.

Krog (1955) made some very interesting temperature measurements of the catkins of pussy willow, *Salix polaris,* growing in Alaska in the early spring. He was impressed with the early appearance of pussy willows in the spring, before the air temperature warmed above freezing. He surmised that the hairy covering of the *Salix* catkin caused solar radiation to heat the catkin. Krog's experiment involved the insertion of fine thermocouples into a catkin in the normal, hairy condition, a catkin covered with lampblack, a catkin with the hairs clipped off, and a small cotton ball the size of a catkin. These were placed in full sun, and their temperatures along with that of the air, were observed. The normal catkin was the warmest, about 12°C above air temperature; the blackened catkin was next warmest, about 9°C above air temperature; and the cotton ball and shaved catkin were only about 4°C warmer than the air. A catkin consists of shiny transparent hairs underlain by an almost black layer of dark scales surrounding the flower bud. Sunlight penetrates the transparent hairs and is absorbed by the dark scales, which causes the catkin to warm. But the outer layer of hairs, which are transparent to sunlight, are probably opaque to longwave infrared radiation and therefore inhibit reradiation and loss of energy. The dense layer of hairs produces a dead air space around the catkin and reduces convection. Krog also found woolly bear caterpillars (Arctiidae) crawling on the snow surface in Alaska to be 15°C warmer than the air near the substrate surface.

It is apparent that pubescence produces no single result in all plant leaves with bloom of *Magnolia virginiana* before and after removal of be gained by the leaf of another species or even by leaves of the same species at different times of the year.

In Fig. 8.27 are shown reflectance, transmittance, and absorptance by a leaves with bloom of *Magnolia virginiana* before and after removal of pubescence. The arrows run from the values with pubescence to the values without pubescence at seven different wavelengths, 500, 550, 660, 700, 720, 730, and 1050 nm. These wavelengths were selected on the following basis: the 500-nm wavelength (in the blue region), because it is a peak of chlorophyll absorption; the 550-nm wavelength because it is in the center of the visible spectrum, in the green region; the 660- and

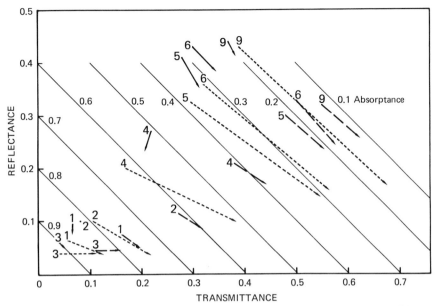

Figure 8.27. Changes of the reflectance, transmittance, and absorptance with removal of pubescence from a magnolia (*Magnolia virginiana*) leaf for light of wavelength (1) 500 nm; (2) 550 nm, (3) 660 nm, (4) 700 nm, (5) 720 nm, (6) 730 nm, and
(9) 1050 nm: (———) leaf 1, 21 May; (——) leaf 2, 1 May; (– – –) leaf 3, 6 May.

730-nm wavelengths because the red form and far-red forms of phy-
tochrome are interconverted by radiation of these wavelengths; and the
1050-nm wavelength because it represents a position on the high, near-
infrared part of the spectrum.

In the visible range, at 500, 550, and 600 nm, the changes corresponding to removal of pubescence are relatively small and consist mainly
in a decrease of absorptance. Hence the presence of pubescence increases absorptance by the leaf, although most of this increase occurs in
the pubescent layer itself. In the infrared, i.e., at wavelengths 700, 720, 730,
and 1050 nm, extremely large changes of reflectance and transmittance
occur without any significant change in the total absorptance. This is seen
in Fig. 8.27, where, despite the enormous changes in ordinate and abscissa values, the arrows run nearly parallel to the lines of constant absorption.

Absorptance changes little during development of the leaves of
Quercus alba. The young leaves have stellate hairs on the undersurface of
the young leaf. As the leaf develops, it loses its pubescence. At first, the
density of hairs decreases substantially as the surface area expands, with
the number of hairs staying the same. In mature leaves, the hairs tend to
break off from the surface. This suggests that pubescence plays a protective role against radiation incident on the young leaf.

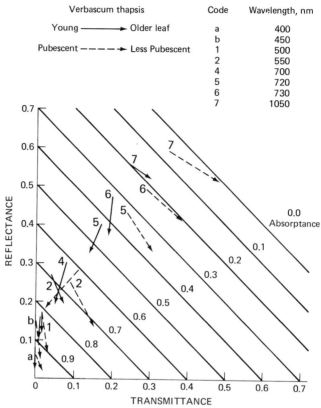

Figure 8.28. Changes of the reflectance, transmittance, and absorptance with removal of pubescence from a mullein (*Verbascum thapsis*) leaf for light of wavelength (a) 400 nm, (b) 450 nm, (1) 500 nm, (2) 550 nm, (4) 700 nm, (5) 720 nm, (6) 730 nm, and (7) 1050 nm.

Verbascum thapsus (mullein) is a plant whose leaves possess extremely heavy pubescence consisting of loosely packed trichomes, as contrasted to plant leaves that have heavy, felted pubescence. The pubescence is on both leaf surfaces. A young mullein leaf has a greater density of pubescence than an older leaf. Figure 8.28 shows the change of reflectance , transmittance, and absorptance of *Verbascum thapsus* leaves with the removal of the pubescence and the change with age for various wavelengths. In each case, the changes are similar and consistent. The presence of pubescence tends to increase reflectance, reduce transmittance slightly, and decrease absorptance at ultraviolet and visible wavelengths. At 700, 720, and 730 nm, there is an increased reflectance, increased transmittance, and decreased absorptance in old (less pubescent) as compared to young (more pubescence) leaves. In the older leaves, however, the presence of pubescence causes increased reflectance, decreased transmittance, and little or no change of absorptance. At 1050 nm, an increased amount of pubescence in each case caused an increased reflectance, decreased transmittance, and little or no change of absorptance. One might suggest that pubescence affords the leaf mesophyll some protection against radiation of ultraviolet and visible wavelengths, particularly in view of the fact that some absorption occurs in the pubescent layer itself. Transmittivity measurements were made of the pubescence alone. Although not highly accurate because the pubescence had to be removed from the leaf and reconstituted on a glass slide, measurements showed an approximate transmittance of 0.40 at 400 nm, 0.52 at 450 nm, 0.56 at 500 nm, 0.59 at 550 nm, 0.62 at 660 nm, and a monotonic increase to 0.70 at 1050 nm. This indicates that substantial amounts of incident radiation do not reach the leaf mesophyll because of the screening effect of the pubescent layer and that screening is greater at short wavelengths.

Seasonal Reflectance Changes

The changes that occur in the spectral properties of plant leaves during the growing season are illustrated in Figs. 8.29 and 8.30 for leaves of white oak, *Quercus alba*. The very young, much folded, compact, underdeveloped leaf of 17 April exhibited a lack of chlorophyll. Absorption in the visible range is caused by protochlorophyll and anthocyanin. The leaf ap-

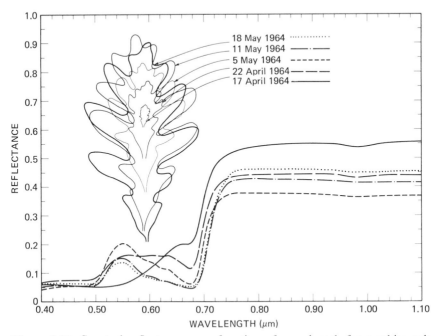

Figure 8.29. Spectral reflectance as a function of wavelength for a white oak (*Quercus alba*) leaf for different times of year.

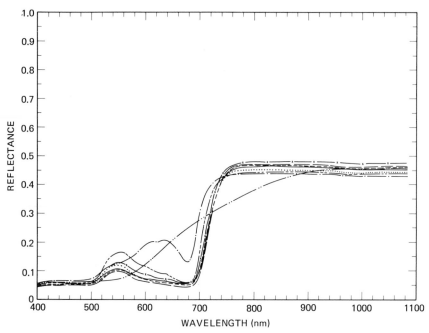

Figure 8.30. Spectral reflectance as a function of wavelength for a white oak (*Quercus alba*) leaf at different times of the year: (———) 19 June; (———) 20 July; (– – –) 18 August; (· · · ·) 18 Sept; (– × –) 7 Oct; (– – – –) 21 Oct; (— · —) 28 Oct; (– · · –) 2 Nov.

peared reddish to the eye. By 22 April, the young leaf began to exhibit green reflection and decreased red reflection. By 5 May, the leaf had developed most of its "normal" spectral characteristics, with the green reflectance strong and the red and blue spectral regions much absorbed. The near-infrared reflectance decreased to its minimum value of about 0.37 from a high of 0.56 on 17 April. This striking decrease appears to be caused by the unfolding and expansion of the leaf and resultant loss of a multitude of reflecting surfaces which existed in the much folded, very young leaf. By 11 May, the color of the leaf, which was bright, light green on 5 May, had darkened; chlorophyll absorption had become well developed. However, the near-infrared reflectance increased from 0.37 to 0.42, possibly as a result of the development of air spaces in the mesophyll and the presence of many reflecting surfaces within the leaf. By 18 May, no change of the reflectance in the visible range had occurred, but an increase (to 0.46) had occurred in the near infrared. Throughout much of the summer period, the white oak leaf exhibited a fairly stable reflectance throughout the visible and near infrared. Variations from leaf to leaf are astonishingly small from 19 June through 18 September. Beginning in mid-September and early October, however, there is evidence portending changes in spectral reflectances. By 21 October, strong changes in pigment, probably chlorophyll, were occurring. Green reflectance increased dramatically as the blue and red absorption bands weakened. The white oak leaf appeared very yellow by 28 October with the disappearance of nearly all chlorophyll absorption, but some of the absorption by anthocyanin remained. By 2 November, senescence was complete, with most of the pigmentation gone; the leaf was dried out and had collapsed cells throughout most of its structure. At this stage, the leaf had a "dead," brown appearance. It is interesting to note that the near-infrared reflectance over the range 700 to 900 nm diminished strongly but that the reflectance in the region beyond 900 nm changed very little.

Reflectance Edge

The dramatic change in reflectance and transmittance which occurs for green plant leaves at 700 nm is of great interest. The precise position of this reflection and transmission edge is largely dependent on the nature of the pigments in the leaf and represents the transition from the strong electronic absorption bands of pigments, which control the nature of reflection and transmission spectra at visible and ultraviolet wavelengths, to the region of the near infrared, where neither electronic nor vibration–rotation bands of molecules are very dominant. Because of the nature of the electronic-energy transitions in molecules, one might expect this edge to move toward wavelengths shorter than about 700 nm, but not to shift to substantially greater wavelengths, for different combinations of pigments. An illustration that this is indeed the case is given in Fig. 8.31, where the reflectance of a white magnolia petal and the reflectance and transmittance of a red rose petal and green cottonwood leaf are shown. The absorption edge, which is also the edge of strong reflection and transmission, is at 700 nm for the green leaf, 600 nm for the red petal, and 400 nm for the white petal. It is apparent from the spectral curves why the magnolia blossom appears white and the rose blossom appears red to the human eye. The red rose petal has a small amount of blue reflection mixed in with the red reflection and therefore appears somewhat purplish. For the magnolia and rose petals, the eye sees light reflected at wavelengths greater than the wavelength of the absorption edge, but for the green plant leaf, the eye is responding to a relatively weak secondary reflection maximum which occurs at wavelengths considerably shorter than the absorption edge. If the chlorophyll pigment plays a dominant role in the wavelength position of the absorption edge at 700 nm, consider what happens when the albino form of the same plant leaf is taken into account. Gates et al. (1965) have given the absorption spectrum of the green and albino forms of leaves of *Hedera helix*. The albino leaf is without pigments that absorb very strongly in the visible, although it is clear from the spectrum that there is some blue absorption but very little green and red absorption. The human eye, with its peak sensitivity in the green, sees the albino *He-*

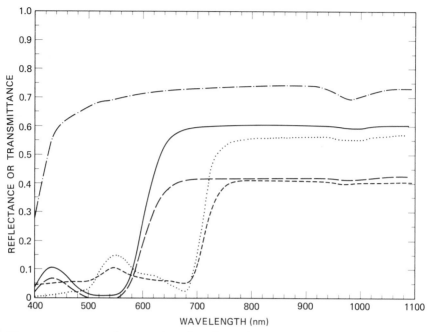

Figure 8.31. Spectral properties of various plant parts as a function of wavelength: (–·–) *Magnolia* petal reflectance, upper surface; (——) red rose petal reflectance, upper surface; (———) red rose petal transmittance; (–––) cottonwood leaf reflectance, upper surface; (· · · ·) cottonwood leaf transmittance.

dera helix leaf as yellow, since more blue than red light is missing from the sunlight reflected by the leaf.

Coniferous Plants

Pine, cedar, spruce, fir, and other conifers have the lowest reflectances and highest absorptances throughout the ultraviolet, visible, and near infrared of any plants measured. Lack of transmittance through the needles and leaves of coniferous trees and low reflectance result from the darker pigmentation of these plants in comparison with the leaves of other plants. In infrared photographs, conifers generally appear as dark areas in contrast to deciduous trees. Gates et al. (1965) give the mean absorption to direct solar radiation (including skylight) and to cloudlight as 0.89 and 0.88 for *Pinus strobus* and 0.88 and 0.88 for *Thuja occidentalis*. Absorptance in the visible is particularly high for conifers, with some values as high as 0.974. The cylindrical geometry of pine needles, as contrasted with the more planar aspect of broad leaves, generally results in less solar radiation being absorbed per unit surface area. The dark pigmentation of conifers may be an adaptive characteristic which has evolved to allow conifers to compensate for reduced absorption resulting from geometry.

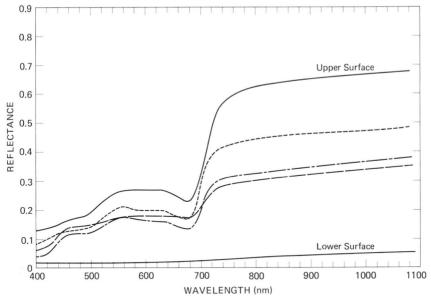

Figure 8.32. Spectral reflectance as a function of wavelength for several lichens: (———) foliose lichen gyrophore; (——) dry crustose lichen *Parmelia* on sandstone; (– – –) wet crustose lichen *Parmelia* on sandstone; (– – –) crustose lichen on decaying hemlock.

Lichens

The spectral reflectances of selected lichens are shown in Fig. 8.32. A lichen is an unusual plant in that it consists of a fungus and an alga living symbiotically. The alga contains chlorophyll; the fungus does not. The chlorophyll is screened by the fungus so that the reflection spectra do not show the strong green reflection peak at 550 nm so characteristic of green plants. The lichen reflection spectra shown in Fig. 8.32 were obtained from dry and wet specimens of the lichen *Parmelia* occurring on a sandstone substrate. The water content of the lichen reduced the reflectance in the visible and increased the reflectance in the near infrared. A lichen transmits very little or no radiation, and therefore the reflectance is a direct measure of absorptance directly. The abrupt change in reflectance and absorption at 700 nm is present in the lichen spectra just as in higher green plants.

Infrared Reflectance

The spectral characteristics of plant leaves at infrared wavelengths are of great importance for the exchange of radiation between a plant and its environment. There are two spectral regions of great importance in the infrared: the region from 700 to 3000 nm, which contains most of the infrared solar radiation, and the region beyond 3000 nm, over which most terrestrial objects radiate heat. The first region, in the very near infrared, is characterized by strong reflectance and transmittance and very low absorptance for coniferous and succulent plant parts. The longwave region of the infrared, beyond 3000 nm, comprises what spectroscopists refer to as the intermediate and far infrared since most objects at temperatures of about 300°K radiate a broad spectrum of wavelengths from 3,000 to 50,000 nm and beyond.

Plant leaves are generally fairly "black" at wavelengths greater than 4 μm. Gates and Tantraporn (1952) measured the specular spectral reflectance of leaves of 27 plant species and found the reflectance to be about 0.05 for most of them. All leaves were opaque. A few had higher reflectance, however, and *Citrus limonia,* in particular, had a reflectance of 0.17 at 10 μm. Most of these measurements were made at an angle of incidence of 65°; a few were made at 20°.

Wong and Blevin (1967) reported diffuse spectral reflectance measurements over the wavelength range from 2 to 14 μm for 15 plant species and at wavelengths of 8, 10, and 12 μm for an additional 32 plant species. Generally, reflectance values throughout the wavelength range 3 to 14 μm were less than 0.05. In the range 2 to 3 μm, quite high reflectances (up to 0.45) were observed for a few species. Reflectances at 8 μm were under 0.04 with the following exceptions: *Atriplex nummularia* lower surface, 0.05; *Gazania* hybrid lower surface, 0.05; and *Acacia aneura* upper surface, 0.05. Also, at 10 μm, reflectances were generally 0.01 to 0.04, and only in a few species did reflectances measure 0.06 or 0.07.

A careful study of the effect of drying on leaves of French bean, *Phaeolus vulgaris,* was made. The oven-dried leaf had much greater reflectances in the regions between the liquid-water absorption bands which are at 1.9, 2.7, and 6.3 μm. The leaves of plants with the highest reflectances in the 8- to 12-μm region had surface structures which could account for the increased reflectances. A thick mat of hairs occurs on the upper surfaces of *Cerastium tomentosum* and *Lychnis coronaria* and the lower surfaces of *Garzania* hybrid and *Spinifex hirsutus;* reflectances at 10 μm were 0.06, 0.07, 0.06, and 0.06, respectively. Stripping off the cuticle and epidermal layer of *Agapanthus umbellatus* and *Gazania* hybrid reduced leaf reflectances markedly. Gates and Tantraporn (1952) found that the lower leaf surface generally had a higher reflectance than the upper surface, a finding that has been confirmed by Wong and Blevin (1967).

The infrared reflectances of black spruce, *Picea mariana,* and sugar maple, *Acer saccharum,* are shown in Fig. 8.33. The liquid-water absorption bands are evident in these spectra, but much more structure is present than that associated with water alone. The reflectances of tap water and an apple leaf are shown in Fig. 8.34. These spectra were obtained at the National Bureau of Standards by Victor Weidner. The sharp reflectance signal at about 2700 cm⁻¹ (3.7 μm) is produced by the CH

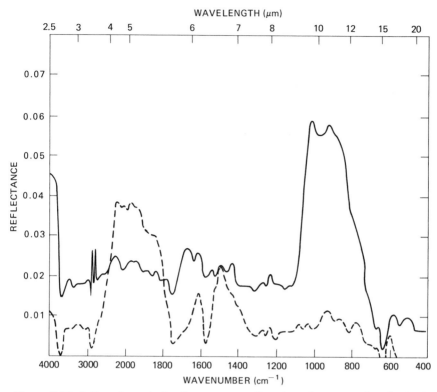

Figure 8.33. Infrared spectral reflectance as a function of wavenumber for (--) black spruce (*Picea mariana*) and (—) sugar maple (*Acer saccharum*) leaves.

stretching modes in the hydrocarbons of the plant material. Most of the features characteristic of any one of these spectral curves are found in all of them. The surprising thing is the variability of certain features, even though variability is small on an absolute basis. For example, in the spectral region around 2,000 cm^{-1} (or 5 μm), there is a very strong reflectance for black spruce, with a substantially lower value for sugar maple. The tulip tree leaf exhibits an unusually strong reflectance at about 2,564 cm^{-1} (or 3.9 μm) that does not appear in the other spectra. All spectral curves show a strong reflectance peak in the region 900 to 1,000 cm^{-1}, but sugar maple exhibits a peak that is substantially stronger and broadened considerably toward both lower and higher frequencies. The spectral reflectance of liquid water of 6.3 mm depth on steel is shown for comparison; the underlying steel is not considered to interfere with the reflectance of liquid water when the depth of the water is this great. The reflectance features of liquid water vary opposite to the reflectance spectra of plant leaves. The complex molecules of leaf cuticle, epidermis, or mesophyll absorb very strongly at wavelengths greater than 12 μm. These curves suggest that the average reflectance of plant leaves in the infrared region

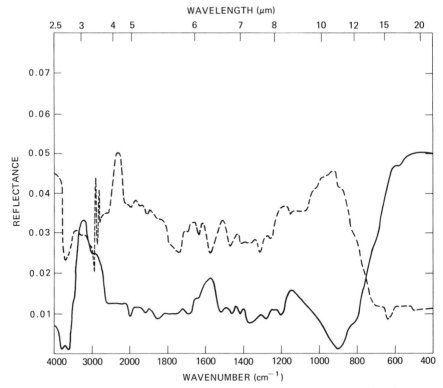

Figure 8.34. Infrared spectral reflectance as a function of wavenumber for (—) tap water and (--) an apple leaf (*Malus* sp.). The water was measured at 6.3 mm depth on steel.

varies between about 0.01, for black spruce, and 0.03, for broad leaf plants, and thus that the emittances vary from 0.99 to 0.97, respectively.

Specular or Diffuse Reflectance

The question is often raised as to whether or not plant leaves are ideal diffuse reflectors of radiation, specular reflectors, or a combination of both. Moss and Loomis (1952) have reported on data collected by Dinger (1941) showing that, for the visible region, the reflection curve for a green *Coleus* leaf conforms fairly well to the cosine law for a perfect diffusing surface, whereas that for a pansy leaf exhibits considerable departure from the cosine law. Wong and Blevin (1967) showed that leaf reflection conforms reasonably well to the cosine law at 2 μm, but shows a very strong specular reflection when the angle of incidence was 10° at 10 μm. One would not expect a leaf to resemble a perfect diffuse reflector since the chloroplasts, mitochondria, peroxisomes, and other particles are not randomly distributed, but located in preferred positions within the cells. Measurements by Woolley (1971) with white and green portions of a *Philodendron* leaf and

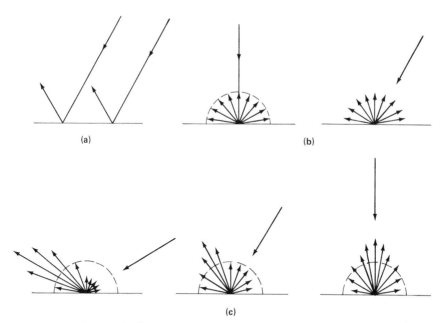

Figure 8.35. Types of reflectance: (a) specular reflectance; (b) diffuse reflectance; (c) specular and diffuse reflectance combined.

also a maize leaf for visible light of angles of incidence 15° and 45° showed that there are specular and diffuse components in most leaf reflection. About 4% of the visible light striking the white portion of the *Philodendron* leaf was specularly reflected and about 3% of the light striking the green portion. The average total reflectance of the leaf was 0.55; hence more than half the total reflectance was specular. Many leaves have a sheen, and this is the specular component of reflectance. Howard (1966) discussed the reflectance of eucalyptus leaves and showed that they were never highly specular, or were they perfectly diffusing, but always a combination. Figure 8.35 illustrates the various types of reflectance. Pure diffuse reflectance, which always produces the same intensity reflected in any direction, irrespective of the angle of incidence, is shown in Fig. 8.35b. Pure specular reflectance, in which light is not reflected in every direction but where the angle of reflection is equal to the angle of incidence, is shown in Fig. 8.35a. Most leaves exhibit a combination of the two reflectance types (Fig. 8.35c).

Most authors seem to feel that a leaf is a reasonably good diffuse reflector or transmitter of radiation. Willstätter and Stoll (1918) proposed that reflectance from a leaf occurs by critical reflection at spongy mesophyll cell wall to air space interfaces. Sinclair et al. (1973) proposed a modification of the Willstätter and Stoll hypothesis, what they termed a "diffuse reflectance hypothesis," based on the idea that cell walls act as diffuse reflectors owing to their microfibril structure. The authors feel that their hypothesis adequately explains the substantial amount of data they obtained on leaf reflectance and transmittance. Meyers and Allen (1968) used the Kubelka–Munk theory of successive diffusing layers to explain reflectance from a leaf. This work was further extended in a paper by Allen et al. (1969). Meyers and Allen used normal-incidence reflection at 35 interfaces along the mean optical path through the leaf. Gausman et al. (1969) noted that if oblique reflections were considered, fewer interfaces were required to explain leaf reflectance. According to Kumar and Silva (1973), who have reviewed this subject, "Knipling (1969) emphasized that the air spaces within the palisade parenchyma layer of a leaf mesophyll may be more important in scattering light than air spaces in the spongy parenchyma layer." Birth (1971) has given a critical review of the various concepts of leaf reflectance and states that Sinclair et al. (1973) demonstrated that the diffuse character of radiation in a leaf begins with the initial interface.

Ray-Tracing Leaf-Reflectance Model

Kumar and Silva (1973) present a leaf-reflectance model which illustrates very succinctly the problems involved in attempting to analyze the interaction of radiation and leaf structure. They present an illustration of the pathway of a light ray through a leaf that is distinctly improved compared to that used earlier by many authors.

The leaf-reflectance model of Kumar and Silva is a ray-tracing technique that uses properties of geometrical optics relating to the indexes of refraction of plant cells. The following assumptions are made:

1. Each leaf element, whether cell wall, chloroplast, cell sap, or intercellular air space, is considered to be homogeneous and isotropic.
2. Geometrical optics is applied to the ray tracing. Diffraction by plastids and chloroplasts is not considered, nor is scattering.
3. Only the near-infrared wavelength region from 0.7 to 1.3 μm is considered here, since absorption is neglected in the first approximation. The same model will work for the visible or intermediate infrared region, but the complex indexes of refraction of the leaf constituents must be known.
4. A two-dimensional leaf cross-section is used for the model.

Standard geometrical optics were used for ray-tracing utilizing the following relationships: Snell's law, Fresnel's equations, and certain boundary conditions.

The index of refraction of air inside or outside the leaf is 1.0. The refractive index of a potato cell wall was shown by Renck (1972) to be about 1.52 in the visible range. The refractive indexes for cell sap and chloroplasts were determined by Charney and Brackett (1961) to be 1.36 and 1.42, respectively, in the visible range. Values are not available for the indexes of refraction of cell walls, cell sap, and chloroplasts in the near-infrared region (0.7 to 1.3 μm) but are almost the same as that for liquid water according to Irvine and Pollack (1968). These latter values were used by Kumar and Silva (1973) in their ray-tracing model.

Light is reflected when a ray travels from one media to another having a different index of refraction. In a leaf, there are many such optical interfaces. The model of Kumar and Silva considers the following eight interfaces: (1) air to cell wall; (2) cell wall to cell sap; (3) cell wall to chloroplast; and (4) cell sap to chloroplast. A light intensity of 1.000 was assumed for each of the incident intensities. An angle of incidence of 5° was used since this is the angle of the light rays on a sample in the DK-2A spectrophotometer with reflectance attachment. The rays were traced by hand by means of a drafting set which measured angles to an accuracy of 5°. By using Snell's law and the appropriate indexes of refraction for each interface, the rays were traced through the leaf. The incident light ray passed through a total of 253 interfaces, out of which there were 18 cases of total reflection at cell-wall-to-air interfaces, 2 at cell-wall-to-chloroplast interfaces, and 1 at cell wall to cell sap interface. A ray whose intensity became less than 0.018 was discounted, but all other rays were followed completely through the leaf until they escaped as either transmitted or reflected light.

When a collimated beam of light irradiates a leaf surface, a nearly infinite number of light rays pass through the leaf structure. Each light ray encounters different internal surfaces and is therefore reflected and transmitted in different directions. In this manner, the collimated beam of light incident at the leaf surface becomes diffuse as it is transmitted through the leaf. Near-infrared radiation (0.7 to 1.3 μm) reflected from a leaf is more diffuse than visible light because near-infrared light passes through more interfaces than visible light before its intensity is reduced to a minimum since the near infrared is absorbed much less than the visible. Both the visible and near-infrared light transmitted by a leaf are fairly diffuse since a light ray must pass through a large number of interfaces before it is transmitted.

Kumar and Silva (1973) compared the amount of reflected and transmitted light for a soybean leaf as calculated by the ray-tracing technique with observations by Sinclair (1968) using a DK-2A spectrophotometer. The calculated and observed reflectances were 0.46 and 0.47, respectively, and the transmittances were 0.54 and 0.53 for the wavelength range 0.7 to 1.3 μm. Zero absorption was assumed for the leaf in this wavelength interval. These results suggest that the ray-tracing technique using a leaf with eight types of interfaces is a good approximation to the reality. Furthermore, when these authors used the ray tracing for air-to-cell-wall and cell-wall-to-air interfaces only, they obtained a reflectance of 0.30 and a transmittance of 0.70. It is clear that the larger number of interfaces used in the eight-interface model is more representative of the true situation in the leaf. These authors conclude that the older leaf model, first suggested by Willstätter and Stoll (1918), is too simple and does not adequately account for leaf reflectance and transmittance. There is some contribution to reflection by Rayleigh and Mie scattering owing to particles in the leaf cells, but the contribution by reflection at interfaces account for most of the observed reflectance and transmittance.

Kubelka–Munk Theory

Allen and Richardson (1968) used the Kubelka–Munk theory of diffuse light reflection and transmission to explain the optical properties of a plant canopy. They then applied this theory to reflectance and transmittance by a single leaf. The theory is sufficiently important to most problems of light reflectance and transmittance to be reproduced here.

A plant leaf or canopy is considered to be made up of a number n of randomly oriented, light-absorbing and light-scattering elements per unit area. The total number of elements between the upper and lower surface is N. The plane $n = 0$ is the illuminated, or upper, surface of the leaf or canopy, and the plane $n = N$ is the under or lower surface of the leaf or canopy. Radiant flux in the positive n direction (downward) is I, and that in the negative n direction (upward) is J. Incident flux on the upper surface is I_0 and is taken to be unity. Since $I_0 = 1.0$, reflected flux R is equivalent to reflectance, and the transmitted flux T is equivalent to transmittance. When $n = 0$,

$$I(0) = 1, \quad J(0) = R, \tag{8.22}$$

and when $n = N$,

$$I(N) = T, \quad J(N) = R_g I(N), \tag{8.23}$$

where R_g is the reflectance of the lower surface of the leaf or of the ground in the case of a plant canopy.

The Kubelka–Munk theory of diffuse reflectance and transmittance is based on the following differential equations of incremental absorption and scattering

$$dI = -(k + s)I \, dn + s \, J \, dn \tag{8.24}$$

$$dJ = (k + s)J \, dn - s \, I \, dn, \tag{8.25}$$

where k is the absorption coefficient and s is the scattering coefficient.

Solving Eq. (8.25) for I and substituting this factor into Eq. (8.24) gives

$$\frac{d^2J}{dn^2} - \alpha^2 J = 0. \tag{8.26}$$

Then, by solving for J in Eq. (8.24) and substituting into Eq. (8.25), one gets

$$\frac{d^2 I}{dn^2} - \alpha^2 I = 0, \tag{8.27}$$

where

$$\alpha^2 = k(k + 2s). \tag{8.28}$$

Equations (8.26) and (8.27) are then solved and the boundary conditions of Eqs. 8.22 and 8.23 are imposed to give the following two-parameter representation of light intensity within a leaf:

$$J = \frac{(1 - \beta^2)[e^{\alpha(N-n)} - e^{-\alpha(N-n)}] - R_g[(1 - \beta)^2 e^{\alpha(N-n)} - (1 + \beta)^2 e^{-\alpha(N-n)}]}{(1 + \beta)^2 e^{\alpha N} - (1 - \beta)^2 e^{-\alpha N} - R_g(1 - \beta^2)(e^{\alpha N} - e^{-\alpha N})} \tag{8.29}$$

$$I = \frac{(1 + \beta)^2 e^{\alpha(N-n)} - (1 - \beta)^2 e^{-\alpha(N-n)} - R_g(1 - \beta^2)[e^{\alpha(N-n)} - e^{-\alpha(N-n)}]}{(1 + \beta)^2 e^{\alpha N} - (1 - \beta)^2 e^{-\alpha N} - R_g(1 - \beta^2)(e^{\alpha N} - e^{-\alpha N})}, \tag{8.30}$$

where

$$\beta^2 = k/(k + 2s). \tag{8.31}$$

Equations (8.29) and (8.30) are the generalized Kubelka–Munk reflectance and transmittance equations and can be applied to the single leaf, an array of stacked leaves, or a plant canopy.

Two special cases are of interest here: zero absorptance in a leaf or canopy, which is approximated by the interaction of near infrared (0.7–1.3 μm) radiation with a leaf or canopy, and the case of zero scattering, which is only of academic interest. In the case of zero absorptance, Allen and Richardson (1968) give the relationship

$$J_0/[s(N - n)] = I_0/[1 + s(N - n)] = 1/(1 + sN) \tag{8.32}$$

and state that in a nonabsorbing medium, the quantity $J - I$ is independent of n and takes the form

$$J - I = -(1 + sN)^{-1} + [TN^2 R_g/(1 - R_g R_0)]. \tag{8.33}$$

This suggests that in a given nonabsorbing leaf or canopy, net radiant flux at any point is a function of the reflectances of the upper and lower surfaces and of the total number of scattering elements in between. Reflectance R and transmittance T for a nonabsorbent canopy or leaf with $R_g = 0$ is obtained by setting $n = 0$ and $n = N$ in Eq. (8.32) to obtain

$$R = sN/(1 + sN) \tag{8.34}$$

$$T = 1/(1 + sN). \tag{8.35}$$

These equations are equilateral hyperbolas with asymptotes $R = 1, T = 0$, and $N = -1/s$. Since only scattering is considered, attenuation is by scattering rather than absorption.

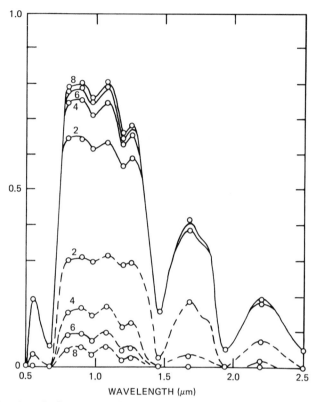

Figure 8.36. (—) Reflectance and (—) transmittance as a function of wavelength for two, four, six, and eight cotton leaves placed in a stacked array. Numbers refer to the number of leaves in the stack. (From Allen and Richardson, 1968.)

If the canopy or leaf has no scattering and finite absorption when $\alpha = k$ and $\beta = 1$, Eqs. (8.29) and (8.30) give the relationships

$$J = R_g\, e^{-k(2N-n)} \tag{8.36}$$

$$I = I_0\, e^{-kn}. \tag{8.37}$$

These equations are essentially the Boguer–Lambert law of attenuation for radiation passing through an absorbing medium.

Allen and Richardson (1968) have applied their theory to the reflectance and transmittance of mature cotton leaves singly and stacked for measurement with a spectrophotometer. The leaves each had a thickness of 0.018 cm. The comparison between theory and observation is shown in Fig. 8.36. The standard deviation between theoretical and experimental values was 1%. Scattering and absorption coefficients derived for a single mature cotton leaf are shown in Figs. 8.37 and 8.38. The absorption coefficient shows relatively large values at wavelengths that correspond to the chlorophyll absorption band at 0.68 nm and to liquid-water absorption

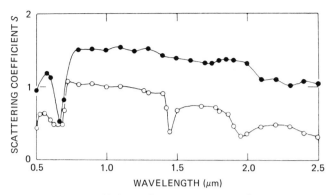

Figure 8.37. Scattering coefficient of (○—○) a normal mature cotton leaf and (●—●) a dehydrated leaf as a function of wavelength. (From Allen and Richardson, 1968.)

bands at 0.97, 1.21, 1.45, and 1.95 nm. When the leaf is dried and shrinks in thickness to 0.010 cm, the water absorption bands nearly disappear. It should be noted that, in Fig. 8.38, the normal leaf has essentially no absorption from 0.7 to 1.3 nm, the wavelength region for which Kumar and Silva (1973) assumed no absorption for their ray-tracing model of leaf reflectance and transmittance. By contrast, the scattering coefficient shown in Fig. 8.37 tends to remain relatively constant over the entire spectral range, except for small dips at 1.45 and 1.95 nm. It is also interesting to note that the scattering coefficient increases with dehydration, since dehydration results in a collapse of leaf mesophyll and palisade cells and

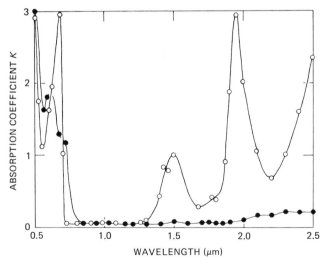

Figure 8.38. Absorption coefficient of (○—○) a normal mature cotton leaf and (●—●) a dehydrated leaf as a function of wavelength. (From Allen and Richardson, 1968.)

would be expected to give more rough surfaces for scattering. This demonstrates the accuracy of the Kubelka–Munk theory of diffuse reflectance and transmittance to plant leaves and, along with the ray-tracing technique demonstrated by Kumar and Silva, seems to explain fully the interaction of light with plant leaves.

Parallel-Plate Theory

Despite the apparent excellence of application of the Kubelka–Munk theory to light reflectance and transmittance by plant leaves, there are considerably simpler models which allow good, consistent empirical matching to observational data. Certainly, the Kubelka–Munk theory is mathematically difficult. Allen et al. (1969) have derived a plane-parallel-plate model of a compact leaf without intercellular air spaces. Allen et al. (1970) have extended this model to plant leaves containing intercellular air spaces.

Using transmittance and reflectance measurements on a corn leaf, Allen et al. (1969) derived the values of the effective index of refraction shown in Fig. 8.39. This dispersion curve seems to agree reasonably well with the refractive index measured by Eglinton and Hamilton (1967) for epicuticular wax found on the surfaces of many leaves. The index of refraction for carnuba wax from the leaf surface of the carnuba palm, *Capernicia cerifera,* is shown as a single value at 0.5 μm and agrees well with the derived curve at this wavelength.

The plate-model theory contains improvements over the Kubelka–Munk theory in that the scattering coefficient *s* has been interpreted in terms of Fresnel reflections from the leaf or cell surfaces, polarization has

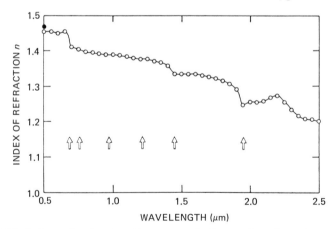

Figure 8.39. Index of refraction for a corn leaf as a function of wavelength. Arrows mark the positions of absorption bands by chlorophyll and liquid water. The solid circle at 0.50 μm is the published index of refraction for carnuba wax. (From Allen et al., 1960.)

been included, and the absorption coefficient derived by the model is in closer agreement with that of pure liquid water than that calculated from the Kubelka–Munk formulation.

Stochastic Model of a Leaf

Tucker and Garratt (1977) have reported on a leaf model based on a random process making up a Markov chain with discrete states. Four radiation states (incident, reflected, absorbed, and transmitted radiation) are represented by six compartments: (1) solar input; (2) initial reflection from the cuticle; (3) absorption in the palisade parenchyma; (4) diffusely reflected radiation; (5) absorption in the spongy mesophyll; and (6) diffusely transmitted radiation. Two cellular aggregates are represented by two compartments: (7) palisade parenchyma and (8) spongy mesophyll. Two scattering mechanisms are represented by two compartments: (9) scattering occurring in the palisade parenchyma and (10) scattering occurring in the spongy mesophyll. Radiative interactions within the leaf are considered as a random-walk process, and probabilities are assigned for each possible interaction based on knowledge of pigment composition and concentration, leaf structure, and leaf water content. All the observed features of spectral reflectance for plant leaves are accounted for by the stochastic model.

Leaf Optical Properties with Angle of Incidence

Leaf absorptance, reflectance, and transmittance to incident radiation remains relatively constant with angle of incidence up to angles of about 50° according to Farrar and Mapunda (1977). Gates and Tantraporn (1952) have reported measurements of longwave infrared reflectance at angles of incidence of 20° and 65°. Generally, the spectral composition of reflected sunlight does not change with angle of incidence. However, a more detailed inspection of reflectance r and absorbtance a as reported by Farrar and Mapunda does show a gradual change occurring at angles of incidence of 20° or greater. The following cubic functions describe these changes fairly well for the leaves of some plants:

$$r = A + B\ i^3 \tag{8.38}$$

$$a = A^1 - B^1\ i^3, \tag{8.39}$$

where i is the angle of incidence, B and B^1 are constants, and A and A^1 are reflectivity and absorptivity, respectively, at $i = 0$. Values for *Phaseolus vulgaris* are $A = 0.174$, $B = 4.82 \times 10^{-7}$, and $A^1 = 0.708$, and $B^1 = 5.93 \times 10^{-7}$.

Transmittance t of the leaf as a function of the angle of incidence is

$$t = 1 - r - a$$
$$= 1 - (A + A^1) - (B - B^1)^3. \qquad (8.40)$$

For *Phaseolus vulgaris,* transmittance is

$$t = 0.118 + 1.11 \times 10^{-7} i^3 . \qquad (8.41)$$

Other plants measured by Farrar and Mapunda followed the same type of change with angle of incidence, with some exceptions; for example, some plants exhibited a sudden drop of transmittance at angles of incidence between 60° and 70°. They also mentioned that Tageeva and Brandt (1960) obtained results for a variety of plants contrary to their own. Clearly, this is a subject that requires further study.

Spectral Properties of Animals

Introduction

Coloration of animals has been a topic of enormous conjecture and considerable speculation for a very long time. Many questions concerning coloration remain unanswered, and some confusion exists in the scientific literature. A treatise concerning this topic by Hamilton (1973) is provocative and challenging, as well as being open to some dispute (see Gates, 1974). A great deal of attention has been given to the reflectance and transmittance of human skin, partially because of the health problems such as sunburn and skin cancer. An excellent review of this topic is contained in a book edited by Urbach (1969), which deals with ultraviolet radiation and its biological effects. The spectral properties of animal hair, fur, and feathers are given in the form of graphs in Dirmhirn (1964). Whereas the spectral reflectances of various plant leaves differ mainly quantitatively, spectral reflectances of animal surfaces often differ both qualitatively and quantitatively.

It is important to consider the interaction of an animal's surface with radiation of the following spectral intervals: ultraviolet, visible, near infrared, and far infrared. Very high-frequency ultraviolet radiation can cause cell damage and genetic mutation, as well as having beneficial effects. Visible radiation may cause very limited molecular or cellular damage, including thermal effects, but produces many extremely beneficial biological reactions and responses. Near-infrared radiation, which comprises nearly 50% of the total incident solar radiation, produces very few photochemical reactions and is largely converted to thermal energy upon absorption by an organism. All objects, including organisms, that have ambient temperatures in the general vicinity of 300°K emit longwave radiation of 4 μm wavelength or greater. The reflectance, transmittance, or absorptance of an organism's surface at ultraviolet wavelengths may be similar to or very different from that at visible, near-infrared, or far-infrared wavelengths. An object that is white at visible wavelengths may be "gray" at other wavelengths. Almost all organism surfaces, no matter what their reflectances at visible wavelengths, are "black" at far-infrared wavelengths, or nearly so. Most organisms reflect less than 5% of far-infrared wavelengths and therefore have emittances of 95% or greater to longwave radiation.

In many cases, "protective," or cryptic, coloration in the form of camouflage is evident among animals as a means to reduce predation or improve their predation strategy. Many animal surfaces of brown, gray, or mottled appearance are examples of protective coloration but may also include white or black against backgrounds or substrates of similar coloration. The green adaptive coloration of the preying mantis, katydid, walkingstick, and leafhoppers are familiar examples. The mottled pattern of the nighthawk makes it difficult to discern against the brown grasses and rocks of its nesting habitat. A white snowshoe hare or white ptarmigan is difficult to detect against the winter snow, and their brown summer garb appears to be cryptic coloration as well. But beyond the vast number of cases of animal coloration which seem to be readily explainable are the more difficult problems of animal energetics and protection against photochemical damage.

Color Matching

The color of an animal's surface may act as a temperature-control mechanism or may serve as protective coloration. In addition, the coloration of an animal may play a social role or act as a sexual attractant. The protective-coloration role would be expected to operate over the spectral range of visual sensitivity for most animals. The eyes of most animals have a sensitivity limited to approximately the same wavelength range (400 to 700 nm) as the human eye, except for certain insects which seem to have ultraviolet sensitivity. Are temperature control and protective coloration mutually exclusive or do they act simultaneously? Many animals, particularly some reptiles and amphibians, are able to undergo color change as the conditions of their environments change. Kenneth Norris (1967) has made extensive studies of reptilian coloration, and in particular into the spectral regions beyond the visible. Most of the solar energy reaching the earth's surface is contained in the spectrum between 300 and 1800 nm, with about 60% of this energy contained in the nonvisible ultraviolet and infrared portions of the spectrum.

It is well known that the transfer of melanophore pigments in the skin may produce subtle or sometimes dramatic changes in the color of reptiles or amphibians. Often, the transfer of these pigments is induced by temperature in the natural environment. Transfer can also be artificially induced by electric shock, or by injections of ACTH or MSH hormones. Dr. Norris devised a special temperature-controlled holding device in order to mount small animals against the reflectance apparatus of his spectrophotometer. In this manner, he was able to measure the influence of temperature on the spectral reflectance characteristics of the animals. He could also induce color change by the methods mentioned above. By measuring the spectral reflectance characteristics of the animals under various conditions and the spectral properties of the soil, sand, or rock backgrounds on which the animals live, he was able to record the degree of matching or mismatching between the two.

Norris posed the question, Does color matching extend into the invisible spectral regions? From some 25 examples of animal and background reflectance, it was found that color matching seldom extends into the ultraviolet or infrared portions of the spectrum. It is also usually the case that good matching does not even extend throughout the entire visible spectrum. Sometimes, near-ultraviolet matching is sufficiently good to suggest that some predator vision occurs in the spectral range 325 to 400 nm.

Norris posed the question: does color matching extend into the invisi-properties fall into three well-defined classes:

1. Those species that match their backgrounds when in their lightest state, such as *Uma* sp. This class is called an *upper activity-level color match*.
2. Those species that match their backgrounds at intermediate-activity temperatures and overshoot background reflectance, becoming much lighter than the background, at higher temperatures. This class is termed a *middle activity-level color match*. Species in this group are *Dipsosaurus dorsalis* and *Callisaurus dracanoides*.
3. Those species that achieve an apparent match with background, such as *Holbrookia maculata ruthveni*. This class is called *pseudomatch*.

The color-match behavior of *Uma scoparia* (fringed-toed sand lizard) is shown in Fig. 8.40, along with the reflectance of sand from its habitat. Animal and sand curves diverge at about 325 nm in the ultraviolet and at 650 nm in the red. This lizard has some of the most precise color-matching at visible wavelengths of any animal observed.

The color habitats of desert regions vary a great deal. Dune sands, which are highly reflective, are the habitat of the desert iguana, *Dipsosaurus dorsalis*. This animal matches the sand with good fidelity when in its active temperature range (around 36°C). At very high body temperatures, 44–47°C, it becomes more reflective than the sand. The desert iguana can reflect as much as 20% of the total solar energy incident upon it and thereby remain active longer than any other lizard in the heat of the day.

Approximately the same kind of color change occurred with the

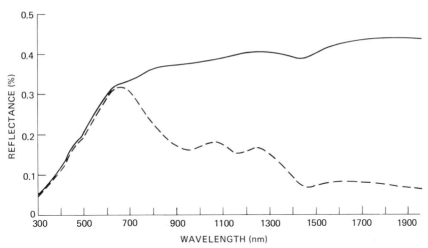

Figure 8.40. Spectral color match of (−−) the fringed-toed sand lizard (*Uma scoparia*) and (—) the sand from its habitat. Temperature of the lizard, 36°C. (Redrawn from Norris, 1967.)

zebra-tailed lizard, *Callisaurus dracanoides*, except that maximum concentration of melanin took place at a considerably lower temperature (up to 44°C). The spectral reflectance curves for this animal are shown in Fig. 8.41. Norris (1967) makes the following comment: "In both of these thermophyllic species, this "superlight" condition has been noted in nature. It is seen during very hot spring and summer days preceding retreat by the lizards into burrows or shade, and at times when nearly every other vertebrate has fled from the heat into various retreats. The animals are gleaming white and rather conspicuously shining against the sand. These two species remain active into thermal periods when predators have probably completely left the scene. May it not be that color matching is more precise when these species are active in an environment filled with other creatures, including predators, and once these associates have retreated, the activity periods of the two lizard species are extended by sacrifice of some measure of protective coloration no longer needed? The extended periods of both forms are probably of use because of the vegetarian proclivities of both species. The adult desert iguana is almost wholly vegetarian and the zebra-tailed lizard is known to eat flower heads, perhaps for their insect content."

For the *pseudomatch* color-matching animals, the substratum background is very reflective. Light seems to saturate the human eye, and presumably predator eyes, the so-called dazzle effect. Quoting Norris on this point: "White Sands, N.M., provides the best example of this phenomenon in any of my records. The stark white gypsum sands so fill the eyes with reflected light that it is very difficult for a person's eyes to accomodate. A human walking amid the dunes may be near visual pain thresholds

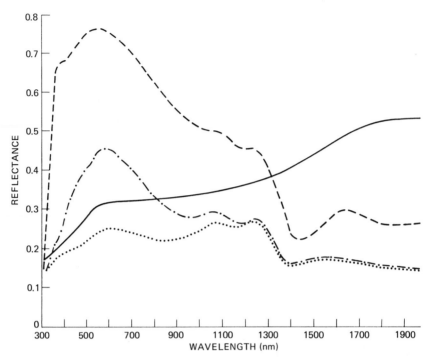

Figure 8.41. Spectral reflectance of the zebra-tailed lizard (*Callisaurus dracon-oides*) before and after treatment with ACTH together with the sand from its habitat: (—) sand; (– · –) dorsal surface, 44°C; (· · · ·) dorsal surface, 1 hr after ACTH; (– –) ventral surface at 36°C. (Redrawn from Norris, 1967.)

continuously on a bright day. The most common lizard on these dunes, *Holbrookia maculata ruthveni*, is all but invisible in the glaring wilderness. Usually, only its eyes and its black lateral auxillary blotch or its movement reveals its presence. Yet, when this lizard is captured and taken from the dunes and placed in a box of White Sands dune sand, it is seen to be slightly yellowish in tone and significantly darker than the sand from which it came. The match is not particularly good, and the eye can detect the failure in such a restricted situation, but on the open dunes no finer color match exists."

The best color matches occur when the backgrounds reflect about 30% of the incident visible light or less. This is seen by comparing the color matching of *Uta stansburiana* (reflectance of 0.6) living on the dark Pisgah Lava Flow in Southern California, where the match is within 1%, and that of *Holbrookia m. ruthveni*, whose reflectance, at 0.45, differs by nearly 30% from the reflectance of the underlying sand (0.75) at 700 nm. A number of species of reptiles are closely color matched to dark lava flows, and this phenomenon has also been reported for some mammals, (Summer, 1921; Blair, 1947). Although rodent populations on lava flows sometimes show melanism, they often do not. Existence of color matching seems to be related to the size of the flow and its degree of isolation from other dark substrata by light-colored alluvium as discussed by Hooper (1941). Norris (1967) reports that the minimum area for color matching is about 0.09 acre, and territorial studies by Tinkle et al. (1962) give the territory of male *Uta stansburiana* in western Texas as 0.06 acre. These areas seem to be too small to support a functional gene pool, and the color matching produced is largely phenotypic. *Uta stansburiana* lizards form an inexact color match with their backgrounds when born but gradually, during the first month or two, become increasingly blended with their background. Animals living on reddish lava become buff to rusty in tone as a result of phaeomelanins in their skin. *Callisaurus dracanoides,* living on an alluvial apron composed of reddish pebbles, was of rusty coloration. James Heath told Norris about hatchling *Utas* living on substrates covered with white dust from a cement plant which were not color matched but eventually became well-matched to the white color with age. Apparently, there is a color-matching ability which is not genetically controlled.

In order to investigate further the adaptive strategy of color matching, Norris (1967) studied a large number of specimens of the side-blotched lizard, *Uta stansburiana stejnegeri,* found along a transect near Pisgah Crater that crossed small stringers of lava, intervening sandy swales, a dry lake bed, lava ridges, and an alluvial fan. When he tested these lizards on the spectrophotometer, he found the following surprising results: "Animals taken from pure lava situations were found to lack, or nearly lack the capacity to color change and were found to be matched to their background all of the time. On the other hand, those lizards taken from adjacent lighter-colored areas all possessed color change potential, the greatest color lability existing in those animals taken from the lightest soils. Such lizards matched their background only when in their lightest phase. Hormonal, electrical, or thermal stimulation of lava-dwelling animals produced little or sometimes no change at all, except in males, in which the characteristic rows of small blue dorsal spots could be made to scintillate with structural blue color or to darken into conformance with the surrounding lava coloration. This blue color pattern was found to be exhibited by all male *Utas* during aggressive display, but not to be present in females. The dorsal blue coloration of male *Uta stansburiana* seems to be concerned with aggression and territoriality."

Spectral Transmission of Body Walls

There has been considerable interest in the damaging effect of solar radiation upon the body walls of vertebrates. Ultraviolet radiation is known to be damaging to organisms and tissues (see Hanawalt, 1966; Urbach, 1969). The presence or absence of black peritoneums in various animals suggests a protective role against solar radiation. The gonads of many diurnal vertebrates are surrounded by pigmented tissue, whereas their nocturnal relatives are not pigmented. Black peritoneums are observed in colorless fishes, and clear peritoneums in colored fishes. Warren Porter became interested in this subject for a Ph.D. thesis at UCLA while working with Kenneth Norris. He wrote that his interest was aroused "because of repeated observations in the literature about the correlation between diurnality and internal pigmentation and its inferred protection from solar radiation (1967, p. 273)." There seemed to be doubt that black peritoneum could afford sufficient protection against the ultraviolet radiation which reaches the surface of the earth. On the other hand, there was considerable question as to whether or not it could absorb sufficient energy to affect the energy budget of the animal and thereby its temperature.

According to Porter (1967), melanin concentration in the peritoneum of *Uta stansburiana,* as well as in other lizards possessing black peritoneums, varies in different parts of the body cavity. Melanin is found in all parts of the body wall, in the skin, muscles, and peritoneum. The distribution of melanin in the body wall greatly affects absorption and transmission of light by the wall. Porter made a detailed determination of the amount of solar energy absorbed by the black peritoneum and showed that, in terms of the total energy incident on the animal, the energy absorbed by the black peritoneum is small (about 4%). The muscle absorbs about 10% and the skin, with its layer of keratin, about 72% of the incident sunlight. Porter also reported similar measurements made on the body walls of aquatic amphibians and fishes which showed the black peritoneum to play an insignificant role in the total energy absorbed for thermoregulation. Porter made measurements of many diurnal lizards with black peritoneum, some lizards without a peritoneum, some nocturnal lizards, diurnal, crepuscular and nocturnal snakes, and amphibians, birds, mammals, and insects. His paper is an exceedingly important contribution to the subject of radiation and the body wall. Because his conclusions are so significant, they are reproduced here (Porter, 1967, p. 294):

> 1. The quantity and quality of light reaching the body cavity of all diurnal animals examined is remarkably consistent irrespective of the presence of a melanized peritoneum. When a black peritoneum occurs, other light-absorbing components of the body wall, e.g., skin or intermuscular melanin deposits, are less well-developed or absent. The black peritoneum in those species which possess it excludes significant amounts of ultraviolet light over the spectral interval of 290–400 nm.
> 2. The pigmented peritoneum seems insignificant in thermoregulation because a) it absorbs usually about 5% of incident solar energy whereas the skin usually absorbs 60–70% and the muscles 10–20%, b) it is present in many aquatic amphibians and fishes which would lose any heat gained to the surrounding water, and c) since all diurnal animals are opaque to light passing completely through the body, they would absorb any penetrating energy anyway. The peritoneum does not seem to be a heat shield for internal organs within the peritoneum; any energy ab-

sorbed by the peritoneum would be conducted very rapidly to the internal organs, thus precluding any significant temperature differential between the peritoneum and the internal organs.

3. Tissue death has no effect on light transmission provided the tissue is kept cool and does not dehydrate.

4. Ultraviolet light of wavelengths as short as 300 nm and with intensities sufficient to induce mutations could penetrate to the body cavity of reptiles which have had the peritoneum removed.

This last conclusion has been disputed by Hamilton (1973). More experiments are required in order to resolve the difference of opinion. Porter's conclusion is based on the fact that the destructive action or mutagenic effect of ultraviolet radiation is an integrative phenomenon with time and, although the instantaneous amount of ultraviolet solar radiation is not sufficient to produce measurable consequences, the total amount integrated over the daylight hours on 22 June at 30° north latitude might be sufficient. Since biological tissue consists largely of water, Porter points out that liquid water does transmit ultraviolet radiation. This is readily seen in the spectra for liquid water presented earlier in this chapter. The transmission spectra of body walls clearly show the liquid-water absorption bands in the infrared. These characteristic absorptions are always easily recognized at about 960, 1100, 1400, 1900, and 2700 nm.

Porter (1967) included measurements of reflectance and transmittance of a black carabid beetle, *Calosoma semilaeva,* because black beetles are often found in desert habitats during the daytime. The black beetle might be imagined to be extremely conspicuous against the light-colored desert sands, where it would absorb a great deal of solar radiation and become overheated. Porter found that *Calosoma* reflected only 8.6% of the total incident sunlight, transmitted 2.4% (all in the infrared), and absorbed 89.0%. The beetle's dorsal body wall and elytra excluded all incident ultraviolet and visible radiation. Although the beetle would appear to become very warm from the absorbed radiation, because of its small size, the beetle has a large convective heat transfer coefficient and exhibits strong convective cooling. This phenomenon was studied and reported by Parry (1951).

Color Change and Energy Budget of Lizards

Color change potentially can influence either absorption of radiation or concealment against predators or both. Cole (1943) and Klauber (1939) have investigated these possibilities. Very few investigations of the non-visible spectral properties of animals were made prior to the early 1950s, when Bodenheimer (1952) and Bodenheimer et al. (1953) reported transmittance measurements for excised back skin of *Ophisops,* and a number of insect tissues from the ultraviolet through 1500 nm. Hutchinson and Larimer (1960) reported reflectance measurements of excised lizard skins for wavelengths from 300 to 1200 nm and gave estimates of the effect on the animal heat budgets. Spectral measurements were needed for the entire solar spectrum of living animals in order to explore properly the possible influence of reflectance or absorptance on the animal's energy budget.

That the spectral properties of certain animals may be of special significance is indicated by the following general observations. Pearson (1954) has reported observations of an Andean iguanid lizard *Liolaemus multiformis,* which has a body temperature as much as 30°C above ambient air temperature when exposed to full sunshine in the high mountains of Peru. The sidewinder rattlesnake, *Crotalus cerastes,* is highly reflective, particularly at infrared wavelengths, as shown in Fig. 8.42. The sidewinder basks in the morning sun when partially buried for many hours and is apparently aided in doing so by its unusually high reflectance.

Animal size may be related to ability to undergo color change that affects the heat budget. The rate at which a poikilotherm changes body temperature is a product of its total insulation, including absorptance to radiation and heat capacity. A small animal tends to change temperature quickly, and absorptance could have considerable influence on this rate. A large animal changes its body temperature very slowly, and the reflec-

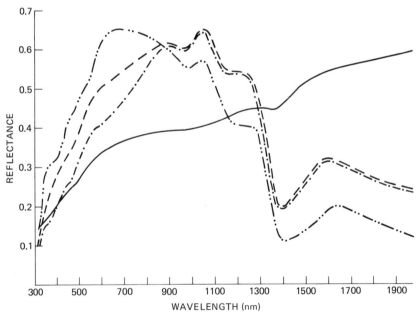

Figure 8.42. Reflectance of sidewinder rattlesnake (*Crotalus cerastes*) and sand from its habitat as a function of wavelength: (——) sand; (– –) dorsal surface, 38°–40°C; (– · –) dorsal surface, 20°–21°C; (– · · –) ventral surface. (From Norris, 1967.)

tance or absorptance of its surface will only affect its equilibrium temperature. Norris (1967) reported that small lizards have less reflectance and greater absorptance than large lizards. The transition point, at which absorptance goes from high to low, is for a lizard of about 40 g and a surface-to-mass ratio of 3.5. The small animal has a distinctly greater convective heat-transfer coefficient than the larger animal. Therefore, there is not much of a temperature disadvantage to being dark-colored, since the tendency to reach a higher temperature is compensated by increased convective cooling, a factor that is particularly important in desert regions, which tend to be windy. The dark-colored smaller lizard is able to make fairly quick temperature adjustments in the radiation field of the desert environment, whereas a large lizard such as the giant spiny chuckwalla (ca. 1000 g), *Sauromalus hispidus,* is sluggish and has a high reflectance and reduced absorptance which allows it to extend its activity period by reducing its heat load during the hotter part of the day.

Considerable color lability shows up in all small heliothermic (solar heated) iguanid genera explored by Norris, including *Uta, Uma, Callisaurus, Urosaurus, Sceloporus, Dipsosaurus, Crotaphytus,* and *Phrynosoma.* Color change was common among small agamid lizards, but was not found in thigmothermic families such as Teiidae, Lacertidae, and Scincidae. The heavy-bodied iguanid genus *Sauromalus* had good color lability in the smallest species, *S. obesus,* and only slight color lability in the larger species *S. varius* and *S. hispidus.* Color lability was very evident in the young of both large species, but diminished during growth. The young of the American alligator, *Alligator mississippiensis,* and the Nile crocodile, *Crocodilus niloticus,* showed some color change. Some pattern change was observed in *Iguana iguana.* The monitor lizards apparently do not change color.

The spectral reflectance curves of the larger lizards are quite different from those of the smaller lizards. For those lizards greater than 30 to 40 g and having a surface-to-mass ratio less than 3.5, near-infrared reflectance is consistently greater than the visible reflectance. The small heliotherms of about 9 to 18 g or less and having a surface-to-mass ratio of 4.0 to 4.8 or more have reflectances which peak in the visible, dip sharply in the red part of the spectrum, and reach low values in the near infrared. Comparison of the lizard reflectances with those of the substrates on which they are found shows that dorsal reflectance continues to rise above substrate reflectance in the red and near infrared in large lizards but drops below that of the substrate, often starting in the red and continuing to drop strongly in the near infrared, in small species.

The effect of color change on the energy budget of *Dipsosaurus dorsalis* was carefully determined by Norris for dorsal and ventral surfaces. He found that this lizard absorbs a total amount of solar radiation of 6.14 W in its lightest state and 7.60 W in its darkened state, a significant difference.

Spectral Reflectance and Absorptance
of Vertebrates

The spectral reflectance of the surface of various vertebrate skins has been measured in my laboratory. For the most part, reflectance of the skins is very close to the reflectance of the living animal. The spectral absorptance is equal to 1.0 minus the spectral reflectance. Typical spectral absorptance curves from 300 to 2500 nm are given for the dorsal surface of mammals in Fig. 8.43 and Fig. 8.44 for birds. As a general rule, the absorptance is above 0.90 in the ultraviolet and at visible wavelengths shorter than 500 nm. Usually, at about 700 nm, absorptance drops and continues to drop until about 1300 nm, when it levels out or begins to increase. A second maximum occurs around 2500 nm (presumably this maximum is at 2700 nm, where liquid water has a strong absorption band). The other absorption bands produced by liquid water are readily identified in all of the spectral absorptance curves at 1450 and 1900 nm. Weaker water bands at 1100 nm or shorter do not usually appear in these spectral curves. Visual coloration of the animal surface occurs at wavelengths between 500 and 700 nm. For example, the sharp dip in absorptance for the cardinal at about 650 nm indicates a reflectance peak at this wavelength; the bird thus appears red to the eye.

The mean value of absorptance to solar radiation is calculated by multiplying the spectral absorptance by the spectral intensity of solar radiation, integrating over all wavelengths, and dividing by the total solar energy incident on the animal. Mean absorptances were determined for about 100 different animal species for a high and low sun. The low sun is relatively rich in red wavelengths but somewhat depleted of blue wavelengths because of the long slant path through the atmosphere. Since spectral absorptance usually begins to drop at red wavelengths in contrast to the blue and ultraviolet, the mean absorptance to radiation from a low sun is always less than that from a high sun. The mean absorptance values for many animals are listed in Table 8.2. Almost invariably, the ventral mean absorptance is less than the dorsal mean absorptance by about 10% and sometimes as much as 30%. Most vertebrates (mammals and birds) have mean absorptances between 0.70 and 0.80, with 0.75 very common. Of the animals on this list, more birds tend to be darker than mammals although the species listed here are not totally representative of the animal kingdom. None of the animals listed here has dorsal absorptances to high or low sun of less than 0.40, except for a juvenile white swan, with absorptances of 0.367 and 0.314 to a high and low sun, respectively. However, many ventral absorptances are between 0.30 and 0.40.

How "black" is black and how "white" is white for various animal surfaces to solar radiation? None of the birds or mammals has dorsal absorptances above 0.90, except the Siamang, *Hylobates syndactlylus*, a very black animal with mean absorptance to solar radiation of 0.946.

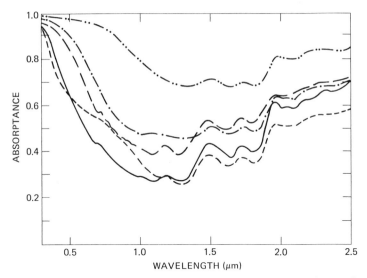

Figure 8.43. Absorptance of the dorsal surface of several mammals as a function of wavelength: (– · · –) grizzley bear, *Ursus horribilis;* (–––) Eastern cottontail rabbit, *Sylvilagus floridanus;* (– · –) red fox, *Vulpes fulva;* (—) Eastern chipmunk, *Tamias striatus;* (——) fox squirrel, *Sciurus niger.*

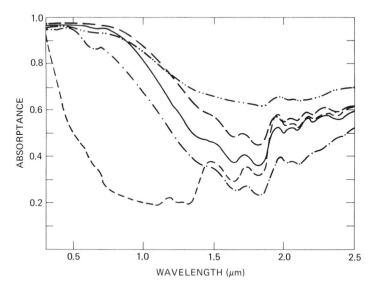

Figure 8.44. Aborptance of the dorsal surface of several birds as a function of wavelength: (– · · –) starling, *Sturnis vulgaris;* (—) pileated woodpecker, *Hylatomus pileatus;* (——) Muscovy black duck, *Cairina moschata;* (–––) whistling swan, *Cygnus columbianus;* (– · –) cardinal, *Richmondena cardinalis.*

Table 8.2. Mean Solar Reflectances and Absorptances of the Surfaces of Animals for a Sun Low and High in the Sky.[a]

Species	Part	R_L	R_H	A_L	A_H
Birds					
Accipiter cooperii, Cooper's Hawk	Dorsal	0.286	0.216	0.714	0.784
	Ventral	0.203	0.165	0.797	0.835
Ajaia ajaja, American roseate spoonbill	Dorsal	0.510	0.455	0.490	0.545
Ammospiza caudacuta, sharp-tailed sparrow	Dorsal	0.334	0.250	0.666	0.750
	Ventral	0.688	0.639	0.312	0.361
Asio otus wilsonianus, long-eared owl	Dorsal	0.451	0.342	0.549	0.658
	Ventral	0.434	0.351	0.566	0.649
Bombycilla cedrorum, cedar waxwing	Dorsal	0.252	0.193	0.748	0.807
	Ventral	0.289	0.236	0.711	0.764
Cairina moschata, Muscovy duck	Wing	0.164	0.121	0.836	0.879
Circus cyaneus hudsonius, marsh hawk	Dorsal	0.189	0.139	0.811	0.861
	Ventral	0.441	0.350	0.559	0.650
	Egg	0.861	0.802	0.139	0.198
Coccyzus americanus, yellow-billed cuckoo	Dorsal	0.300	0.233	0.700	0.767
	Ventral	0.707	0.658	0.293	0.342
Colaptes auratus, flicker	Dorsal	0.313	0.239	0.687	0.761
	Ventral	0.516	0.461	0.484	0.539
Colinus virginianus, quail	Dorsal	0.366	0.280	0.634	0.720
	Ventral	0.477	0.405	0.523	0.595
	Egg	0.869	0.818	0.131	0.182
Corvus conix, hooded crow	Egg	0.626	0.514	0.374	0.486
Cyanocitta cristata, blue jay	Dorsal	0.268	0.214	0.732	0.786
	Ventral	0.403	0.347	0.597	0.653
	Egg	0.638	0.527	0.362	0.473
Cyanocitta stelleri, Stellar's jay	Dorsal	0.156	0.120	0.844	0.880
Cygnus columbianus, white swan (juvenile)	Dorsal	0.686	0.633	0.314	0.367
Falco sparverius, sparrow hawk	Egg	0.769	0.665	0.231	0.335
Florida coerulea coerulea, little blue heron	Egg	0.733	0.647	0.267	0.353

Table 8.2. (*continued*)

Species	Part	R_L	R_H	A_L	A_H
Fregata ariel, lesser frigate bird	Egg	0.891	0.865	0.109	0.135
Fregata magnificens, Magnificent frigate bird	Egg	0.857	0.819	0.143	0.181
Geococcyx californianus, roadrunner	Wing	0.250	0.208	0.750	0.792
	Egg	0.829	0.818	0.171	0.182
Geothylpis trichas, yellow throat	Dorsal	0.352	0.265	0.648	0.735
	Ventral	0.680	0.609	0.320	0.391
Grus americana, whooping crane (chick)	Dorsal	0.415	0.331	0.585	0.669
Guara rubra, scarlett ibis	Wing	0.532	0.427	0.468	0.573
Haliaetus leucocephalus, bald eagle	Egg	0.792	0.708	0.208	0.292
Hylatomus pileatus, pileated woodpecker	Red Head	0.410	0.321	0.590	0.679
	Dorsal	0.243	0.177	0.757	0.823
	Middorsal	0.225	0.165	0.775	0.835
	Middorsal	0.211	0.154	0.789	0.846
	Egg	0.825	0.812	0.175	0.188
Ixobrychus exilis exilis, least bittern	Egg	0.783	0.733	0.217	0.267
Leucophoyx thula thula, snowy egret	Egg	0.799	0.714	0.201	0.286
Meleagris gallopavo, wild turkey	Egg	0.719	0.651	0.281	0.349
Megaceryle alcyon, kingfisher	Middorsal	0.136	0.108	0.864	0.892
	Dorsal	0.264	0.208	0.736	0.792
	White neck	0.686	0.662	0.314	0.338
Mniotilta varia, black-and-white warbler	Dorsal	0.321	0.247	0.679	0.753
	Ventral	0.684	0.652	0.316	0.348
Nycticora nycticora hoactli, black-crowned night heron	Dorsal	0.175	0.142	0.825	0.858
	Ventral	0.355	0.299	0.645	0.701
	Egg	0.795	0.709	0.205	0.291
Otus asio, screech owl	Middorsal	0.292	0.227	0.708	0.773
	Dorsal	0.419	0.330	0.581	0.670
	Ventral	0.590	0.519	0.410	0.481
	Egg	0.920	0.892	0.080	0.108

Table 8.2. (*continued*)

Species	Part	R_L	R_H	A_L	A_H
Phasianus colchicus torquatus, ring-necked pheasant	Dorsal	0.232	0.180	0.768	0.820
	Ventral	0.289	0.221	0.711	0.779
Prophyrula martinica, purple gallinule	Egg	0.796	0.717	0.204	0.283
Pygoscelis adeliae, Adelie penguin	Right front	0.199	0.153	0.801	0.847
	White breast	0.735	0.676	0.265	0.324
Quisalus versicolor, bronzed grackle	Egg	0.619	0.506	0.381	0.494
Rallus elegans elegans, king rail	Dorsal	0.200	0.147	0.800	0.853
	Ventral	0.318	0.256	0.682	0.744
	Egg	0.729	0.655	0.271	0.345
Richmondena cardinalis, cardinal	Dorsal	0.320	0.246	0.680	0.754
	Ventral	0.496	0.408	0.504	0.592
Seiurus aurocapillus, ovenbird	Dorsal	0.330	0.247	0.670	0.753
	Ventral	0.622	0.576	0.378	0.424
Setophaga ruticilla, redstart	Dorsal	0.335	0.253	0.665	0.747
	Ventral	0.635	0.561	0.365	0.439
Speotyto cunicularia, burrowing owl	Egg	0.899	0.898	0.101	0.102
Spheniscus demersus, black-foot penguin	Egg	0.846	0.812	0.154	0.188
Sterna albifrons, least tren	Egg	0.724	0.601	0.276	0.399
Sterna fuscata fuscata, sooty tern	Egg	0.744	0.674	0.256	0.326
Strix varia, barred owl	Dorsal	0.412	0.315	0.588	0.685
	Ventral	0.499	0.405	0.501	0.595
	Egg	0.843	0.802	0.157	0.198
Sturnella magna, eastern meadowlark	Egg	0.874	0.824	0.126	0.176
Sturnis vulgaris, starling	Dorsal	0.163	0.127	0.837	0.873
	Ventral	0.149	0.112	0.851	0.888
	Egg	0.760	0.721	0.240	0.279
Struthio camelus, ostrich (chick)	Middorsal	0.175	0.143	0.825	0.857
Turdus migratorius, robin	Dorsal	0.337	0.258	0.663	0.742
	Ventral	0.479	0.409	0.521	0.591

Table 8.2. (*continued*)

Species	Part	R_L	R_H	A_L	A_H
Vireo olivaceus,	Dorsal	0.336	0.258	0.664	0.742
red-eyed vireo	Ventral	0.661	0.614	0.339	0.386
Mammals					
Ammotragus lervia,	White middorsal	0.349	0.296	0.651	0.704
audad, or Barbary sheep	White-tan	0.438	0.358	0.562	0.642
	Brown dorsal	0.476	0.392	0.524	0.608
Bison bison,	Dorsal	0.298	0.223	0.702	0.777
American bison					
Blarina brevicauda,	Dorsal	0.337	0.264	0.663	0.736
short-tailed shrew	Ventral	0.342	0.280	0.658	0.720
	Gray-black				
	middorsal	0.237	0.193	0.763	0.807
	Brown middorsal	0.266	0.218	0.734	0.782
Canis familiaris,	White flank	0.557	0.541	0.443	0.459
pointer (pup)	White dorsal	0.597	0.577	0.403	0.423
Canis latrans,	Dorsal	0.393	0.311	0.607	0.689
coyote					
Canis latrans,	Yellow-brown				
coyote (female pup)	flank	0.313	0.260	0.687	0.740
Canis lupus,	Dorsal	0.380	0.304	0.620	0.696
gray wolf					
Canis lupus,	Flank	0.243	0.197	0.757	0.803
wolf (pup)					
Castor canadensis,	Dorsal	0.311	0.237	0.689	0.763
beaver					
Cercocebus torquatus,	Dorsal	0.169	0.131	0.831	0.869
sooty mangabey					
Citellus beldingi,	Dorsal	0.230	0.169	0.770	0.831
Belding ground squirrel	Side	0.407	0.342	0.593	0.658
Didelphis marsupialis,	Dorsal	0.560	0.458	0.440	0.542
opossum	Ventral	0.545	0.463	0.455	0.537
Eptesicus fuscus,	Dorsal	0.352	0.264	0.648	0.736
big brown bat					
Erethizon dorsatum,	Dorsal	0.188	0.147	0.812	0.853
porcupine					
Felis catus	Dorsal	0.585	0.555	0.415	0.445
cat (white)					

Table 8.2. (*continued*)

Species	Part	R_L	R_H	A_L	A_H
Felis pardalis,	Light middorsal	0.267	0.223	0.733	0.777
ocelot	Dark middorsal	0.173	0.135	0.827	0.865
Giraffa camelopardalis,	Dorsal	0.321	0.248	0.679	0.752
Masai giraffe	Dark brown cheek	0.210	0.163	0.790	0.837
	Light gray cheek	0.408	0.367	0.592	0.633
Glaucomys volans,	Dorsal	0.387	0.294	0.613	0.706
southern flying squirrel	Ventral	0.694	0.640	0.306	0.360
Hylobates syndactlylus,		0.070	0.054	0.930	0.946
siamang (gibbon)					
Lynx rufus,	Dorsal	0.377	0.301	0.623	0.699
bobcat					
Marmota monas,	Dorsal	0.433	0.346	0.567	0.654
groundhog	Dorsal	0.335	0.281	0.665	0.719
Mephitis mephitis,	Black dorsal	0.210	0.156	0.790	0.844
striped skunk	White dorsal	0.545	0.507	0.455	0.493
Microtus pinetorum,	Dorsal	0.343	0.251	0.657	0.749
pine vole					
Mus musculus,	Dorsal	0.305	0.228	0.695	0.772
house mouse					
Mustela frenata,	Dorsal	0.331	0.258	0.669	0.742
weasel					
Myotis grisescens,	Dorsal	0.388	0.293	0.612	0.707
gray myotis bat					
Neotoma floridana,	Dorsal	0.323	0.246	0.677	0.754
eastern woodrat	Ventral	0.594	0.541	0.406	0.459
Ochotona princeps,	Dorsal	0.247	0.198	0.753	0.802
pika					
Odocoileus virginianus,	Neck	0.260	0.204	0.740	0.796
white-tailed deer	Red-tan middorsal	0.349	0.274	0.651	0.726
Ondatra zibethica,	Dorsal	0.355	0.270	0.645	0.730
muskrat	Ventral	0.504		0.496	
Panthera leo,	Tan thigh	0.350	0.272	0.650	0.728
lion (female)					
Saimiri sciureus,					
squirrel monkey		0.219	0.171	0.781	0.829
Sciurus niger,	Dorsal	0.267	0.213	0.733	0.787
fox squirrel	Ventral	0.450	0.379	0.550	0.621

Table 8.2. (*continued*)

Species	Part	R_L	R_H	A_L	A_H
Sigmodon hispidus, cotton rat	Dorsal	0.271	0.201	0.729	0.799
Spilogale putorius, spotted skunk	Dorsal	0.338	0.256	0.662	0.744
	Ventral	0.328	0.248	0.672	0.752
Sus scrofa, pig	Black middorsal	0.200	0.183	0.800	0.817
	White middorsal	0.328	0.322	0.672	0.678
Sylvilagus floridanus, eastern cottontail rabbit	Dorsal	0.351	0.270	0.649	0.730
	Ventral	0.540	0.485	0.460	0.515
Talpa europaea, common mole	Middorsal	0.317	0.264	0.683	0.736
Tamias striatus, eastern chipmunk	Dorsal	0.323	0.242	0.677	0.758
	Ventral	0.591	0.529	0.409	0.471
Tragelaphus strepsiceros, greater kudu	Dorsal	0.338	0.294	0.662	0.706
Uncia uncia, Siberian snow leopard	Dark brown	0.156	0.117	0.844	0.883
	Lightest	0.244	0.212	0.756	0.788
Urocyon cinereoargenteus, fox	Dorsal	0.245	0.195	0.755	0.805
	Black brown	0.218	0.165	0.782	0.835
	Flank	0.231	0.182	0.769	0.818
Urocyon cinereoargenteus, gray fox	Dorsal	0.333	0.279	0.667	0.721
Ursus horribilis, grizzly bear	Dorsal	0.151	0.115	0.849	0.885
Vulpes fulva, red fox	Dorsal	0.408	0.314	0.592	0.686
	Dorsal	0.400	0.325	0.600	0.675
Reptiles					
Alligator mississippiensis, alligator	Living dorsal	0.119	0.100	0.881	0.900
	Living ventral	0.413	0.426	0.587	0.574
	Dead dorsal	0.112	0.091	0.888	0.909
Cnemidophorus tigris, lizard	Shoulder	0.111	0.111	0.889	0.889
	Dorsal	0.154	0.150	0.846	0.850
Scelloporous malachiticus, fence lizard (female)	Dorsal	0.068	0.063	0.932	0.937
	Dorsal	0.058	0.052	0.942	0.948
Scelloporous malachiticus, fence lizard (male)	Dorsal	0.112	0.107	0.888	0.893
	Tail	0.076	0.085	0.924	0.915
Amphibians					
Desmognathus fuscus brimleyorum, central dusky salamander	Dorsal	0.265	0.293	0.735	0.707

Table 8.2. (*continued*)

Species	Part	R_L	R_H	A_L	A_H
Rana temporaria, frog	White foam nest	0.312	0.297	0.483	0.518
Plethodon caddoensis, Caddo Mountain salamander	Dorsal	0.113	0.094	0.887	0.906
Plethodon glotinosus, slimy salamander	Dorsal	0.099	0.081	0.901	0.919
Plethodon ovachitae, Rich Mountain salamander	Dorsal	0.101	0.086	0.899	0.914
	Dorsal	0.109	0.093	0.891	0.907
	Dorsal	0.091	0.077	0.909	0.923
Humans					
Homo sapiens, Caucasian	Back of hand	0.345	0.321	0.655	0.679
		0.311	0.289	0.689	0.711
		0.355	0.324	0.645	0.676
		0.305	0.272	0.695	0.728
		0.316	0.283	0.684	0.717
Homo sapiens, Negro	Back of hand	0.247	0.207	0.753	0.793
		0.248	0.209	0.752	0.791

[a] R_L, reflectance for low sun; R_H, reflectance for high sun; A_L, absorptance for low sun; A_H, absorptance for high sun. All animals are adult males unless otherwise indicated.

Black-appearing fur or feathers have solar absorptances of about 0.85. The pileated woodpecker has values of 0.835 and 0.846; a Muscovy duck, 0.879; and the dark, middorsal region of a kingfisher, 0.892. Several salamanders were very dark and had absorptances from 0.906 to 0.923. The fence lizard, *Scelloporus malachiticus,* had values from 0.893 to 0.948. Another lizard, *Cnemidophorus tigris,* had mean solar absorptances of 0.850 and 0.889. An alligator had values of 0.90 on a living hide and 0.909 on the hide of a dead animal. White-appearing feathers had mean solar absorptances as follows: white neck feathers of a kingfisher, 0.338; white breast feathers of an Adelie penguin, 0.324; and a white swan, 0.367. Often, white is not as "white" as it appears. A white domestic cat had a mean solar absorptance of 0.445, and the white stripe on a shunk, *Mephitis mephitis,* had an absorptance of 0.493 (its black fur, 0.844). The ventral surfaces of several birds had absorptances of 0.34 to 0.46, as did the southern flying squirrel.

Problems

1. A piece of glass of thickness 0.02 m has a transmittivity to green light of 0.90 at wavelength 550 nm. What are the monochromatic "deci-

mal'' and "Napierian" absorption coefficients for glass at this wavelength?

2. How much greater is the intensity of Rayleigh scattered skylight at a wavelength of 300 nm than that at 500 nm?

3. How much greater is the intensity of Rayleigh scattered skylight at a wavelength of 500 nm than that at 1000 nm? Consider that the sky is full of dust. Now, using the Ångstrom relationship, determine the expected ratio of scattered skylight of wavelength 500 nm to that at wavelength 1000 nm.

4. Using the monochromatic intensity of the extraterrestrial solar spectrum, obtain the ratio of the monochromatic intensities at 550 and 1000 nm. Now convert these monochromatic intensities from a wavelength basis to a frequency basis, that is, instead of per nm of wavelength to per cm^{-1} (or m^{-1}) of frequency and take the same ratio. What does the result indicate?

5. If the index of refraction of liquid water to visible light is $n = 1.33$, calculate the reflectance of the water surface to light when the angle of incidence is 20°, 40°, and 80°.

6. If the "Napierian" absorption coefficient of a lake is 0.20 m^{-1} for green light, at what depth is 50% of the light absorbed? At what depth is 98% absorbed?

7. Two leaves designated A and B have absorptances as a function of the wavelength as given in the table below. The incident intensity of direct sunlight has the spectral distribution given. Determine the amount of direct solar energy absorbed by each leaf. The leaves have emissivities of 1.0. Compute the leaf temperatures if the leaves are in radiative equilibrium only. Does your answer tell you something about the advantages of the spectral properties of real leaves? If the leaf absorptances (and hence emittances) in the wavelength range 2.5 to 50 μm were 0.5 instead of 1.0, what would the radiative equilibrium temperature be?

Wavelength internal (μm)	Absorptance		Solar energy W m^{-2} $(\mu m)^{-1}$
	A	B	
0–0.3	1.0	1.0	0
0.3–0.5	1.0	1.0	250
0.5–0.7	1.0	1.0	500
0.7–1.0	0.1	0.8	400
1.0–1.5	0.1	0.8	160
1.5–2.5	0.5	0.9	100
2.5–50.0	1.0	1.0	0

Chapter 9

Conduction and Convection

Introduction

An organism immersed in a medium such as air, water, or soil exchanges heat by conduction and possibly by convection, radiation, and evaporation. In order to understand convective heat transfer, we should first understand the somewhat simpler mechanism of conductive heat transfer. Conduction is the transfer of heat along a temperature gradient from a region of higher to lower kinetic activity. Heat is transferred by conduction without mass movement of the medium. An example is the conduction of heat along the handle of a spoon in a cup of hot coffee. The process of heat transfer in soil is primarily one of conduction, although some mass movement of gases or water may occur. The process of heat transfer between an organism and a fluid is a combination of conduction and mass movement. Convection is simply the conduction of heat in a moving fluid. In convection, heat is transferred between a surface and a fluid by direct conduction through the boundary layer of fluid. The boundary layer is a layer of fluid which adheres to a surface because of the viscosity of the fluid. Fluid motion transfers energy to or from the boundary layer into the general mass of fluid outside the boundary layer. The effectiveness of energy transfer by convection depends upon the mixing action of the fluid.

Definition of Free and Forced Convection

There are two types of convection. In free, or natural, convection, flow is created by density differences resulting from temperature differences within the fluid. In forced convection, fluid is forced to move by an external force, such as a fan, blower, or pressure difference, which creates a "wind." Currents of warm air rise from a warm radiator in free convection, and so also warm air rises from a hand surrounded by cool air. As parcels of warm air rise from our hand or a radiator they carry away heat. Warm air rises because it is less dense than the ambient air nearby. Likewise, cold air drains off of a cold surface, and heat is transferred to the cold surface from the warmer ambient air that replaces the air draining off. Open the door of the freezer compartment of the refrigerator and a draft of cold air immediately pours forth and drops toward the floor. This is free convection. Cold air drains off of a window pane on a cold winter night. A wind produces forced convection of heat to or from trees if there is a temperature difference between the trees and the air. A hand placed in a cold rushing mountain stream experiences the loss of heat to the water by means of forced convection.

Definition of a Fluid

In order to understand convection, one must understand the properties of a fluid. An ideal fluid is defined as a substance that will not transmit a shearing stress. An ideal solid, in contrast, has rigidity, transmits a shearing stress well, and retains its shape when acted upon by a force. A fluid does not retain its shape when acted upon by a force and generally tends to occupy or take the form of the vessel containing it. Fluids are subdivided into gases and liquids. A gas is a fluid that has no free surface and, by expanding, will fill completely any vessel containing it. A liquid, on the other hand, may present a free surface. Liquids are slightly compressible, i.e., possess large bulk moduli (comparable to those of solids). Gases are highly compressible and possess small bulk moduli.

Rather than referring to the mass m of a fluid, one usually refers to the density ρ, which the mass of the fluid per unit volume v:

$$\rho = \frac{m}{v}. \tag{9.1}$$

The quantity ρ has a value of 1.2 kg m^{-3} for air at 20°C and 1000 kg m^{-3} for water at 20°C.

Viscosity

Although a perfect or ideal fluid is defined as a substance that does not transmit a shearing stress, all actual fluids transmit shearing stresses and possess the property known as "viscosity." Fluids are viscous substances, and some fluids are more viscous than others. Honey, glycerin, molasses, and tar, for instance, are much more viscous than water, alcohol, or air.

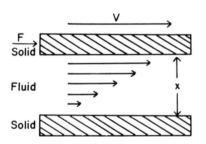

Figure 9.1. Shearing stress exerted on a fluid between two parallel plates separated by a distance x. If a force f is applied to one plate, it will accelerate to a velocity v when the other plate is held stationary.

Coefficient of Viscosity

Imagine two surfaces, shown in Fig. 9.1, separated a distance x, with the space between the two plates filled with a fluid such as oil, water, or air. If we apply a constant force f to the upper plate as shown, the upper plate will accelerate to a velocity V, moving at a uniform rate relative to the lower plate. If the force applied to the upper plate is doubled, then the velocity of the upper plate relative to the lower plate will double. If the distance between the plates is doubled, then the velocity of the upper plate will double. If each plate has a surface area A, it is found experimentally that the velocity of the upper plate for a given amount of force f goes down as A increases. These observations can be expressed mathematically as follows:

$$V \propto f\, x\, \frac{1}{A}. \tag{9.2}$$

Solving for f gives

$$f = \mu\, A\, \frac{V}{x}, \tag{9.3}$$

where the proportionality constant μ is the coefficient of viscosity, sometimes called the dynamic viscosity. The viscosity coefficient is clearly the tangential force per unit area of surface required to maintain a certain velocity gradient in the fluid. Hence

$$\mu = \frac{f}{A}\frac{x}{V}, \tag{9.4}$$

where μ is in units of Newton seconds per square meter. The tangential force per unit area is referred to as the shearing stress.

Since the effect of a given force upon an object is inversely proportional to the mass of the object, the affect of a force on a fluid is inversely proportional to the density of the fluid. Hence another coefficient, called the kinematic viscosity ν, is often used. This quantity is defined as follows:

$$\nu \equiv \frac{\mu}{\rho}. \tag{9.5}$$

The units of ν are square meters per second. There is often confusion between the terms dynamic and kinematic viscosity, and it is necessary to be careful in the use of these terms. Note that the units for these two co-efficients of viscosity are different.

At 20°C, air has a dynamic viscosity μ of 18.2×10^{-6} N s m^{-2} and a kinematic viscosity ν of 15.2×10^{-6} m^2 s^{-1} since the density of air is 1.2 kg m^{-3}. For water at 20°C, $\mu = 10.05 \times 10^{-2}$ N s m^{-2}, $\nu = 10.05 \times 10^{-2}$ m^2 s^{-1}, and $\rho = 1.00$ kg m^{-3}.

The viscosity of a liquid is caused by the cohesive forces of attraction of the molecules, which are relatively close together. In a gas, the molecules are too far apart to exert cohesive forces on one another, and the viscosity is the result of momentum transfer of the moving molecules from one part of the gas to another. For a gas, the coefficient of viscosity increases with the temperature just as one would expect from the momentum-transfer theory of gas viscosity with increased collision frequency with temperature. For a liquid, however, viscosity decreases with an increase in temperature precisely as one would expect intuitively since the theory of viscosity of liquids is based on the notion of molecular attraction which weakens with increased temperature.

The Boundary Layer

When fluid flows along a surface, the viscosity of the fluid produces a drag force, so that fluid particles near the surface are slowed down. In fact, the fluid particles next to the surface stick to the surface and have zero velocity relative to it. Full velocity is achieved only at some distance from the surface. The transition zone, within which the velocity is increasing in the direction away from the surface, is referred to as the "boundary layer." This is illustrated in Fig. 9.2. The boundary layer is sometimes less than a millimeter and may be more than a centimeter in width. The boundary layer is also a transition zone for temperature; there may exist not only a velocity profile but also a temperature profile adjacent to the surface (see Fig. 9.2). A velocity profile may occur without a temperature profile if the surface is at the same temperature as the fluid.

Each person can experience the boundary layer attached to his or her own body, and at times it is possible to see the boundary layer attached to other surfaces. Sometimes, when the light is just right, one can see beneath the shimmering currents of air rising from the hot hood of an automobile the thick adhering layer of air next to the metal surface. Within this boundary layer, there is little evident air movement for a space of about 1 cm above the surface. There are numerous other instances in which boundary layers near surfaces can be seen. The observation is clearest when there is a dramatic temperature difference between the surface and the air, giving rise to contrasts of air density.

When a person is immersed in warm water in a bathtub, the establish-

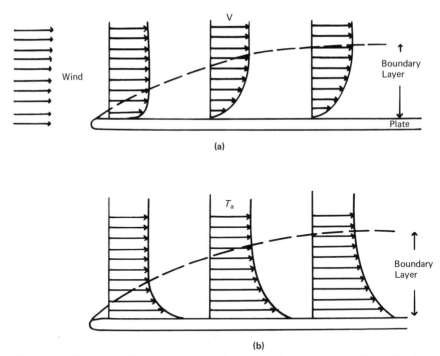

Figure 9.2. The transition zone, or boundary layer, between the surface of a plate and the free air beyond the plate's influence as illustrated by the (a) wind-speed and (b) temperature profile when the plate is warmer than the air.

ment of a boundary layer of water adhering to the skin occurs during the first few seconds. At first, the water feels warm, but within seconds, seems cooler. This is the result of the water adjacent to the skin having reached skin temperature. If one stirs the water in the bathtub, thereby putting it into motion, the boundary layer is partially broken down and made thinner, warm water gets close to the skin, and more heat is conducted to the skin itself. The temperature gradient is much greater when the boundary layer is small; the warmth of the water is conducted more effectively to the skin in this case. This is an experiment that anyone can do.

Laminar and Turbulent Flow

Since transport of heat by convection is intimately related to fluid flow along a surface, it is important to understand the properties or modes of fluid flow. Fluid dynamicists recognize two primary types of flow in fluids, laminar flow and turbulent flow.

Laminar flow, illustrated in Fig. 9.3, is a streamlined flow in which one fluid particle follows another in an orderly fashion. If the surface over which the fluid is flowing is at a temperature different from that of the

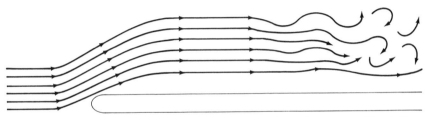

Figure 9.3. Air flow over a flat plate. Initially, there is laminar (streamlined) flow. Further along the plate, turbulent flow develops.

fluid, then heat is transferred by molecular conduction within the fluid and at the interface of the fluid and the surface. Temperature and velocity profiles for laminar flow at various positions along a surface are shown in Fig. 9.2.

Turbulent flow is in distinct contrast to the streamlined nature of laminar flow. Under conditions of turbulent flow, the smooth character of the flow breaks down, and there are numerous eddies and irregularities generated as the fluid passes over the solid surface. This type of flow is also illustrated in Fig. 9.3. The actual behavior of a fluid in turbulent flow over a surface is much more complex, however. A more detailed illustration of turbulent flow appears in Fig. 9.4, which indicates that near the surface of a plate there is a laminar sublayer (which may be very thin), an intermediate buffer layer, and the fully turbulent layer above this. Heat is transferred in turbulent flow by the innumerable eddies which carry parcels of fluid outward.

It is relatively easy to demonstrate to oneself both laminar and turbulent flow in air. Hold a cigarette in the air and watch the plume of smoke rising from it. Warm air rising from a cigarette is buoyant and gains speed as it rises. At first, the flow is streamline, regular, and smooth; then, as the smoke rises more rapidly, the flow pattern becomes less regular, and turbulence sets in.

The same phenomena is observed with fluid flow along the surface of a flat plate. There is always laminar flow near the leading edge, but this breaks down because of viscous forces between the plate surface and the fluid, and flow becomes turbulent. The phenomena of fluid flow is complex, but it is possible to predict approximately where the transition from laminar to turbulent flow will occur. There are always small disturbances and waves in a flowing fluid; however, if the viscous properties of the fluid are large, they will prevent the disturbances from growing in magnitude. As fluid flows over the plate back from the leading edge, the boundary

Figure 9.4. Turbulent flow over a plane surface illustrating the existence of a laminar sublayer, an intermediate buffer layer, and the fully turbulent layer.

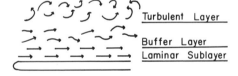

Turbulent Layer

Buffer Layer

Laminar Sublayer

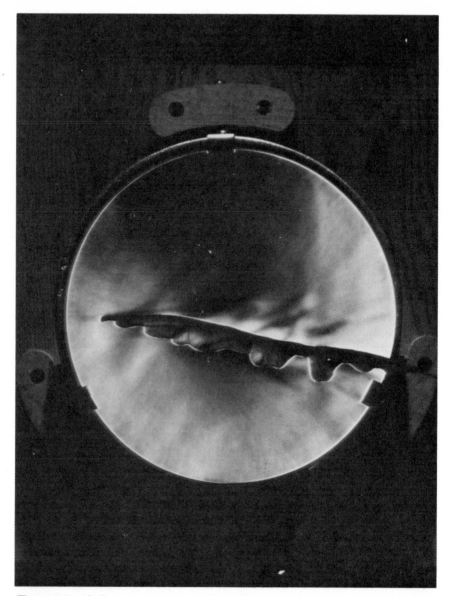

Figure 9.5. Air flow over a bur oak leaf (*Quercus macrocarpa*). Air is flowing from left to right at a speed of about 1.33 m s⁻¹ (3 mph). Note the laminar flow near the leading edge, with turbulent flow occurring over much of the leaf surface.

layer thickens, the ratio of viscous to inertial forces decreases, and, eventually, a position is reached where the disturbances are no longer damped out but grow with time. When this occurs, the boundary layer becomes unstable, and the transition from laminar to turbulent flow begins.

The coefficients of heat and mass transfer for turbulent flow are different from those for laminar flow. Flow of air along a plant leaf at a wind speed of approximately 1.33 m s^{-1} is shown in Fig. 9.5. The laminar flow of the air back from the leading edge at the left is seen clearly. Turbulent flow is evident further along the leaf, at the place where eddies form and the motion becomes very irregular. Air flow over a small leaf is usually laminar, and the rate of heat and mass transfer is distinctly greater than that of a large leaf. The heat-transfer coefficient of a large leaf averages the laminar flow back from the leading edge and the turbulent flow occurring across the remainder of the leaf. If the leaf is of considerable size, the nature of the flow is primarily turbulent and the laminar condition may be inconsequential at most air speeds. The rate at which heat is transferred by convection is increased by increasing the velocity and turbulence of the fluid since this decreases the boundary-layer thickness.

Heat Transfer by Conduction

It is useful to describe the nature of heat transfer by conduction before taking up the subject of convection. The simplest heat-flow problem is the steady flow of energy from a region of higher to lower temperature across a slab of material with plane, parallel faces. The geometry of plants and animals, however, is often cylindrical, spherical, or some combination (often asymmetrical) of these forms. Most heat-transfer analysis for animals is done using a plane-parallel-plate analog, but more often a cylindrical or spherical geometry should be used for more precision. The distinctions among these models will become apparent in the following discussion.

Heat Conduction for a Plate

The quantity of heat q transferred across a layer of material having plane, parallel faces maintained at temperatures T_1 and T_2 is directly dependent on the temperature difference $T_1 - T_2$, the surface, or cross sectional, area A, the time t, and the conductivity k of the material and inversely dependent on the thickness d of the material. Hence

$$q = k \, A \, t \frac{T_1 - T_2}{d}, \qquad (9.6)$$

where T_1 is greater than T_2.

If the rate of heat flow per unit area per unit time Q with any temperature gradient $\partial T / \partial x$ is constant, then

$$Q = \frac{q}{At} = k \frac{\partial T}{\partial x} \qquad (9.7)$$

and is a constant. The solution to this equation is

$$T = T_1 - \frac{x}{d}(T_1 - T_2). \qquad (9.8)$$

Thus the temperature gradient is constant throughout the slab, and the temperature change with depth is strictly linear.

Heat Conduction for a Cylinder

The flow of energy per unit of time through a hollow cylindrical surface of length ℓ and radius r is

$$\frac{q}{t} = k(2\pi r)\,\ell\,\frac{\partial T}{\partial r}, \tag{9.9}$$

and takes a constant value.
Integration gives

$$T = A \ln r + B, \tag{9.10}$$

where A and B are constants of integration and are given by the boundary conditions. If $T = T_1$ at $r = r_1$, and $T = T_2$ at $r = r_2$, then equating the flow of heat through each of these cylindrical surfaces gives

$$T = \frac{T_1 - T_2}{\ln r_1/r_2} \ln r + \frac{T_2 \ln r_1 - T_1 \ln r_2}{\ln r_1/r_2}. \tag{9.11}$$

Whereas the temperature distribution across a uniform infinite slab is linear, that across a cylinder falls off exponentially with increasing radius. Differentiating $\partial T/\partial r$ and putting this quantity into Eq. (9.9) gives

$$Q = \frac{q}{2\pi r\,\ell t} = \frac{k\,(T_1 - T_2)}{r \ln r_1/r_2}. \tag{9.12}$$

Heat Transfer for a Sphere

The flow of energy q per unit of time t through a spherical surface of radius r is

$$\frac{q}{t} = k\,4\pi r^2\,\frac{\partial T}{\partial r}, \tag{9.13}$$

which is a constant value. Integration gives

$$T = \frac{A}{r} + B, \tag{9.14}$$

where A and B are constants of integration and are given by the boundary conditions. If $T = T_1$ at $r = r_1$ and $T = T_2$ at $r = r_2$, then

$$T = \frac{T_1 - T_2}{r_2 - r_1}\frac{r_1\,r_2}{r} + \frac{r_2\,T_2 - r_1\,T_1}{r_2 - r_1}. \tag{9.15}$$

Differentiating $\partial T/\partial r$ and substituting into Eq. (9.13) gives

$$Q = \frac{q}{4\pi r^2\,t} = \frac{k\,(T_1 - T_2)}{r_2 - r_1}\frac{r_1\,r_2}{r^2}. \tag{9.16}$$

Theory of Convection

Heat is transferred from a surface to a fluid at a rate dq/dt proportional to the area A of the surface and the temperature difference ΔT between the surface and the fluid. Hence

$$\frac{dq}{dt} = \overline{h}_c \, A \, \Delta T, \qquad (9.17)$$

where $\overline{h}_c$ is the average convective heat-transfer coefficient, or thermal convective conductance, in watts per square meter per degree centigrade.

Next to the plate, leaf, or animal surface, heat is transferred only by conduction. In this case, the rate of heat transfer is proportional to the temperature gradient dT/dy at $y = 0$, hence

$$\frac{dq}{dt} = kA \left.\frac{dT}{dy}\right|_{y=0}, \qquad (9.18)$$

where y is the direction of the normal to the surface and k is the thermal conductivity of the fluid. The amount of heat conducted across the boundary layer is equal to the amount of heat convected to or from the free air beyond the boundary layer. Hence

$$\frac{dq}{dt} = kA \left.\frac{dT}{dy}\right|_{y=0} = \overline{h}_c \, A(T_s - T_a), \qquad (9.19)$$

where T_s is the temperature of the surface and T_a is the temperature of the fluid (usually air or water). Consider that the temperature gradient is within a relatively thin boundary layer δ (Fig. 9.6). Although the actual

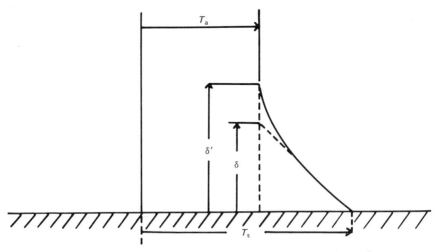

Figure 9.6. The temperature profile near a plane surface when the surface temperature is T_s and the air temperature is T_a. The actual boundary-layer thickness is δ', and the approximate boundary-layer thickness is δ.

boundary layer is of thickness δ', one may, to a first approximation, consider a linear relation between the temperature difference $T_s - T_a$ and the equivalent boundary-layer thickness δ. Then $dT = T_s - T_a$ and $dy = \delta$, so that the rate of heat transfer per unit area is

$$\frac{1}{A}\frac{dq}{dt} = k\frac{T_s - T_a}{\delta} = \overline{h}_c(T_s - T_a). \tag{9.20}$$

Therefore,

$$\overline{h}_c = \frac{k}{\delta}. \tag{9.21}$$

It is seen from Eq. (9.21) that there is a definite relation between boundary-layer thickness and the mean convective heat-transfer coefficient. At 20°C, thermal conductivity k for air is 25.7×10^{-3} W m^{-1}°C^{-1} and for water is 58.6×10^{-2} W m^{-1}°C^{-1}. For a boundary layer of air at 20°C 0.01 m in depth, then $\overline{h}_c$ is about 2.51 W m^{-2}°C^{-1} and for a boundary-layer 3.6×10^{-3} m, $\overline{h}_c = 6.98$ W m^{-2}°C^{-1}. It will be shown later that the convective heat-transfer coefficient for many objects of ordinary size is about 4.19 W m^{-2}°C^{-1}. This corresponds to a boundary-layer thickness of about 6×10^{-3} m.

The convective heat-transfer rate varies over the surface from point to point. An increment of surface area dA will have a local coefficient given by the amount of heat dq_c transferred from it as follows:

$$\frac{dq_c}{dt} = \overline{h}_c\, dA\, \Delta T. \tag{9.22}$$

The average convective heat-transfer coefficient is

$$\overline{h}_c = \frac{1}{A}\int\!\!\int_A \overline{h}_c\, dA. \tag{9.23}$$

The theory of convective heat transfer is complex, and to develop it fully is difficult. Engineers and physicists have had to combine four general techniques in order to obtain convective heat-transfer coefficients: (1) dimensional analysis combined with observations; (2) exact mathematical solutions of the boundary-layer equations; (3) approximate boundary-layer analysis by integral equations; and (4) the analogy between heat and mass transfer.

The engineer gets basic information concerning convective heat-transfer coefficients from direct laboratory measurements. The theory is developed as far as possible in order to provide a model within which to fit the data, but basically, the heat- and mass-transfer process is sufficiently complex for objects of unusual shape that one must resort to direct measurement. Certainly, this is true for the shapes of organisms, what we are concerned with here.

The leaves of many plants are sufficiently flat that one can approximate their convective heat-transfer coefficients by means of engineering theory developed for heated flat plates. Later, coefficients of convective heat transfer will be derived from observations in a wind tunnel using real leaves and models of real leaves. Tree trunks, limbs, branches, and so on may be approximated by cylinders for purposes of heat-transfer analysis. Most animals can be approximated by spheres, and the heat-transfer properties of a sphere are well known. Except for special geometrical arrangements, such as an array of cylinders, a collection of cooling fins, or fluid flow through pipes, the engineering literature contains information about heat transfer for plates, cylinders, or spheres. The biologist, in trying to understand the flow of air across the surface of organisms, is confronted with surfaces far more complex than those represented in the engineering literature.

Dimensionless Groups

The convective heat-transfer coefficient for any object immersed in a fluid is a function of many different variables, including size, shape, and orientation of the object, as density, viscosity, specific heat, and thermal conductivity of the fluid, velocity of the flow, and occurrence of laminar or turbulent flow. The functional relationship among these variables is complex and difficult to establish. Fortunately, there is a somewhat simplified scheme for establishing the proper relationships which helps significantly. Many of the variables involved in heat transfer can be combined in dimensionless groups, and there are functional relationships among these dimensionless groups. The basic character of heat transfer between an object and the fluid in which it is immersed is the same irrespective of the nature of the fluid. The actual rates of heat transfer differ from fluid to fluid, of course, but the fundamental character of the heat-transfer process is essentially invariant. This suggests that there are basic functional relationships among certain dimensionless groups which give a general description of the rate of heat transfer between an object and a fluid. Irrespective of whether the measurements are made in air, water, oil, molasses, or any other fluid, the functional relationships between certain sets of dimensionless groups have exactly the same form.

The procedure used for forming the proper dimensionless groups is known as dimensional analysis. It is a purely algebraic procedure and is straightforward but lengthy. For this reason, an explanation is not given here (see Kreith, 1966).

It is helpful to treat the formulation of convective heat transfer in forced convection separate from that for free convection. For convenience, the case of forced convection is described first. The convective heat-transfer coefficient h_c in forced convection is a function of one variable involving the object, i.e., the characteristic dimension D of the plate, cylinder, sphere, or organism. It is a function of five variables involving the fluid; these variables are thermal conductivity k, velocity of flow V, density ρ, viscosity μ or ν, and specific heat at constant pressure c_p. It is possible to define three dimensionless groups from these seven variables describing heat transfer for forced convection and derive functional relationships among them. Only one additional dimensionless group is required for the description of free convection.

Nusselt Number

For a fluid in motion, such as a stream current or wind, convective heat exchange occurs between an object and the fluid. It is necessary to have a dimensionless group containing the convective heat-transfer coefficient. Such a group is known as the Nusselt number (Nu) and is defined by

$$Nu \equiv \frac{h_c D}{k}. \tag{9.24}$$

From Eq. (9.21), k/h_c is equal to the boundary layer thickness δ and has the primary dimension of length. Since D also has the primary dimension of length, it is evident that Nu is dimensionless. Since Nu $= D/\delta$, the Nusselt number is the ratio of the characteristic dimension of the object or organism to its boundary-layer thickness. The Nusselt number also describes the ratio of the convective conductivity of the surface to the thermal conductivity per unit of dimension.

Reynolds Number

It is necessary to have a dimensionless group that expresses the dynamic properties of the fluid. Such a group is known as the Reynolds number (Re) and is defined by

$$Re \equiv \frac{VD\rho}{\mu} = \frac{VD}{\nu}. \tag{9.25}$$

The Reynolds number is an indication of whether flow over a surface is laminar or turbulent in character. When the Reynolds number is small, flow is laminar; when it is large, flow is turbulent. When the character of flow changes from laminar to turbulent, the functional relation between the Nusselt and Reynolds numbers changes considerably.

There is a principle of fluid dynamics known as the "model law" which states the following (according to Kreith, 1966): "The behavior of two systems will be similar if the ratios of their linear dimensions, forces, velocities, etc., are the same. Under conditions of forced convection in geometrically similar systems, the velocity fields will be similar provided the ratio of inertial forces to viscous forces is the same in both fluids. The Reynolds number is the ratio of these forces, and consequently we expect similar flow conditions in forced convection for a given value of the Reynolds number.''

Prandtl Number

A third dimensionless group, which expresses the static properties of the fluid, is also useful. This group is known as the Prandtl number (Pr) and is defined by

$$Pr \equiv \frac{c_p \mu}{k}. \tag{9.26}$$

Physically, the Prandtl number represents the ratio between the molecular diffusivities of momentum and heat, or is the relative efficiency of the conducting system for the molecular transport of momentum and energy.

Forced-Convection Function

Although the convective heat-transfer coefficient was a function of six variables, three dimensionless groups have been formed from them. Heat-transfer theory shows that, in the case of forced convection, the following relationship holds:

$$Nu = f(Re, Pr). \tag{9.27}$$

Experimental data can now be correlated in terms of three dimensionless groups instead of the original seven variables. This is an enormous reduction in complexity.

Grashof Number

In the case of free convection, one must arrange a new set of variables to represent the ability of a parcel of fluid warmer or colder than the surrounding fluid to rise against or fall with the attractive force of gravity. The set of variables includes the temperature difference ΔT between the object's surface and the fluid, the coefficient of volumetric thermal expansion β, and the acceleration due to gravity g. The fluid velocity V, does not enter into free convection, although the fluid is in gentle motion because of density differences set up by temperature differences. The results of the dimensional analysis for free convection suggest the use of the following dimensionless group, known as the Grashof number (Gr):

$$Gr \equiv \frac{\rho^2 \, g \, \beta \, \Delta T \, D^3}{\mu^2} = \frac{g \, \beta \, \Delta T \, D^3}{\nu^2}. \tag{9.28}$$

This quantity represents the ratio of the buoyant and inertial forces to the square of the viscous force. The Grashof number accounts for the "chimney" effect in free convection, by which warm air rises and cool air sinks. For Grashof numbers greater than about 2×10^7, the flow for a warm square plate facing upward or a cool plate facing downward is turbulent rather than laminar. For large Grashof numbers, buoyancy is sufficiently great that free convection is strong and inertial forces dominate over viscous forces, with the result that flow is turbulent. For horizontal cylinders, the transition from laminar to turbulent flow occurs at Grashof numbers greater than 10^9. This is readily demonstrated by smoke rising from a warm cigarette. The flow immediately above the cigarette is smooth and streamlined, or laminar, during the interval in which the viscous forces damp out instabilities produced by the fluid velocity but becomes turbulent above this region, as the momentum of the fluid overcomes viscous damping.

Free-Convection Function

Dimensional analysis for free convection, as well as detailed heat-transfer theory, shows that, for free convection,

$$\overline{\text{Nu}} = f(\text{Gr}, \text{Pr}). \tag{9.29}$$

Once again, what initially was an extremely complex problem is, in the case of free, or natural, convection, reduced to the functional relationship between three dimensionless groups.

Values of Dimensionless Groups in Air

It is now possible to evaluate the dimensionless groups for the fluids of particular interest here, i.e., air and water. For air at 20°C, thermal conductivity k is 25.7×10^{-3} W m^{-1}°C^{-1}, kinematic viscosity ν is 15.3×10^{-6} m^2 s^{-1}, dynamic viscosity μ is 18.2×10^{-6} N s m^{-2}, specific heat at constant pressure c_p is 1.01×10^3 J kg^{-1}°C^{-1}, and coefficient of volumetric expansion β is 3.67×10^{-3}°C^{-1}. Acceleration by gravity g is 9.80 m s^{-2}. Values for the Nusselt, Reynolds, Prandtl, and Grashof numbers are as follows:

$$\text{Nu} = \frac{h_c D}{k} = 38.9 \, h_c D \tag{9.30}$$

$$\text{Re} = \frac{VD}{\nu} = 6.54 \times 10^4 \, VD \tag{9.31}$$

$$\text{Pr} = \frac{c_p \mu}{k} = 0.72 \tag{9.32}$$

$$\text{Gr} = \frac{g \, \beta \, \Delta T \, D^3}{\nu^2} = 15.4 \times 10^7 \, \Delta T \, \Delta D. \tag{9.33}$$

Values of V for wind speed range from 0.1 to 10 m s^{-1}, and values of characteristic dimension D for most organisms range from 10^{-3} to 1 m. Therefore, the Nusselt number ranges from 0.04 h_c to 40 h_c, the Reynolds number from 6.6 to 6.6 $\times$ 10^5 (a 100,000-fold range), and the Grashof number from 1.6 $\times$ 10^{-2} to 4.6 $\times$ 10^9 if ΔT varies from 0.1° to 30°C. The Prandtl number is nearly constant in air.

The constancy of the Prandtl number in air allows one to write the equations

$$\overline{\text{Nu}} = f(\text{Gr}), \qquad (9.34)$$

for free convection, and

$$\overline{\text{Nu}} = f(\text{Re}), \qquad (9.35)$$

for forced convection.

Values of Dimensionless Groups in Water

For water at 20°C, the thermal conductivity k is 59.9 $\times$ 10^{-2} W m^{-1}°C^{-1}, kinematic viscosity ν is 10.05 $\times$ 10^{-2} m^2 s^{-1}, dynamic viscosity μ is 10.05 $\times$ 10^{-2} N m^{-2} s, specific heat c_w is 41.8 $\times$ 10^2 J kg^{-1}°K^{-1}, and coefficient of volumetric expansion β is 41.9 $\times$ 10^{-4}°C^{-1}. Values for the Nusselt, Reynolds, Prandtl, and Grashof numbers are therefore

$$\text{Nu} = \frac{h_c D}{k} = 1.67\, h_c D \qquad (9.36)$$

$$\text{Re} = \frac{VD}{\nu} = 99.5\, VD \qquad (9.37)$$

$$\text{Pr} = \frac{c_p \mu}{k} = 701 \qquad (9.38)$$

$$\text{Gr} = \frac{g\, \beta\, \Delta T\, D^3}{\nu^2} = 4.07\, \Delta T\, D^3. \qquad (9.39)$$

The Prandtl number varies considerably with temperature in water and cannot be approximated by a constant value as was done for air.

Free or Forced Convection?

There is a wind speed V_c below which free convection exists and above which forced convection exists. One can obtain an estimate of this value by comparing the Grashof number with the square of the Reynolds number. Since the Grashof number is the ratio (buoyancy) (inertial forces)/(viscous forces)^2 and the square of the Reynolds number is (inertial forces)^2/(viscous forces)^2, Gr/Re^2 is the ratio of buoyancy forces to inertial forces in fluid flow. Experimental evidence shows that the free-

convection function for heat transfer is to be used when Gr $>$ 16 Re^2 and
the forced-convection function when Gr $<$ 0.1 Re^2. For intermediate values, the larger number of $\overline{\text{Nu}}$ should be used. Using Eqs. (9.31) and (9.33)
for air at 20°C when Gr equals 16 Re^2, one gets $\Delta T\, D = 444\ V_c^2$. If $D =$
0.01 m and $\Delta T = 3°C$, $V_c = 8.2 \times 10^{-3}$ m s^{-1}, and if $D = 0.1$ m and $\Delta T =$
10°C, $V_c = 0.47$ m s^{-1}. As a general rule, we have usually considered air
flow of 0.1 m s^{-1} or greater to be forced convection and values less than
this to be free convection.

Forced Convection with Laminar Flow

Convective heat-transfer coefficients for fluid flow over a flat plate in
which the flow is parallel to the surface and of such a speed that only the
laminar condition prevails are derived by theory and confirmed by measurements. Two somewhat idealized situations are used by the heat-
transfer engineer. One is when a flat plate has its surface maintained at a
uniform temperature but exchanges heat with the medium at a nonuniform rate over its surface. The other is when a flat plate is uniformly
heated, or heat is transferred at a uniform rate over its surface, but temperature varies over the surface.

Flat Plate at Uniform Temperature

For forced laminar flow across a flat plate of uniform surface temperature,

$$\overline{\text{Nu}} = 0.595\ \text{Re}^{0.5}. \tag{9.40}$$

The Reynolds and Nusselt numbers as a function of air speed and characteristic dimension are given in Table 9.1.

Hence, from Eqs. (9.24) and (9.40), the convective heat-transfer coefficient for forced laminar flow across a flat plate of uniform surface temperature is given by

$$_1\overline{h}_c = 0.595\ \frac{k}{D}\ \text{Re}^{0.5}. \tag{9.41}$$

In air at 20°C,

$$_1\overline{h}_c = 3.93\left(\frac{V}{D}\right)^{0.5}. \tag{9.42}$$

In water at 20°C,

$$_1\overline{h}_c = 357\left(\frac{V}{D}\right)^{0.5}. \tag{9.43}$$

The quantity $_1h_c$ is in watts per square meter per degree Centigrade, V is
in meters per second, and D is in meters. Values of $_1\overline{h}_c$ in air at 20°C are
given in Table 9.1.

Table 9.1. Reynolds Numbers, Nusselt Numbers, and Convective Heat-Transfer Coefficients (W m^{-2}C^{-1}) for Forced Convection in Laminar Flow across a Flat Plate in Air at 20°C as a Function of Wind Speed V and Characteristic Dimension D.[a]

	V				
D(m)	0.1 ms^{-1}	1.0 ms^{-1}	2.0 ms^{-1}	5.0 ms^{-1}	10.0 ms^{-1}
	Re				
0.01	20	660	1320	3300	6600
0.05	330	3300	6600	16500	33000
0.10	660	6600	13200	33000	66000
0.20	1320	13200	26400	66000	132000
	$_1$Nu				
0.01	5.0	15.3	21.6	34.1	48.3
0.05	10.8	34.1	48.3	76.3	108
0.10	15.3	48.3	68.4	108	153
0.20	21.6	68.4	96.7	153	216
	$_2$Nu				
0.01	6.8	20.8	29.5	46.5	65.9
0.05	14.7	46.5	65.9	104	147
0.10	20.8	65.9	93.3	147	209
0.20	29.5	93.3	132	209	295
	$_1h_c$				
0.01	12.6	39.1	55.1	87.2	123.5
0.05	5.6	17.4	25.1	39.1	55.1
0.10	4.2	12.6	17.4	27.9	39.1
0.20	2.8	9.1	12.6	19.5	27.9
	$_2h_c$				
0.01	17.4	53.7	75.4	119.3	168.9
0.05	7.7	23.7	34.2	53.7	75.4
0.10	5.6	16.7	23.7	37.7	53.7
0.20	3.5	11.9	16.7	26.5	37.7

[a] $_1$Nu and $_1\bar{h}_c$ is for a plate with uniform temperature, and $_2$Nu and $_2\bar{h}_c$ is for a plate with uniform heat flux from the surface. The quantities $_1h_0$ and $_2h_c$ are in watts per square meter per degree Centigrade.

Flat Plate at Uniform Heat Flux

For forced laminar flow over a flat plate with a uniform heat flux across the surface,

$$_2\overline{Nu} = 0.812 \, Re^{0.5}. \tag{9.44}$$

In the case of Eq. (9.29), if the plate is maintained at constant surface temperature throughout [Eq. (9.29)], the boundary layer thickens and the heat transferred by convection drops off markedly away from the leading edge as flow proceeds across the surface. Equation (9.33) describes the case in which the rate of heat transfer by convection remains uniform at all positions across the plate.

Hence, from Eqs. (9.24) and (9.44), the convective heat-transfer coefficient for forced laminar flow across a flat plate with uniform heat flux is given by

$$_2\bar{h}_c = 5.37 \left(\frac{V}{D}\right)^{0.5}, \tag{9.45}$$

for air at 20°C, and

$$_2\bar{h}_c = 488 \left(\frac{V}{D}\right)^{0.5} \tag{9.46}$$

for water at 20°C.
Units are watts per square meter per degree Centigrade; values in air at 20°C are listed in Table 9.1.

Boundary-Layer Thickness

It is of interest to know the thickness of the laminar boundary layer for air flow over a flat plate. Careful experimental measurements, together with the use of dimensionless-variable analysis, allows one to define the hydrodynamic boundary-layer thickness as that distance from the surface at which the local velocity reaches 99% of the free-stream value. The boundary-layer thickness for laminar flow along a flat plate in forced convection is then given by

$$\delta = 5\, d\, Re^{-0.5}, \tag{9.47}$$

where d is the distance from the leading edge. Hence, at a Reynolds number of 1600, $\delta = 5d/40$ (for $d = 0.01$ m, $\delta = 1.25 \times 10^{-3}$ m, and for $d = 0.1$ m, $\delta = 1.25 \times 10^{-2}$ m). If the Reynolds number is 10,000, $\delta = 5d/100$ (for $d = 0.01$ m, $\delta = 5 \times 10^{-4}$ m, and for $d = 0.1$ m, $\delta = 5.0 \times 10^{-3}$ m). For most leaves, this is probably an overestimate of the boundary-layer thickness since any roughness along the leading edge of the leaf will act as a "spoiler" to the air flow and cause some erosion of the boundary layer.

Forced Convection with Turbulent Flow

Air flow along a smooth surface is always laminar near the leading edge, continues in laminar flow for some distance downstream, and becomes turbulent farther along the surface. The position along a flat plate at which the transition from laminar to turbulent flow occurs depends upon the amount of turbulence in the mainstream of the fluid, the shape of the leading edge of the object, and the velocity of flow. If the air is disturbed upstream, as would occur with wind through tree branches or leaves, and if the leading edge is abrupt and irregular, then much of the flow along the object will be turbulent. If the main flow is comparatively smooth, the air velocity is relatively low, and the object is reasonably smooth, then flow will continue in laminar form far along the surface. In fact, for small objects, including small leaves or animals, most of the flow encountered is laminar for relatively low wind speeds.

Transition Distance from Laminar to Turbulent Flow

The Reynolds number is used as a measure of the conditions under which the transition between laminar and turbulent flow occurs. Experimental observations show that for smooth surfaces of plates, cylinders, spheres, etc., the transition from laminar to turbulent flow occurs at a Reynolds number of 5×10^5 but may occur at values as low as 8×10^4 or as high as 10^6. However, observations with plant leaves using Schlieren photography and results of Pearman et al. (1972) and Wigley and Clark (1974) indicate that turbulent flow may develop at Reynolds numbers as low as 10^3. Measurements of Gates et al. (1968), however, suggest that the transition occurs at Reynolds numbers of about 10^4. Adequate information is not available to establish the transition value of the Reynolds number for leaves under field conditions, where the small-scale turbulence is quite different from that generated in a laboratory wind tunnel. Therefore, the value of 10^4 is an estimate, to be used only until more definitive experiments are done. Using the definition of the Reynolds number and letting $D = d_c$, the critical distance from the leading edge at which the transition from laminar to turbulent flow occurs, one gets

$$d_c = \frac{\nu \, \text{Re}}{V} = \frac{15.2 \times 10^{-2}}{V}. \tag{9.48}$$

If $V = 1.0$ m s^{-1}, $d_c = 0.152$ m; if $V = 2.0$ m s^{-1}, $d_c = 0.076$ m; and if $V = 3.0$ m s^{-1}, $d_c = 0.051$ m. These distances are generally consistent with the characteristics of air flow over plant leaves as shown in our Schlieren photographs. At low wind speeds and low Reynolds numbers, flow over a flat broad leaf is mostly laminar, but turbulence begins to dominate the flow at higher wind speeds. In general, small leaves will usually experience laminar flow, and large leaves ($D > 0.05$ m) will experience turbulent flow for wind speeds of 3 m s^{-1} or greater.

Heat Transfer for Turbulent Flow

Turbulent flow along the surface of a flat plate is described by the following heat-transfer relation:

$$\overline{\text{Nu}} = 0.032 \, \text{Re}^{0.8}. \tag{9.49}$$

Table 9.2. Reynolds Numbers, Nusselt Numbers, and Convective Heat-Transfer Coefficients (W m^{-2} C^{-1}) for Forced Convection in Turbulent Flow across a Flat Plate in Air at 20°C as a Function of Wind Speed V and Characteristic Dimension D.

			V		
D(m)	0.1 ms^{-1}	1.0 ms^{-1}	2.0 ms^{-1}	5.0 ms^{-1}	10.0 ms^{-1}
			Re		
0.01	66	660	1320	3300	6600
0.05	330	3300	6600	16500	33000
0.10	660	6600	13200	33000	66000
0.20	1320	13200	26400	66000	132000
			Nu		
0.01	0.91	5.75	1.0	20.8	36.3
0.05	3.30	20.8	36.3	75.5	131
0.10	5.75	36.3	63.2	131	229
			h_c		
0.01	2.1	14.7	25.8	53.7	93.5
0.05	2.1	12.6	22.3	46.7	80.9
0.10	1.4	10.5	18.8	39.1	67.0
0.20	1.4	9.1	16.0	33.5	58.6

Converting from Nusselt number to the convective heat-transfer coefficient for air and water gives

$$\bar{h}_c = 5.85 \ V^{0.8} \ D^{-0.2} \tag{9.50}$$

in air at 20°C and

$$\bar{h}_c = 76.00 \ V^{0.08} \ D^{-0.2} \tag{9.51}$$

in water at 20°C.

The Reynolds numbers, Nusselt numbers, and convective heat-transfer coefficients for turbulent flow along a flat plate are given in Table 9.2 as a function of wind speed and characteristic dimension of the plate. Comparison of the laminar-flow and turbulent-flow equations shows that at low wind speeds, less than approximately 1.0 m s^{-1}, the laminar convective heat-transfer coefficient is greater than the turbulent convective coefficient for all except very large leaves and plates. Heat-transfer phenomena in leaves will be discussed in more detail later in the chapter.

Thickness of the Turbulent Boundary Layer

It is possible to derive from experiments an estimate of the thickness of the turbulent boundary layer for flow over a plane surface. Neglecting to a first approximation the laminar sublayer, then, for Reynolds numbers between 10^4 and 10^7, one gets

$$\delta = 0.376d \ \text{Re}^{-0.2}, \qquad (9.52)$$

where d is the distance along the plate from the leading edge. By comparing Eq. (9.47) with Eq. (9.52), one can see that at any given value of d, a turbulent boundary layer will be smaller than a laminar boundary layer at low Reynolds numbers and larger at high Reynolds numbers. Despite its greater thickness, the turbulent boundary layer offers less resistance to heat flow than a laminar layer because the eddies of turbulence produce a more rapid mixing between warmer and cooler regions.

Forced Flow over a Cylinder

Many organisms are cylindrical in shape. Fluid flow across a cylinder should be a good approximation of the flow across tree trunks, limbs, branches, twigs, and the bodies of animals. Therefore, the behavior of fluid flow across a cylinder at right angles to the axis is of special interest. Fluid flow across a circular cylinder at two different Reynolds numbers is illustrated in Fig. 9.7. As air flows across a cylinder, flow changes from streamlined to turbulent. This change in the character of the flow is called "separation." Beyond the point of separation, fluid near the surface of the cylinder flows in a direction opposite to the main stream. This local reversal of the flow results in eddies being formed and turbulence being established. In the lee of a cylinder, eddies form on both sides and create a turbulent wake. The most-forward point of a cylinder in fluid flow is known as the stagnation point. Fluid particles striking this point are brought to rest, and the pressure here rises over that in the immediate vicinity.

When the Reynolds number is very small, less than unity, flow across a cylinder adheres to the surface, and the streamlines follow the cylinder contour. At Reynolds numbers of about 10, inertial forces become appreciable, and two weak vortices develop in the rear of the cylinder. When the Reynolds number becomes about 100, laminar flow prevails from the leading edge to the point of separation, which lies at an angular position about 80° and 85° from the apex or direction of flow. At Reynolds numbers greater than 10^4, flow becomes turbulent, and, curiously, the separation point moves to the rear, with the turbulent wake being reduced in size. At these high Reynolds numbers, flow over the forward part of the cylinder is laminar, and heat transfer and fluid flow resemble those for laminar flow along a flat plate. Kreith (1966) on p. 373 writes as follows:

> The local conductance is largest at the stagnation point and decreases with distance along the surface as the boundary layer thickness increases. The conductance reaches a minimum on the sides of the cylin-

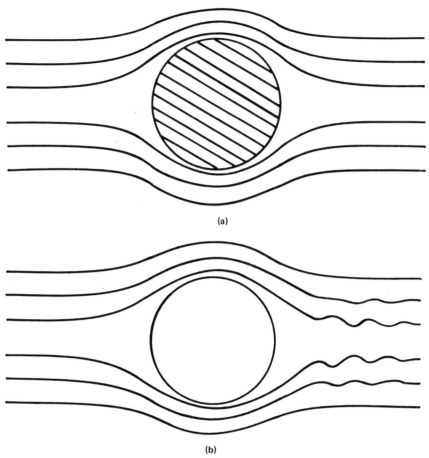

Figure 9.7. Fluid flow around a circular cylinder for a Reynolds number of (a) about 1.0 and (b) greater than 10^4. Flow is from left to right.

der near the separation point. Beyond the separation point the local conductance increases because considerable turbulence exists over the rear portion of the cylinder where the eddies of the wake sweep the surface. However, the conductance over the rear is no larger than over the front, because the eddies recirculate part of the fluid and, despite their high turbulence, are not as effective in mixing the fluid in the vicinity of the surface with the fluid in the main stream as a turbulent boundary layer. At Reynolds numbers large enough to permit transition from laminar to turbulent flow in the boundary layer without separation of the laminar boundary layer, the unit-surface conductance has two minima around the cylinder. The first minimum occurs at the point of transition. As the transition from laminar to turbulent flow progresses, the unit conductance increases and reaches a maximum approximately at the point where the boundary layer becomes fully turbulent. Then the unit-surface conductance begins to decrease again and reaches a second maximum at about

130 degrees, the point at which the turbulent boundary layer separates from the cylinder. Over the rear of the cylinder the unit conductance increases to another maximum at the rear stagnation point.

Angles are clockwise in degrees from the direction of flow.

Convective Heat Transfer

The relationship between the Nusselt and Reynolds numbers for fluid flow across a cylinder has been shown to have the form

$$\text{Nu} = c \, \text{Re}^n, \tag{9.53}$$

where c and n are empirical constants whose numerical values vary with the Reynolds number (see Table 9.3). For cylinders, the characteristic dimension is the diameter of the cylinder. Reynolds numbers for most organisms and fluid-flow conditions generally range from 6.6 to 6.6×10^5.

By placing the values given in Table 9.3 in Eq. (9.53) and converting them to $\bar{h}_c$ for a cylinder in air at 20°C and standard pressure, one gets the following results (where $\bar{h}_c$ is in units of watts per square meter per degree Centigrade):

$0.4 < \text{Re} < 4$	$\bar{h}_c = 0.923 \, V^{0.333} \, D^{-0.666}$	(9.54)
$4 < \text{Re} < 40$	$\bar{h}_c = 1.507 \, V^{0.385} \, D^{-0.615}$	(9.55)
$40 < \text{Re} < 4{,}000$	$\bar{h}_c = 2.77 \, V^{0.466} \, D^{-0.534}$	(9.56)
$4{,}000 < \text{Re} < 40{,}000$	$\bar{h}_c = 4.23 \, V^{0.168} \, D^{-0.382}$	(9.57)
$40{,}000 < \text{Re} < 400{,}000$	$\bar{h}_c = 4.64 \, V^{0.805} \, D^{-0.195}$	(9.58)

Table 9.4 gives values of the Reynolds number, Nusselt number, and convective heat-transfer coefficient as a function of the cylinder diameter D (in meters) and the air flow speed V (in meters per second) for a smooth cylinder in cross flow. It is clear that Reynolds numbers change in magnitude one-thousand-fold just for the range of air speeds and cylinder diameters exhibited in Table 9.4. The majority of values for small objects and low wind speeds fall between Reynolds numbers of 40 and 4,000, how-

Table 9.3. Coefficients for Calculation of Average Heat-Transfer Coefficient of a Circular Cylinder in a Fluid Flowing Normal to the Axis of the Cylinder.

Re	c	n
0.4–4.0	0.891	0.330
4–40.0	0.821	0.385
40–4000	0.615	0.466
4,000–40,000	0.174	0.618
40,000–400,000	0.024	0.805

Table 9.4. Reynolds Numbers, Nusselt Numbers, and Convective Heat-Transfer Coefficients (W m^{-2} °C^{-1}) for a Cylinder of Diameter D in Fluid Flow at Speed V Perpendicular to the Axis of the Cylinder.[a]

	V			
D(m)	0.1 ms^{-1}	1.0 ms^{-1}	2.0 ms^{-1}	5.0 ms^{-1}
		Re		
0.01	66	660	1320	3300
0.05	330	3300	6600	16500
0.10	660	6600	13200	33000
0.20	1320	13200	26400	66000
		Nu		
0.01	4.3	12.7	17.5	27.0
0.05	9.2	27.0	40.0	69.6
0.10	12.7	40.0	60.9	108
0.20	17.5	60.9	94.0	182
		(W m^{-2} °C^{-1})		
0.01	11.2	33.5	46.1	70.5
0.05	4.9	14.0	20.2	36.3
0.10	2.8	10.5	15.4	27.9
0.20	2.1	7.7	11.9	20.9

[a] The convective heat-transfer coefficient is computed for an air temperature of 20°C and standard atmospheric pressure.

ever, so that Eq. (9.56) is most appropriate for computation of the convective heat-transfer coefficient. At wind speeds of 2.0 m s^{-1} or greater, and for objects of diameters 0.05 m or greater, Reynolds numbers are in the range 4,000 to 40,000, and Eq. (9.57) should be used.

Boundary-Layer Thickness

It is possible to use Eq. (9.12) to define the thickness δ of a thermal boundary layer surrounding a cylinder of radius r in air or water when a quantity of heat Q is conducted across it according to the equation

$$Q = \frac{k (T_s - T_a)}{r \ln \dfrac{r + \delta}{r}}. \tag{9.59}$$

Heat transfer is described in terms of the convection coefficient such that

$$Q = h_c (T_s - T_a). \tag{9.60}$$

Then,

$$\frac{k}{h_c r} = \ln \frac{r + \delta}{r} = \frac{2}{\mathrm{Nu}} \tag{9.61}$$

since $\mathrm{Nu} = h_c D / k$, where $D = 2r$ is the diameter of the cylinder. Therefore,

$$\delta = r[e^{2/\mathrm{Nu}} - 1]. \tag{9.62}$$

Nobel (1974) has investigated experimentally fluid flow across cylinders and cylindrically shaped plant parts such as stems, branches, and trunks. Using the relationship of Eq. (9.53) in Eq. (9.62), he was able to obtain the following relationship for VD between 1 and 3 m² s⁻¹:

$$\delta = 0.74\, D\, (e^{0.01/(VD)^{1/2}} - 1), \tag{9.63}$$

where δ is in meters if V is in meters per second and D is in meters. Eq. (9.63) holds at 20°C, and δ can be corrected to other temperatures by multiplying by $(T + 273)/273$.

Forced Flow over a Sphere

Convective Heat Transfer

For Reynolds numbers from about 25 to 100,000, the range most frequently encountered in nature, the relation between the average Nusselt number and the Reynolds number for fluid flow over a sphere is

$$\overline{\mathrm{Nu}} = 0.37\, \mathrm{Re}^{0.6}. \tag{9.64}$$

The rate of convective heat transfer from a sphere is always greater than that from a cylinder of the same diameter, usually by a factor of about 1.5 to 1.8. Some animals or animal parts (such as a head) can, for purposes of convective heat transfer, be represented by a sphere.

Boundary-Layer Thickness

Plant fruits and sometimes whole plants are of spherical shape, as are the heads of some animals. For purposes of understanding heat and mass transfer from these plant and animal parts, it is useful to know the approximate boundary-layer thickness.

One can define the thickness δ of a thermal boundary layer surrounding a sphere of radius r in air or water when a quantity of heat Q [as given by Eq. (9.16)] is conducted across it as follows:

$$Q = \frac{(r + \delta)\, k(T_s - T_a)}{\delta r}. \tag{9.65}$$

Heat transfer is described in terms of the convection coefficient such that

$$Q = h_c(T_s - T_a).$$ (9.66)

One gets from the definition of the Nusselt number [Eq. (9.24)] the following relationship:

$$\text{Nu} = \frac{h_c D}{k} = \frac{D + 2\delta}{\delta}.$$ (9.67)

For forced fluid flow over a sphere, by equating Eq. (9.64) with Eq. (9.67) and solving for δ, one gets

$$\delta = 0.28 \sqrt{D/V} + \frac{0.025}{V},$$ (9.68)

where V is the wind speed in meters per second and D is in meters.

Nobel (1975) reported laboratory measurements of the boundary-layer thickness which confirmed the accuracy of Eq. (9.68). He also derived an empirical relationship to fit his data with a 5% accuracy. Boundary-layer thicknesses ranged from 6.2×10^{-5} m for a sphere of diameter 10^{-3} m and V equal to 10 m s^{-1} to 2.9×10^{-2} m for a sphere of diameter 1.0 m and V equal to 3.2×10^{-2} m s^{-1}.

Forced Flow over Animals

Although many workers have assumed that forced flow over animals can be represented by convective flow over cylinders or spheres, there have been few detailed tests of such an assumption. Mitchell (1976) compared a variety of observations involving forced convective heat transfer for animals, including man, and derived some "best fit" conclusions that are both simplifying and accurate.

According to Mitchell, convective heat transfer for animal shapes is best estimated by using the convection relationship for a sphere and proceeding according to the following steps:

1. The volume of an animal is determined from its total mass divided by the density.
2. The characteristic dimension of an animal is evaluated as the one-third power of the volume.
3. The Nusselt number is computed using Eq. (9.64). The convection coefficient is then derived from the Nusselt number.
4. Enhancement of the convection coefficient owing to outdoor turbulence is estimated as a function of height z above the surface. Enhancement varies from 1.7 when the animal is resting on the ground surface (for which $z/D = 0.5$) to 1.0 when the animal is far above the surface

$(z/D > 100)$. For many situations, an enhancement factor of 1.3 is a good approximation.

Mitchell demonstrates that the convective heat transfer as determined by the above procedure is within 20% for the following animals: cow, man, sheep, frog, lizard, insect, and spider. His results are sufficiently convincing that use of this procedure is recommended.

Free Convection

Free convection occurs when an object is warmer or colder than the fluid that surrounds it. As a result of the temperature difference, heat flows between the body and the fluid and produces a change in the density of the fluid in the vicinity of the surface. Warmer, less dense fluid rises, and cooler, more dense fluid falls. This movement is referred to as free convection, and heat is transferred in the process. Warm air rises from a warm object, and heat is transferred from the object to the air by free convection. Cool air sinks from a cool object, and heat is transferred from the air to the object by free convection. Because of the gentleness of the flow during free convection, the heat-transfer coefficients for free convection are generally smaller than those for forced convection.

Dimensional analysis showed that for free convection, the Nusselt number is a direct function of the Grashof and Prandtl numbers [see Eq. (9.29)]. Different constants of proportionality are involved depending upon whether the surface of a heated flat plate is facing upward or downward or is vertical.

Functional Relationships

The formulas given below express the functional relationships between the Nusselt, Grashof, and Prandtl numbers for free convection for several characteristic geometrical surfaces with various orientations. Generally, it is necessary to distinguish between laminar and turbulent flow in order to specify the range of Grashof numbers for each case.

Horizontal cylinder
   Laminar range

$$Nu = 0.530 \, (Gr \cdot Pr)^{0.250} \tag{9.69}$$

Vertical cylinder
   Laminar range

$$Nu = 0.726 \, (Gr \cdot Pr)^{0.250} \tag{9.70}$$

   Turbulent range

$$Nu = 0.067 \, (Gr \cdot Pr^{1.29})^{0.333} \tag{9.71}$$

Plane square flat surface
  Laminar range
    Horizontal warm plate facing upward

$$Nu = 0.710 \ (Gr \cdot Pr)^{0.250} \tag{9.72}$$

Horizontal warm plate facing downward

$$Nu = 0.350 \ (Gr \cdot Pr)^{0.250} \tag{9.73}$$

Vertical

$$Nu = 0.523 \ (Gr \cdot Pr)^{0.250} \tag{9.74}$$

Turbulent range
  Horizontal warm plate facing upward

$$Nu = 0.170 \ (Gr \cdot Pr)^{0.333} \tag{9.75}$$

Horizontal warm plate facing downward

$$Nu = 0.080 \ (Gr \cdot Pr)^{0.333} \tag{9.76}$$

Vertical

$$Nu = 0.130 \ (Gr \cdot Pr)^{0.333} \tag{9.77}$$

A warm horizontal plate facing upward has the same rate of convective heat transfer as a cold horizontal plate facing downward. Also, a warm horizontal plate facing downward has the same rate of convective heat transfer as a cold horizontal plate facing upward. One should note that the Nusselt number for a vertical plate is approximately the average of that for the upward- and downward-facing horizontal plane surfaces.

Convection Coefficients

Since the free convection usually encountered by organisms is of the laminar form, we shall derive as an example the convection coefficients for various surfaces. For example, using $Pr = 0.72$ and converting Eqs. (9.69) and (9.72) from Nusselt number to $\bar{h}_c$ by means of the definition of Nu, one gets the following relationships for the laminar range at 20°C and standard atmospheric pressure:
  Horizontal cylinder

$$\bar{h}_c = 1.40 \ (\Delta T/D)^{0.250} \tag{9.78}$$

Plane square flat surface
  Horizontal warm plate facing upward

$$\bar{h}_c = 1.87 \ (\Delta T/D)^{0.250} \tag{9.79}$$

The other examples can be converted to convection coefficients in a similar manner.

The magnitude of free convection is generally small in terms of the overall heat budget of a leaf, and the influence of free convection on leaf temperature is also small. Amount of radiation absorbed or emitted by a leaf is often 280 to 560 W m^{-2}. Using Eq. (9.79), one finds for a warm upward-facing plate in laminar flow, that $\overline{h}_c$ varies from about 1.05 for $\Delta T = 1°C$ and $D = 0.1$ m to 3.32 W m$^{-2}°C^{-1}$ for $\Delta T = 10°C$ and $D = 0.01$ m. The quantity of heat transferred is $\overline{h}_c \Delta T$, which gives 1.05 and 33.2 W m^{-2}, respectively, for these two values of $\overline{h}_c$. In the first case, the quantity of heat transferred is very small; however, in the second case, for a small leaf and a large temperature difference, the heat transferred is about 6 to 12% of the absorbed radiation.

A comparison of the convective heat-transfer values for free convection with those for forced convection at an air speed of 0.1 m s^{-1} shows them to be of the same magnitude. In other words, forced convection at an air flow of 0.1 m s^{-1} is nearly equivalent to free convection. Generally, it is simplest to apply the forced-convection formulation to all conditions and consider the lower limit of forced convection to be the amount of air movement at which the convective heat-transfer coefficients for forced convection equal those for free convection.

Characteristic Dimension

There is always a question regarding determination of characteristic dimension for a leaf of irregular shape, or, for that matter, a rectangular plate of unequal dimensions. This problem is discussed in a later section. The characteristic dimension D of a flat, rectangular plate is taken to be its width, or narrowest dimension. Equations describing convection for horizontal flat plates will approximate heat transfer from horizontal circular disks if the characteristic dimension is chosen as 0.9, the diameter of the disk. For cylinders, the characteristic dimension is the diameter of the cylinder. In the case of spheres, when the Grashof number approaches zero, the Nusselt number approaches a value of 2. This condition corresponds to pure conduction through a stagnant layer of air adjacent to the surface of the sphere.

Actual Broad Leaves

History of Research

Once it is recognized that the heat transfer from a broad leaf may be approximated by heat transfer from a flat plate of a certain characteristic dimension, it is necessary to experimentally confirm or deny the validity of the approximation. Numerous investigators recognized very early that convection plays a significant role in the transfer of heat between plant leaves and the air. Brown and Escombe (1900, 1905), Brown and Wilson (1905), Walter (1926), and others were pioneers in this field and made significant contributions to our understanding. Martin (1943) wrote a remarkable paper in which he thoroughly explored the influence of leaf size and wind speed on the convective heat-transfer coefficient. During World War II, Koriba (1943) published a magnificent monograph on the subject of convection and transpiration, in which he reviewed all of the earlier work concerning both convection and transpiration. This monograph is illustrated with many photographs of currents of air flowing around various surfaces and leaves. The currents of convection are made visible with smoke introduced into the air flow. I do not believe this monograph has received attention commensurate with its importance, and for those who can read German, it is worthwhile looking it up.

One can consider the heat-transfer coefficient to have the following functional relation:

$$h_c \propto D^m \ V^n, \tag{9.80}$$

where D is the characteristic dimension and V is the wind speed. Walter's results showed the exponent m to be -0.2 to -0.4; n was not determined. Martin's results gave

$$h_c \propto W^{-0.2} \ L^{-0.3} \ V^{0.5}, \tag{9.81}$$

where W is the dimension of the leaf at right angles to the wind direction and L is the dimension of the leaf parallel to the wind direction. Clearly, if $W = L = D$, then $h_c \propto (V/D)^{0.5}$.

Raschke (1954, 1956a, 1956b, 1960) published several important papers describing heat transfer by convection. He found that n is equal to 0.5 for thin leaves, but that with thick leaf models made of copper plates, the wind function was matched better by $n = 0.7$. Raschke commented that this finding was similar to those made by Hilpert (1933), in studies made on cylinders with square cross sections of dependency $n = 0.675$. Raschke concluded that convective heat transfer depends on leaf size and wind as follows:

$$h_c \propto D^{-0.3} \ V^{0.5}. \tag{9.82}$$

For *Canna* leaves with the wind blowing across them transversely, Raschke found the convection coefficient to be given by

$$h_c = 136.0 \ D^{-0.3} \ V^{0.5} \tag{9.83}$$

when the ratio of leaf length to width was $9:5$. If the wind was blowing along the leaf length, he found that

$$h_c = 118.3 \ D^{-0.3} \ V^{0.5}. \tag{9.84}$$

Raschke compared the heat-transfer coefficients between a circular disk, a *Canna* leaf, and a compound leaf of *Delonix* and found the ratio to be

0.91:1.00:1.23. Certainly, on basic principles, one would expect that *Delonix* to have a greater heat-transfer coefficient than *Canna,* even though both leaves may occupy about the same total area, since the *Delonix* leaf is finely divided into many segments.

Knoerr and Gay (1965) have reported measurements of the convection coefficients of broad leaves as deduced from energy-balance determinations. Leaves of *Quercus velutina, Magnolia grandiflora,* and *Prunus caroliniana* were studied. Convective heat losses for these leaves were determined as a residual term in the energy-balance equation; net radiation and transpiration rates were measured. Coefficients for forced convection agreed very well with the value given by Eq. (9.45). Coefficients for free convection are close to the theoretical values only for the larger leaves and increase to about twice the predicted values as leaf size decreases. This is also the experience of other workers. Gates and Benedict (1963) reported convection from small leaves, as observed by means of Schlieren photography, to be about twice the theoretical value. These workers felt that part of the problem was due to the fact that there was additional air movement past the leaf caused by rising air currents from the warm ground or floor beneath the leaves. This mild air movement probably increased convection over that predicted for free convection only. The smaller leaves, having the greatest perimeter-to-area ratio, would be affected most by these rising currents of air. In addition, turbulence of air in the natural environment can produce a greater heat-transfer coefficient than one might get from low-turbulence wind-tunnel experiments. Pearman et al. (1972) made measurements of heat-transfer coefficients on metal disks of diameters 1.5 to 12.7 cm placed above several agricultural- and nonagricultural-type surfaces with wind velocities from 0.1 to 3.5 m s^{-1}. Heat-transfer coefficients averaged approximately 1.5 times greater than was predicted from transfer theory for laminar flow, but there was considerable scatter to the data, with most estimates falling between the theoretically predicted values and twice the theoretical. Kestin (1966) has discussed the fact that free-stream turbulence enhances the heat-transfer coefficient for flow around cylinders and spheres. He has also shown that for a flat plate with zero attack angle, the critical length at which the transition from laminar to turbulent flow occurs is shortened with increased free-stream turbulence. Kestin observes that heat transfer is enhanced only about 10% for turbulence intensities of about 5%. Sogin and Subramanian (1961) have shown that for a flat plate at normal incidence, heat transfer increases 50% with an increase in free-stream turbulence of an unspecified amount. Pearman et al. (1972) have reviewed many of these ideas and make the following statement:

> The scale of turbulence at a leaf under natural winds would vary from large to small with components normal to the leaf. Normal wind components, similar to wedge flow, could produce a favorable pressure gradient of uncertain magnitude on the lead side of the leaf, but the effect on the

wake side is not known. With normal wind components, the situation also is similar to that of Sogin and Subramanian on the side presented to normal flow, but no data exist for estimating the effect of turbulence at the rear, if any.

Pearman et al. (1972) gave the following relationship as the best-fit regression equation for 268 separate experiments under natural field conditions:

$$\overline{Nu} = 1.08 \ Pr^{.33} \ Re^{.05}. \tag{9.85}$$

This equation compares favorably to that suggested by Monteith (1965), using the data of Raschke (1956a), which was

$$\overline{Nu} = 1.13 \ Pr^{0.33} \ Re^{0.5}. \tag{9.86}$$

Parlange et al. (1971) found that the mean heat-transfer coefficient of tobacco leaves of dimension 0.2×0.2 m under uniform heat flux in turbulent air was about 2.5 times greater than that predicted for a flat plate at uniform temperature in laminar flow. Wigley and Clark (1974) using data from Parlange et al. (1971) obtained the following relationship:

$$\overline{Nu} = 1.66 \ Pr^{0.333} \ Re^{0.5}. \tag{9.87}$$

When Pearman et al. (1972) compared their results with these, they concluded that "due to the scatter of the present data, we only can say the convective heat loss from uniform-temperature metal plates under natural conditions generally exceeds that predicted from laminar transfer theory, but less than twice the theorietical value, averaging about 1.5 times greater and that the increase is due to the turbulence of natural winds." All the data from these various investigators agree very well with the $Re^{0.5}$ relationship and do not appear to be represented by a higher power like $Re^{0.8}$, as suggested by turbulent flow over flat plates.

Gates et al. (1968) used a convection coefficient which was 1.8 times greater than that given by laminar flow across a flat plate at uniform temperature when the leaf dimensions were greater than 0.05 m and the cross-flow dimension of the leaf was greater than the dimension in the direction of flow. However, when the cross-flow dimension was less than the dimension in the wind direction or when leaf dimensions were less than 0.05 m, the convection coefficient used was 2.8 times the coefficient given by laminar flow at uniform temperature. If, however, one compares these coefficients with those for a uniformly heated plate in laminar flow, they are only 1.2 and 2.0 times greater, respectively. These coefficients were established by means of a large number of careful experiments in a wind tunnel which used rectangular strips of dry and wet blotting paper of various sizes. The convective heat-transfer coefficient was given as

$$h_c = 6.98 \ (V/D)^{0.5}, \tag{9.88}$$

when $W > D$ or $W = D$ and is greater than 0.05 m, where D is the characteristic dimension in the direction of flow and W is the dimension across the flow. The coefficient is

$$h_c = 11.30 \ (V/D)^{0.5} \tag{9.89}$$

when $W < D \le 0.1$ m or $W = D$ and is ≤ 0.05 m.

It was clear during the course of these experiments that an abrupt change ocurred in the nature of the air flow and the convective heat-transfer rates when the plate or leaf size became less than about 0.05 m or when the ratio of longitudinal to cross-flow dimensions of the leaf changed from $W > D$ to $W < D$. This is probably a change from fairly turbulent average flow conditions over the surface to generally laminar flow conditions on the average. The formulations given in Eqs. (9.88) and (9.89) are equivalent to the following in terms of the dimensionless groups:

$$\begin{aligned} Nu &= 1.06 \ Re^{0.5} \\ &= 1.18 \ Pr^{0.33} \ Re^{0.5} \end{aligned} \tag{9.90}$$

and

$$\begin{aligned} Nu &= 1.67 \ Re^{0.5} \\ &= 1.86 \ Pr^{0.33} \ Re^{0.5}. \end{aligned} \tag{9.91}$$

It will be shown later that Eq. (9.90) is the recommended relationship based on all information available.

It is interesting to note that Eq. (9.90) is almost identical to Eq. (9.86), which was derived by Monteith (1965a) from Raschke's data. Certainly, it would be reasonable to use Eq. (9.88) for most calculations and to use a higher value of the convective heat-transfer coefficient for small leaves. Gates and Papian (1971) used Eqs. (9.88) and (9.89) or their equivalent, Eqs. (9.90) and (9.91), to calculate the energy budgets for plant leaves given in that compilation.

Parkhurst et al. (1968a) decided to simulate "real" leaves by means of models made with an electrical heating element of Nichrome wire embedded between thin copper sheets the same shape as the "real" leaf. A cast of the "real" leaf was made in plaster of paris, and the copper sheet was pressed into the imprint in order to cut out the same precise figure. In addition, copper leaves of arbitrary and varied size and shape were made with heating elements within them. Tests of these at low Reynolds numbers showed that the Nusselt numbers fell between the functions given by Eqs. (9.40) and (9.44). Much of the data could be well matched by the relationship

$$Nu = 0.650 \ Re^{0.5} \tag{9.92}$$

for Reynolds numbers between 4000 and 10^5. For lower Reynolds numbers, Eq. (9.40) fits the observations very well.

A very thorough study of heat-transfer coefficients of broad leaves was made by Wigley and Clark (1974), who compared much of the earlier work referred to above and made exceedingly precise measurements of local convection coefficients over the leaf surface. They showed that considerable temperature differences exist over a leaf surface which is strongly irradiated and undergoing both laminar and turbulent air flow. When a leaf is inclined at 30° to the flow direction, major temperature differences exist for both laminar and turbulent flow. The greatest temperature variations over a leaf surface occur when flow is laminar. In general, local leaf temperature increases with the square root of the distance from the leading edge. This is especially true when the leaf surface is parallel to the direction of flow. When a leaf is tilted 30° to the direction of flow, the leaf in laminar flow has a maximum temperature near the center rather than near the trailing edge, but the maximum occurs near the trailing edge in turbulent flow.

Wigley and Clark obtained integrated values of the Nusselt numbers for electrically heated models of real leaves. They found that for an isothermal leaf model in parallel laminar flow,

$$Nu = 0.664 \ Pr^{0.333} \ Re^{0.50}, \tag{9.93}$$

and for a uniform heat flux model in parallel laminar flow,

$$Nu = 0.68 \ Pr^{0.333} \ Re^{0.50}. \tag{9.94}$$

For a condition of turbulent flow over a parallel leaf surface, they obtained the relationship

$$Nu = 0.052 \ Pr^{0.333} \ Re^{0.84}. \tag{9.95}$$

The model used by Wigley and Clark was geometrically similar to a primary leaf of bean, *Phaseolus vulgaris*. These leaves were greater than 0.05 m and appeared to range up to 0.1 m in diameter. The average Nusselt numbers obtained for these leaves were thus in the regime represented by Eq. (9.90) rather than Eq. (9.91). Hence there develops a dilemma as to whether the coefficient in Eq. (9.91) is too large or that in Eq. (9.90) is too low. We know from Wigley and Clark that leaf temperatures are not uniform over a leaf surface and that a leaf, for heat-transfer purposes, is better represented by a surface of uniform heat flux.

Wigley and Clark indicate that for leaves in the field, there is sufficient small-scale turbulence to justify the general use of Eq. (9.95). However, their experiments were done in a wind tunnel and not in the field. Data do not exist which demonstrate unequivocally the relationship to use for leaves in a variety of field conditions.

A thorough discussion of convection coefficients and the influence of wind on plants is given in Grace (1978). A critical evaluation of the heat-transfer characteristics of simple and pinnate leaf models is given by Balding and Cunningham (1976).

Recommended Relationships for Convective Heat Transfer

In comparing the convective heat-transfer coefficients of a flat plate in laminar flow, as given by either Eqs. (9.42) or (9.44), with that for a flat plate in turbulent flow, as given by Eq. (9.50), one finds that at low Reynolds numbers, the laminar-flow relationships give larger Nusselt numbers than the turbulent-flow relationships and that the reverse is true at large Reynolds numbers. For most leaves with dimensions from 0.01 to 0.1 m, and for air speeds from 0.5 to 5 m s^{-1}, Reynolds numbers are between 329 and 3.29×10^4. This suggests that fully tubulent conditions are not so likely to exist over a leaf surface. When wind speeds are above 5 m s^{-1}, calculated leaf temperatures are extremely close to air temperature no matter which convective heat-transfer relationship is used. Whether the distinction between one relationship or another is important for leaves in strong wind depends on the reason for determining leaf temperature. For most purposes, the small temperature differences with different coefficients at high wind speeds are negligible. It is important to note that the majority of all reports mentioned above show the Nusselt number to depend on the square root of the Reynolds number. This suggests that laminar flow prevailed over the leaves or their models for most of the experiments reported and that the square-root law should be used. As a result of all these deliberations, I recommend the use of Eq. (9.90) generally. For small leaves, a larger coefficient should be used, although the laminar-flow relationship still holds. Perhaps the coefficient of Eq. (9.91) is somewhat too high and a compromise between the coefficients of Eqs. (9.90) and (9.91) would be more correct. The temperature of small leaves tends to be close to air temperature, and once again, the difference between the two methods is probably not significant for ecological purposes.

Leaf Adaptation to Environment

Plant leaves differ enormously in size, shape, color, pubescence, stomate size and density, thickness, and internal anatomy. Many people speculate that leaf characteristics have developed in response to the environment of the leaf and that specific size, shape, color, and so on, is the direct result of adaptation to environmental conditions. There is now some evidence that this is indeed the case, although it is not yet clear why certain particular leaf features have developed in the way they have.

Parkhurst and Loucks (1972) have presented the most complete discussion of these important questions. They discuss all environmental variables as they relate to leaf temperature, water use, photosynthesis, and water-use efficiency (defined as the ratio of photosynthetic rate to transpiration rate). Water-use efficiency is affected by leaf size through the convection coefficient. Greater leaf size means not only a thicker boundary layer and increased boundary-layer resistance to the diffusion of water vapor and carbon dioxide but also a reduced convective heat-transfer coefficient, which in turn means higher leaf temperatures, other things being equal. These authors have shown that, for certain environmental conditions, an order of magnitude increase in leaf size, which produces a decrease of h_c by $10^{-0.5}$, can result in a 40% increase in transpiration and a 50% decrease in water-use efficiency.

Sun leaves tend to be small and lobed, whereas shade leaves are large and less lobed, and analysis by Parkhurst and Loucks (1972) has shown that water-use efficiency is greatest for small leaves (or high wind speeds) when absorbed radiation is high and for larger leaves (or low wind speeds) when absorbed radiation is low. Leaf size and wind speed are related through the convection coefficient. These authors also concluded that "the range of radiation–absorption levels for which efficiency increases with increasing leaf size (decreasing convection coefficient) is much smaller at lower air temperatures than at high air temperatures." This statement implies that small leaves will be selected for in cold climates no matter what the radiation level. This argument is, of course, based on the premise that water-use efficiency is a determining factor in evolution. It is certainly not the only factor but may, nevertheless, be important. Parkhurst and Loucks also state that at high air temperatures, large leaves would be at an advantage where there is little light and small leaves where light is intense, because water-use efficiency is increased.

Vogel (1968) has shown that for radiantly heated sun and shade leaves of white oak (*Quercus alba* L.) in still air or a gentle updraft, the difference between ambient and leaf temperature in degrees Centigrade is about 20% less for sun leaves than for shade leaves. In another paper, Vogel (1970) has shown that lobed leaves and plates are more effective convectors than circular plates. In addition, circular plates (disks) dissipated about one-fourth more heat when vertical than when horizontal in still air with free convection, but lobed plates and leaves showed no difference with orientation. In forced convection, lobed plates had maximum heat dissipation by convection when oblique to the air stream under certain circumstances. Vogel states the following (1970, p. 99): "While general differences between lobed and unlobed plates are apparent and consistent, identification of the specific significant features of the different models is difficult. No single model proved superior for all situations which were investigated." He concludes, "In short, while the beneficial effect of lobing in general is clear, the relative advantages and advantageous features of the specific geometries tested are not." Vogel's copper-plate models each had the same surface area but ranged in shape from a regular six-pointed star to greatly serrated disks and exact replicas of sun and shade white oak leaves. This work followed the method of Parkhurst et al. (1968) in using heated copper models in a wind tunnel to determine the effective characteristic dimension for convective heat exchange of real leaves of various shape, size, and orientation.

Thom (1968) used metal models of thin aluminum sheet for heat-transfer and drag experiments and attached filter paper to this surface for experiments with the evaporation of volatile fluids. These experiments are discussed in the chapter concerned with evaporation and transpiration.

Problems

1. A flat plate has a surface area of 0.08 m² and a thickness of 0.01 m. If its thermal conductivity k is 0.01 W m^{-1}°C^{-1}, what is the total rate of heat loss across the plate for a temperature differential of 20°C. What is the rate of heat loss per unit surface area (one side)?

2. A hollow cylinder has an internal radius of 0.04 m, an external radius of 0.05 m, and a length 0.20 m. If the thermal conductivity of the cylinder is $k = 0.01$ W m^{-1}°C^{-1}, what is the total rate of heat loss through its surface for a temperature differential of 20°C (neglecting loss through the ends)? What is the rate of heat loss per unit surface area? Calculate the surface area, volume, and ratio of area to volume.

3. A hollow sphere has an internal radius of 0.04 m and an external radius of 0.05 m. If the thermal conductivity of the sphere is $k = 0.01$ W m^{-1}°C^{-1}, what is the total rate of heat loss through its surface for a temperature differential of 20°C? What is the rate of heat loss per unit surface area? Calculate the surface area, volume, and ratio of area to volume. Now compare the rates of heat loss and the area-to-volume ratios of this sphere and the cylinder of Problem 2.

4. Repeat the calculations for the heat loss from a cylinder, putting in end corrections for the cylinder. Assume the ends to be hemispheres (treated as half-spheres) of the same radius as the cylinder. Compute the surface area, volume, and area-to-volume ratio and compare the rates of heat loss (total loss and loss per unit area) with the results of Problems 2 and 4.

5. Determine the ratio of the convection coefficient in air at 10°C with that in air at 40°C at standard sea level pressure for (a) a flat plate with uniform temperature in laminar flow, (b) a flat plate with uniform heat flux in laminar flow, and (c) a flat plate in turbulent flow.

6. Determine the convection coefficient for a cylinder of diameter 0.1 m in air at 20°C and standard sea-level pressure for wind speeds of 1.0 and 5.0 m s^{-1}. Also determine the boundary-layer thickness for the cylinder at these two wind speeds.

7. Determine the convection coefficient for a sphere of diameter 0.1 m in air at 20°C and standard sea-level pressure for wind speeds of 1.0 and 5.0 m s^{-1}. Also determine the boundary-layer thickness for the sphere at these two wind speeds.

8. Determine the convection coefficients for small leaves, using Eq. (9.89), for the ratios V/D of 1.0, 10, 50, and 100.

9. Determine the convection coefficients for large leaves using Eq. (9.88) for the ratios V/D of 1, 10, 50, and 100.

Chapter 10

Evaporation and Transpiration

Introduction

Water is ubiquitous and extremely important to life throughout the biosphere. Evaporation of water from both physical and organic surfaces is of great significance not only for cooling the surface and surrounding media but also in adding water to the atmosphere, where it is a significant meteorological factor. Water is found in all plant and animal cells, where it plays many roles, including acting as a solvent and transporting solutes of various kinds. The water molecule is small and therefore diffuses easily through cell walls. Because of its highly polar character, water readily accepts other polar substances.

General Properties of Water

Water is truly a remarkable compound, and fortunately so because of its facultative role in the living system. It is easy to take the presence of water for granted and forget its tremendously varied role throughout the living world. The high dielectric constant of the water molecule makes it an excellent solvent for ions and as a result many nutrients are transported through biological systems. The intermolecular attraction between water molecules is extremely great, and thermal agitation does not disrupt these easily. Water has an unusually high boiling point (100°C) compared to other closely related electronic compounds and also possesses a comparatively high melting point (0°C). Strong intermolecular forces give water an extremely high surface tension and tensile strength. These properties make it possible for water to rise in the xylem vessels of woody plants and allow some insects to walk on a water surface. Water has a large heat of vaporization, a fairly substantial heat of fusion, a high heat capacity, and good thermal conductivity. These thermal properties make it possible for water to act as a temperature equalizer and buffer temperature changes throughout the biosphere. The density of liquid water is greatest at 4°C, and the bottom waters of deep lakes are usually about this temperature. Water is less dense in its solid form than in its liquid form, which makes it possible for ice to float.

Liquid water, as well as water vapor, is highly transparent to visible radiation but absorbs in the ultraviolet and infrared. This optical property of water is highly significant in the biosphere. The earth's atmosphere contains a great deal of water vapor. The broad spectral transparency of water vapor allows sunlight to reach the earth's surface, and its infrared opacity acts as a thermal screen against longwave radiation loss to space. The optical transparency of liquid water allows light to penetrate deep into oceans, lakes, ponds, and streams, as well as into the cells of plants and animals. But the high infrared opacity of liquid water suggests that longwave radiation is exchanged at the surface of all bodies of water and of biological cells. The heat-transfer implications of infrared absorption and emission by liquid water surfaces are highly significant.

Water forms a continuum for plants, from the soil through the roots, stems, trunks, and leaves to the atmosphere. Water plays an important role in photosynthesis and is the source of oxygen evolved from plants and of hydrogen involved in the reduction of CO_2. The role of water in the energy budgets of plants is highly significant. Nearly all animal cells and tissues are permeable to water, and evaporation of water from animal surfaces is important in determining surface temperatures. Loss of water from a plant or animal has considerable effect on the osmotic potential of cells and concentrations of solutes within cells.

The properties of water which relate to the evaporation and transpiration of water from plant and animal surfaces, as well as from bodies of liquid water, are described in this chapter.

Latent Heat

The quantity of heat necessary to produce a change of state from solid to liquid is called the latent heat of fusion, and that quantity necessary to produce a change from liquid to vapor is the latent heat of vaporization. The latent heat of fusion of water is about 3.33×10^5 J kg^{-1} at 0°C. The latent heat of vaporization of water is 2.260×10^6 J kg^{-1} at 100°C, 2.356×10^6 J kg^{-1} at 60°C, 2.406×10^6 J kg^{-1} at 40°C, 2.454×10^6 J kg^{-1} at 20°C, and 2.501×10^6 J kg^{-1} at 0°C. The value often used for biological purposes is 2.427×10^6 J kg^{-1} at 30°C since many organisms exist at about this temperature much of the time. The error involved with using $2.427 \times$

10^6 J kg^{-1} is only 2.5% at 40°C and 2.0% at 10°C, so for many purposes it is scarcely necessary to put in a temperature function.

Ideal Gas Law

Water vapor may be treated, to a first approximation, as an ideal gas, provided condensation is not taking place. An ideal gas of n moles at pressure p and temperature T occupying a volume V obeys the law

$$pV = nRT, \qquad (10.1)$$

where R is the gas constant. It is known that 1 mole (the mass in grams, equal to the molecular weight) of any ideal gas occupies 22.4×10^{-3} m^3 at 0°C and a pressure of 1 atm (101 N m^{-2}). Therefore,

$$R = 8.31 \text{ J deg}^{-1} \text{ mole}^{-1}. \qquad (10.2)$$

The number n of gram-moles of a gas equals the mass m of the gas divided by its molecular weight M. Hence,

$$n = \frac{m}{M}, \qquad (10.3)$$

and the ideal gas law is written

$$pV = \frac{m}{M} RT. \qquad (10.4)$$

The density ρ of a gas is the mass m per unit volume V, and the ideal gas law becomes

$$p = \rho \frac{R}{M} T, \qquad (10.5)$$

or

$$\rho = \frac{p}{RT} M. \qquad (10.6)$$

Water Vapor

Now one can apply the gas law to water vapor, considered as an ideal gas. If e is the pressure of water vapor, ρ_w its density, and M_w its molecular weight (18), then

$$\rho_w = \frac{e}{RT} M_w. \qquad (10.7)$$

The density of water vapor ρ_w is also referred to as the absolute humidity. Table 10.1 lists values of the density of water vapor as a function of temperature. Dry air also obeys the ideal gas law to a first approximation. If ρ_a is the density of dry air at a pressure p_a and temperature T, then

$$\rho_a = \frac{p_a}{RT} M_a, \tag{10.8}$$

where M_a is the molecular weight of dry air (28.97).

It is desirable to show how the density of moist air ρ relates to the density of dry air ρ_a. The total pressure p of moist air is the sum of the pressure of dry air and water vapor:

$$p = p_a + e. \tag{10.9}$$

Also, the density ρ of moist air is the sum of the density of dry air and water vapor

$$\begin{aligned}
\rho &= \rho_a + \rho_w \\
&= \frac{p_a}{RT} M_a + \frac{e}{RT} M_w \\
&= \frac{p - e}{RT} M_a + \frac{e}{RT} M_w \\
&= \frac{pM_a}{RT}\left(1 - 0.379 \frac{e}{p}\right),
\end{aligned} \tag{10.10}$$

where $(M_a - M_w)/M_a = 0.379$.

Equation (10.10) demonstrates that the density of moist air is always less than the density of dry air at the same temperature and pressure.

Table 10.1. Density of Pure Water Vapor at Saturation over Water[a]

	Relative humidity				
	0.20	0.40	0.60	0.80	0.100
Temperature (°C)					
0	0.969	1.839	2.908	3.678	4.847
5	1.359	2.719	4.078	5.438	6.797
10	1.880	3.760	5.639	7.519	9.399
15	2.566	5.132	7.698	10.264	12.83
20	3.460	6.920	10.380	13.840	17.30
25	4.610	9.220	13.830	18.440	23.05
30	6.076	12.152	18.228	24.304	30.38
35	7.926	15.852	23.778	31.704	39.63
40	10.238	20.476	30.714	40.952	51.19
45	13.100	26.200	39.300	52.400	65.50
50	16.612	33.224	39.836	66.448	83.06

[a] Units are grams per cubic meter.

This, of course, is the reason why moist air rises in the atmosphere. Moist air coming off vegetation, ponds, lakes, or the ground surface will rise.

Definitions

Absolute humidity. Absolute humidity is the density of water vapor and depends on temperature and pressure according to Eq. (10.7).

Relative humidity. Relative humidity is the ratio of the actual vapor pressure p of the moist air to the saturation vapor pressure e_s. Hence

$$h = \frac{p}{e_s}. \tag{10.11}$$

Saturation of moist air at a temperature T and total pressure p occurs if the liquid and vapor coexist when the vapor is above a plane surface of pure liquid water. Table 10.1 shows these values. At constant temperature, the ratio of the vapor pressure is, according to Eq. (10.5), also the ratio of the vapor densities. Therefore,

$$h = \frac{\rho_w}{\rho_s}. \tag{10.12}$$

Specific humidity. Specific humidity is the ratio of the density of water vapor to the density of moist air. Hence, from Eqs. (10.7) and (10.10), one gets

$$q = \frac{\rho_w}{\rho} = \frac{eM_w}{pM_a\left(1 - 0.379\,\dfrac{e}{p}\right)}$$

$$= 0.622\,\frac{e}{p - 0.379\,e}, \tag{10.13}$$

where $M_w/M_a = 0.622$. The specific humidity q is usually given in grams per kilogram. For cold, relatively dry air, the values of q are from 1 to 3, and for warm, relatively moist air from 12 to 25.

Mixing ratio. Mixing ratio is the ratio of the density of water vapor to the density of dry air:

$$w = \frac{\rho_w}{\rho_a} = \frac{e\,M_w}{p_a M_a} = 0.622\,\frac{e}{p - e}. \tag{10.14}$$

Usually, p is about two orders of magnitude greater than e, or $p - e \simeq p$. Then, comparing Eqs. (10.13) and (10.14), one gets

$$w = 0.622\,\frac{e}{p} = q. \tag{10.15}$$

This equation shows that mixing ratio and specific humidity can be used interchangeably. It also leads to the result that relative humidity is the ratio of the mixing ratio of the moist air to the mixing ratio for the air at saturation w_s. Hence

$$h = \frac{w}{w_s}. \tag{10.16}$$

The vapor pressure of water vapor in moist air at a total pressure p is given by rewriting Eq. (10.14) as follows:

$$e = \frac{w}{0.622 + w} p. \tag{10.17}$$

Saturation deficit. Saturation deficit is the difference in vapor pressure between the saturation and actual water-vapor pressure. Therefore, one can write

$$sd = e_s - e. \tag{10.18}$$

Dew point. Dew point is the temperature to which air must be cooled at constant pressure and constant water vapor content in order to become saturated. At temperatures below freezing, it is called the frost-point temperature.

Wet-bulb temperature. Amount of moisture in the air and relative humidity are often measured using a thermometer having wet and dry bulbs. The dry bulb measures the air, or dry-bulb, temperature. The wet-bulb temperature is the temperature the wet cloth assumes at steady state as moisture is evaporated into the air around it when it is well ventilated. If the air is initially saturated with water vapor, then there will be no evaporative cooling of the cloth, and the wet-bulb temperature will equal the dry-bulb temperature. If the water-vapor content of the air is below saturation, water will evaporate from the wet cloth and latent heat loss will reduce its temperature below the dry-bulb temperature. The air is initially at temperature T and pressure p and has a vapor pressure e. The saturation vapor pressure of the air in equilibrium with the liquid water is e_s. Since e is less than e_s, water will evaporate into the air, and both e and p will increase. In the process of producing more water vapor, latent heat λ is removed from the air, which is cooled to a lower temperature T'. Since the system is well insulated, the only place from which the latent heat (used to evaporate the water) can come is from the liquid–air system itself. The saturation vapor pressure at the new temperature is $e_s(T')$.

Initial water-vapor density is given by Eq. (10.14) as

$$\rho_w = \rho_a(0.622) \frac{e}{p} \tag{10.19}$$

since $p \gg e$ and $p - e \approx p$. The water-vapor density of moist air at saturation is

$$\rho'_{ws} = \rho_a(0.622) \frac{e_s(T')}{p}. \tag{10.20}$$

As the water-vapor pressure of the air rises from $e(T)$ to $e_s(T')$, the change of sensible heat is given by

$$\lambda(\rho'_{ws} - \rho_w) = \lambda\rho_a(0.622)\frac{[e_s(T') - e(T)]}{p}. \tag{10.21}$$

If this amount of heat comes from cooling the air (whose specific heat is c_p) from temperature T to temperature T', then

$$\lambda\rho_a(0.622)\frac{[e_s(T') - e(T)]}{p} = \rho_a\,c_p(T - T'). \tag{10.22}$$

The quantity $T - T'$ is referred to as the wet-bulb depression.
Solving for $e(T)$, one gets

$$\begin{aligned}
e(T) &= e_s(T') - \frac{c_p\,p}{(0.622)\lambda}(T - T') \\
&= e_s(T') - \gamma\,(T - T'),
\end{aligned} \tag{10.23}$$

where the quantity γ is known as the psychrometer constant and has a value of 0.66 mbar°C^{-1} at a temperature of 20°C and a pressure of 1,000 mbar. The quantity γ varies from 0.65 at 5°C to 0.67 at 40°C.

Plotting water-vapor-pressure difference against temperature difference, one gets a slope of $-\gamma$.

If the psychrometer is not adequately ventilated to remove the cool, moist air from around the bulb as quickly as it evaporates, then the value of γ will be lower than that given above.

Mass Transfer

Just as the transfer of heat between an object and the air around it is described by convection and use is made of dimensionless numbers such as the Nusselt or Reynolds number, so also can the exchange of mass such as water vapor be described. First, however, we must describe mass flow by diffusion.

Fick's Law of Diffusion

In relatively still air, gases move along a concentration gradient from regions of higher to lower concentration. In this manner, carbon dioxide diffuses into the leaf mesophyll from the air surrounding the leaf, and water vapor and oxygen diffuse from within the leaf outward to the atmosphere. The concentration gradient is the driving force which produces mass movement, or current flow. However, there is resistance to the flow, or drag, owing to the viscosity of the fluid and its interaction with the surfaces or walls past which it is flowing. Water must be pushed through a hose in order to overcome the drag of the walls, and the gases diffusing in or out of a leaf must likewise overcome the drag produced by the cell walls and stomate aperture. These relationships may be expressed by a one-dimensional equation known as Fick's first law of diffusion, first derived in 1855. If there is a concentration or density difference $\Delta\rho_j$ along an increment of distance Δz, then the amount of a particular species j (water vapor, oxygen, or carbon dioxide, for example) flowing per unit area per unit time is

$$J_j = \mathcal{D}_j \frac{\Delta\rho_j}{\Delta z}, \tag{10.24}$$

where $\mathcal{D}_j$ is a proportionality constant known as the diffusion coefficient and is pressure- and temperature-dependent. The value of $\mathcal{D}$ for water vapor is 2.57×10^{-5} m^2 s^{-1} at 20°C and varies from 2.26×10^{-5} at 0°C to 2.88×10^{-5} m^2 s^{-1} at 40°C. The value of $\mathcal{D}$ for carbon dioxide is 1.47×10^5 m^2 s^{-1} at 20°C and varies from 1.29×10^{-5} at 0°C to 1.65×10^{-5} m^2 s^{-1} at 40°C.

If Fick's law is written in terms of the resistance r_j to flow offered by the diffusion pathway, it becomes

$$J_j = \frac{\Delta\rho_j}{r_j}, \tag{10.25}$$

and therefore

$$r_j = \frac{\Delta z}{\mathcal{D}_j}. \tag{10.26}$$

Units are as follows: z is in meters, ρ_j is in kilograms per cubic meter, $\mathcal{D}_j$ is in square meters per second, r_j is in seconds per meter, and J_j is in kilograms per square meter per second. The reciprocal of resistance is conductance, measured in meters per second, and some people prefer to use this terminology. By analogy the resistance to convective heat transfer is given by Eq. (9.20) as $r_h = 1/\overline{h_c}$ and using Eq. (9.24) it becomes $\mathcal{D}/k$ Nu.

It is now possible to write the diffusion of water vapor, or transpiration, from a leaf in terms of Fick's law. Let the water vapor concentration or density in the leaf at saturation at the leaf temperature T_ℓ be $_s\rho_\ell(T_\ell)$. Let water-vapor saturation of the air at the air temperature T_a be $_s\rho_a(T_a)$. Then, if the air has a relative humidity h, the actual water-vapor density of the air is $h\ _s\rho_a(T_a)$. The internal structure of a leaf offers resistance r_ℓ to the flow of water vapor from the inside to the outside of the leaf. The boundary layer of viscous air adhering to the leaf surface offers additional resistance r_a to diffusive flow of water vapor. Amount of water vapor leaving a leaf per unit area per unit time is E, given by

$$E = \frac{_s\rho_\ell(T_\ell) - h\ _s\rho_a(T_a)}{r_\ell + r_a}. \tag{10.27}$$

The units for E are kilograms per square meter per second. When E is multiplied by the latent heat λ of vaporization for water (2.427×10^6 J kg^{-1}), the product λE is in joules per square meter per second, or watts per square meter.

Sherwood Number

Dimensionless analysis for various objects of diferent shape, size, and orientation in a variety of media leads to a set of equations for mass transfer analogous to that achieved for convective heat transfer. Flow conditions will influence the boundary-layer thickness and change the rate of gas diffusion from J_j to J_j'. It is possible to define a dimensionless group known as the Sherwood number, which represents the ratio of J_j' to J_j,

$$\text{Sh} = \frac{F'}{\mathscr{D}(\rho_1 - \rho_2)/\delta} = \frac{J_j'}{\mathscr{D} \, \Delta \rho_j/\Delta z}. \tag{10.28}$$

The Sherwood number is the ratio between the actual flux density of water vapor (or other gas) and the flux density which would occur by pure diffusion across a thickness of still air Δz.

Schmidt Number

Another dimensionless group useful in defining mass flow is the Schmidt number (Sc). It is analogous to the Prandtl number used in the analysis of convective heat exchange. The Schmidt number is

$$\text{Sc} \equiv \nu/\mathscr{D}, \tag{10.29}$$

where ν is the coefficient of kinematic viscosity.

Mass- and Heat-Flow Analog

The result of dimensionless analysis gives, for a flat plate in forced flow,

$$\text{Sh} = 0.66 \, \text{Re}^{1/2} \, \text{Sc}^{1/3}, \tag{10.30}$$

as compared with heat exchange for forced convection, which is

$$\text{Nu} = 0.66 \, \text{Re}^{1/2} \, \text{Pr}^{1/3}. \tag{10.31}$$

The above equations tell us that the fundamental properties of heat and mass transfer by molecular diffusion are nearly identical and that the dimensionless variables Sc or Pr simply take account of differences in mass or heat transfer through the boundary layer.

Lewis Number

When forced convection occurs, one can obtain the relationship between the Sherwood and Nusselt numbers by dividing Eq. (10.30) by Eq. (10.31):

$$\text{Sh} = \text{Nu}(\text{Sc}/\text{Pr})^{1/3} = \text{Nu}(k/\mathscr{D})^{1/3}. \tag{10.32}$$

The ratio $k/\mathscr{D}$ is called the Lewis number (Le). In air, $(k/\mathscr{D})^{1/3}$ is 0.96 for water vapor and 1.14 for CO_2. Monteith (1973) gives the following ratios between the resistances to water-vapor transfer r or carbon dioxide transfer R and the resistance to heat transfer r_h:

$$r/r_h = (k/\mathscr{D})^{2/3} = 0.93 \tag{10.33}$$

$$R/r_h = (k/\mathscr{D}_c)^{2/3} = 1.32, \tag{10.34}$$

where $\mathscr{D}$ and $\mathscr{D}_c$ are the diffusion coefficients for water vapor and carbon dioxide, respectively.

First Law of Thermodynamics

Some useful energy definitions and relationships are derived by making use of the first law of thermodynamics, which states that in a closed system, the sum of all forms of energy remains constant. When a quantity of heat dQ is added to a body, a portion of the heat may go into increased internal energy dU and another portion into increased potential energy if an expansion of the system, given by PdV, has taken place. Therefore,

$$dQ = dU + PdV = c_v\, dT + PdV, \tag{10.35}$$

where P, V, and T are pressure, volume, and temperature and c_v is the specific heat of the substance at a constant volume. One can rewrite Eq. (10.35) as

$$\frac{dQ}{T} = \frac{dU}{T} + \frac{PdV}{T} = c_v\frac{dT}{T} + R\frac{dV}{V} \tag{10.36}$$
$$= d(c_v \log T + R \log V) = dS.$$

By definition, the quantity S, whose variation is represented by an exact differential, is called the entropy of the system. Rewriting Eq. (10.36), one gets

$$dU = dQ - PdV = TdS - PdV. \tag{10.37}$$

Also,

$$d(PV) = PdV + VdP. \tag{10.38}$$

Since these are exact differentials, one can write

$$d(U + PV) = TdS + VdP \tag{10.39}$$

$$d(TS) = TdS + SdT \tag{10.40}$$

$$d(U - TS) = -PdV - SdT. \tag{10.41}$$

The quantity $U + PV$ is called the "total heat," or enthalpy, of a substance, and is denoted by H. The quantity $U - TS$ is called the Helmholtz free energy of the system, or the maximum available energy, and is usually denoted by F. The Helmholtz free energy is useful in cases in which no external work is done by a system. Most biological systems undergo changes, such as cell growth and expansion, which involve small volume changes against a constant external pressure. In this situation, it is convenient to define a function G, which is known as the Gibbs free energy and is defined by

$$G = U - TS + PV = F + PV. \tag{10.42}$$

A change in the Gibbs free energy is often referred to as the maximum useful work a system undergoes, and a change in the Helmholtz free energy as representing the maximum total work done. Usually, ΔG for a physiological or biological process is almost exactly the same as ΔF.

Enthalpy and Equivalent Temperature

In studying the evaporation and condensation of moisture in a parcel of air in contact with liquid water, it is useful to consider how the total quantity of heat H is distributed when the temperature of the system is changed from T_1 to T_2 and the vapor pressure changes from e_1 to e_2. The energy budget of the air is made up of a sensible-heat term and a latent-heat term. The total heat content, or enthalpy, of the air per unit volume is

$$H = \rho_a\, c_p(T_2 - T_1) + \lambda\rho_a(0.622)(e_2 - e_1)/p. \tag{10.43}$$

When adiabatic changes are made in a parcel of air, there is no change in the total quantity of heat, and the gain in one term equals the loss in the other, as shown in Eq. (10.23). There the psychrometer constant γ, which is equal to $c_p\, p/(0.622)\lambda$, was defined. Using this constant, one can rewrite Eq. (10.43) as

$$\begin{aligned} H &= \rho_a\, c_p(T_2 - T_1) + \rho_a\, c_p(e_2 - e_1)/\gamma \\ &= \rho_a\, c_p(\theta_2 - \theta_1), \end{aligned} \tag{10.44}$$

where θ is known as the equivalent temperature of the air and is defined by the equation

$$\theta = T + e/\gamma. \tag{10.45}$$

The equivalent temperature θ is the temperature which relates to the change of total heat H, or enthalpy, a nonadiabatic change involving both sensible- and latent-heat terms as given in Eq. (10.44). By comparison, the ordinary temperature T determines the change of sensible heat only.

The equivalent temperature is the temperature air assumes when the pressure is decreased adiabatically until all the water in the air is condensed. It should be noted that, because of changes in the value of e, θ changes in value much more than T.

An adiabatic system is one which is completely insulated from its surroundings and therefore is only approximated by real-world conditions. However, the concept of an adiabetic change, in which energy cannot flow into or out of the system, gives relationships between temperature and moisture which are highly useful.

Temperature–Enthalpy Diagram

A very useful diagram is a temperature–enthalpy diagram, referred to as a TED, in which the ordinate θ (equivalent temperature) is a measure of the enthalpy and the abscissa is the temperature T. Monteith (1973) gives many such diagrams for various applications.

A TED can be developed for the behavior of a wet surface with respect to its energy balance. By letting Q equal q/At, or total heat loss per unit area per unit time, as we have done throughout this book, one can write the behavior of a wet evaporating surface in terms of the diffusion resistance to heat r_h and diffusion resistance to water vapor r. If the wet surface is at T_2, e_2, and θ_2, and the free air above it is at T_1, e_1, and θ_1, then the total heat loss from the surface is the sum of sensible- and latent-heat changes as follows:

$$Q = \frac{\rho \, c_p (T_2 - T_1)}{r_h} + \frac{\rho \, c_p}{\gamma} \frac{(e_2 - e_1)}{r}. \tag{10.46}$$

In order to rewrite Eq. (10.46) in terms of equivalent temperatures θ_1^* and θ_2^*, it is convenient to substitute $\gamma^* = \gamma \, r/r_h$. One then obtains

$$Q = \frac{\rho \, c_p (T_2 - T_1)}{r_h} + \frac{\rho \, c_p (e_2 - e_1)}{\gamma^* \, r_h}$$

$$= \frac{\rho \, c_p}{r_h} (\theta_2^* - \theta_1^*). \tag{10.47}$$

Therefore,

$$\theta^* = T + e/\gamma^*. \tag{10.48}$$

This equation describes the temperature–enthalpy properties of an evaporating surface. If the evaporating surface is a wet bulb, one can see from Eq. (10.23) that the equivalent temperature θ of the air is uniquely related to the air temperature T' by the relationship

$$\theta = T' + e_s(T')/\gamma. \tag{10.49}$$

Wet Surface of Vegetation

Theory

The rate at which water evaporates from a lake, river, pond, or other broad flat surface, as well as from a leaf, is of enormous interest. It is assumed that there is a net amount of energy available for evaporation and convection. The net energy available would normally be the net radiation at the surface and also that available as storage in the water.

Let there be a net radiation R_n at a wet evaporating surface and some heat going into or out of storage, represented by X. Then, to a first approximation, if λE is the energy consumed by evaporation,

$$Q = R_n - X = \lambda E + C, \tag{10.50}$$

where C is the sensible heat exchanged through convection. From Eq. (10.46) or (10.47), it is evident that

$$C/\lambda E = \beta = \gamma \frac{r}{r_h} \frac{T_s - T_a}{e_s - e_a} = \gamma^* \frac{T_s - T_a}{e_s - e_a}. \tag{10.51}$$

The ratio β is often referred to as Bowen's ratio. If the storage term is $X = 0$, then Eq. (10.50) can be written

$$R_n = \lambda E(1 + \beta). \tag{10.52}$$

The very simplest expression for the evaporation from a free water surface was given by Dalton in 1802. The basic relationship, which has been used ever since by most investigators, is

$$E = (e_s - e_a) f(V), \tag{10.53}$$

where e_s and e_a are the vapor pressures at the evaporating surface and in the air, respectively, and $f(V)$ is a function of the wind speed. There have been various modifications of this formula, but in general they remain about the same. Penman (1948) has reviewed the measurements of Rohwer (1931) and writes

$$E = 0.40(e_s - e_a)(1 + 0.17 \, V_2), \tag{10.54}$$

where V_2 is the mean wind speed at a height of 2 m above the water surface and E is in units of millimeters per day. Variations of this formula are given in Penman (1948). Of particular significance is the equation given for evaporation from a wet rectangular strip of length x and width y,

$$E = c(e_s - e_a) \, V_2^{0.76} \times x^{0.88} \, y, \tag{10.55}$$

where c is a parameter that depends on the temperature.

It is evident from Eq. (10.52) that to obtain the evaporation rate, one must know β, which, from Eq. (10.51), requires knowledge of T_s and e_s as well as T_a and e_a. Since it is difficult to obtain values for T_s and e_s, it is desirable to eliminate these and use instead the slope of the saturation vapor curve to give an approximation to the exact formulation.

Let

$$\Delta = \frac{d\,e_s}{dT},\tag{10.56}$$

where Δ, the change of saturation vapor pressure with temperature.

The following procedure, which is used to derive an expression for the evaporation rate of a water surface in terms of other readily determined parameters, is taken from Monteith (1973), with some modification. Let T_0, e_0, and θ_0^* be the temperature, vapor pressure, and equivalent temperature of a wet surface. This position is represented in the TED diagram of Fig. 10.1 by point Y on the saturation vapor-pressure curve. Let T_a, e_a, and θ_a^* be the temperature, vapor pressure, and equivalent temperature of the air above the water. The heat budget of the water–air interface is, according to Eq. (10.50),

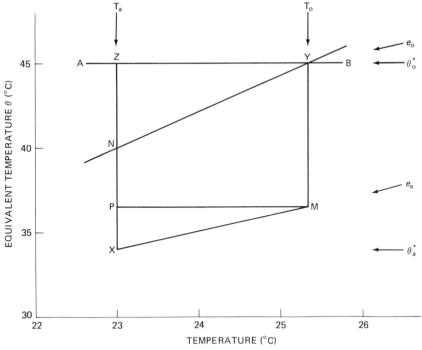

Figure 10.1. Temperature–enthalpy diagram (TED) for evaporation (Y) from a wet surface (X) into air X. The saturation vapor-pressure curve and its chord is NY. (Redrawn from Monteith, 1973.)

$$Q = R_n - X = \frac{\rho \, c_p (T_0 - T_a)}{r_h} + \frac{\rho \, c_p (e_0 - e_a)}{\gamma^* \, r_h}$$

$$= \frac{\rho \, c_p}{r_h} (\theta_0^* - \theta_a^*).$$

(10.57)

Therefore,

$$\theta_0^* = \theta_a^* + \frac{Q \, r_h}{\rho \, c_p}.$$

(10.58)

The lines AB and PM of Fig. 10.1 represent constant equivalent-temperature conditions. The line XM is of constant vapor pressure, e.g., of slope equal to 1.0, according to Eq. (10.45). (Remember that the TED diagram is scaled with the T-axis expanded four times the θ^*-axis; therefore, a line of constant vapor pressure appears to have a slope of 1/4.) Then, PM = ZY = $T_0 - T_a$ and is proportional to the sensible heat change. The vertical lines YM and ZP are at constant temperature and, according to Eq. (10.46) or (10.48), represent a change of latent heat since YM = ZP = $(e_0 - e_a)/\gamma^*$. The slope of the chord YN is $\partial(T + e/\gamma^*)/\partial T = 1 + \partial e/\partial T$, or $1 + \Delta$, where $\Delta = \partial e/\partial T$ [Eq. (10.56)].

In order to determine the latent heat in terms of other quantities, it is convenient to write from Fig. 10.1 the expression

$$YM = ZP = ZN + NX - PX$$

$$= ZY \left(1 + \frac{\Delta}{\gamma^*} \right) + NX - PX.$$

(10.59)

Therefore,

$$\frac{e_0 - e_a}{\gamma^*} = (T_0 - T_a)\left(1 + \frac{\Delta}{\gamma^*} \right) + \frac{[e_s(T) - e]}{\gamma^*} - (T - T_a).$$

(10.60)

Multiplying throughout by $\rho \, c_p / r_h$ gives

$$\lambda E = \frac{\Delta}{\gamma} C + \frac{\rho \, c_p [e_s(T) - e]}{\gamma^* \, r_h},$$

(10.61)

where $C = \rho \, c_p (T_0 - T_a)/r_h$.

From Eq. (10.50), it is possible to write $C = Q - \lambda E$. Substituting into Eq. (10.61) and rearranging gives

$$\lambda E = \frac{Q\Delta + \rho \, c_p [e_s(T) - e]/r_h}{\Delta + \gamma^*}.$$

(10.62)

This expression was first derived by Penman (1948) and has been modified by Slatyer and McIlroy (1961). It is a useful expression and has been applied to estimate rates of evaporation from water surfaces and vegetation. The quantity Q may represent just the net radiation R_n, or the net radiation plus or minus the heat to or from storage $R_n \pm X$, or whatever energy is available for sensible and latent heat. Thus by knowing the net radiation over a period of time in which $X = 0$, the saturation deficit $e_s(T) - e$, the temperature, and the wind speed, it is possible to estimate, by means of Eq. (10.62), the rates of evaporation for a free water surface. In order to get r_h, Penman used measurements of evaporation from a small tank of water to determine a linear relationship between r_h^{-1} and the wind speed at a height of 2 m, in a procedure that follows naturally from aerodynamic analysis of evaporation from a surface.

Also from Eq. (10.50), it is possible to write $\lambda E = Q - C$. Substitution of this quantity into Eq. (10.61) gives

$$C = \frac{\gamma^* Q - \rho\, c_p [e_s(T) - e]/r_h}{\Delta + \gamma^*}. \tag{10.63}$$

Since C can be determined in this manner, then from the definition of C, one can write the surface temperature T_0 of the wet evaporating surface as

$$T_0 = T_a + \frac{C\, r_h}{\rho\, c_p}. \tag{10.64}$$

Observational Results

One of the earlier papers concerning rates of evaporation from open bodies of water was by Cummings and Richardson (1927). In fact, the formula derived by these authors was based on Bowen's (1926) work and is equivalent to Eq. (10.52). The purpose of this formulation was to arrive at a relationship which is not dependent upon wind speed. Penman (1948) was interested in establishing the relationship between the evapotranspiration rate from vegetated surface and that from an open water surface. He found that evaporation from a turf was a certain fraction of the evaporation from a free water surface, depending on the time of year, for southern England. The ratio of turf to open water evaporation E_t/E_0 was 0.6 in midwinter, 0.7 in spring and autumn, and 0.8 in midsummer, and the average for the whole year was 0.75. According to Monteith (1973), these values are generally valid for any form of vegetation, as long as the roots have free access to water, and are not restricted to short green grass that is freely watered. Evapotranspiration in arid zones is often higher than that for turf because of the very dry air and perhaps because of an "oasis" effect. The Penman formula, Eq. (10.62), has been tested under a wide variety of conditions for reservoirs, lakes, pans, and crops throughout the world and found to be valid. In fact, Stanhill (1961) writes that "measurements of crop evapotranspiration with a non-limiting soil moisture supply have shown close agreement with the amount of evaporation from an open water surface, when this was either measured with a standard British Meteorological Office evaporation tank or calculated by Penman's method." However, when crops are subjected to stress caused by water deficits, the evapotranspiration rate is reduced 40% or more below that of a well-watered crop.

What sort of values prevail for the evaporation rates from open bodies of water at different times of year? Evaporation rates from an 85-acre reservoir in Boston ranged from 10.7 mm/day in June to 2.5 mm/day in October. A reservoir at Molato, Italy, had values of 7.9 mm/day in July, 5.7 mm/day in August, and 1.3 mm/day in December. Measurements on Sevan Lake in Soviet Armenia gave values from 4.5 mm/day in September to 0.2 mm/day in May. Evaporation from rye grass at Davis, Calif., varied from a minimum of about 1 mm/day in December and January to a maximum of about 7.2 mm/day in June according to Pruitt and Angus (1961). In the same study, a U.S. Weather Bureau class-A pan located within a large grass field usually evaporated nearly 40% more water than the rye grass; maximum pan evaporation exceeded 14 mm/day during clear, dry, north-wind days.

Monteith (1965) derived a combination formula for leaves and also for a plant canopy that resembles Eq. (10.62) in form, but Tanner and Fuchs (1968) offer considerable criticism of Monteith's formulation. In fact, Tanner and Fuchs have derived a general relation between actual evaporation from a surface and potential evaporation from the surface when there is no saturation deficit at the surface and a condition of nonlimiting water supply exists.

Fuchs and Tanner (1967) give the evaporation E from drying soil in terms of the potential evaporation E_p as

$$E = \frac{\Delta + \gamma}{\Delta} E_p + \frac{\rho c_p}{\Delta} h(e_0^* - e_z), \tag{10.65}$$

where h is a transfer coefficient which is essentially the reciprocal of resistance. The units of h are meters per second. The notation of Fuchs and Tanner is used here. The quantity e_0^* is the saturation vapor pressure at the surface, and e_z is the actual vapor pressure at height z above the surface.

Potential evaporation is given by

$$E_p = \frac{\Delta(R_n - G) + \rho c_p h(e_z^* - e_z)}{\Delta + \gamma}, \tag{10.66}$$

where e_z^* is the saturation vapor pressure at height z.

The derivation of Eq. (10.65) involves no assumption concerning the saturation deficit at the surface or in the soil. This suggests not that the evaporation rate is independent of the saturation deficit at the surface but that its effect is accounted for by the surface temperature. The saturation vapor pressure e_0^* at the surface is directly dependent on the surface temperature. As a soil dries, the surface temperature increases, and e_0^* increases greatly even though e_0 may decrease.

Fuchs and Tanner (1967), following earlier work by others, describe the transfer coefficient in terms of wind speed V and height above the surface z as

$$h = k^w V[\Phi + \ell n(z + z_0)/z_0]^2, \qquad (10.67)$$

where k is the von Karman constant $(= 0.428)$ and Φ is a profile diabatic-influence function which depends on height above the surface in a complex manner. It is not worthwhile to describe this function in more detail here; reference to the original paper will give the reader the necessary information.

Transpiration by Leaves

The process of transpiration by a plant leaf may be the inadvertent consequence of the leaf taking in of carbon dioxide through the stomates. The CO_2 molecule has a molecular weight of 44, compared with 18 for H_2O. This means that H_2O molecules diffuse through membranes much more readily than CO_2 molecules. However, a plant leaf must take in CO_2 in order to supply it to the chloroplasts at a reasonable rate for photosynthesis. Membranes have not evolved that pass CO_2 through from outside the leaf into the chloroplasts without passing H_2O molecules out to the external environment. By means of valves, or stomates, the loss of water vapor is controlled while still allowing CO_2 to diffuse in to the leaf. These two conflicting requirements demand that some sort of control mechanism exist which can open and close stomates in response to both internal and external conditions. The control of stomate opening is one of the more vital of all plant functions. Land plants have evolved an epidermis covered with a waxy cuticle which has a low permeability to water and carbon dioxide. Diffusion of these gases through the leaf cuticle is from 10 to 100 less than that occurring through the stomates of a leaf when they are wide open. The lowest resistances of stomates are about 100 s m^{-1} or a little less. Minimum diffusion resistance depends on stomatal size, density, and distribution, which in turn are dependent on species, developmental conditions of light, temperature, water potential, nutrients, and such things as age.

Water enters a plant from the soil through root hairs, epidermis, or breaks in the periderm tissue. Once water has reached the xylem, it moves up through the stems to the leaves, where it passes into the mesophyll cells and is vaporized. It is evaporation of water from the leaf that creates a drop in water potential in the leaf, acts as a driving force along an entire sequence of dropping water potential, and drives water from root to stem to leaf to atmosphere. Jarvis (1975) has reviewed the mechanisms of water transfer in plants. It is not the purpose here to concern ourselves with the transfer of water within the plant or to describe the physiology of plant water relations. Descriptions of this sort can be found in such physiology texts as Nobel (1974) or Salisbury and Ross (1969).

All parts of the pathway for water transport, from the soil through the roots, xylem, mesophyll, stomates, and boundary layer to the atmosphere, offer resistance to the flow of water. Roots are thought to offer the largest resistance to the flow of water from the soil to leaf to atmosphere, but Boyer (1969) has shown that the resistances of stem and leaves together nearly match that of the roots. The ratio of resistances of root to stems to leaves for sunflower was 2:1:1, with the moist soil contributing negligible resistance. When normalized on a per-unit-length basis, the ratio of root to stem to leaf resistances is 1:0.0005:1 for sunflower.

Water Potential

The chemical potential, or energy status, of water in cells determines the direction of net water movement. At equilibrium, the chemical or water potential is the same across a membrane between two cells or across any interface. An internal hydrostatic pressure within plant cells produces stresses in cell walls, and a certain deformation of the cell wall results. Turgor pressure exists within the cells. The water potential of a cell or a parcel of air containing water vapor is its free energy per mole. The free energy per mole of any particular chemical species is known as the chemical potential of that species. Hence the water potential is simply the chemical potential of water, written

$$\mu_w = \mu_w^* - \overline{V}_w \, \pi + \overline{V}_w \, P + m_w g h, \tag{10.68}$$

where μ_w^* is an arbitrary reference level, π is the osmotic pressure, $\overline{V}_w$ is the partial molal volume of water, the volume of 1 mole of water, P is a pressure taken to be the pressure in excess of atmospheric pressure, and $m_w g h$ is the work done in raising a mass of water equal to its gram-molecular weight (18.016×10^{-3} kg/mole) through a height h (with g the gravitational constant).

Water potential is defined as

$$\Psi = \frac{\mu_w - \mu_w^*}{\overline{V}_w} = P - \pi + \rho_w g h, \tag{10.69}$$

where $\rho_w = m_w / \overline{V}_w$.

The gravitational term is often very small in going from one cell to another within a plant and can be neglected. If water moves 10 m vertically up a tree, the water potential is increased by about 1 bar since

$\rho_w g = 0.098$ bar/m. A pressure of 1 bar is defined as 10^6 dynes cm^{-2}, which is equivalent to 0.987 atm, or 0.76 mm of Hg at sea level.

The water potential for water vapor in air is given by

$$\Psi_{wv} = \frac{\mu_{wv} - \mu_{wv}^*}{\overline{V}_w} = \frac{RT}{\overline{V}_w} \ell n h + \rho_w g \hbar. \qquad (10.70)$$

It is important that $\overline{V}_w$ and not $\overline{V}_{wv}$ be used in the definition of water potential since one wishes to compare the chemical potential of water in the vapor phase with that in the cells of plants. The term $\rho_w g \hbar$ is small and can be dropped.

The quantity $RT/\overline{V}_w$ takes the value 1350 bars at 20°C. If the relative humidity of air is 1.0, then ℓn 1.0 $= 0$, and the water potential of saturated air is 0. Pure liquid water at atmospheric pressure has a water potential of 0, and the saturated air is, of course, in equilibrium with it. At a relative humidity of 0.99, the water potential is 1350 ℓn 0.99, or -13.6 bar; at 0.95, 1350 ℓn 0.95, or -69.2 bars; and at 0.50%, 1350 ℓn 0.50, or -936 bars. It is clear that small changes in the relative humidity of a parcel of air have enormous effects on the water potential. The air inside plant leaves, in the substomatal cavity for example, is usually very close to saturation and is in the range of -3 to -30 bars. A few desert plants have water potentials as low as -50 bars, which corresponds to a relative humidity of 0.97. Since the air outside a leaf is usually at a relative humidity lower than 0.97, there is almost always loss of water from the leaf. The assumption which is made in the transpiration formula that the air inside the leaf is saturated is therefore a very good approximation.

Boyer (1969) reported measurements of water potential and resistances to the transfer of free energy, and therefore water, for soil, roots, stems, and leaves. Sunflower leaves with a water deficit had water potentials of -15 bars, and well-watered leaves had potentials of -2 bars.

Theory of Leaf Resistance

Water vapor diffuses from the wet mesophyll cell walls surrounding the intercellular cavities of a leaf through the intercellular air spaces, the stomates, and finally the boundary layer of adhering air next to the leaf surface. In addition, there is a parallel circuit by which water is lost from a leaf through the cuticle and across the boundary layer. These diffusion pathways are shown in Fig. 10.2. By using Fick's law [Eq. (10.69)], it is possible to define diffusion resistances for the intercellular pathway, stomatal pathway, cuticular pathway, and boundary layer. From electrical circuit theory, we know that series resistances are always additive and that for parallel circuits, the reciprocal of the total resistance is the sum of the reciprocals of the resistances. Therefore, two resistances r_1 and r_2 in series would be equivalent to $r_1 + r_2$ and in parallel would be equal to

$r_1 r_2/(r_1 + r_2)$. Therefore, the resistances of a leaf may be written

$$r_\ell = \frac{(r_i + r_s)r_c}{r_i + r_s + r_c},\qquad (10.71)$$

where r_i is intercellular air-space resistance, r_s is stomatal resistance, and r_c is cuticular resistance. Total resistance r of a leaf and its adhering boundary layer is

$$r = r_\ell + r_a = \frac{(r_i + r_s)r_c}{r_i + r_s + r_c} + r_a,\qquad (10.72)$$

where r_a is the boundary-layer resistance. All resistances are average values for the leaf as a whole.

Actually, there are two leaf surfaces, and the diffusion of water vapor through the upper surface is in parallel with the exchange through the lower surface. If the superscript u indicates the upper surface and the superscript ℓ the lower surface, then the total resistance of a leaf is written

$$r = \frac{r^u\, r^\ell}{r^u + r^\ell},\qquad (10.73)$$

where

$$r^u = \frac{(r_i^u + r_s^u)\, r_c^u}{r_i^u + r_s^u + r_c^u} + r_a^u \qquad (10.74)$$

$$r^\ell = \frac{(r_i^\ell + r_s^\ell)\, r_c^\ell}{r_i^\ell + r_s^\ell + r_c^\ell} + r_a^\ell.\qquad (10.75)$$

Usually, one does not need to use these fairly complicated equations for the upper and lower leaf surfaces individually, but can use average values as given in Eq. (10.72) for the whole leaf. Usually, $r_i^u = r_i^\ell$, $r_a^u = r_a^\ell$ for a vertical leaf and a leaf of any orientation in wind, and $r_a^\ell = 2r_a^u$ for a horizontal leaf in still air. For most leaves, $r_c^u \gg r_s^u$ and $r_c^\ell \gg r_s^\ell$. If r_c is very

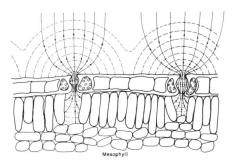

Figure 10.2. Diffusion pathways for water vapor out of a leaf from wet cell wall surfaces in the mesophyll and parenchyma.

large, then Eqs. (10.74) and (10.75) are given by

$$r^u = r_i^u + r_s^u + r_a^u \tag{10.76}$$

$$r^\ell = r_i^\ell + r_s^\ell + r_a^\ell, \tag{10.77}$$

and, of course, Eq. (10.73) still applies for the whole leaf.

A leaf with stomates on both sides (an amphistomatous leaf) has an equivalent total leaf resistance of $r^u/2 = r^\ell/2$ if $r^u = r^\ell$, whereas a leaf with stomates on the lower side only (a hypostomatous leaf) has an equivalent resistance of $r^\ell r^u/r^\ell + r^u$ as given by Eq. (10.73). Throughout this book, we use the average resistance of the whole leaf and the boundary layers.

Resistance Values

It is useful to know the minimum leaf resistances to the diffusion of water vapor. The degree of stomatal opening is a function of many factors, including water potential and turgor pressure in the leaf, relative humidity of the air, temperature of the leaf, level of illumination, and the carbon dioxide concentration inside and outside the leaf. If one knows the minimum resistance, then, depending on the values of all other factors, one can estimate the resistance for any actual condition. There seems to have been no systematic effort to determine diffusion resistance for a large number of plant species.

Until relatively recently, it has been difficult to measure the diffusion resistance of a leaf with any degree of accuracy. Now, with the development of various porometers, it can be done, and considerable amounts of resistance data are accumulating. For porometer design, one is referred to Kanemasu et al. (1969), Monteith and Bull (1970), Bierhuizen et al. (1965), Kenny and McGruddy (1972), Stigter (1972), Stigter et al. (1973), van Bavel et al. (1965), and Byrne et al. (1970). Taylor and Gates (1970) have described three methods to obtain resistance values from a porometer and energy-budget combination method whose accuracy is $\pm 5\%$, a quick-weigh method accurate to $\pm 9\%$, and a severed-leaf technique accurate to $\pm 14\%$.

Minimum diffusion resistances to water are generally as follows: r_a, 30–120 s m^{-1} for small and moderate size leaves in still air or light wind; r_i, 20 s m^{-1}; r_s, 100 to 200 s m^{-1} for many mesophytes (some leaves have values as low as 30 or as large as 500 s m^{-1}, and xerophytes have values up to 6,000 s m^{-1}; and r_c, 2,000 to 10,000 for mesophytes and 4,000 to 60,000 for xerophytes. For example, if $r_a = 50$ s m^{-1}, $r_i = 20$ s m^{-1}, $r_s = 200$ s m^{-1}, and $r_c = 10,000$ s m^{-1}, then according to Eq. (10.71), $r_\ell = 215$ s m^{-1}, and according to Eq. (10.72), $r = 265$ s m^{-1}. Usually, r_a is a variable resistance which depends upon leaf size and wind speed.

The actual leaf resistance in the above example will vary from 265 s

m^{-1} in still air to 215 s m^{-1} in a strong breeze. Although increased wind removes the boundary layer and reduces resistance, it is likely that the wind-agitated leaves will close some stomates and stomatal resistance will increase. When a leaf is in a condition of minimum leaf resistance (high light intensity, high humidity, optimum temperature, etc.), then a change of almost any environmental factor will cause some stomatal closure, or partial closure, and total resistance will increase.

Table 10.2 shows values of r_a, r_s, and r_c for various plants as reported by Holmgren et al. (1965). Often, the leaf boundary layer has a relatively small resistance compared with stomatal resistance. *Helianthus annuus* (sunflower) is an exception since $r_s = 40$ s m^{-1} and $r_a = 50$ s m^{-1}. The stomatal-resistance value for sunflower is so low as to be suspect. Ehrler and van Bavel (1968) obtained measurements for sunflower of $r_\ell = 120$ s m^{-1}, and Whiteman and Kohler (1967) reported $r_s = 360$ s m^{-1} at 25°C. With moderate amounts of wind, r_a will reduce to 10 or 20 s m^{-1}. Wind movement in the canopy of vegetation occurs with sufficient frequency to make the influence of a leaf boundary-layer resistance on water loss and productivity negligible compared with other resistances. Baker and Myhre (1969) compared carbon fixation in deeply lobed, "okra" cotton leaves, which have relatively thin boundary layers, to that in normal cotton leaves and found no difference whatsoever. These authors concluded that the leaf boundary-layer resistance under field conditions is small compared with the total CO_2 diffusion resistance.

The resistances of various other plant species have been reported. For cotton, values of $r_\ell = 180$ s m^{-1} and $r_a = 140$ s m^{-1} (Ehrler and van Bavel, 1968) and $r_\ell = 110$ s m^{-1} and $r_c = 3200$ s m^{-1} (Slatyer and Bierhuizen (1964) have been recorded. Slatyer and Bierhuizen have also shown that $r_a = 350$ to 160 s m^{-1} when the wind speed is 0.006 and 0.031 m s^{-1}, respectively, and $r_a < 50$ when the wind speed is 0.25 m s^{-1}. Ehrler and van Bavel (1968) made the following measurements of leaf resistance r_ℓ:

Table 10.2. Average Values of Minimum Stomatal Resistance with Low Amount of Wind, Cuticular Resistance, and Boundary-Layer Resistance[a]

	r_s	r_c	r_a
Populus tremula	230	–	60
Betula verrucosa	120	7000	80
Quercus robur	1160	29000	90
Acer platanoides	790	10100	80
Circaea lutetiana	1240	9000	50
Lamium galeobdolon	940	3000	80
Helianthus annuus	40	–	50

[a] Units are seconds per meter. All plants are hypostomatous.
(From Holmgren et al., 1965.)

lemon, 180 s m^{-1}; alfalfa, 120 s m^{-1}; and snap bean, 120 and 300 s m^{-1}. van Bavel et al. (1965), using wet blotting paper cut in the shape of cotton leaves of size 0.05 × 0.10 m, found $r_a = 138$ s m^{-1}, which is consistent with the findings of Slatyer and Bierhuizen.

Parkhurst and Gates (1966) reported transpiration measurements with *Populus sargentii* and obtained a minimum r_ℓ of 400 s m^{-1}. Miller and Gates (1967) and Gates (1966a) reported energy-budget determinations using a potometer and energy-budget combination method with various plants in the field. Typical values were *Acer rubrum*, 1100 s m^{-1}; *Ammophila breveligulata*, 300 s m^{-1}; *Arctostaphylos uva ursi*, 500 s m^{-1}; *Betula papyrifera*, 550 s m^{-1}; *Populus tremuloides*, 200 s m^{-1} (wet site) and 360 s m^{-1} (dry site); and *Thuja occidentalis*, 700 s m^{-1}. It is not certain that these are minimum resistance values for all plants because of water stress which may have occurred in the field in northern Michigan in July. *Chamaedaphne calyculata*, a bog plant, was growing in water at all times and had a resistance of 850 s m^{-1}. The values for *Pinus resinosa* (2000 s m^{-1}) and *P. strobus* (3000 s m^{-1}) were high, but not out of line with calculations by Lee and Gates (1964), which gave 934 s m^{-1} for *P. resinosa* from the basic anatomy of the needles.

In the alpine tundra, *Bistorta bistortoides* occurs on dry, snow-free meadows or slopes and occasionally on wet meadows. It has a minimum diffusion resistance of $r_\ell = 160$ early in the season when well supplied with water and 210 s m^{-1} during the drier midseason period, according to Ehleringer and Miller (1975). These workers also studied *Caltha leptosepala*, an herbaceous species growing in marshy areas, and found r_ℓ to be 220 s m^{-1} during the early summer and 310 s m^{-1} in midseason. Janke (1970) reported on the transpiration resistance of *Vaccinium myrtillus*, an understory plant of the subalpine forest in Colorado, and found r_ℓ to be 220 and 400 s m^{-1} for sun and shade plants, respectively. The ratio of number of stomates on shade plants to that on sun plants was 17,700/9,860, or 1.80, and this is the same ratio as the resistances, i.e., 1.82.

Stomates

The stomates of plant leaves are critical controllers of gas exchange between the plant and its environment. A stomate is simply an intercellular opening between two specialized epidermal cells, known as "guard cells." Changes in turgor pressure of the guard cells cause opening or closing of the stomate. During daylight hours, most stomates open as the guard cells become crescent-shaped. The opening is an elliptical-shaped pore from 1 to 10 μm wide and up to 20 μm in length.

There is an enormous diversity of stomate size, density, distribution, and physiological behavior among various plants. Leaves with stomates on both the upper and lower surfaces are called amphistomatous leaves; those with stomates on the lower surface only are hypostomatous leaves. Rate of water loss from a leaf is controlled by the available energy and the stomate structure.

The stomates of most grasses are distributed fairly evenly over both upper and lower leaf surfaces. In maize, for example, there are approximately 6,000 stomates per square centimeter. The broad leaves of most trees have few or no stomates on their upper surfaces. Most herbaceous broad-leaved plants have stomates on both the upper and lower surfaces which are quite densely distributed. The number of stomates on upper and lower surfaces of a few selected species are as follows: alfalfa (*Medicago sativa*), 16,900 and 13,800; bean (*Phaseolus vulgaris*), 4,000 and 28,100; black oak (*Quercus velutina*), 0 and 58,000; cabbage (*Brassica oleracea*), 14,100 and 22,600; lilac (*Syringa vulgaris*), 0 and 33,000; tomato (*Lycopersicon esculentum*), 1,200 and 13,000; and sunflower (*Helianthus annuus*), 8,500 and 15,600. Total number of stomates per plant is enormous. For example, it is estimated that a fully grown corn plant has a total of 140 to 240 million stomates. A single pumpkin leaf may have 50 to 60 million stomates.

Stomates respond extremely rapidly to changes in light, temperature, carbon dioxide, water, and other environmental factors. Very often, the stomate opening changes from wide open to nearly closed in a matter of a few seconds. Raschke (1975) has reviewed the physiology of stomatal action. Much is known concerning the ecological significance of stomatal action, but a great deal remains to be learned concerning the basic mechanisms involved. Stomatal opening must synchronize with the demand for CO_2 by the metabolic centers in the leaf. Stomates seem to open in response to a depletion of intercellular CO_2, whether produced by photosynthesis or by fixation in the dark as with crassulacean acid metabolism. In this way, stomatal conductance is proportional to the assimilatory activity of the mesophyll. The guard cells must respond not only to the need for CO_2 by the mesophyll but also to the water potential of the leaf. When there is water stress, guard cells may override opening caused by CO_2 depletion and close the stomates. Guard cells may lose water to the surrounding tissue and thus lose turgor and produce stomatal closure. Generally, stomates are relatively insensitive to reductions in water potential until a certain threshold is exceeded, at which point they close rapidly and fairly completely. Stälfelt (1961) recognized the negative-feedback effect of water loss on guard cells through changes in solute content. He conceived of two reactions, one of which he termed hydroactive and the other hydropassive. The hydroactive reaction initiates or increases output of water from the guard cells and seems to involve the metabolism of the cells, whereas the hydropassive mechanism leaves the solute content of the guard cells unchanged and does not involve metabolism. Lange et al. (1971) showed that stomates respond to changes in the humidity of the ambient air. Guard cells appear to function as "humidity sensors" to the ambient air by responding to the difference in water potential of the air inside and outside the leaf. Direct negative feedback between transpiration and stomatal opening is of enormous value for the water balance of plants. A decrease in humidity produces a strong reduction in stomatal aperture. These interesting modern observations were preceded by some remarkable observations by LaRue (1930), used by Williams (1950) to show that stomatal behavior depends more on humidity of the external atmosphere than on internal water content of the leaf and that the water supply of the epidermis of many plants is obtained primarily from the main veins as disseminated to the epidermal cells by lateral diffusion.

Transpiration cools a leaf and exerts a negative-feedback effect upon water loss by reducing the water vapor pressure in the leaf. Leaf temperature affects CO_2 assimilation, and stomates respond through several feedback loops involving CO_2 and water vapor. Transpiration and assimilation are affected to different degrees because the temperature functions of these two processes are quite different. According to Raschke (1975), "If stomata respond to a change in an environmental factor, it is uncertain which feedback loop will determine the ultimate stomatal aperture because CO_2 exchange and water vapor loss are always simultaneously affected by stomatal movement. Stomata will rarely be in a steady state, either in the field or in the laboratory." A very comprehensive review of oscillations of stomatal opening, with particular attention to bean leaves, is given by Hopmans (1971). He shows the way diffusion resistance and leaf temperature oscillate with the same frequency. Farquahar and Cowan (1974) have also discussed stomatal oscillations and water loss.

Stomates appear to respond to light by reacting indirectly to the concentration of CO_2 in the leaf, which is affected by light acting on assimilation or respiration. Both mesophyll cells and guard cells have their CO_2 concentrations affected by light acting on the chloroplasts in the cells. Light controls stomatal opening by changing both intercellular and intracellular amount of CO_2. Guard cells respond to changes in intracellular amount of CO_2, exchange of CO_2 in and out of the cells, and CO_2 consumption by photosynthesis and, when it occurs, dark fixation. According to Zeiger and Hepler (1977), however, there is evidence that a blue-light receptor, bound to the plasmalemma or tonoplast, or both, initiates a flow of electrons when illuminated with blue light as a normal component of sunlight. A resulting charge separation of OH^- and H^+ across the membrane would establish a pH gradient which would result in a K^+ uptake and stomatal opening. Although this mechanism is conjectural, these authors find that the stomates of onion swell and open only when illuminated with blue light.

Stomate Resistance and Geometry

Since stomates control the loss of water from a leaf, a great deal of interest has centered on understanding the detailed relationships of stomate geometry and transpiration rates. The diffusion of water vapor from the intercellular air spaces through the stomates and out through the adhering surface boundary layer is shown in Fig. 10.2. Once the physics of gaseous diffusion through small channels and porous membranes was understood, application of these principles to diffusion of water vapor out of a leaf was straightforward. In the process, however, some serious mistakes were made and invalid conclusions drawn. Stefan derived the basic evaporation law for a liquid disk surface of finite radius in 1881, and from this was developed the so-called "diameter law" for diffusion through stomates.

The first workers to investigate loss of water from leaves as a diffusion process through pores were Brown and Escombe (1900). They simulated stomates and the leaf epidermis by placing small apertures in thin membranes of metal, mica, or celluloid placed above reservoirs of water. The pores were separated by 10 or more pore diameters, a very important fact relating to the interpretation of their data. Lee and Gates (1964) reviewed the whole matter of diffusion through porous membranes and its application to plants. The following theory will explain the Brown and Escombe results.

The diffusion resistance of a single hollow cylinder of length ℓ and radius s is given by

$$r = \frac{\ell + \frac{\pi}{2} s}{\mathscr{D}}, \tag{10.78}$$

where $\mathscr{D}$ is the diffusion coefficient for water vapor.

The quantity $(\pi/2)s$ is an end correction term for the effect of the end of the cylinder in extending its effective length. For the case in which the pore is elliptical in cross-section, instead of cylindrical, Penman and Schofield (1951) derived the equation

$$r = \frac{0.9\left(\ell + \frac{\pi}{2} s\right)}{\mathscr{D}}, \tag{10.79}$$

where $s = s_1(s_1/s_2)^{1/2}$, with s_1 is the radius at the narrowest part of the stomate and s_2 is the radius at the mouth. Since the cross section is elliptical, s_1 and s_2 are the geometrical means of the axes of the ellipse.

Each stomate loses moisture through a cross-sectional area of πs^2. If the total number of stomates on a leaf of surface area A is N, then the fraction of the total surface through which moisture is lost is $N \pi s^2/A$. Let n be the number of stomates per unit area; the fraction then becomes $n \pi s^2$.

Amount of water vapor per unit area per unit time transpired by a leaf is

$$E = \frac{\mathscr{D} \, n \, \pi s^2}{\ell + \dfrac{\pi}{2} \, s} \, \Delta c, \tag{10.80}$$

where Δc is the concentration difference of water vapor inside and outside the leaf. Therefore,

$$r_s = \frac{\Delta c}{E} = \frac{\ell + \dfrac{\pi}{2} \, s}{n \, \mathscr{D} \, \pi \, s^2}. \tag{10.81}$$

One can estimate the stomate resistance of a leaf to water vapor from Eq. (19.83) as follows. If $n = 2 \times 10^8$ stomates m^{-2}, $\ell = 10 \ \mu$m, and $s = 2 \ \mu$m, then $r_s = 210$ s m^{-1}. If all other values are the same and $s = 4 \ \mu$m, then $r_s = 184$ s m^{-1}. If $n = 2 \times 10^8$ stomates m^{-2}, $\ell = 5 \ \mu$m, and $s = 4 \ \mu$m, then $r_s = 105$ s m^{-1}. If the number of stomates increases to 3×10^8 stomates m^{-2} and if $\ell = 5 \ \mu$m and $s = 2 \ \mu$m, then $r_s = 130$ s m^{-1}. It is clear that for a plant leaf to have a stomatal resistance near 100 s m^{-1}, it must have not only a large number of stomates per unit area but also large stomates of short length. *Zebrina pendula* has among the largest stomates ($35 \times 20 \ \mu$m) but only 1.62×10^7 stomates m^{-2}, with a stomate length of about 10 μm. *Medicago sativa* has 1.54×10^8 stomates m^{-2} but a maximum opening of $9 \times 4 \ \mu$m and a stomate length of 12 μm.

For many years, investigators of transpiration from leaves described a "diameter law" by which the rate of diffusion of water vapor was proportional to the diameter of the stomates. Theoretical studies, however, have established that a "diameter law" would be observed only under fairly restrictive conditions. When $\ell \ll s$, then, from Eq. (10.81),

$$r_s = \frac{1}{2 \, n \, \mathscr{D} \, s}, \tag{10.82}$$

which shows that transpiration is proportional to $2s$ when membrane thickness or pore length is small compared to the diameter.

One can rewrite Eq. (10.81) as follows:

$$r_s = \frac{\left(1 + \dfrac{\pi}{2} \dfrac{s}{\ell}\right) \ell}{n \, \mathscr{D} \, \pi \left(\dfrac{s}{\ell}\right)^2 \ell^2}. \tag{10.83}$$

If s/ℓ is a constant, then

$$r_s \propto \frac{1}{\ell} \propto \frac{1}{s}. \tag{10.84}$$

This equation is known as the "diameter law."

If $\ell \gg s$, then

$$r_s = \frac{\ell}{n \, \mathcal{D} \, \pi \, s^2},$$ (10.85)

and conductance or transpiration is proportional to πs^2.

The geometry of stomates is such that $\ell \gg s$; rarely is $\ell = s$ or $\ll s$. Application of Stefan's original "diameter law" to stomates is a misappropriation and a use to which Stefan probably never intended it be put.

Ting and Loomis (1963) tested the "diameter law" by an experiment in which they prepared holes in brass shim stock 50 μm thick, with the holes of diameter 100, 200, 400, and 800 μm. They also used shim stock of 25-μm thickness with holes of diameter 20 and 50 μm. Only in one instance was ℓ large compared with s. Ting and Loomis fortuitously selected combinations of ℓ and s in their experiments with perforated membranes which gave them results which fit the "diameter law." The validity of Eq. (10.83), in distinction to the "diameter law," is confirmed by the data of Verduin (1947), Sayre (1926), and Brown and Escombe (1900). These authors found diameter proportionality only when the thicknesses of the membranes were small compared to the pore diameters. These details are discussed by Lee and Gates (1964).

Substomatal Resistance

The resistance offered by the intercellular air spaces is relatively small. Bange (1953) and Lee and Gates (1964) have shown that although substomatal cavities offer some resistance to the diffusion of water vapor from the cell walls to the stomates, neglect of substomatal cavity resistance involves an error in the total resistance between 2 and 5%. Bange in particular gives a theoretical derivation of the formulas for calculating diffusion resistance of cavities with a variety of geometries; this work has been extended and applied to further examples by Lee and Gates.

Boundary-Layer Resistance

A layer of air adheres to a leaf surface because of the viscosity of air. This layer is known as a "boundary layer" and represents a transition zone between the still air next to the leaf surface and moving air near the leaf. This cushion of air offers resistance to the diffusion of air from the leaf surface to the free air. Gates et al. (1968), from a large number of experiments in a low-velocity wind tunnel with rectangular leaf models made of wet blotting paper showed the boundary-layer resistance to be

$$r_a = k_2 \frac{W^{0.20} \, D^{0.35}}{V^{0.55}},$$ (10.86)

where D and W are the leaf dimensions (in meters) in the direction of air flow and across the air flow, respectively, V is the wind speed (in meters per second), and $k_2 = 180 \, s^{0.45} \, m^{-1}$ such that r_a is in seconds per meter.

Environmental Influences on Leaf Resistance

Several environmental factors influence the degree of opening of stomates and the diffusion resistance of the leaf. As shown in the review by Raschke (1975), the mechanisms by which environmental factors influence stomatal aperture are not entirely understood. Nevertheless, there is sufficient empirical information to allow a description of the general influence of environmental factors.

Water Potential

The effect of leaf water potential and turgor potential on leaf diffusion resistance was studied by Turner (1974) for maize, sorghum, and tobacco. Leaf resistance remained unchanged with decreasing turgor pressure, as shown in Fig. 10.3, in all three species. A marked increase in leaf resistance occurred at a turgor pressure of about 2 bars in maize, -1 bar in sorghum, and 1 bar in tobacco. At lower values of turgor pressure, leaf resistance went to very high values very quickly as a result of stomatal closure.

Leaf resistance remained unchanged with decreasing water potential until a "critical" value of water potential was reached, at which point r_ℓ increased sharply. The critical water potential varied with the species and the position of the leaf in the canopy. For leaves at an intermediate height, the effect of water potential on leaf resistance is shown in Fig. 10.3. The critical water potential was -17, -20, and -13 bars in maize, sorghum, and tobacco, respectively. Maize and sorghum are tropical grasses and C_4 species, whereas tobacco is a C_3 species. The greater yielding ability of sorghum over corn may be the result of the fact that sorghum keeps its stomates open at a lower water potential than maize. The critical water potential was higher (less negative) for leaves low in the canopy and lower (more negative) for leaves at the top of the canopy. The critical value of water potential varies with species, being -6 bars for onion and -14 bars for a vine and sunflower. In snap beans, the critical water potential was higher for adaxial stomates than abaxial stomates. Field-grown plants generally have lower (more negative) critical water potentials than those grown in a growth chamber. Turner (1974), like LaRue (1930), concluded that the turgor balance between guard cells and subsidiary cells is largely independent of changes in mesophyll turgor.

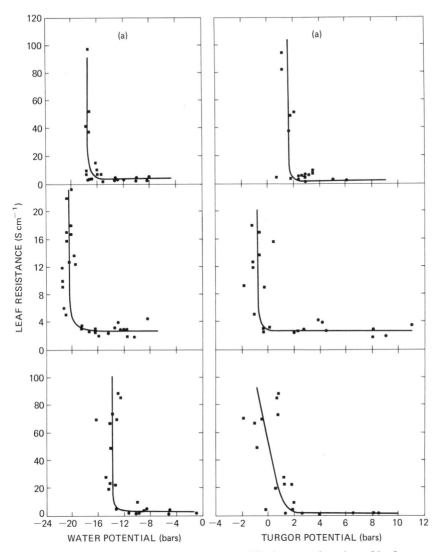

Figure 10.3. Leaf resistance to water vapor diffusion as a function of leaf water potential and turgor potential for (a) maize, (b) sorghum, and (c) tobacco at (●) high, i.e., greater than -2 bars, and (■) low i.e., -4 to -7 bars, soil water potentials for leaves irradiated at greater than 419 W m^{-2}. (Redrawn from Turner, 1974.)

The relationship between leaf resistance and leaf water potential has been applied by Ehleringer and Miller (1975) to a simulation model of water relations and production by plants in the alpine tundra of Colorado. These authors write: "Physiologically, it is through the daily course of leaf resistances that plants regulate the trade-off between water loss and carbon gain in their respective environments and thus determine plant performance.''

Illumination

Given adequate leaf water potentials, normal amounts of carbon dioxide, and moderate temperatures, the stomatal resistance of a leaf is related to light intensity in a hyperbolic fashion, as shown in Fig. 10.4. The data shown are from Tenhunen and Gates (1975) for *Asclepias syriaca*. Very similar curves were obtained by Kuiper (1961) for bean (*Phaseolus vulgaris*) and by Gaastra (1959) for onion. The light curves for leaf resistance given by Ehrler and van Bavel (1968) for leaf diffusion resistance included the following plant species: alfalfa, *Medicago sativa;* snap bean, *Phaseolus vulgaris;* fava beans, *Vicia faba;* cotton, *Gossypium barbadense;* sunflower, *Helianthus annuus;* lemon, *Citrus hybrida;* corn, *Zea mays;* and sorghum, *Sorghum vulgare*. Most of these curves had the hyperbolic shape, but all terminated at a very finite value of leaf resistance for darkness. This value probably represents the cuticular resistance of the leaf since leaf resistance was used rather than stomatal resistance.

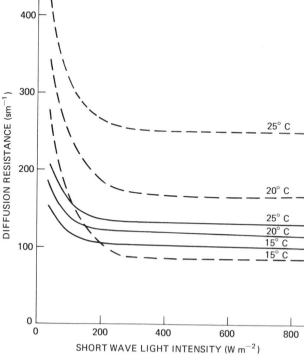

Figure 10.4. Leaf diffusion resistance to water vapor of *Asclepias syriaca* as a function of light intensity at three temperatures for a (—) wet and (––) dry plot.

These authors noted that complete stomatal opening occurred for most of these plants only when illumination reached 50 klux. Each plant had a different saturation illumination and minimum resistance.

The functional dependence of stomatal resistance on light intensity is given by

$$r_s = \frac{L + K_{rL}}{L} \, {}_{min}r_s, \qquad (10.87)$$

where ${}_{min}r_s$ is the minimum values of r_s reached at high light intensities and K_{rL} is the value of L such that $r_s = 2({}_{min}r_s)$.

Stomates have an endogenous rhythm of a phase such that stomates are open during daylight hours and closed during darkness. A plant leaf has very high resistance to gas exchange with its environment during darkness and a low resistance during daylight, modulated by leaf water potential, leaf temperature, carbon dioxide concentration, and other factors. The endogenous behavior of leaf resistances for *Bistorta bistortoides* and *Caltha leptosepala* in the alpine tundra are given by Ehleringer and Miller (1975).

In Fig. 10.4, one can see some of the influence of water potential (since plants from both wet and dry plots were used) and leaf temperature on leaf resistance. When working with plants in the field, one must take all environmental factors into account.

Temperature

As a general rule, leaf resistance changes with changing leaf temperature, but whether resistance increases or decreases with increasing temperature depends upon the status of other factors. For a long time, it was thought that resistance increased with temperature. This was shown for stomatal aperture by Heath and Orchard (1957) and is further discussed by Heath and Meidner (1957). Wuenscher and Kozlowski (1971) measured the change in transpiration rates and leaf resistances with leaf temperature for *Quercus macrocarpa, Q. velutina, Q. alba, Q. rubra,* and *Acer saccharum.* They found leaf resistance to increase monotonically with temperature, as did the other authors referred to above, and surmised that increased resistance with increased temperature may be significant in preventing rapid desiccation of leaves under the large evaporative stress imposed by high leaf temperature.

Hall and Kaufmann (1975) report that they, as well as others, find a decrease of leaf resistance with increasing leaf temperature under some circumstances. Hall and Kaufmann find that if the humidity gradient between the leaf and the air is kept constant, then stomatal resistance will remain constant or decrease with increasing leaf temperature, depending on the magnitude of the gradient and the species.

If the difference in water-vapor concentration between leaf and air is kept constant and only the leaf temperature is permitted to vary, then, according to Schulze et al. (1974), leaf resistance when under light saturation for net photosynthesis is given by

$$r_\ell = a + b\, T_\ell, \tag{10.88}$$

where a is the intercept and b is the slope of the line, which turns out to be nearly independent of the water-vapor concentration difference as long as the stomates are not completely open. The slope of the regression line becomes less negative only for very humid conditions, when the stomates are already wide open.

If leaf temperature changes, the change of leaf resistance is given by

$$\frac{\partial r_\ell}{\partial T_\ell} = b. \tag{10.89}$$

Values of b were determined by Lange et al. (1975) for various desert plant species as a function of leaf water potential. All species exhibited stomatal opening with increasing leaf temperature, i.e., had b negative at high (least negative) plant water potentials and b positive at low (most negative) water potentials.

Humidity

Experiments reported by Lange et al. (1971) show very clearly that stomates react to changes in the humidity of the ambient air. Degree of turgor of the guard cells determines stomatal movement. It seems that a certain equilibrium between water recharge of the guard cells and peristomatal transpiration corresponds to a certain degree of stomatal aperture. The guard cell acts as a humidity sensor for the ambient air by measuring the difference in water potential of the air inside and outside the leaf.

As the outside air becomes drier, the stomatal aperture decreases very markedly, and, as a result, stomatal resistance greatly increases. This direct negative feedback between transpiration and stomatal aperture may be of enormous importance for the water balance of plants. The advantage would be that, with a decrease in humidity, water loss could be restricted by the stomatal diffusion resistance before the water potentials of other leaf tissues are affected seriously. Lange and his colleagues were able to show that every stomate functions independently of every other stomate.

Further evidence that stomatal aperture, stomatal resistance, and water loss are influenced strongly by the vapor-pressure gradient between the leaf and the air is given by Hall and Kaufmann (1975) and O'Leary (1975). Schulze et al. (1972) have shown, for some xerophytic species and one mesophytic species, that increases in leaf resistance resulting from increased evaporative demand reduce the level of water stress in leaves. The evidence now seems strong that stomates of many plants may respond to humidity independent of the more-traditional hydroactive mechanism. However, it is also well-known that many plants have stomatal response nearly unaffected by humidity, as stated by Meidner and Mansfield (1968) and Whiteman and Koller (1967).

Schulze et al. (1974) derived the following empirical formula to fit the dependence of leaf resistance on the difference in water-vapor concentration between the evaporating sites of the leaf and the surrounding air:

$$r_\ell = a' + b'(\rho_\ell - \rho_a) + c' \, e^{-d'} \, (\rho_\ell - \rho_a), \qquad (10.90)$$

where a', b', c', and d' are constants and ρ_ℓ and ρ_a are the water-vapor densities in the leaf and in the air, respectively. For example, for the plant *Prunus armeniaca* at Avdat, Israel, this relationship was

$$r_\ell = -325 + 51.5(\rho_\ell - \rho_a) \times 10^3 + 532 \, e^{-178}(\rho_\ell - \rho_a) \quad (10.91)$$

in August and

$$r_\ell = -188 + 32.7(\rho_\ell - \rho_a) \times 10^3 + 396 \, e^{-165}(\rho_\ell - \rho_a), \quad (10.92)$$

in May, where ρ_ℓ and ρ_a are in kilograms per cubic meter of H_2O.

One is reminded that the water-vapor density of the air ρ_a is the absolute humidity and is the relative humidity times the saturation water-vapor density for a given air temperature, and similarly for the water-vapor density of the air in the intercellular air spaces, ρ_ℓ. Therefore, leaf resistance depends on the humidity inside and outside the leaf, as well as on the temperatures.

At constant water difference but with varying leaf temperatures, Schulze et al. derived the relationship

$$r_\ell = a'' + b'' T_\ell \qquad (10.93)$$

for a temperature range from 24° to 43°C. The parameter b'' is almost independent of $\rho_\ell - \rho_a$, as long as the stomates are not completely open. The slope of the line (b'') becomes less negative only when the air is extremely humid and the stomates are completely open. For *Prunus armeniaca*, when the water vapor difference was 28×10^{-3} kg m^{-3}, the regression line was

$$r_\ell = 1762 - 31.0 \, T_\ell. \qquad (10.94)$$

When $\rho_\ell - \rho_a = 22 \times 10^{-3}$ kg m^{-3}, it was

$$r_\ell = 1631 - 32.0 \, T_\ell, \qquad (10.95)$$

and when $\rho_\ell - \rho_a = 11 \times 10^{-3}$ kg m^{-3}, it was

$$r_\ell = 665 - 13.0 \, T_\ell. \qquad (10.96)$$

For a change in the water-vapor-density difference, there is a change in leaf resistance of

$$_1\Delta r_\ell = b'\,\Delta(\rho_\ell - \rho_a) + c'\,d'\,e^{-d'(\rho_\ell - \rho_a)}\,\Delta(\rho_\ell - \rho_a). \qquad (10.97)$$

Change of leaf resistance with leaf temperature is

$$_2\Delta r_\ell = b''\,\Delta T_\ell. \qquad (10.98)$$

Then, total change of leaf resistance with a simultaneous change of vapor-density difference and leaf temperature is

$$
\begin{aligned}
\Delta r_\ell &= {}_1\Delta r_\ell + {}_2\Delta r_\ell \\
&= b'\,\Delta(\rho_\ell - \rho_a) + c'\,d'\,e^{-d'(\rho_\ell - \rho_a)}\,\Delta(\rho_\ell - \rho_a) + b''\,\Delta T_\ell. \quad (10.99)
\end{aligned}
$$

Schulze et al. (1974) have used these relationships to simulate the changes of leaf resistance throughout a day for certain desert plant species. They applied this empirical model to leaves of *Prunus armeniaca* for a dry, hot day of 17 August and a cool, moist day of 21 August 1971 at Avdat, Israel. During the dry day, leaf temperature went from 19°C at sunrise to 45°C in the afternoon. During this time, the water-vapor difference $\rho_\ell - \rho_a$ increased from about 7 to 52×10^{-3} kg m^{-3} (relative humidity 0.19). The moist, cool day was about 8°C cooler at noon than the dry, hot day. Leaf temperatures rose to only 37°C, and the water-vapor difference increased from about 2 to 32×10^{-3} kg m^{-3}. According to calculation, diffusion resistance during the dry day should increase 2100 s m^{-1} because of humidity controlled stomatal closure and decrease by 700 s m^{-1} because of the temperature increase, to give a net change of 1400 s m^{-1}. In the morning of the dry day, the predicted values of r_ℓ attain 98% of the observed values and in the afternoon about 81%. There is an additional stomatal control during the afternoon, which produces a higher leaf resistance than predicted.

For the moist, cool day (21 August), the change in leaf resistance resulting from the humidity control is an increase of 900 s m^{-1} and a decrease of 400 s m^{-1}, to give a net change of 500 s m^{-1}. The predicted change of leaf resistance is better than 98% of the observed change during the morning and about 96% during the afternoon.

Photosynthesis is low in the morning of the dry, hot day and at noon decreases to the compensation point. On the moist, cool day, there is a higher rate of apparent photosynthesis in the morning, which continues strong at noon. Stomatal opening seemed to have little effect on the rate of net photosynthesis during the late afternoons of both days, probably because light intensity was getting low. On the dry day, there was a small second peak of assimilation just before sunset.

The daily course of transpiration was higher on the moist day than on the dry day, and the difference was greatest during the afternoon. This highly significant ecological fact shows that on a moist day that followed a dry day, there were sufficient water reserves in tree and soil to permit considerable water loss.

The authors make another important ecological observation. After the midday depression, the stomates were closed until late in the afternoon. The reopening which then occurred at low light intensities had little effect on the gas exchange. At the end of the dry season, when water supply from the soil was most limited, stomates opened in the afternoon to such an extent that there was increased transpirational water loss even at the end of a dry desert day.

There is no doubt that the predictive model could be improved with a better understanding of stomatal control mechanisms. The methodology demonstrated by these investigators is encouraging, however, and there is no reason why quite accurate predictions of water loss and carbon fixation cannot be achieved for any plant in any environment.

Carbon Dioxide

There is considerable evidence that increased internal carbon dioxide concentration produces a reduction of stomatal aperture and decreased carbon dioxide concentration causes stomatal opening. Heath and Orchard (1957) and Heath and Meidner (1957) have shown that onion stomates respond to increased carbon dioxide concentration by closure and that high-temperature closure is accounted for by accumulation of carbon dioxide inside the leaf. They also show that if accumulation of intercellular carbon dioxide is prevented, high leaf temperature causes opening of stomates.

Schulze et al. (1975) have explored the significance of leaf water potential and carbon dioxide concentration on stomatal resistance for apricot, *Prunus armeniaca,* growing at Avdat, Israel. They found that daily changes of water potential had little effect on the daily course of stomatal resistance. The daily changes in stomatal resistance had little correlation with changes of internal carbon dioxide concentration. In the morning, a decreasing internal carbon dioxide concentration was inversely correlated with stomatal response. In the afternoon, the effect of external climate and increasing internal carbon dioxide concentration on stomatal response may have been additive. Generally, it seems that for apricot, external conditions override most internal factors in the control of stomatal opening.

Problems

1. The air near the ground is saturated with moisture early in the morning when the air temperature is 10°C. The sun warms the air to a temperature of 20°C. What was the original density of the saturated air and the relative humidity after warming? What is the mixing ratio of the air at 10°C and at 20°C? What is the saturation deficit for the air at 20°C?

2. Using the value of the diffusion coefficient for water vapor at 20°C, determine the resistance to diffusion of a boundary layer of air 0.005 m thick.

3. If the relative humidity of air is 0.90 at an air temperature of 20°C, determine the water potential of the air. If the relative humidity is 0.90 at an air temperature of 40°C, determine the water potential of the air.

4. If the water vapor is saturated inside a leaf of temperature 25°C and the air outside the leaf is at 20°C and relative humidity 0.60, what is the water potential difference between inside and outside the leaf?

5. A leaf has a stomatal resistance of 100 s m^{-1}, a cuticular resistance of 1000 s m^{-1}, and a boundary-layer resistance of 200 s m^{-1} for each leaf surface. Determine the total leaf resistance if the leaf is (a) hypostomatous (stomates on underside only) and (b) amphistomatous (stomates on both sides).

6. Determine the rate of water loss from a leaf using Eq. (10.46) for the following conditions: air temperature, 20°C; relative humidity, 0.60; leaf temperature, 25°C; total resistance to water vapor diffusion, 300 s m^{-1}; and resistance to convective heat exchange, 200 s m^{-1}.

Chapter 11

Energy Budgets of Plants

Introduction

All parts of plants have temperatures determined by their environmental conditions. The energy status of a plant or a plant leaf is manifested by its temperature. The leaves and stems of plants have small masses and are generally of low heat capacity. Because of this, the temperatures of these plant parts respond quickly to changes of environmental conditions. Trunks of trees, tree limbs, and other massive plant parts have much greater heat capacities and respond slowly to environmental variations, sometimes with a thermal lag of several hours, as shown in the study of tree trunks by Derby and Gates (1966).

Energy is exchanged between a plant leaf and the environment by radiation, convection, conduction, and transpiration. The environment in the proximity of a plant as it affects the plant's energy status is described by the variables of radiation flux, air temperature, wind speed, and water-vapor density. Although these variables are all very different properties, with each measured in terms of its own set of dimensions, the one common mechanism by which they interact with the plant is the exchange of energy. One of the first workers to give a thorough analysis of the energy exchange of plant leaves was Raschke (1956b). Gates (1962) published a description of the energy environment of plants which included a discussion of the radiation regime in which plants live.

The radiation flux of the environment is coupled to the energy status of the plant by means of the plant's absorptance to radiation. If a plant reflected all incident radiation, it would be completely decoupled from the radiation flux. If a plant were black and absorbed all incident radiation, it would have its internal energy status, and thereby its temperature, strongly coupled to the external radiation field. Many plant leaves absorb only about 50% of the incident solar radiation, and some absorb as little as 20% of shortwave radiation. At the same time, most plant parts absorb about 95% of incident thermal, or longwave, radiation. Therefore, plants are moderately coupled to incident shortwave radiation and strongly coupled to longwave radiation.

Coupling between a plant and the air temperature is through the convection coefficient. Because many plant parts are relatively small, the convection coefficient tends to be large, and there is strong coupling between plant temperature and air temperature. This is true for all small branches, twigs, leaves, needles, thorns, spines, etc. Many small leaves cannot depart from air temperature by more than 2° or 3°C, as shown by Gates et al. (1968). The trunks of trees, large branches, very large leaves, and massive cushion plants are weakly coupled to air temperature, and their temperatures are affected much more strongly by radiation flux.

The concentrations of various gases in a leaf are coupled to the concentrations in the air beyond the surface boundary layer by the diffusion resistance. Oxygen, carbon dioxide, and water vapor are the main gases exchanged between the interior and exterior of a leaf, but other gases may be involved, including pollutants. Exchange of water vapor affects the energy status of a leaf since it requires approximately 2.428×10^6 J kg^{-1} to vaporize liquid water. Actually, the latent heat of water varies with the temperature of the water. Through diffusion of water vapor from a leaf, evaporative cooling occurs, and the temperature of the leaf is reduced. Coupling of leaf temperature to water-vapor density of the air is through the diffusion resistance of the stomates. Some leaves have a very low diffusion resistance, and their leaf temperatures are strongly influenced by the vapor density of the air outside of the leaf. Other leaves have a high diffusion resistance, and their temperatures are decoupled from the water-vapor density. A large leaf with very low diffusion resistance can have its temperature reduced as much as 30°C when transpiring vigorously under intense radiation in still air. Other plants with high diffusion resistances may have their leaf temperatures affected less than 5°C under intense radiation in still air. When a plant leaf is not exposed to intense radiation, the transpiration rate is much reduced, and its effect on leaf temperature is small, on the order of 2° or 3°C. Two conditions are required for loss of water from a leaf: energy must be available, and there must be a diffusion gradient from the inside to the outside of the leaf.

Some energy is stored by a leaf by means of photosynthetic chemical events, and some energy is released chemically through respiration. The effect of these processes on the temperature of leaves is generally negligible. However, the heat released during short periods of intense respiration in the spadix of eastern skunk cabbage (*Symplocarpus foetidus*) produced temperature increases of up to 15° to 35°C above air temperature for periods up to 14 days when the air temperature was −15° to

+15°C (see Knutson, 1974). The spadix consumes oxygen at a rate comparable to that of homeothermic animals of comparable size. Respiration of bacteria in mounds of decaying vegetation may also produce considerable temperature increases.

There are other processes which may produce energy changes in plants, including electrical discharges, sudden magnetic-field variations, and gravitational-field effects, but, as a general rule, these processes produce negligible consequences.

Time-Dependent Energy Budgets

The time-dependent energy budget for a plant part, including a term for heat stored or lost, is given by

$$Q_a - P + W - \epsilon\sigma(T_\ell + 273)^4 - k_1\left(\frac{V}{D}\right)^{0.5}(T_\ell - T_a)$$

$$- \lambda(T_\ell)E - C\frac{dT_\ell}{dt} = 0, \quad (11.1)$$

where Q_a is the amount of absorbed radiation, P is the energy consumed in photosynthesis, W is the energy released by respiration, E is the rate of transpiration, $\lambda(T_\ell)$ is the latent heat of vaporization as a function of leaf temperature, C is the heat capacity of the plant part, T_ℓ and T_a are leaf and air temperatures, respectively, t is the time, V is the wind speed (in meters per second), D is the characteristic dimension (in meters), k_1 is a constant, ϵ is the leaf emissivity, and σ is the Stefan–Boltzmann constant (5.673×10^{-8} Wm^{-2} °K^{-4}). All terms of the equation are in watts per square meter.

Equation (11.1) can be linearized in the same manner shown in Chapter 13 for Eq. (13.3). Therefore,

$$(T_\ell + 273)^4 = 4(T_\ell + 273)(\overline{T}_\ell + 273)^3 - 3(\overline{T}_\ell + 273)^4. \quad (11.2)$$

If $\mathcal{R} = 4\epsilon\sigma(\overline{T}_\ell + 273)^3$, $H = k_1(V/D)^{0.5}$, and $Q_n = Q_a + 3\epsilon\sigma(\overline{T}_\ell + 273)^4 - 273\mathcal{R}$, then Eq. (11.1) becomes

$$\frac{C}{\mathcal{R} + H}\frac{dT_\ell}{dt} + T_\ell - \frac{Q_n + HT_a}{\mathcal{R} + H} - \frac{W - P - \lambda E}{\mathcal{R} + H} = 0. \quad (11.3)$$

This equation can be written in the abbreviated form:

$$\frac{C}{\mathcal{R} + H}\frac{dT_\ell}{dt} + T_\ell - T_e - T_\Delta = 0, \quad (11.4)$$

where

$$T_e = \frac{Q_n + HT_a}{\mathcal{R} + H} \quad (11.5)$$

is the *operative environmental temperature* and

$$T_\Delta = \frac{W - P - \lambda E}{\mathcal{R} + H} \tag{11.6}$$

is the *physiological offset temperature*. Eq. (11.4) can be written

$$\frac{dT_\ell}{dt} + \frac{T_\ell}{\tau} = \frac{T_e + T_\Delta}{\tau}, \tag{11.7}$$

where the time constant τ is $C/(\mathcal{R} + H)$. The solution to this equation has the form

$$T_\ell = T_\infty + (T_0 - T_\infty)e^{-t/\tau}, \tag{11.8}$$

where T_0 is the initial temperature and T_∞ is the final temperature which is approached asymptotically with time. The time constant τ for small and intermediate-sized plant leaves is generally from 5 to 20 s. However, large leaves such as *Monstera* or *Heliconia* have time constants of several minutes. The blades of cacti also have time constants of several minutes. Very small twigs have time constants on the order of minutes, branches several minutes, and trunks and limbs of considerable size may have time constants of an hour to a half-day or longer.

Observed Transients

The temperatures of leaves in nature may change frequently if environmental conditions are variable. Gates (1963) reported the effect of a changing radiation environment and varying amount of wind on the leaves of *Populus acuminata* in Boulder, Colo., on 22 and 29 July 1961. These results are shown in Figs. 11.1 and 11.2. A leaf irradiated by the sun had a leaf temperature 5° to 7°C above air temperature when in still air. When a cloud covered the sun, the leaf temperature dropped to 2° or 3°C below air temperature. Temperature changes occurred in about 6 s or less. A change from no wind to wind resulted in a drop in leaf temperature of as much as 10°C, and again the changes were nearly instantaneous. The frequent variation in leaf temperature which is characteristic of most leaves in natural habitats may have some physiological consequences, but exactly what they are we do not know.

Watson (1933) described the cooling rates of leaves in air, and Shull (1930) estimated rates from the mass per unit surface area. Ansari and Loomis (1959) studied the heating and cooling rates of leaves in a series of experiments designed to reveal the significance of transpiration on leaf temperature. They concluded that leaf temperature was not significantly affected by transpiration. Typical heating and cooling rates, as determined by Ansari and Loomis, are shown in Fig. 11.3. The initial rate of

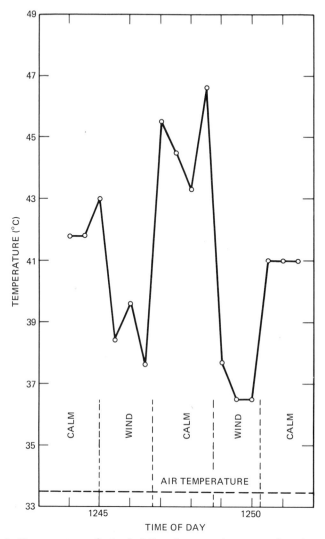

Figure 11.1. Temperature of a leaf of *Populus acuminata* as a function of time of day during alternate calm and windy conditions. Measurements were made on 29 July 1961.

temperature change according to these authors was 17°C min⁻¹. The time constant of this pepper leaf is the time it takes for its temperature to change to $1/e$ of its final value. Letting $t = \tau$ in Eq. (13.8) gives

$$T_\ell = T_\infty + (T_0 - T_\infty)e^{-1}. \tag{11.9}$$

The pepper leaf has a time constant of approximately 20 s, as determined from the data given in Fig. 11.3. For example, if $T_0 = 20.5°C$ at

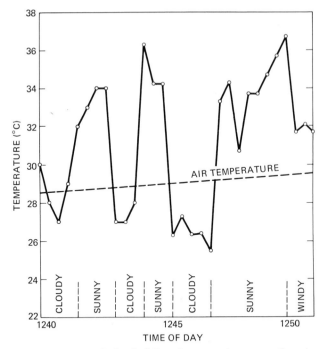

Figure 11.2. Temperature of a leaf of *Populus acuminata* as a function of time of day during alternate sunny and cloudy conditions. Measurements were made on 22 July 1961 in Boulder, Colo.

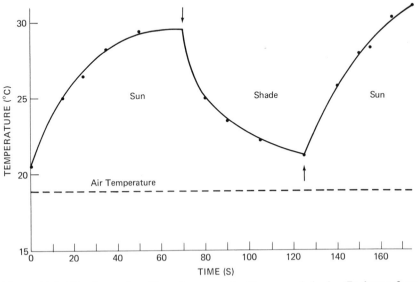

Figure 11.3. Heating and cooling of a pepper leaf in sun and shade. (Redrawn from an example by Ansari and Loomis, 1959.)

time $t = 0$ and $T_\infty = 29.5°C$ at time $t = \infty$, then $T_\ell = 26.2°C$, at a time of 20 s.

Mellor et al. (1964) report studies of leaf temperatures under controlled-environment conditions and show heating and cooling curves for a number of selected species. They obtained the following time constants: *Kalanchoe tomentosa*, 540 s; *Saintpaulia lonantha*, 252 s; *Nicotiana glutinosa*, 102 s; *Pelargonium hortorum*, 72 s; and *Xanthium pennsylvanicum*, 42 s.

Steady-State Energy Budget

It is convenient and practical to consider plant leaves as existing in a steady-state condition. When environmental conditions are constant for periods of time greater than the time constant of a leaf, then steady-state conditions prevail. The steady-state energy budget of a leaf is

$$Q_a - P + W - \epsilon\sigma(T_\ell + 273)^4 - k_1\left(\frac{V}{D}\right)^{0.5} (T_\ell - T_a) - \lambda E = 0. \quad (11.10)$$

The terms P and W do not normally produce a significant effect on leaf temperature or transpiration rate. Neglecting P and W and writing the transpiration rate explicitly, one gets

$$Q_a = \epsilon\sigma(T_\ell + 273)^4 - k_1\left(\frac{V}{D}\right)^{0.5} (T_\ell - T_a)$$

$$+ \lambda(T_\ell)\frac{{}_sd_\ell(T_\ell) - h_sd_a(T_a)}{r_\ell + k_2\left(\dfrac{W^{0.2}\,D^{0.3}}{V^{0.5}}\right)} = 0. \quad (11.11)$$

This is the form of the steady-state energy-budget equation which has been generally used and which Gates and Papian (1971) used for their tabulation of leaf temperatures and transpiration rates for leaves of various characteristics. Values of k_1 and k_2 are such that each term in the energy-budget equation is watts per square meter. The following values are averages of values for k_1 and k_2 first derived by Gates et al. (1968) in some very thorough wind-tunnel experiments: $k_1 = 9.14$ J m^{-2}, and $k_2 = 200$ s$^{1/2}$ m^{-1}.

There is no doubt that although these values are quite good, better values might be attained by means of additional wind-tunnel measurements of leaves and simulated leaves (in the form of wet blotting paper). It would also be desirable to include both leaf dimensions in the convection term. However, the amount of effort involved in establishing a more precise and refined relationship is enormous and is not deemed necessary since the consequences to calculations of leaf temperatures or transpiration rates generally would be small.

If the energy-budget analysis given by Eq. (11.11) is applied to tree trunks, limbs, or branches, then the convection coefficients characteristic of cylinders should be used.

The publication by Gates and Papian (1971), "Atlas of Energy Budgets of Plant Leaves," was designed to make readily available the solutions to Eq. (11.11). The dependent variables are leaf temperature T_ℓ and transpiration rate E. The independent variables are air temperature T_a, wind speed V, radiation absorbed Q_a, and relative humidity h. The plant parameters are the leaf dimensions D and W, diffusion resistance r_ℓ, emissivity ϵ, and absorptance, which is part of the Q_a term. Values of each of the independent variables and of the plant parameters must be selected in order to obtain values for T_ℓ and E. Typical values of these variables and parameters are given elsewhere in this book.

The tables in Gates and Papian (1971) are set up to cover the following sets of environmental conditions: Q_a = 400, 600, 800 and 1,000 W m^{-2}; V = 0.1, 0.2, 0.5, 1.0, 2.0, 4.0, and 8.0 m s^{-1}; T_a = 0°, 10°, 20°, 30°, and 40°C, and h = 0 and 1.0. The leaf dimensions D and W represented are 0.01 × 0.01, 0.05 × 0.01, 0.01 × 0.05, 0.05 × 0.05, 0.10 × 0.05, 0.2 × 0.05, 0.01 × 0.10, 0.05 × 0.10, 0.10 × 0.10, and 0.20 × 0.10 m, and the values for internal diffusion resistances to water vapor are 0, 100, 200, 500, 1000, 2000, and ∞ s m^{-1}. In the tables, cgs units were used rather than mks units as given here. These tables should be used whenever possible. They were calculated using a computer. Also presented are graphs showing the response of one or both dependent variables to changes in several of the independent variables. The following examples will illustrate the use of the tables.

Example 1. Let Q_a = 800 W m^{-2}, T_a = 30°C, h = 0.50, V = 0.20 m s^{-1}. The diffusion resistance of the leaf is 200 s m^{-1}.

For a leaf of size 0.01 × 0.01 m, T_ℓ = 32.6°C and E = 8.6 × 10^{-5} kg m^{-2} s^{-1}. For a leaf 0.05 × 0.05 m, T_ℓ = 34.6°C and E = 8.9 × 10^{-5} kg m^{-2} s^{-1}. And for a leaf 0.20 × 0.10 m, T_ℓ = 39.4°C and E = 9.2 × 10^{-5} kg m^{-2} s^{-1}.

In this sequence, the only change was in the size of the leaf, which changed from 0.01 × 0.01 to 0.05 × 0.05 to 0.20 × 0.10 m. Leaf temperature ranged from 32.6° to 34.6° to 39.4°C, and transpiration rate from 8.6 to 9.4 to 9.2 × 10^{-5} kg m^{-2} s^{-1}. As the leaf became larger, both the leaf temperature and the rate of transpiration increased. Can you show why this happened?

Example 2. Let Q_a = 800 W m^{-2}, T_a = 40°C, h = 0.50, and V = 0.20 m s^{-1}. Compute T_ℓ and E for the same leaf sizes as above and a leaf resistance of 200 s m^{-1}.

One gets, for a leaf 0.01 × 0.01 m, T_ℓ = 40.1°C and E = 11.5 × 10^{-5} kg m^{-2} s^{-1}; for a leaf 0.05 × 0.05 m, T_ℓ = 40.9°C and E = 10.7 × 10^{-5} kg m^{-2} s^{-1}; and for a leaf 0.20 × 0.10 m, T_ℓ = 43.6°C and E = 9.7 × 10^{-5} kg m^{-2} s^{-1}. In this example, leaf temperature increases with leaf size, but transpiration rate systematically decreases.

Example 3. Let $Q_a = 800$ W m^{-2}, $T_a = 40°C$, $h = 0.50$, and $V = 0.20$ m s^{-1}. Let the leaf size sequence be the same as above, but let the diffusion resistance to water vapor be 2000 s m^{-1}.

One gets, for a leaf 0.01×0.01 m, $T_\ell = 44.2°C$ and $E = 1.9 \times 10^{-5}$ kg m^{-2} s^{-1}; for a leaf 0.05×0.05 m, $T_\ell = 47.7°C$ and $E = 2.3 \times 10^{-5}$ kg m^{-2} s^{-1}; and for a leaf 0.20×0.10 m, $T_\ell = 54.3°C$ and $E = 3.4 \times 10^{-5}$ kg m^{-2} s^{-1}. This time, with all environmental conditions the same as previously but the diffusion resistance 2000 s m^{-1} rather than 200 s m^{-1}, leaf temperature and transpiration rate both increase monotonically with increasing leaf size. It should be noted that leaf temperatures are 4.2°, 7.7°, and 14.3°C above the air temperature for the three leaf sizes.

Example 4. Let $Q_a = 1000$ W m^{-2}, $T_a = 40°C$, $h = 0.20$, and $V = 1.0$ m s^{-1}. These are conditions which exist in a desert environment under clear skies during midsummer at noon in the middle latitudes. Compute T_ℓ and E for leaf sizes of 0.01×0.01, 0.05×0.05, and 0.10×0.10 m if the diffusion resistance is 1000 s m^{-1}.

For a leaf 0.01×0.01 m, $T_\ell = 43.1°C$ and $E = 4.9 \times 10^{-5}$ kg m^{-2} s^{-1}; for a leaf 0.05×0.05 m, $T_\ell = 46.0°C$ and $E = 5.5 \times 10^{-5}$ kg m^{-2} s^{-1}; and for a leaf 0.10×0.10 m, $T_\ell = 50.8°C$ and $E = 7.2 \times 10^{-5}$ kg m^{-2} s^{-1}.

Example 5. Let $Q_a = 600$ W m^{-2}, $T_a = 20°C$, $h = 0.80$, and $V = 0.10$ m s^{-1}. These are very moderate conditions which may exist frequently at midday during spring and autumn in the middle latitudes or early in the day during midsummer. Compute T_ℓ and E for leaf sizes of 0.01×0.01, 0.05×0.05, and 0.10×0.10 m if the diffusion resistance is 500 s m^{-1}.

For a leaf 0.01×0.01 m, $T_\ell = 24.1°C$ and $E = 1.5 \times 10^{-5}$ kg m^{-2} s^{-1}; for a leaf 0.05×0.05 m, $T_\ell = 27.1°C$ and $E = 2.0 \times 10^{-5}$ kg m^{-2} s^{-1}; and for a leaf 0.10×0.10 m, $T_\ell = 31.1°C$ and $E = 2.6 \times 10^{-5}$ kg m^{-2} s^{-1}.

Example 6. Let $Q_a = 400$ W m^{-2}, $T_a = 10°C$, $h = 0.60$, and $V = 2.0$ m s^{-1}. These are cool temperate conditions of early spring or late autumn during midday in the middle latitudes or early summer at higher latitudes. Compute T_ℓ and E for leaf sizes of 0.01×0.01, 0.05×0.05, and 0.10×0.10 m if the diffusion resistance is 1000 s m^{-1}.

For a leaf 0.01×0.01 m, $T_\ell = 10.8°C$, $E = 0.5 \times 10^{-5}$ kg m^{-2} s^{-1}; for a leaf $0.05 = 0.05$ m, $T_\ell = 11.2°C$ and $E = 0.9 \times 10^{-5}$ kg m^{-2} s^{-1}; and for a leaf 0.10×0.10 m, $T_\ell = 12.8°C$ and $E = 0.5 \times 10^{-5}$ kg m^{-2} s^{-1}. Here once again is an example in which leaf temperature increases monotonically with leaf size but transpiration rate increases at first and then decreases. In fact, the transpiration rates of the 0.01×0.01-m leaf and the 0.10×0.10-m leaf are the same, although the temperatures of the leaves differ by 2.4°C.

Example 7. Let $Q_a = 400$ W m^{-2}, $T_a = 20°C$, $h = 0.20$, and $V = 0.10$ m s^{-1}. These are conditions which a leaf might encounter when shaded from the sun but exposed to skylight so that the stomates are at least partly open. Let diffusion resistance be 500 s m^{-1}. Calculate T_ℓ and E for leaf sizes of 0.01×0.01, 0.05×0.05, and 0.10×0.10 m.

For a leaf 0.01×0.01 m, $T_\ell = 18.6°C$ and $E = 2.3 \times 10^{-5}$ kg m^{-2} s^{-1}; for a leaf 0.05×0.05 m, $T_\ell = 17.9°C$ and $E = 1.9 \times 10^{-5}$ kg m^{-2} s^{-1}; and for a leaf 0.10×0.10 m, $T_\ell = 17.1°C$ and $E = 1.6 \times 10^{-5}$ kg m^{-2} s^{-1}. All leaf temperatures are below air temperature and decrease with increasing leaf size. Transpiration diminishes with increasing leaf size. Leaf temperatures as much as 5°C below air temperature have been observed by Gates (1963) and are shown in Fig. 11.4.

These examples should suffice to show the reader how to use the "Atlas of Energy Budgets of Plant Leaves." Values of T_ℓ and E can be found in the atlas for realistic values of the environmental variables and

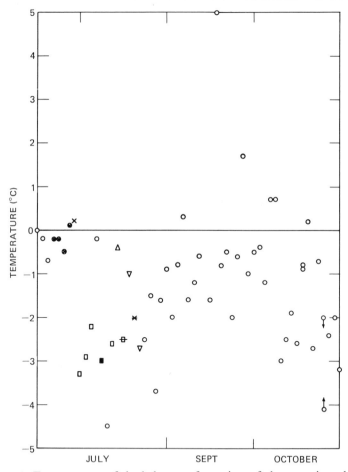

Figure 11.4. Temperatures of shade leaves of a variety of plant species relative to air temperature during the growing season: (○) *Quercus macrocarpa;* (×) Convolvulaceae; (■) ivy; (△) *Ulmus glabra;* (▽) *Ulmus pumila;* (⊟) *Aesculus glabra;* (✳) grape; (□) *Populus deltoides.*

plant parameters. All these calculations were done on the basis of steady state for a given set of values of variables and parameters. If quantity changes in value, one can look up the consequences of such a change on the dependent variables as listed in the stable.

Observed Leaf Temperatures

A large number of measurements of leaf temperatures have been made by various investigators. One of the earliest investigators of leaf temperatures was Askenasy (1875). Other early works that report leaf-temperature measurements include the classical papers of Brown and Escombe (1905) and Brown and Wilson (1905). The winter temperatures of pine needles were measured by Ehlers (1915). Modern measurements of needle temperatures are reported by Perttu (1971). The effect of transpiration and environmental factors on leaf temperatures was reported by Clum (1926). Another early classical work involving the heat exchange of plants was by Huber (1935, 1937). Leaf temperatures of cotton were reported by Eaton and Belden (1929). Curtis (1936) discussed leaf temperatures and the cooling of leaves by radiation. Miller and Saunders (1923) published an early paper on the leaf temperatures of crop plants. Shreve (1919) should be recognized as one who very early realized the importance of leaf temperatures in transpiration. The work by Shull (1930) was referred to earlier in this chapter. Waggoner and Shaw (1952) reported a large number of temperature measurements for potato and tomato leaves.

The temperature of leaves relative to air temperature for any set of conditions is now well understood. The investigator should always keep in mind the reasons for determining leaf temperature. Is temperature significant because of its effect on the biochemical reactions of photosynthesis, respiration, cell enlargement and expansion, diffusion of nutrients, or permeability of membranes? Is temperature important because of cold or heat damage to plant tissue, and if so, what is the nature of the damage and the degree of temperature sensitivity? If temperature is important for biochemical reasons, then what is the degree of temperature sensitivity? Is a difference of 1°C important or is only a difference of 3° or 5°C of real significance? It is relatively easy to measure leaf temperature to an accuracy of about 0.3°C, but to do better than that is quite difficult. In fact, differences over the surface of a leaf are usually much greater than this, as shown by Cook et al. (1964). For some events, such as irreversible heat damage, an increase of leaf temperature of 1° or 2°C can result in thermal death of leaf tissue, whereas for some biochemical events, temperature differences less than about 5°C may not be of great importance.

Daytime Temperatures

The following are various examples of leaf temperature measurements
made in the field (other measurements are described in Chapter 3). Typical
of leaf temperatures during the day are the values shown in Fig. 11.5 for
various plants growing in my yard in Boulder, Colo., during July, 1961.
Leaf temperatures were frequently 6° to 10°C above air temperature, with
a few leaves being very much warmer. For example, some *Quercus macrocarpa* (bur oak) leaves were 20° to 22°C above air temperature at 1445,
although many were only 16° or 17°C above air temperature. The leaves of
Vitis had consistently lower temperatures than the other species. When
clouds obscured the sun, leaf temperatures dropped to within about 2°C of
air temperature, and those leaves in the shade had temperatures about 3°
to 4°C below air temperature. Measurements of leaf temperatures of
Quercus macrocarpa on 4 October showed some fully sunlit leaves at 20°
to 22°C above air temperature. The upper and lower surfaces of *Quercus
macrocarpa* differed in temperature by 1.5° to 2.0°C; the upper surface
was exposed to the sun and was warmer, and the lower surface, which
contained all the stomates, was undergoing transpiration, was shaded
from the sun, and was cooler.

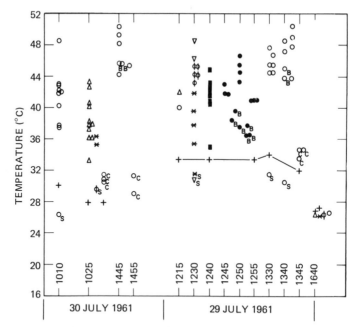

Figure 11.5. Leaf temperatures of a variety of plant species as a function of time of
day: (△) *Ulmus glabra;* (○) *Quercus macrocarpa;* (▽) *Ulmus pumila;* (✳) *Vitis;*
(⏀) *Acer saccharum;* (●) *Populus acuminata;* (■) *Populus tremuloides;* (+) air
temperature. Subscripts: s, shaded leaves; c, cloudy conditions; B, breeze or the
presence of wind.

Gates (1969b) reported the temperatures of various sunlit leaves of plants growing at the Missouri Botanical Garden. The day of the measurements was 15 June 1968. It was sunny, but with a hazy overcast. Measurements reported for 14 plant species were made at noon. The air temperature was 31°C, the relative humidity 0.65, and the incident sunlight between 768 and 907 W m^{-2}. There was no wind. A wet soil surface was at 40°C, and dry leaves on the ground were at 57°C. Leaf temperatures of various sunlit plants varied from a high of 48°C for *Canna* to a low of 33°C for *Corylus* and *Mahonia repens*. A juniper had temperatures around 39°C.

Nighttime Temperatures

Nighttime leaf temperatures are usually below air temperature in the vicinity of the leaf. This is illustrated by the measurements shown in Fig. 11.6. Most of the leaf temperatures were 2° to 4°C below air temperature for leaves exposed to a cold, clear sky, whereas those leaves not exposed and shielded from the sky by other leaves were generally within 1° to 2°C of air temperature. Turrell et al. (1962) reported nocturnal leaf temperatures of *Citrus limonia* and *Citrus paradisi* in Riverside, Calif. They found leaves to be about 1°C cooler than their air during warm nights and 1° to 1.4°C cooler than the air during cool nights when exposed to some sky. Turrell et al. worked out in great detail for citrus leaves their heat-transfer coefficients and view factors to radiation.

Heat-Storm Temperatures

During mid-July 1966, a period of severe heat occurred in St. Louis, Mo. MacBryde et al. (1971) reported leaf temperatures, transpiration rates, and energy budgets of 10 species of trees, shrubs, and herbs during this period. Air temperatures reached highs of about 42°C but were generally about 37° to 40°C during midday and 30°C at night. Various plants used different strategies to survive heat stress. Leaves of *Plantago lanceolata* were always near ambient air temperature, whereas those of *Plantago major* were near air temperature except in wilted leaves, which had temperatures about 5°C above ambient temperatures. High transpiration rates and low diffusion resistances (130 to 200 s m^{-1}) were responsible for the maintenance of low leaf temperatures in *Plantago*. These plants grew close to the ground, where there was little or no air movement and radiation intensity was great. The upper surfaces of *Plantago* had the lowest absorptance (0.497 to 0.514) to solar radiation of any of the plants measured except *Zea mays*, which had an absorptance of 0.442. A small decrease in relative turgidity in the leaves brought about a marked negative water potential, thereby facilitating the movement of water from soil to leaves as a result of the potential difference generated. The high degree of

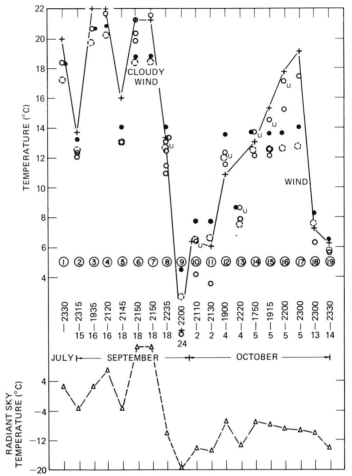

Figure 11.6. Nighttime temperatures of leaves of bur oak (*Quercus marcocarpa*) exposed to the night sky at 19 discrete times: (○) *Quercus macrocarpa*; (●) ground temperature; (+) air temperature; (◌) leaf temperatures predicted from energy budget calculations. Unexposed leaves that are shielded from the cold sky are indicated by the subscript u. The bottom graph gives the sky temperature for the corresponding dates.

success of plantains as weeds seems to be related to their ability to transpire freely and survive heat stress.

During the heat storm, the leaves of *Quercus palustris* were considerably (5° to 10°C) above air temperature during most of the day. The highest leaf temperature measured was about 45°C. Minimum diffusion resistance was 440 s m⁻¹, but very high resistances, up to 31,100 s m⁻¹, were found during midday periods. *Quercus palustris* is well adapted for restricting water loss to a minimum. The upper leaf surface has a thick cuticle. Large water deficits developed in *Quercus* as compared with *Plantago*, but unlike the plantains, the oak leaf did not develop strong negative water potentials. However, the oak leaves exhibited an unusual ability to recover from low relative-turgidity values. For example, leaves of *Q. palustris* could sustain a relative-turgidity drop to below 8%, *P. lanceolata* and *P. major* to 50%, and *Zea mays* to 80%. In the event of heat stress for *Zea*, if high transpiration rates were unable to be maintained because of water shortage, then not only did leaf temperature rise, but low resistance did not restrict water loss and, consequently, the water deficit increased in the tissue. At the time of the heat storm, the corn crop in Missouri was severely damaged. *Quercus palustris* had one distinct advantage over shrubs and ground plants during the heat storm: its leaves were high in the air, where convective cooling was effective because of increased air movement.

Leaf damage occurred to the leaves of the following plant species: *Viburnum hanceanum, Hamamelis vernalis,* and *Tilia heterophylla.* Minimum diffusion resistances were 3580 s m^{-1} for *Viburnum hanceanum,* 740 s m^{-1} for *Hamamelis vernalis,* and 18,800 s m^{-1} for *Tilia heterophylla.* The leaf resistances of virburnum were always quite high, and the transpiration rates were low. Convection would have played an important role in cooling these leaves, except for the fact that during the heat storm, wind speeds were often near zero between ground level and 2m above the ground. Therefore, some shrubs like *V. hanceanum,* which had low transpiration rates, did not receive the convective advantage of *Q. palustris,* which grew to much greater heights and received some light air movement. These plants could take considerable water stress and recover turgidity, but the high leaf temperatures did cause burning. The percent relative turgidity below which recovery did not occur was 18% for *H. vernalis,* 19.5% for *V. hanceanum,* and 20% for *Tilia heterophylla.*

Cool, Humid Elfin Forest

Leaf temperatures were measured of dominant tree species growing in the elfin forest at the summit of Pico del Oeste, Luquillo Mts., P.R., during 7 and 8 January 1968 (Gates, 1969a). The climate in the forest is very wet, with frequent rain and mist. It is cool and windy, and light levels are low. Incident global radiation varied between a maximum of 558 Wm^{-2} and a minimum of 209 Wm^{-2} at noon. Mean wind speed at tree-top level was 6 m s^{-1}, and mean relative humidity was about 0.98.

The dominant trees were *Tabebuia rigida* and *Ocotea spathulata.* Their leaves are approximately 0.06 to 0.08 m in width. Leaf temperatures were 16°C when air temperature was 17°C, relative humidity 0.85, wind speed 1.0 m s^{-1}, and amount of absorbed radiation about 398 Wm^{-2}. The energy budget for these leaves was calculated, and the relationship

between leaf temperature and transpiration rate was determined for estimated values of diffusion resistance. Although transpiration rates were
low because of the high humidity, there nonetheless was definite transpiration going on. Maximum rates were about 1.3×10^{-5} kg m^{-2} s^{-1}.

The leaf temperatures of plants on Pico del Oeste probably never drop
below 15°C or rise above 25°C. It would be interesting to know the optimal temperature for photosynthesis in these plants. It may be just above
these maximal temperatures, and if so, these temperatures would reduce
photosynthesis very little. However, the light level is probably very much
suboptimal, and photosynthesis is strongly reduced. Further measurements are required to confirm or deny these speculations.

Desert Plant Temperatures

Two broad categories of plants exist in dry, desert regions of the world:
plants which as a general rule have very small leaves, such as creosote
bush, paloverde, sage, cat's-claw, etc., and succulent plants such as cacti
or euphorbs. Each of these two types responds in a very different manner
to heat stress and water deficits. The plants with very small leaves have
leaf temperatures strongly coupled to air temperature by convection, as
described in Chapter 3. Temperatures of these leaves never exceed air
temperature by more than 2° or 3°C. Often, these plants shed their leaves
when confronted with extended dry periods. These plants generally have
either C_3 or C_4 metabolism. The succulent plants have a very different
metabolic physiology, referred to as crassulacean acid metabolism; they
are known as CAM plants. Their stomates open primarily at night, during
which period they take up CO_2, store it in the form of acids, and metabolize it during the day by means of photochemical events. The succulents
have thick, waxy cuticles which help in the conservation of water. These
plants often have vertical forms which reflect sunlight effectively during
midday. Also, they may have a dense crown of thorns on top which effectively reflects sunlight and retards penetration of solar heat to the underlying tissues.

I measured the surface and internal temperatures of various succulent plants during brief visits to the desert near Phoenix, Ariz., in June,
1963. A typical set of measurements are those for a barrel cactus in full
sun under clear sky at 1150 on 26 June when the air temperature was
37.7°C. The top surface of the cactus was 41°C, a side surface was 37.7°C;
the temperature was 30°C 0.13 m inside the sunlit side and 27°C 0.10 m inside the shady side. At noon, when the air temperature was 40°C, a
saguaro had a top surface temperature of 47°C and an internal temperature near the center of 37.5°C. On another saguaro, the surface temperature at the top was 54°C and the inside center temperature 35°C when the
air temperature was 40.7°C. At 1510, measurements of a very large
saguaro showed that when the air temperature was 41.9°C, the tempera-

ture was 34°C 0.10 m inside the shady side and 32°C 0.15 m inside the sunlit side. *Opuntia* usually has its broad blades oriented in a vertical plane. When the air temperature was 45°C, the internal temperature was 46.5°C for a vertical blade and 52°C for a horizontal blade. The vertical orientation of the blades of *Opuntia* gives them a decided advantage in avoiding over-heating. At 1440, when the air temperature was 41.9°C, an *Echinocereus* with many thorns had an internal temperature of 50°C. At night, the temperatures inside a saguaro reversed relative to air temperature. At 2310, the air temperature was 32.6°C, whereas the temperature was 38°C 0.18 m inside the saguaro and 33°C 0.04 m inside one of the ridges. At 0800, the air temperature was 29.4°C, and the temperature 0.18 m inside the saguaro was 34.5°C (the plant was still cooling from the previous day). At 1005, when the ambient temperature was 35.6°C, the temperature 0.18 m inside the plant was 33°C. At 1440, when the temperature was 39.7°C, the temperature at the same depth was 32°C. The outside top of the saguaro was at 50°C, and an *Opuntia* blade was at 49°C at this same time.

It is clear that the interior of a large object such as a saguaro or a barrel cactus exhibits a temperature lag relative to the air. Surface temperatures vary enormously from the sunlit to the shaded side, whereas interior temperatures are moderate and nearly 12 hr out of phase with the external temperature. This was shown by Derby and Gates (1966) for tree trunks, by means of both direct measurement and calculation from theory. The saguaro interior at a depth of 0.18 m is coolest in early afternoon and warmest after midnight. It takes 8 to 12 hr for the heat of the day to penetrate to the central portion of the saguaro, and it takes a similar length of time for heat to escape during the night and morning hours. Desert birds which nest in the interior of the saguaro must realize that temperatures there are much more modulated and uniform than those outside—cool in the day-time and a little warmer at night.

Sunlit leaves of an *Agave americana* were at 47°C when the air temperature was 41.9°C at 1420 on 26 June. Shade leaves were at 41°C. *Agave* leaves which were fairly vertical seemed to survive the heat without undue stress, but leaves bent to a horizontal position exhibited much denaturation of tissues. It is interesting to note that the soil surface temperatures were as high as 70°C, although much of the time they were only 5° to 16°C above air temperature during midday hours. At night, soil and air temperatures were about equal.

New Zealand Plant Temperatures

A trip to New Zealand in February, 1968, during the Austral summer offered an opportunity to measure many plant temperatures in the field. Temperature measurements were made in the Wanganui State Forest on the west side of the south island of New Zealand. Plants measured were

Dacrydium cuppressinum, Pseudopenax crassifolium, Leptospernum scoparium, Coprosma parviflora, and *Griselinia littoralis.* Leaf temperatures were within about 5°C of air temperature with a few exceptions (some leaves were 10° to 17°C above air temperature in full sunshine). During cloudy conditions, leaf temperatures were always close to air temperature.

Alpine Tundra Plants

The alpine tundra is a harsh environment of persistent winds, intense solar radiation, air that is often very dry, moisture in the form of rain, hail, sleet, and snow, and temperatures varying from intense cold to temperate. All life tends to hug the ground in the alpine. Trees grow horizontally at the edges of the alpine in the form of *Krumholtz,* and all plants are short and stunted except the cushion plants, which grow snug against soil and rocks, staying out of the wind as well as possible. During summer and autumn, a micro "desert" may exist a few feet from a sodden micro "marsh." Gates and Janke (1965) described the climate of the alpine tundra and reported the temperatures of plants growing on Niwot Ridge above Boulder, Colo., at an altitude of 3350 m, giving a detailed energy analysis for *Polygonum bistortoides.* Salisbury and Spomer (1964) reported leaf temperatures of alpine plants on Mount Evans, Colo., at 4300 m, and along Trail Ridge Road in Rocky Mountain National Park at 3800 m.

The cushion plants were the warmest of all plant forms. The following are representative temperatures in the alpine during conditions with wind less than 1 m s^{-1} and clear sky with scattered cumulus clouds. The time was 1145 to 1215 on 22 July 1963 at Niwot Ridge. Air temperature was 18°C. Leaves of *Mertensia viridis* were 27.3°C; *Polygonum bistortoides,* 29°C, an aster flower, 29°C, *Silene acaulis* (moss campion), 34°C; and various grasses, 29°C. Soil containing seedlings of *Draba streptocarpa, Orioxous alpina, Sedum stenopetalum, Polygonum bistortoides, Artemesia,* and *Loidia seratina* was at 53°C. Dr. William Osburn, who studied these seedlings over a period of several years, first called my attention to the harshness of the seedling environment. These is a high mortality among the seedlings because of leaf burn and denaturation of stem tissue next to the soil surface. The intense solar radiation of the alpine tundra produces very high surface temperatures at times of clear sky and low wind.

On 25 July 1962 at Niwot Ridge, the air temperature was 10.4°C. The sun had been out earlier, but at 0945, the soil surface was 18°C, cushion plants of *Arenaria obtusilobia* were 17.5°C, and various rock surfaces were 17.2°C. Salisbury and Spomer (1964) reported temperatures up to 22°C above air temperature for leaves of alpine plants fully exposed to sunshine, but when plants were shaded by rocks or cliffs, leaf temperatures dropped to as much as 3.5°C below ambient temperature. These workers recorded plant temperatures as high as 32.5°C during the day and as low as 0°C at night during 14 and 15 July 1961 on Mt. Evans, Colo. Wind had enormous influence on leaf temperature. Cushion plants generally have higher temperatures than erect plants, as shown in Fig. 11.7 (from Salisbury and Spomer, 1964).

Gates and Janke (1965) measured and evaluated the complete energy budget for plants of *Polygonum bistortoides*, bistort, growing on the alpine tundra of Niwot Ridge, Colo., during 26 and 27 July 1963. Because the day was overcast, the amount of radiation received was not particularly strong, and the bistort leaves were very close to air temperature. At times, the air was warmer than the leaves, and convection added a small amount of energy to the leaf; at other times, convection cooled the leaves. There was nearly always some wind. Amount of radiation absorbed was calculated from the radiation fluxes for an absorptivity of 0.52 to shortwave radiation and 0.96 to longwave radiation. For comparison, measurements were made during two clear days, 2 and 3 September 1963, during which there were only small amounts of cloud cover. The wind speed was considerably less during this period than in July. Although the midday air temperature was nearly the same for the two periods, leaf temperatures were substantially higher on the clear, calmer day, often being

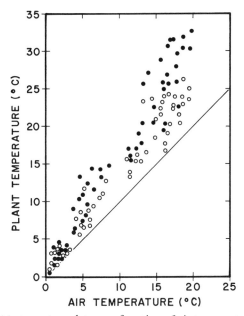

Figure 11.7. Plant temperature data as a function of air temperature obtained at an altitude of 3800 m near Trail Ridge Road in Rocky Mountain National Park, Colo.: (●) cushion plant; (○) erect plant. (After Salisbury and Spomer, 1964.)

5° to 8°C above air temperature. Wind speed and air temperature were always measured at the level of the plant.

Janke (personal communication) measured the needle temperatures of *Pinus contorta* on Niwot Ridge on these same days. Needle temperatures were often about 4°C, and never more than 6°C, above the air temperature during midday. At night, needle and air temperatures were nearly identical. One would expect needle temperatures to be closely coupled to air temperature. This coupling would be very pronounced for single needles; the clusters of needles on branches produce a greater trap for radiation and warmer temperatures. Gates and Janke reported the rapid and frequent variation of leaf and needle temperatures, and Salisbury and Spomer also commented on the lack of steady state. This is particularly true of plants growing in a windy, turbulent environment such as the alpine tundra.

I have measured the temperatures of clusters of needles in Englemann spruce (*Picea engelmanni*) and alpine fir (*Abies lasiocarpa*) at timberline in the *Krumholtz* stands. For full sunshine, the needles are at air temperature on the windward side of a tree and up to 9°C above air temperature on the leeward side. I believe that considerably greater warmth of the plant tissues on the leeward side is in part responsible for the accelerated growth of the tree on this side. All *Krumholtz* trees have the habit of growing downwind. This amount of temperature difference could have notable influence on their growth. Wilson (1957) reports a similar phenomena for clumps of certain plants in the Arctic.

Arctic Plants

Arctic regions are also characterized by extreme environments in which the growth of plants is often temperature-limited. Arctic environments tend to be windy and are unusual because of the long periods of daylight during the summer and long periods of darkness during the winter. Wilson (1957) reported extensive measurements of the temperatures of arctic plants made at Resolute (75°N, 95°W), on Cornwallis Island in the Canadian Arctic Archipelago. Most of his measurements were with *Salix arctica*, but he also measured temperatures of *Parrya arctica, Saxifraga cernua, Ranunculus sulphureus*, and *Arctagrostis latifolia*. He found that all species had about the same temperature response to environmental conditions and therefore reported most extensively the leaf temperatures of *Salix*. Leaf temperatures were generally 3° to 5°C above air temperature during clear, sunny conditions when there was slight wind. The leaves were growing about 1 cm above the ground surface. Measurements were made during days in August when air temperatures were about 7°C. During overcast days, leaf temperatures were always very close to air temperature. It is interesting that Wilson refers to some very early temperature measurements made by T. Wulff during an expedition to Spitzbergen, an island north of Norway, and published in 1902. Wulff found that the surface temperature of a *Silene acaulis* clump in Spitzbergen reached 15.5°C when the air temperature at 1 m above the ground was 4.7°C.

Soil-surface temperatures in the Arctic during sunny days may greatly exceed air temperatures. Extreme values of soil temperature are as high as 68°C, and the mean daily maxima of soil temperatures may exceed air temperature by 15°C. When air temperatures were 16° to 21°C, soil-surface temperatures of 50°C were observed. The soil-surface temperature was of particular interest since many arctic plants are prostrate on the ground surface.

Tikhomirov et al. (1960) reported measurements of the temperatures of 40 species of flowering plants at Tiksi, Yakutsk, USSR, at 71°31′N. They found that the temperatures of plant parts exceeded air temperature by 2° to 5°C during sunny days but often were below air temperature during overcast days. During sunny days, white flowers had temperatures 0.7° to 2.0°C above air temperature and blue flowers 3.4 to 4.2°C above. The interesting work of Krog (1955) is described in Chapter 15. He found that willow catkins with hairs had considerably higher temperatures than the same catkins painted black. He surmised that the natural hairs let solar radiation pass through to the underlying dark scales and at the same time reduce reradiation and the exchange of heat by convection. The microclimates of alpine and arctic tundras have been described in detail by Bliss (1956).

One would expect the temperatures of arctic and alpine plant leaves in full sunshine to be within about 3°C of air temperature because of their generally small size. However, grouping of leaves into tightly knit clusters produces greater differences between leaf temperature and air temperature. Cushion plants are exceptions to the rule concerning the temperatures of single, small leaves.

Leaf Temperatures of Plants in Tadzhikistan

The region of Tadzhikistan is characterized by a continental climate with warm temperatures and dry air. Bregetova and Popova (1963) reported very detailed leaf temperature measurements both day and night. They studied *Paspalum digitaria,* barley, cotton, grape, corn, and tomato. Unfortunately, they did not measure radiation, humidity, or wind speed while they measured leaf temperatures and only give a comparison with air temperatures. However, the distinctive thing about these measurements was the persistence of leaf temperatures below air temperature. *Paspalum digitaria* had leaf temperatures consistently below air temperature, as much as 6°C below. Cotton had leaf temperatures lower than air temperature between the hours of 1030 and 1730 and above air temperature during the early morning and late afternoon. Clearly, this is a transpiration phenomena. At 1400, the temperature difference was zero for a short while because of water stress limiting the transpiration rate. The leaves of corn remained above air temperature until 1330 and below air temperature during the afternoon. Tomato leaves were below air temperature all day, with a maximum difference of 7°C. Grape was slightly below air temperature during the morning, above during the afternoon, and below late in the day. The leaves of all plants were close to or below air temperature at night. Bregetova and Popova (1963) reported work by Kleshnin and Shulgin showing that leaf temperatures of plants growing in the Crimea had temperature gradients of $T_\ell - T_a$ from $-2°$ to $+13°C$. They also discussed plant temperatures as measured in other parts of the Soviet Union. The interesting thing about most of these papers is that leaf temperatures were described with very little explanation as to why they were above or below air temperature. Every one of the observations could be fully understood by means of energy exchange if the environmental conditions had been measured.

Leaf Temperatures in Controlled Environments

The research of Mellor et al. (1964) is concerned with leaf temperatures, transpiration rates, and energy budgets of the leaves of several plant species as measured in controlled environments. In this paper, these authors give an energy balance sheet for *Xanthium pennsylvanicum,* as shown in Table 11.1. They used a shortwave absorptivity of 0.46 to an incoming amount of shortwave radiation of 1284 Wm^{-2}. The ceiling of their chamber emitted at a temperature of 25°C and the floor at 22°C. They used a thermal emissivity for the leaf of 0.95. The convection term in Table 11.1 is adjusted to balance the heat budget and so is the most inaccurate of all the terms. The transpiration rate was measured. It is seen that radiation dissipates about 63% of the total energy absorbed by the

Table 11.1. Energy Balance Sheet for *Xanthium pennsylvanicum* under Controlled Laboratory Conditions[a]

Energy absorbed	
Shortwave from light source	600
Longwave from ceiling and floor	837
Reflected shortwave	28
Total	1465
Energy lost	
Transpiration	467
Thermal radiation emitted	921
Convection	77
Total	1465

[a] Units are watts per square meter.

leaf, transpiration 32%, and convection only 5%. This energy balance sheet is given here to illustrate that it is relatively easy to account for all of the energy exchanged by a plant leaf, particularly under controlled-environment conditions. Analysis of measurements made in the field is only slightly more difficult.

Optimal Leaf Size

The size, shape, and orientation of leaves is very important to the productivity of a plant. Many people have noted the distribution of leaf sizes and shapes among the leaves of a single plant or an entire stand, as well as variations from one habitat to another. Sun leaves of *Citrus, Olea,* and *Quercus* were reported by Bergen (1904) to be narrower than shade leaves in proportion to length. He found that the surface area of individual leaves was 50 to 100% greater for shade leaves than sun leaves. Hanson (1917) reported on leaf structure as related to environment for *Acer saccharum, Fraxinus pennsylvanica, Ulmus americana, Tilia americana, Quercus alba, Quercus rubra, Quercus macrocarpa, Acer negundo, Celtis occidentalis,* and *Plantanus occidentalis.* He found exposed leaves to be smaller, thicker, more deeply lobed, and more prominently toothed than shaded center leaves from the same tree. Hesselbring (1914) found the leaves of shade-grown tobacco plants to be much larger than the leaves of plants grown in the open. Cain and Miller (1933) report that *Rhododendron cataubiense* growing in exposed habitats in the Great Smoky Mountains has smaller leaves than plants of the same species growing in the shade. Whitehead (1962) reports that *Helianthus annuus* has leaves smaller and more xeromorphic when grown under conditions of drying wind than when grown in relatively calm air. Taylor (1971) reports interesting observations in his Ph.D. thesis which seem to be environmentally related. He also reports the apparent effects of environment on leaf thickness, color integument, concavity, or edge turning, orientation, and stomatal density and size. According to Taylor (1975), "Bailey and Sinnott (1916) concluded that the form and size of leaves was more a result of environment than of genetic history, although the latter was certainly an influence." Benson et al. (1967) reported studies of the hybrid oak *Quercus douglasii* × *Q. turbinella* which showed that the leaves of individuals found on the northeast slope were larger than those from the more-arid, southwest slope.

Raunkiaer (1934) studied leaf size and was convinced that it was an indicator of climatic conditions. A modification of Raunkiaer's leaf-size classification is given in Taylor (1975). The small leaves of desert, alpine, and arctic plants, the very large leaves of the understory plants of tropical forests, and the various gradations of leaf size along transects from dry to wet, cold to warm, and shaded to light lead one to feel relatively certain that leaf size and environment are intimately coupled.

"Natural selection leads to organisms having a combination of form and function optimal for growth and reproduction in the environments in which they live." This is Parkhurst and Louck's (1972) statement of the principle of optimal design and the basis of Parkhurst's Ph.D. thesis study. These authors go on to state, "The fundamental axiom is simply that, in a given population in a given environment, those genotypes having phenotypes closest to optimum for growth and reproduction in that environment will tend to survive and reproduce most often." Parkhurst was trained in engineering, worked at the University of Colorado on biophysical ecology, and recognized the importance of theory and analytical modeling in understanding biological systems. The following statement is taken from Parkhurst and Loucks (1972):

> In biology . . . one may attempt to understand or explain a given form or process of an organism by "designing" an organism to fill the same role. The procedure used is to choose some cost variable thought to be of most importance to the survival of the organism, and to minimize that cost. If the resulting design is sufficiently similar to typical forms of the actual living organism, it is likely that one understands reasonably well the factors which have influenced its evolution.

With this statement I concur completely, and it is the principle according to which this book is written and the fundamental reason I have attempted to establish an analytical subdiscipline within ecology.

The paper by Parkhurst and Loucks (1972) is the most significant paper written in this field during recent years and merits careful study. The following is a summary of their procedures and analysis. The energy budget for a leaf is

$$Q_a = \epsilon\sigma(T_\ell + 273)^4 + h_c(T_\ell - T_a) + E. \qquad (11.12)$$

The transpiration rate E, in energy units, is

$$E = \lambda(T_\ell)\frac{{}_sd_\ell(T_\ell) - h\,{}_sd_a(T_a)}{r_\ell + r_a}. \qquad (11.13)$$

The internal diffusion resistance r_ℓ to water vapor is made up of a mesophyll resistance r_m and a stomatal resistance r_s. The rate of photosynthesis is

$$P = \frac{C_a - C_c}{r'_\ell + r'_\ell} f(T)\,g(I). \qquad (11.14)$$

The quantities C_a and C_c are the concentrations of CO_2 in the air and at the chloroplasts, respectively, where $f(T)$ is the temperature function of photosynthesis, $g(I)$ is the light function of photosynthesis, r'_a is the boundary-layer resistance to CO_2 diffusion, and the internal diffusion resistance r'_ℓ to CO_2 diffusion is made up of a mesophyll resistance r'_m and a stomatal resistance r'_s.

The rate of heat transfer by convection is given by

$$h_c(T_\ell - T_a) = \frac{c_p\, \rho(T_\ell - T_a)}{r''_a}, \tag{11.15}$$

where c_p is the specific heat of air at constant pressure, ρ is the density of air, and r''_a is the boundary-layer resistance to exchange of heat between leaf and air. I have used a notation for the resistances different from that of Parkhurst and Loucks. The reader of their paper should take note of these differences.

Because the concentration C_c of CO_2 inside the leaf is unknown, Parkhurst and Loucks defined what they called the normalized net photosynthesis P' as follows:

$$P' = \frac{P}{C_a - C'_\ell} = \frac{1}{r'_\ell + r'_a}. \tag{11.16}$$

They defined water-use efficiency as the ratio $R = P/E$, following an earlier definition by Slatyer (1964). Since normalized net photosynthesis P' and actual net photosynthesis P behave in approximately the same manner and P' is simpler to use than P, Parkhurst and Loucks use a normalized water-use efficiency, $R' = P'/E$, for their optimization calculations.

Optimizing Water-Use Efficiency

Parkhurst and Loucks (1972) made predictions of leaf sizes based on optimizing the normalized water-use-efficiency ratio for as many combinations of the following seven independent variables as they wished. The independent variables are (a) the convection coefficient, (b) air temperature, (c) relative humidity, (d) absorbed radiation, (e) stomatal resistance, (f) mesophyll resistance, and (g) stomatal distribution, i.e., whether the leaf is hypostomatous (undersurface only) or amphistomatous (both surfaces). Although the functional dependence of photosynthesis on leaf temperature and light is indicated in Eq. (11.14), these functions are not taken into account during the first approximation to a solution of the problem. The influence of these functions on water-use efficiency is brought in later.

The first example given by Parkhurst and Loucks is for the following fixed independent variables: air temperature, 20°C; relative humidity, 0.50, stomatal resistance, 2000 s m^{-1}; and mesophyll resistance, 1000 s m^{-1}. The leaf is assumed to be hypostomatous. The authors used two values of absorbed radiation, 419 and 558 Wm^{-2}, which correspond to the probable absorption by shade and sun leaves, respectively, in a southern

Wisconsin oak forest. They determined the response of the three dependent variables (leaf temperature, transpiration rate, and normalized photosynthetic rate) as a function of the convection coefficient. Since the convection coefficient is proportional to the square root of the wind speed and inversely proportional to the square root of the leaf dimension in the direction of air flow, this procedure was tantamount to determining the response of the dependent variables to wind speed or leaf dimension. The same results for leaf temperature and water use can be read from the tables and charts given by Gates and Papian (1971), but the introduction of photosynthesis into the analysis is new. Parkhurst and Loucks found that the normalized water-use efficiency of shade leaves is higher when leaves are large, whereas that of sun leaves is higher when the leaves are small.

Parkhurst and Loucks then tested leaf temperature, transpiration, and water-use efficiency as a function of number of leaf surfaces bearing stomates, relative humidity, stomatal resistance, air temperature, convection coefficient, and amount of absorbed radiation. Water-use efficiency is in general much higher at 0.95 relative humidity than at 0.20, other factors being equal. In a three-dimensional plot of a habitat space with air temperature, relative humidity, and amount of radiation absorbed as axes, Parkhurst and Loucks show a surface profile of $\partial R'/\partial h_c = 0$. In the region above the surface, $\partial R'/\partial h_c < 0$, i.e., water-use efficiency increases with increasing leaf size or decreasing wind speed, and the amount of absorbed radiation is low. In the region below the surface, $\partial R'/\partial h_c > 0$, i.e., water-use efficiency increases with decreasing leaf size or increasing wind speed, and the amount of absorbed radiation is high. This is consistent with the observations by Clements (1904), Graner (1942), Shields (1950), Blackman (1956), and Talbert and Holch (1957) that sun leaves tend to be small and lobed, whereas shade leaves are large and less lobed.

Also noted from their three-dimensional diagram is that the range of amounts of absorbed radiation for which water-use efficiency increases with increasing leaf size and decreasing wind speed is much smaller at low air temperatures than at high air temperatures. If natural selection tends to increase water-use efficiency, then small leaves will be selected for in cold climates regardless of the amount of radiation absorbed. This certainly is consistent with the leaf sizes seen in alpine and arctic tundra regions. At high temperatures, the diagram shows that where light is intense, i.e., with high amounts of absorbed radiation, small leaves are at an advantage. This is consistent with the small leaves observed in all desert plants except the succulents, which operate on a very different physiological basis. Also, at high temperatures, the diagram shows that where there is little light, large leaves have the greatest water-use efficiency. This is consistent with the large leaves found in the interiors of dense tropical forests. The diagram also predicts that in hot, dry, low-light environments, large leaves will evolve. However, such an environment does not occur in nature, since when there is little moisture, the vegetation is not sufficiently dense to produce a shaded, low-light condition for the understory vegetation.

Factorial Approach

Parkhurst and Loucks next use a very valuable technique to explore the sensitivity of response of the dependent variables to the independent variables. This technique is the 2^N factorial-design method used in statistical experimental design as described by Davies (1967) and Simpson et al. (1960). In such a design, each dependent variable is determined for each possible combination of N independent variables each at two allowable levels. There are 2^N such combinations. The ith independent variable is designated by F_i. The main effect of the ith independent variable is defined as the difference between the mean response for the 2^{N-1} independent variable combinations including F_i at its high level and the mean response for the 2^{N-1} independent variable combinations including F_i at its low level. The procedure used is illustrated in Parkhurst and Loucks (1972) and explained in Davies (1967). The seven independent variables listed previously lead to 2^7, or 128, possible combinations of levels to be tested.

The results of the 2^N factorial-design method of main effects and factor interactions on calculated leaf temperatures show that leaf temperatures vary most with air temperature over the temperature range from 15° to 35°C. Leaf temperature varies by 15.5°C over this air-temperature range and is less than air-temperature variation because of transpiration and radiational cooling. After air temperature, absorbed radiation has the largest effect on leaf temperature, followed by stomatal resistance, convection coefficient, and relative humidity. These results are illustrated in Fig. 11.8. Lesser influences on leaf temperature are produced by the first-order interactions of the various independent variables. Thus it is seen that next in order is the influence of the convection coefficient–absorbed radiation reaction, then the stomatal resistance–relative humidity interaction, etc.

Also shown in Fig. 11.8 are the results of the sensitivity influence of the independent variables on transpiration rate and water-use efficiency. The most important variables with respect to transpiration are stomatal resistance, relative humidity, stomatal resistance–relative humidity interaction, air temperature, absorbed radiation, etc. The most important variables with respect to normalized water-use efficiency are relative humidity, absorbed radiation, absorbed radiation–relative humidity interaction, mesophyll resistance, convection coefficient, stomatal resistance, etc. The values listed in Table 11.2 (shown in Fig. 11.8) were used for the "Experiment I" calculations; "Experiment II" calculations used convection coefficients which were 10 times those listed in Experiment I. The re-

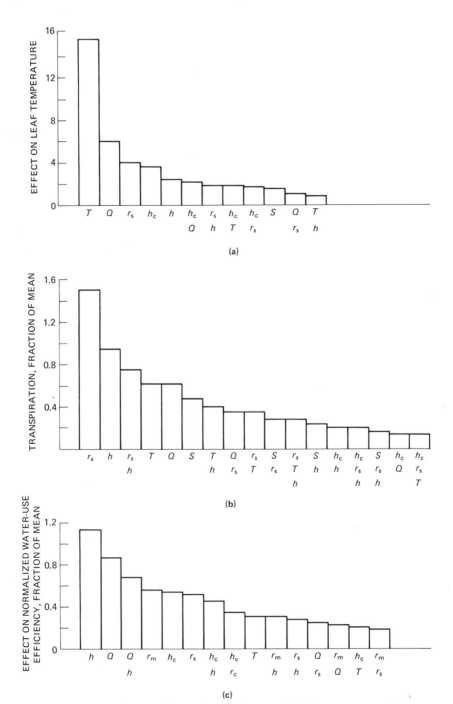

Figure 11.8. Main effects and factor interactions on (a) leaf temperatures, (b) transpiration rates, and (c) normalized water-use efficiencies in a 2^7 factorial design

Table 11.2. Factors, Symbols, Units, and Quantities used by Parkhurst and Loucks (1972)[a]

	Symbol	Units	Low level	High level
Experiment I				
Absorbed radiation	Q	Wm^{-2}	558	837
Air temperature	T	°C	15	35
Relative humidity	U	%	20	95
Number of surfaces bearing				
Stomates	D	–	1	2
Stomatal resistance	S	s m²	200(100)	3000(1500)
Mesophyll resistance	M	s m⁻²	400(200)	2000(1000)
Convection coefficient	H	$Wm^{-2} °C^{-1}$	19.5	62.1
Experiment II				
Convection coefficient	H	$Wm^{-2} °C^{-1}$	195	621

[a] Quantities in parentheses are for an amphistomatous leaf; other quantities are for a hypostomatous leaf.

sults are essentially the same, and the diagrams are similar except for a stronger air-temperature dependence.

Water-use efficiency increases with increasing stomatal resistance as also predicted by Slatyer (1964). According to Parkhurst and Loucks, this finding may explain why stomatal resistances of wild plants, such as oak trees, are usually higher than those of crop plants which have been bred for high CO_2 assimilation under irrigated conditions of ample water. Water-use efficiency increases with decreasing mesophyll resistance as shown by the negative sign in the histogram. Parkhurst and Loucks state as follows: "The optimality principle then implies that mesophyll resistances will tend through natural selection to become as low as is consistent with the structure and chemistry of leaves otherwise adapted to the environment concerned."

Water-use efficiency increases with the convection coefficient, and hence with diminishing leaf size, except in shaded conditions. Water-use efficiency for hypostomatous leaves is generally greater than that for amphistomatous leaves. Higher amounts of absorbed radiation lead to lower water-use efficiency in these calculations. However, this relationship should change at very low radiation levels, where light is limiting to photosynthesis, and at very high radiation levels, where leaf temperatures and transpiration rates become high.

---

analysis. Mean values areas follows: leaf temperature, 29.2°C; transpiration rate, 1.52 g dm⁻¹ hr⁻¹; normalized water-use efficiency, 39 dm³ g⁻¹· Symbols: T, air temperature; h_c, convection coefficient; Q, flux density of absorbed radiation; r_s, stomatal resistance to water-vapor transfer; h, relative humidity; S, solar radiation. (Redrawn from Parkhurst and Loucks, 1972.)

Parkhurst and Loucks comment that the 2^N factorial-design approach using synthetic data produced from a system of equations is extremely useful. First, it provides a means to judge the sensitivity of the dependent variables to the independent variables. Second, relative effects of single variables and interactions among two or more variables are easily compared, a technique of especial value in ecology. Third, results involving many variables are easily graphed and their interactions visualized. However, the n-dimensional hyperspace with sections taken through it has some advantages over the factorial method. Therefore, a combination of the two methods, which are complementary, is probably best for representing the complex dimensional interactions of the many variables found in ecology.

Photosynthesis; Taking into Account Light and Temperature

In the analysis described in the preceding section, Parkhurst and Loucks did not take the influence of temperature and light on photosynthesis directly into account. However, it is straightforward to do so. Using the water-use efficiency $R = P/E$, one can determine the variation of R with P and E, and through the convection coefficient, the dependence on leaf temperature. Hence,

$$\frac{\partial R}{\partial h_c} = \frac{\partial R}{\partial E} \frac{\partial E}{\partial h_c} + \frac{\partial R}{\partial P} \frac{\partial P}{\partial h_c}. \tag{11.17}$$

From the definition of $R = P/E$, one gets $\partial R/\partial E = 1/E$ and $\partial R/\partial E = -P/E^2$. The factors P and E each depend on h_c through the effect of convection on leaf temperature and boundary-layer resistance. Therefore,

$$\frac{\partial E}{\partial h_c} = \frac{\partial E}{\partial T} \frac{\partial T}{\partial h_c} + \frac{\partial E}{\partial r} \frac{\partial r}{\partial h_c}, \tag{11.18}$$

where $r = r_\ell + r_a$, and

$$\frac{\partial P}{\partial h_c} = \frac{\partial P}{\partial T} \frac{\partial T}{\partial h_c} + \frac{\partial P}{\partial r'} \frac{\partial r'}{\partial h_c}, \tag{11.19}$$

where $r = r_m' + r_s' + r_a'$. Combining these results and substituting into Eq. (11.17), one gets

$$\frac{1}{R} \frac{\partial R}{\partial h_c} = \left[\left(\frac{1}{P} \frac{\partial P}{\partial T} - \frac{1}{E} \frac{\partial E}{\partial T} \right) \frac{\partial T}{\partial h_c} \right] + \left[\frac{1}{h_c} \left(\frac{r_a}{E} \frac{\partial E}{\partial r} - \frac{r_a'}{P} \frac{\partial P}{\partial r'} \right) \right]. \tag{11.20}$$

The effects represented by the second bracketed term are those considered earlier when the influence of h_c on water-use efficiency through the transpiration term was taken into account. Therefore, only the first term is considered here.

Since water-vapor concentration in a leaf is roughly an exponential function of leaf temperature, the term $(1/E)$ $(\partial E/\partial T)$ increases with T. The term $(1/P)(\partial P/\partial T)$ is large and positive at low temperatures, zero at the temperature optimum for net photosynthesis, and large and negative at high temperatures according to the slope of the photosynthesis–temperature curve.

At temperatures higher than the optimum for photosynthesis, both $\partial P/\partial T$ and $-\partial E/\partial T$ are negative and therefore additive. The term $\partial T/\partial h_c$ is negative for sunlit leaves with substantial amounts of absorbed radiation. The net result is that at temperatures above the temperature optimum, $\partial R/\partial h_c$ is positive. This means that water-use efficiency increases with an increase in the convection coefficient and therefore with a decrease in leaf size. At low leaf temperatures, the P and E effects on $\partial R/\partial h_c$ are in opposite directions, and the $\partial P/\partial T$ term may dominate over the $\partial E/\partial T$ term. In this case, since $\partial T/\partial h_c$ is still negative, $\partial R/\partial h_c$ is negative, and water-use efficiency increases with increasing leaf size. In other cases, $\partial P/\partial T$ at low temperatures may be less than $\partial E/\partial T$, with the result that $\partial R/\partial h_c$ is positive and water-use efficiency increases with decreasing leaf size.

Parkhurst and Loucks (1972) reviewed the scanty evidence available concerning leaf sizes in different microclimates. Brown (1919) studied the vegetation and the environment of Mt. Maquiling in the Philippines in great detail. His data show a distinct reduction of leaf size from the base at 200 m to an upper altitude of 1100 m, with the mean leaf widths 0.049 and 0.031 m, respectively. Taking first, second, and third stories in a virgin dipterocarp rain-forest stand at 450 m altitude, leaf sizes increased from 0.044 to 0.061 m on the average. This was consistent with observations by Cain et al. (1956) in a Brazilian rain forest, where the taller trees tend to have smaller leaves. The ground herbs in general had quite variable leaf sizes, but the mean leaf size was 0.085 m, which is consistent with the prediction of Parkhurst and Loucks that in more shaded habitats, leaf size will be greater. It is likely that for plants growing on the forest floor, light may limit photosynthesis to such an extent that CO_2 is not limiting. In fact, through respiration and low wind movement, it is likely that CO_2 concentrations will be fairly large near the forest floor. Evaporation stress is probably low here, and leaf sizes may have undergone a genetic drift since they are not under a continuing environmental pressure. This could account for the large variability of leaf size in the understory.

It is generally recognized that both desert and Mediterranean vegetation has mostly small leaves. These are habitats of intense radiation. Alpine and arctic habitats are cold and, as mentioned earlier, are characterized by plants with small leaves. Cushion plants of arctic and alpine environments have apparently violated this rule and have low convection coefficients, since they have rather substantial boundary layers near the ground surface. However, these plants make use of a strategy by which they increase their temperatures by increasing the amount of absorbed radiation, thereby increasing their productivity under predominantly cold conditions. Parkhurst and Loucks comment that the desert succulents have crassulacean acid metabolism, which allows CO_2 uptake at night when the stomates are open and photosynthesis to assimilate the stored carbon during the day when the stomates are closed. In this respect, the succulent plants act effectively as shade plants, for which the model predicts large photosynthesizing surfaces.

I should like to quote from Parkhurst and Loucks (1972) concerning the enormous utility of this analysis for the breeding of crop species:

> Finally, the results of this work may be useful to plant breeders, who should be able to improve water-use efficiencies of crop species by producing plants with a given total leaf area divided into smaller units. Also, when crops are grown under glass or plastic, ventilation rate should be high (because high air speeds have effects similar to small leaves). Exceptions would include crops such as coffee, with its large leaf adapted to shade, and tobacco, when production of large leaves is desirable.

Optimizing Leaf Properties

It is possible to optimize leaf properties other than leaf size. One can study the variation of T, P, E, and P/E with respect to diffusion resistance or amount of absorbed radiation. The best way to do this is mathematically, in terms of partial derivatives, by means of a procedure similar to that just outlined with respect to the convection coefficient (which includes the leaf-dimension effect and the wind speed). Rather than using this approach, S. E. Taylor (1975), who did his Ph.D. work in biophysical ecology at the Missouri Botanical Garden and at Washington University, preferred to create two-dimensional diagrams of the relationship between some of the dependent variables and two of the independent variables. He selected leaf dimension as ordinate and leaf diffusion resistance as abscissa. The diagram is then drawn for a fixed set of values of Q_a, T_a, V, and h. Such a graph is a cross section through a leaf parameter–environmental variable hyperspace. In this case, the cross section is only cut through the leaf-parameter portion of the hyperspace.

From the energy budget, the leaf temperature and transpiration rate are determined for a given set of environment conditions. From the quadratic equation which expresses the photosynthetic rate for a leaf (see Chapters 3 and 14), one can calculate P. Taylor (1971) did this using the simple model of a nonrespiring leaf. He derived the diagrams shown in Fig. 11.9 for a warm, moist, sunny tropical environment with a moderate amount of wind. The conditions were $T_a = 30°C$, $h = 0.80$, $V = 0.50$ m s^{-1}, and

$Q_a = 840$ Wm^{-2}. It is clear from Fig. 11.9 that transpiration and photo-synthesis are influenced very little by leaf dimension, but strongly by leaf resistance. Photosynthesis is slightly more influenced by leaf resistance than transpiration. Leaf temperature is strongly affected by leaf dimension, particularly when diffusion resistance is large. Diffusion resistance

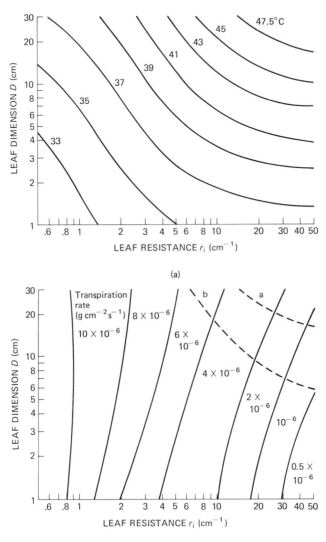

(a)

(b)

Figure 11.9. The dependent variables (a) leaf temperature, (b) transpiration rate, (c) (next page) photosynthetic rate, and (d) (next page) water-use efficiency ratio P/E for hot, humid, low wind, and high radiation conditions as a function of leaf dimension and diffusion resistance. Air temperature, 30°C; relative humidity, 80%; wind speed, 0.5 m s^{-1}; amount of radiation absorbed, 837 Wm^{-2}. (Redrawn from Taylor, 1971.)

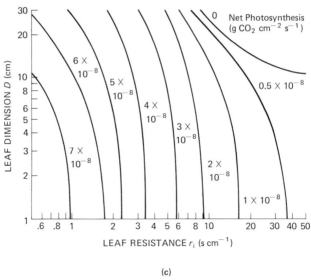

(c)

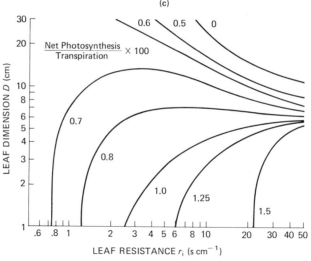

(d)

Figure 11.9. (*continued*).

has less influence for small leaves than for large leaves. The leaf-temperature contour at 47.5°C is considered to be the thermal death limit. In the warm, moist, sunny tropical environment, the water-use efficiency ratio P/E is strongly diffusion-resistance influenced for leaves of small dimensions, with increased efficiencies at high resistances. For leaves of intermediate size, there appears to be no change of water-use efficiency with changing diffusion resistance. This graph suggests that, for greatest water-use efficiency, leaves of exposed, humid tropical habitats should have relatively high resistance and small-to-intermediate leaf size; however, this would result in reduced photosynthesis. If ample moisture is available, selection for increased photosynthesis might occur; low resistance and small leaf size would be indicated in this case, although small-to-intermediate leaf size probably would result.

A similar analysis indicates that for a typical desert situation, both photosynthetic productivity and the water-use efficiency ratio produce pressure for small-sized leaves. When water is unavailable, the effective leaf resistance is high and maximum water-use efficiency ensues. Since the leaves of desert plants are small, leaf temperatures are tightly coupled to air temperature. However, desert air temperatures are usually between 40° and 47°C, and leaf temperatures generally do not exceed 50°C. Usually, under very hot conditions, desert plants shed their leaves until more favorable moisture and temperature conditions occur.

It is clear that diagrams such as these are exceedingly revealing and, in addition, are consistent with the results predicted by Parkhurst and Loucks (1972). Neither Taylor (1971, 1975) nor Parkhurst and Loucks used the complete mathematical description of photosynthesis as described in Chapter 14. However, Taylor's model was somewhat more refined than that of Parkhurst and Loucks. There is no doubt that the general results presented here are correct and consistent with field observations; a more elaborate description of photosynthesis would result only in refinement of the model.

Adaptation to High Radiation

Many plants exhibit a remarkable ability to cope with intense amounts of radiation by means of leaf orientation. Once one learns to recognize this behavior, it is astounding how frequently it is observed. This phenomenon was mentioned in Chapter 3 for the leaves of *Erythrina indica,* which were observed and measured by Gates (1968) in Australia. The leaves of *Arctostaphylos-uva-ursi* on the sand dunes of Michigan are always vertical during the day. Although these leaves are very small, they also utilize the vertical advantage to reduce the amount of absorbed solar radiation. The leaves of various species of *Manzonita* growing in arid western habitats of the United States are vertical during daylight hours, yet often have diameters of a few centimeters. The leaves of many desert species, which are of sizes less than 0.01 m, are usually vertical during midday. These include creosote bush (*Larrea divericata*), cat's-claw (*Mimosa biuncifera*), mesquite (*Prosopis glandulosa*), and many *Acacias*. There are certainly many exceptions to the vertical habit among desert plants, and yet the fact that the leaves of all desert plants are small is the first-order adaptation. All of the succulents have essentially a vertical habit. The blades of various prickly pears (*Opuntia*) are nearly always vertical. The various barrel cacti have vertical sides and a dense layer of thorns on top. The saguaro (*Carnegiae gigantea*) is perfectly vertical and protects its rounded top with a dense layer of thorns. The lechuguilla (*Agave lechuguilla*) is a fibrous, swordlike plant with a needle on the end of the leaf. These long leaves are always nearly vertical or within 30° of vertical. So also are the parts of the century plant (*Agave scabra*). The flower stalks of all of these plants are strictly vertical. Other plants of dry regions have geometries of vertical habit in adaptation to intense radiation. A Mexican juniper, known as drooping juniper (*Juniperus flaccida*) has its leaves hanging very nearly vertical.

Taylor (1971) studied the vertical behavior of several different plant species. He observed the leaf temperatures of *Erythrina berteroana* growing in a greenhouse at the Missouri Botanical Garden when the air temperature was 26°C, relative humidity 0.35, wind 1.12 m s^{-1}, solar radiation 730 Wm^{-2}, floor temperature 48°C, and plate-glass-ceiling temperature 32°C. The plant naturally orients its leaves vertically with strong daylight illumination. He forced some leaves to be horizontal while in full sunlight. Other plants he kept in the shade. He found the horizontal leaves to be 35.3°C, the naturally oriented vertical leaves 33°C, and the shade leaves 24°C. The vertically oriented leaves were almost parallel to the sun's rays. The vertical sunlit leaves transpired nearly eight times less water than the horizontal leaves.

A well-known plant in gardens is the redbud (*Cercis canadensis*). Its sunlit leaves hang nearly vertical, whereas its shade leaves are horizontally oriented. Taylor (1971) investigated the temperatures and transpiration rates of redbud at the Missouri Botanical Garden for moist and dry soil conditions. He forced some of the vertical sunlit leaves to a horizontal position. During moist conditions, with the air temperature 31.9°C, the diffusion resistance to water vapor was 400 and 790 s m^{-1} for vertical and horizontal sunlit leaves. Leaf temperatures were 35.8° and 42.1°C, and transpiration rates were 5.68 and 4.84 $\times$ 10^{-5} kg m^{-2} s^{-1}. During dry conditions, with the air temperature 30.1°C, diffusion resistances of vertical and horizontal sunlit leaves were 3420 and 3380 s m^{-1}, leaf temperatures were 38.6° and 43.5°C, and transpiration rates were 1.0 and 0.9 $\times$ 10^{-5} kg m^{-2} s^{-1}, respectively.

Taylor (1971) observed an interesting response for the leaves of *Erythrina alitansis* when well-watered plants in the greenhouse were moved from shade to sun with a wind speed of 1.0 m s^{-1}. Within 4 min, the most fully exposed leaves rotated to near-vertical orientation. Another plant was placed in the sun, but under a neutral gray-plastic filter. The temperatures of all leaves of both plants were 42°C when the air temperature was 34°C. When the speed of the wind, created by a fan, was reduced from 1.0 to 0.5 m s^{-1}, the leaves were immediately warmer, and yet within 3 min had all returned to 42°C. Sunlit leaves held horizontal by the experiment then rose to 47°C. This indicates once again that leaves have a remarkable ability to control their temperatures and usually do so. Taylor tried the same experiment, placing plants in direct and filtered sunlight, with burdock (*Arctium minus*). Within 5 min, the sunlit leaves, which had initially heated above the temperature of the partially shaded leaves, had cooled to nearly the same temperature, where they remained. The internal diffusion resistance of a sun leaf was about one-half that of a shade leaf, and transpiration was almost twice as great. The plants were well watered. Clearly, these plants were regulating their leaf temperatures through the use of water. We are reminded of the example of *Mimulus* described in Chapter 3.

Problems

1. A leaf has a time constant τ of 10 s for a given set of environmental conditions of net radiation and convection. If its initial temperature is 18°C and its final temperature 26°C, how long does it take the leaf to reach a temperature of 22°C?

2. A leaf absorbs an amount of radiation of 1000 Wm^{-2}. The air temperature is 40°C, the relative humidity 0.60, and the wind speed 0.1 m s^{-1}. The leaf is 0.05×0.05 m and has a leaf diffusion resistance of 1000 s m^{-1}. Determine the leaf temperature and the rate of water loss. Use the tables of Gates and Papian (1971) if available.

3. For the same leaf and environmental conditions as in Problem 2, determine the leaf temperature and rate of water loss if the leaf opens its stomates more fully and total leaf resistance is 200 s m^{-1}.

4. For the same leaf and environmental conditions as in Problem 2, let the relative humidity be 0.20 instead of 0.60 and determine the leaf temperature and rate of water loss with a leaf diffusion resistance of 1000 s m^{-1}. Now make the same determination for a leaf resistance of 200 s m^{-1}.

5. For the same leaf parameters and environmental conditions as in Problem 2, have the wind speed change from 0.1 to 1.0 m s^{-1} and determine the leaf temperature and rate of water loss.

6. Determine the rate of water loss from a leaf as a function of wind speed for the following environmental conditions: amount of radiation absorbed, 400 Wm^{-2}; air temperature, 30°C; and relative humidity, 40%. The leaf is 0.05×0.05 m and has a leaf resistance to water vapor diffusion of 100 s m^{-1}. Use the full range of wind speeds given in Gates and Papian (1971), i.e., 0.1, 0.2, 0.5, 1.0, 2.0, 4.0 and 8.0 m s^{-1}. Plot transpiration rate versus wind speed. Repeat for Q_a equal to 600 and 800 Wm^{-2}. Why does the rate of water loss increase with increasing wind speed in some instances and decrease in other instances?

Chapter 12

Energy Budgets of Animals

Introduction

Animals have evolved and adapted to live in all parts of the world. Lower forms of life, such as bacteria, are found in extreme high-temperatures habitats such as hot springs (Brock, 1967; Bott and Brock, 1969) and places of intense cold such as Antarctica (Cameron, 1971). An interesting discussion of microbial growth under extreme conditions is given by Brock (1969). Insects are found in all habitats of the world, including very high elevations on mountains and very dry desert regions, with the exception of the Antarctic interior. Vertebrates are also widely distributed throughout the world, but are somewhat more restricted in occurrence along the edges of extreme habitats than invertebrates. Birds have been seen flying over the South Pole (probably skuas) and over the highest peaks of the Himalayas. These sightings are rare, however, and vertebrate animals are not normally found in such extreme habitats. Nevertheless, one is immensely impressed with the adaptability of animals and their ability to fill vacant niches wherever they occur.

One is also impressed by the fact that animals, whether living in the desert, tundra, rain forest, boreal forest, savanah, or taiga, are often subjected to considerable extremes of climate and other environmental factors. How do animals cope with extreme conditions? What physical or physiological properties permit them to survive environmental extremes? How does an animal behave when subjected to harsh conditions? Where does it go or which microhabitats does it seek? The answers to these questions are sought by the physiologist and ecologist. There are several methods of approach. No matter which is used, the scientist must have considerable information concerning the animal's habitat and the environmental conditions, as well as information about the animal itself. In addition, animal–animal interactions must also be considered, whether predator–prey, male–female, or host–parasite. This book does not concern itself with the biology of animals per se, but with animal–environmental interactions and animal–animal interactions to the extent that they are affected by the environment. The book is written on the premise that environmental factors are very important force functions on an animal and that a considerable amount of animal distribution, behavior, movement, and development relates to the interactions, or coupling, between an animal and its environment.

Emphasized here is a largely mechanistic approach to animal ecology based on analysis of mass and energy transfer between an animal and its environment. Physical factors such as wind, water, radiation and temperature have been described in great detail, along with the manner by which these factors transfer energy or mass to or from an animal. The task in this chapter is to bring these factors together in an integrated fashion since they act simultaneously in affecting the energy status and responses of an animal. Analysis is in terms of the energy budget of an animal, which also includes some aspects of a mass budget. This treatment is based on the premise that a thoroughly analytical, mechanistic approach will provide a useful framework for understanding these complex processes involving many variables and parameters acting simultaneously and also the conviction that without such an analytical framework, certain important questions will never be answered and certain insights never gained. The material contained in this chapter represents only the first step toward an analytical treatment of these difficult questions. Improved understanding will evolve in the future. It is the establishment of a sound methodology that is particularly necessary at this time.

The analysis of animal energy budgets presented in this chapter uses fundamental physical principles, combined with physiological information, to create a mathematical formulation concerning heat flow between an animal and its environment. It is useful to quote from the introduction to a recent paper by Bakken and Gates (1975) concerning the philosophy of this kind of analysis:

> Mathematical modelling studies in science are not an end in themselves, but rather a tool. The fundamental purpose of mathematical analysis is to provide generalized, intellectually tractable insight into the operation and interaction of the complex factors involved in the physical and biological process under study. Too often contradictory requirements must be met. First, the model must include all relevant factors, with sufficient detail to give predictive precision for validation and practical applications of the analysis. Second, the analysis must be simple enough to be intellectually tractable and give a clearer subjective understanding than the raw data and knowledge of the individual processes involved. The second requirement implies that the model be constructed so that it is analytically soluble. For the complex nonlinear processes in biology, this is seldom possible without sacrificing precision. This is certainly true in thermal analysis of animals.

Early in the book, the concept of animal energy budgets was introduced and the energy budget was reduced to its most simplistic form. Energy budgets are simplified in two ways. The first method considers non-time-dependent, steady-state situations; the second treats an animal in terms of a lumped-parameter model. Both of these simplifications, although inherently wrong in detail, provide a means to make a reasonable approximation to a difficult problem. Yet once the simplistic analysis is accomplished, it is relatively straightforward to build onto it the more detailed and difficult aspects of the analytical model. Animals are frequently moving about their habitats or else environmental conditions are changing about the animal. In either case, animal–environmental interactions are time dependent. Time-dependent analysis will be treated later.

An animal has a torso, head, appendages, and other anatomical parts. To consider an animal as a single geometrical object of one size, shape, color, insulation, and so on is clearly not correct, but it is a useful way to begin an analysis. A more detailed treatment of the energy-budget analysis of an animal will be given later in the chapter. If the fully realistic analysis is made in the beginning, it is somewhat overwhelming in its complexity. The success of the simplified approach of a steady-state, lumped-parameter model, as tested against observations of animals in nature have been great. It is hoped that additional success will be achieved as time-dependent, distributed-parameter models are fully developed. Any science must tolerate the inadequacies of its methodology while it strives for greater precision and predictability. It is with this in mind that the simplistic analytical methods are set down here. More detailed and elaborate methods will follow.

Steady-State Energy Budgets

For purposes of heat transfer, an animal may be thought of as cylindrical in shape, with a central core at a body temperature T_b surrounded by an insulating layer of fat, fur, or feathers of insulation quality I, or thermal conductance $K = 1/I$. Inside the body core, there is net heat production at the rate $M - \lambda E_b$, where M is the metabolic rate, E_b is the evaporative water loss by respiration, and λ is the latent heat of water. The cylindrical surface is the surface between the animal's exterior and interior. Radiation is exchanged at the surface, where direct sunlight, scattered skylight, reflected light, and thermal radiation from ground or sky are absorbed and thermal radiation is emitted by the skin or pelage according to the fourth power of the absolute temperature T_r. The skin or pelage surface is also the transfer point for the convective exchange of heat between the animal and the surrounding air or water. The body core may lose or gain heat, and its temperature T_b will change correspondingly. Heat flow through fur or feather pelage is a fairly complex process and will be discussed later in the chapter. In this simple cylindrical/mechanical model, heat flow includes conduction, convection, and radiation.

It is easy to conceive of an electrical analog to the mechanical heat-flow model. Such an analog is shown in Fig. 12.1. The body core of heat capacity C is shown, with net metabolic heat production $M - \lambda E_b$ and temperature T_b. The skin surface is shown at temperature T_r, with an evaporative water loss E_r. In this simple model, the surface temperature T_r represents the external surface of the animal, whether it is skin or pelage. The external streams of heat flow are shown as incoming radiation absorbed Q_a, emitted radiation $\epsilon\sigma T_r + 273)^4$, and heat exchanged by convection $H(T_r - T_a)$. The convection coefficient is H. By Kirchoff's law, all heat flow (or electric current) at any point in the circuit must add up to zero; in other words, there must be conservation of energy.

At the body core, there is heat lost or gained from storage $C\, dT_b/dt$, heat conducted out through the body wall $K(T_b - T_r)$, and heat production $M - \lambda E_b$. Hence

$$M - \lambda E_b = K(T_b - T_r) + C\, dT_b/dt. \tag{12.1}$$

Conservation of energy at the body surface requires that

$$Q_a + K(T_b - T_r) = \epsilon\sigma(T_r + 273)^4 + H(T_r - T_a) + \lambda E_r \tag{12.2}$$

It is now possible to solve Eq. (12.1) for T_r and substitute into Eq. (12.2) to get

$$Q_a + M - \lambda E_b - C\, dT_b/dt = \epsilon\sigma(T_b - \frac{M - \lambda E_b - C\, dT_b/dt}{K} + 273)^4$$

$$+ H\left(T_b - T_a - \frac{M - \lambda E_b - C\, dT_b/dt}{K}\right) + \lambda E_r. \tag{12.3}$$

This differential equation is a lumped-parameter, isothermal model of the time-dependent energy budget of an animal. It is possible to solve this equation for body temperature as a function of time, but because of the nonlinearity of the equation, the procedure is difficult [the reason given

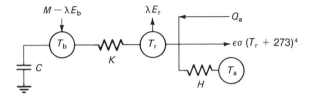

Figure 12.1. Electrical analog of the heat-flow model for an animal: C, body heat capacity; T_b, body temperature; M, metabolic rate; E_b, evaporative water loss rate by respiration; K, thermal conductivity of body; T_r, skin surface temperature; E_r, evaporative water loss from skin; Q_a, incoming radiation absorbed; $\epsilon\sigma T_r^4$, radiation emitted from skin surface; H, convection coefficient.

later in the chapter for linearizing Eq. (12.3)]. This expression contains all of the environmental factors which explicitly or implicitly affect the energy status of an animal in any given environment. The convection coefficient H as a function of the wind speed and the characteristic dimension of the animal is discussed in Chapter 9. The radiation term Q_a contains all the incident fluxes of radiation, the absorptance of the animal's surface, and the geometrical exposure factors. Air temperature enters the convection term explicitly. Air humidity is contained implicitly in the evaporative-cooling terms and will be discussed later in this chapter. The various physiological properties of the animal (M, E_b, E_r, C, K, and T_b) are all contained explicitly in the equation. The quantities M, E_b, and E_r may be time dependent, in which case they complicate the solution of the equation.

For the moment, the mathematical analysis may be simplified by assuming a steady-state condition for which no terms are time dependent and $C \, dT_b/dt = 0$. In this case, Eq. (12.3) becomes

$$Q_a + M - \lambda E_b = \epsilon\sigma\left(T_b - \frac{M - \lambda E_b}{K} + 273\right)^4$$

$$+ H\left(T_b - T_a - \frac{M - \lambda E_b}{K}\right) + \lambda E_r. \quad (12.4)$$

This is the steady-state equation used by Porter and Gates (1969), Gates (1970), Heller and Gates (1971), Spotila et al. (1972), and Morhardt and Gates (1974) in a series of papers published concerning animal energy budgets except that insulation $I\,(= 1/K)$ is used rather than K. This is the same as Eq. (4.4), in which $H = kV^{0.5}/D^{0.5}$ and $E = E_b + E_r$.

The steady-state equation is easy to solve for an unknown if the values of all other quantities are given. For a given animal, the values of M, E_b, E_r, T_b, K, and H are known or can be estimated reasonably well for various conditions. Then, for any value of Q_a in Eq. (12.4), there is a specific value of T_a which will balance the equation. The quantities M, E_b, and E_r vary with the so-called "environmental" temperature of an animal (usually the air and wall temperature of a metabolic chamber into which an animal is placed).

The ability to predict any combination of Q_a and T_a for the values of M, E_b, E_r, K, and H at thermal maximum and thermal minimum makes it possible to define the environmental limits in a climate-space diagram for any given animal. This type of diagram was described in Chapter 4. The climate-space for an animal is that portion of a dimensional space consisting of radiation absorbed, air temperature, wind speed, and humidity within which an animal can survive. The earlier description of climate-space limited it to three dimensions, i.e., radiation absorbed, air temperature, and wind speed. The humidity factor will be brought into the analysis later. After a brief discussion of metabolic rates, evaporative water loss, insulation, convection, and radiation absorbed, a series of examples of climate-space diagrams for various animals will be given.

In the scientific literature, various linearized expressions of the energy budgets of animals are given which are approximations to Eq. (12.4). In another section, Eq. (12.4) will be linearized and these approximations discussed. For the time being, however, it is important to work with Eq. (12.4) as given, since it shows explicitly the mechanisms by which energy is transferred between an animal and its environment. Some of these processes are distinctly nonlinear functions. When a linearized analysis is used without reference to the nonlinear functions, there is a tendency to forget the physics involved. In some of the physiological and ecological literature, linearized heat exchange is used to the point of giving invalid results and the true mechanisms of heat transfer are not addressed. In the steady-state situation, it is quite easy to work with Eq. (12.4) for the solution of certain types of problems. For other situations, there is justification for linearizing the energy-exchange expression. These situations will be discussed later in the chapter.

Before applying energy-budget analysis to specific cases, it is important to consider the various animal parameters that enter the analysis. The important parameters are body temperature, metabolic rate, evaporative water-loss rates, thermal conductance or insulation of fat, skin, fur, or feathers, characteristic dimension in the convection term, geometrical-shape factors, and absorptance and emittance of the animal's surface. Discussion of physiological parameters such as metabolic or evaporative-water-loss rates will be minimal since information concerning these parameters is readily available elsewhere (Kleiber, 1961; Schmidt-Nielsen, 1964; Whittow, 1970; Maloiy, 1972; and others).

Metabolic Rate

Animals fall into several classes with regard to control of body temperature. Poikilotherms are those so-called cold-blooded animals whose body temperatures or energy status follows closely the temperature or energy regime of the environment. Among these animals are all invertebrates (including insects), reptiles, fish, and amphibians (with a few exceptions). Homeotherms are the warm-blooded animals; these animals maintain a reasonably constant temperature over a wide range of environmental conditions. Some animals are warm-blooded part of the year and hibernate during cold periods. Then there are the altricial birds, which are naked when emerging from the egg and do not possess much ability to thermoregulate, or the India python, *Python moluris*, which can raise its metabolism and body temperature while incubating its eggs. The metabolic rates of various animals differ enormously and depend on size, environmental conditions, environmental history, class of thermoregulation, and so on. Certain generalizations are possible, however, and these are important to our discussion concerning animal energy budgets.

Poikilotherms

Metabolic rates of poikilotherms increase monotonically with body temperature and, like many biochemical processes, are described by the rule of van't Hoff. Poikilotherms have metabolic rates, or rates of oxygen consumption, that increase with increasing temperature, unlike those of homeotherms, which decrease with increasing temperature until thermoneutrality is reached and increase thereafter. The rate at which a chemical reaction proceeds at any given temperature is

$$k_T = B \, e^{-A/RT}, \tag{12.5}$$

where B is a constant (except in the cases where it varies with temperature), A is the activation energy of reacting molecules, and k_T is the velocity constant, or rate, of the particular reaction. When applied to a complex biological process, the coefficients do not have specific chemical significance except that the exponential formula empirically describes the overall reaction. From this expression, biologists have found it convenient to define a temperature coefficient Q_{10}, which is the factor by which the rate of activity (in this case, oxygen consumption) increases with each 10°C rise in temperature:

$$Q_{10} = \frac{k_T + 10}{k_T} = \frac{B(T + 10)}{B(T)} \, e^{[10A/RT(T+10)]}. \tag{12.6}$$

If the coefficient B varies with the square root of T, then one gets

$$Q_{10} = \sqrt{\frac{T + 10}{T}} \, e^{[10A/RT(T+10)]}. \tag{12.7}$$

The Q_{10} values for the metabolic rate of poikilotherms vary from about 1.3 to 4.0. In general, chemical reactions have a Q_{10} between 2.0 and 4.0, whereas physical processes such as gas diffusion and osmosis have values between 1.1 and 1.4. Generally, Q_{10} values decline with increasing temperature.

Alligators have a Q_{10} of about 1.3, whereas the rattlesnake *Crotalus atrox* has about 3.9. Many lizards have Q_{10} values from 2.5 to 3.3. Among reptiles, metabolic rates vary enormously with temperature. According to Templeton (1970), the metabolic rates of some lizards are altered by exposing them to constant but different temperatures. Lizards acclimated at higher temperatures, e.g., 35°C, had lower oxygen consumption than those acclimated at lower temperatures, e.g., 15°C. Thermophilic lizards, such as the desert iguana and collared lizard, consumed less oxygen than the skink and the alligator lizard at 39° or 40°C. The alligator lizard is relatively insensitive to low temperatures and has a Q_{10} of 3.1 at 5°C, whereas the collared lizard has a Q_{10} of 7.7 at the same temperature. An overwintering beetle, *Melasoma populi,* has a very low Q_{10} value, which would be to its advantage since it must subsist on fat reserves yet sustain considerable temperature variation. Overwintering potato beetles, living in deep burrows where temperature variations are very small, had a higher Q_{10}.

Metabolic heat production by insects can raise their body (thoracic) temperatures by as much as 25°C. According to Heinrich (1974), most flying insects have a higher rate of metabolism than other animals (on the basis of body weight). Insect metabolism is discussed in more detail under body temperatures.

Homeotherms

Metabolic rates in homeotherms vary enormously and depend upon several factors. Nevertheless, for any given animal, the metabolic rate under specified conditions can be determined, and its variation with changing conditions can be described. According to Gessaman (1973), four kinds of metabolism are distinguishable on the basis of air temperature, level of animal activity, absorptive state of the animal, and time duration of the measurement. Basal metabolism and standard metabolic rate represent the minimal value of heat production as measured within the zone of thermoneutrality with the animal at rest and in a postabsorptive state. The resting metabolic rate is measured at temperatures below the zone of thermoneutrality when the animal is at rest but not postabsorptive and is therefore greater than the basic metabolism rate owing to heat liberated from the digestion of food and the occurrence of some thermoregulation. In addition, there are measures of the average daily metabolic rates which include basal metabolism, thermoregulatory metabolism, specific dynamic action, and activity metabolism. Finally, there is what is known as existence metabolism, which is used as an estimate of free-living metabolism. It is not the purpose here to get involved in the complications of metabolic physiology or with definitions, except that a general notion of the metabolic rates important to problems of energy exchange, particularly as they pertain to free-living animals.

Figure 12.2a is an idealized diagram of the metabolic rate of a typical homeotherm as a function of environmental temperature. The horizontal portion AB between the lower critical temperature $_lT_c$ and the upper critical temperature $_uT_c$ is the thermoneutral range, over which heat production is constant with variable environmental temperature. Body temperature may vary throughout the thermoneutral range. Within the thermal tolerance range (TOL), there are no lethal effects of temperature, no matter what the exposure time. Above and below the thermal tolerance range are zones of resistance (RES), where death occurs as a function of time. Acclimation of an animal to a lower or higher environmental temperature

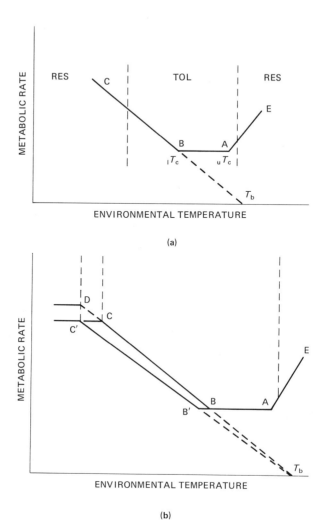

Figure 12.2. Idealized diagrams of metabolic rate versus environmental temperature for homeotherms when (a) normally acclimated, (b) cold acclimated, and (facing page) (c) heat acclimated. The symbols ℓT_c and $_u T_c$ represent the lower and upper critical temperatures. TOL is the thermal tolerance range; RES are zones of resistance, in which thermal death occurs given sufficient time.

will shift the metabolism–environmental temperature response curve. This is shown in Fig. 12.2b. When an animal acclimates to cold, there is a change in slope of the line BC to something resembling B′C′ as a result of a change in thermal insulation owing to increased fat, fur, or feathers (down). An increase in maximum heat production (as shown by the horizontal line at C and C′) may not accompany this change. The other adjustment an animal may make to cold acclimation is to maintain the same conductance (the slope of line BC) and achieve a greater maximum heat

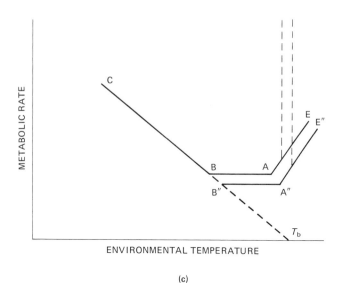

ENVIRONMENTAL TEMPERATURE

(c)

Figure 12.2 (*continued*).

output (shown by the horizontal line at D). Cold acclimation produces a shift to colder temperatures of the tolerance zone TOL. When there is acclimation to heat, some homeotherms elevate the upper critical temperature $_uT_c$ and lower the metabolic rate of the thermoneutral zone from AB to A″B″, as shown in Fig. 12.2c. Heat acclimation produces a shift to warmer temperatures of the tolerance zone TOL. Sometimes, heat acclimation produces greater resistance to hyperthermia and is accompanied by an increase in body temperature. There is a considerable amount of variability from animal to animal within a given species, and of course various species differ in their responses, adaptability, acclimation, acclimatization, and so on.

Animal physiologists have long taken a simplified view of heat transfer between animals and their environments. If an animal is exchanging heat with its environment in a purely linear manner proportional to the difference between body temperature T_b and environmental temperature T_e, then

$$M - \lambda E = K(T_b - T_e). \tag{12.8}$$

This expression was often used in the early physiological literature to suggest that the slope of the line BC in Fig. 12.2 is the total thermal conductance K of the animal. This can only be true if the environmental conditions are such that the surface temperature of the animal is equal to the environmental temperature; otherwise, Eq. (12.8) provides only a very crude approximation to total conductance. Total conductance for heat flowing from the body core to the external environment is made up of conductance between the body core and the animal's surface and that from the surface to the environment. Conductance of the body shell, which involves layers of fat, skin, fur, or feathers, may be quite linear. Conductance from the animal's surface to the environment is usually very nonlinear, with temperature difference as given by Eq. (12.4). Tracy (1972) has discussed this problem in some detail, showing, in particular, the effect of wind on surface conductance. The surface conductance contains transfer of heat by radiation and by convection, a strongly nonlinear function. The radiation term involves the fourth power of the absolute temperature of the surface. The convection component involves the square root or some other power of the wind speed. McNab (1970) recognized some of the difficulties involved in the application of a linearized heat-transfer law but then combined it with various parametric expressions which add complications. Usually, metabolic rates are determined with an animal in a chamber with walls of uniform temperature. The animal's surface temperature is usually slightly above the wall temperature, with the air temperature intermediate. Porter (1969) has discussed the problems associated with metabolic chambers and shown that, for meaningful metabolic measurements to be obtained, the chamber wall should be "black" to infrared radiation or else be as large as possible compared to the size of the animal. Otherwise, radiant heat emitted by the animal reflects from the chamber walls and adds to the energy budget of the animal, thereby causing it to reduce its metabolic output.

Thermal Conductance

The thermal conductance of the animal shell from body core to surface can be determined accurately if the surface temperature of the animal is measured. Surface temperature can be measured with an infrared radiometer. Animal surface temperature often vary considerably because of the irregular and nonhomogeneous nature of an animal's body and the nonisotropic and variable environmental conditions that may prevail around the animal. In a metabolic chamber, however, the environment can be maintained in a uniform and steady state for long periods of time, and animal surface temperatures can be reasonably uniform. As suggested by Morhardt and Gates (1974), the term conductance is appropriately applied to the transfer of heat between an animal's body core and its surface since this transfer occurs primarily by conduction and depends on the thickness and conductivities of the tissue and fur or feathers. Porter and Gates (1969) discuss these conductances in some detail. The conductance of an animal shell to a first approximation is given by the equation

$$M - \lambda E = K_1(T_b - T_r). \qquad (12.9)$$

Conductance and Body Temperature

The more complete treatment of heat transfer from an animal as partitioned among various flow processes will be given later in the chapter. According to Eq. (12.9), if the dry metabolic heat production $M - \lambda E$ is plotted against surface temperature T_r, the slope of a line similar to BC of Fig. 12.2a will be the shell conductance K_1, and the intercept with the abscissa will be the body temperature T_b. If only the metabolic rate M is known, then the intercept will be an overestimate of body temperature, but the overestimation will be small if the evaporative water loss E is small. Often, when determining conductances, the body temperature is measured, and the line BC is drawn through the known body-temperature intercept. Unfortunately, many workers do not determine body temperature when measuring metabolic rates. Usually, when conductance is determined from measurements of metabolic rates, only the lowest or min-

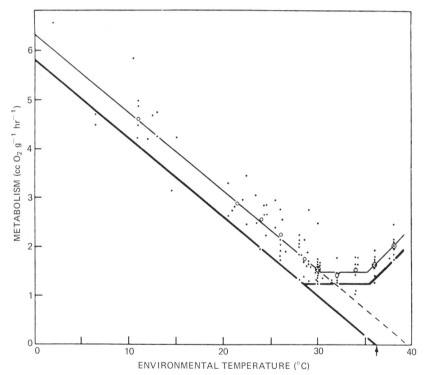

Figure 12.3. Minimal rates of metabolism of *Peromyscus eremicus eremicus* in relation to environmental temperature. Points represent the minimal rate from an individual experiment; circles represent the mean of these minimal values at the specified temperature. The arrow below the abscissa is the mean of the body temperature taken after experiments below thermoneutrality. Mean weight of animals, 21.5 g; N, 109. (Redrawn from McNab and Morrison, 1963.)

imum values are used. Then a line equivalent to BC in Fig. 12.2a is drawn such that it intercepts the body temperature and has less than about 10% of the points falling below it. An example from an excellent paper by McNab and Morrison (1963) is shown in Fig. 12.3. The thin line represents the average of all the points and does not pass through the body-temperature intercept. The heavy line has the same slope as the light line and passes through the body-temperature intercept. (The literature of animal metabolism, however, is variable with respect to credibility, and the reader will need to keep this in mind.) Morhardt and Gates (1974) discuss the determination of surface temperature and its use in the determination of conductance.

Series Conductances

Series conductances K_1 (internal) and K_2 (external) can be treated as follows. The amount of heat flowing from the body core to the surface is given by Eq. (12.9). The same amount of heat is flowing from the surface to the environment:

$$M - \lambda E = K_2(T_r - T_e). \tag{12.10}$$

If Eq. (12.9) is divided by K_1 and Eq. (12.10) is divided by K_2 and the resulting equations are added, one gets

$$\frac{M - \lambda E}{K_1} + \frac{M - \lambda E}{K_2} = T_b - T_e, \tag{12.11}$$

or

$$M - \lambda E = \frac{K_1 K_2}{K_1 + K_2} (T_b - T_a). \tag{12.12}$$

If $K_1 \gg K_2$, this equation becomes

$$M - \lambda E = K_2(T_b - T_e). \tag{12.13}$$

If $K_2 \gg K_1$, it becomes

$$M - \lambda E = K_1(T_b - T_e). \tag{12.14}$$

If Eq. (12.12) were written in terms of specific resistance to the flow of heat, or insulation, instead of conductances, the resistance of the body shell would simply be additive to the external specific resistance. It is a matter of convenience whether conductance or specific resistance (insulation) is used for the expression of heat transfer.

Biot Number

The ratio K_2/K_1 is known as the Biot number. If the Biot number is 10 or greater, then the overall conductance is approximately equal to the internal conductance, and Eq. (12.14) represents the rate of heat transfer. Most homeotherms have Biot numbers considerably lower than 10, so the combined conductances as given by Eq. (12.12) must be used. Usually, to obtain $K_2 \gg K_1$, the surface of the animal must be ventilated by strong wind. According to Eq. (9.45), the conductance of the surface boundary layer by convection for animals whose Reynolds numbers (Re) are between 40 and 4000 is

$$_cK_2 = 2.77 \ V^{0.466} \ D^{-0.534} \tag{12.15}$$

and for animals whose Reynolds numbers are between 4,000 and 40,000 is

$$_cK_2 = 4.23 \ V^{0.618} \ D^{-0.382}. \tag{12.16}$$

The quantity $_cK_2$ is in units of watts per square meter per degree Centigrade. The Reynolds number is $6.54 \times 10^4 \ VD$ in air. Hence, if $V = 1$ m s^{-1} and $D = 0.05$ m, Re $= 3270$ and Eq. (12.15) gives the conductance. In this case, $_cK_2 = 13.7$ Wm^{-2}°C^{-1}. If, instead, $V = 10$ m s^{-1}, then Re $= 32,700$ and $_cK_2 = 55.1$ Wm^{-2}°C^{-1}. For the Belding ground squirrel, Morhardt and Gates (1964) found the average internal thermal conductance $_cK_1$ to be 8.58 Wm^{-2}°C^{-1}. Although $_cK_2 > {_cK_1}$, the ratio $_cK_2/_cK_1$ ranges from 1.6 to 6.4 for wind speeds from 1.0 to 10.0 m s^{-1}. These Biot numbers are not close to 10, and therefore both conductances must be used.

Metabolism and Body Size

Weight Relationship

Heat exchange between an animal and the environment takes place through the animal's surface. One might expect, therefore, that the rate of metabolic heat production would be proportional to the surface area. Kleiber (1961) has reviewed the various theories concerning metabolic rates and surface area and analyzed a large amount of the available data. He makes the following conclusions (Kleiber, 1961, pp. 214–215):

1. Among homeotherms, from mice to cattle, metabolic rate and body size are correlated, and this correlation is specifically high when the metabolic rates are measured under standard conditions.
2. The metabolic rates of large and small homeotherms are more nearly proportional to the areas of their respective body surfaces than to their body weights. This relationship is known as the *surface law*.
3. In natural selection, those animals prove to be better fit whose rate of oxygen consumption is regulated so as to permit the more efficient temperature regulation as well as the more efficient transport of oxygen and nutrients.

4. Relatively recent results on homeotherms, ranging from mice to cattle, indicate that the metabolic rate per unit of the surface area is greater the larger the animal. A linear correlation between the logarithm of metabolic rate and the logarithm of body weight shows that metabolic rate is proportional to a given power function of body weight. The metabolic rate divided by the three-fourth power of body weight is independent of body size. The body weight in kilograms, raised to the three-fourth power, measures the metabolic body size of an animal in $kg^{3/4}$.

5. The metabolic level of an animal may be characterized as the metabolic rate per $kg^{3/4}$ per day or about 3 kcal per $kg^{3/4}$ per hour.

Kleiber wrote the relationship between "standard" metabolic rate M and body weight W as

$$M = 70 \ W^{3/4}, \tag{12.17}$$

where W is in kilograms and M is in kilocalories per kilogram$^{3/4}$ per day.

From the standpoint of heat transfer, one would not expect the rate of heat loss to be strictly proportional to the animal's total surface area. Kleiber expresses his view of this as follows (1961, p. 199): "The surface law is unreliable mainly because the definition of an animal surface is vague. But even if the surface area could be defined and measured accurately, there is no theoretical basis for the hypothesis that the metabolic rate of homeotherms should be exactly proportional to their particular surface area rather than to a more general function of body size." With the first statement I agree, but I disagree with the second statement for the following reasons. The definition and determination of animal surface area is difficult and vague, but this does not rule out the fact that there is an appropriate surface area for each process of heat transfer, radiation, convection, conduction, and evaporation. The fact that a weight-specific relationship is established as given by Eq. (12.17) and that a surface-area-specific relationship does not seem to generally apply only suggests that the surface-area relationship is more difficult to establish. One expects a priori that metabolic heat production is proportional to body weight. One also expects respiratory evaporative water loss to be proportional to body volume (lung tidal volume capacity) and hence to body weight. However, the dry metabolic heat produced ($M - \lambda E_b$) is heat which must be transferred to the environment through the surface area of an animal. Heat-transfer theory shows us the mechanisms by which this heat is dumped, i.e., radiation, convection, conduction, and evaporation. Radiation exchange between an animal surface and the environment involves only a certain fraction of the total surface area of an animal. Bartlett and Gates (1967) have given a determination of the radiant surface areas for the lizard *Sceloporus occidentalis,* using a technique established earlier by

Tibbals et al. (1964) for the radiant surface areas of conifers. The convective heat exchange for an animal involves a different "effective" surface area and is usually determined in terms of a "characteristic" dimension for the animal. Mitchell (1975) described a procedure based on the heat-transfer relationship for a sphere that gives a good description of the convective transfer of heat from an animal. This procedure is described in Chapter 9 [Eq. (9.48)].

Finally, for many animals, there is sweating, which provides evaporative water loss from a fraction of the total surface area of an animal. These evaporative areas can be readily determined. In many cases, the evaporative area is not significant. In other cases, however, such as for some amphibians, it is important. If the proper surface areas for radiative, convective, conductive, and evaporative heat exchange are properly determined, I am quite certain that the proper "law" or relationship between dry metabolic heat production and an "effective" surface area will ensue.

There is an interesting accounting for the weight-specific law for metabolic rates as given by McMahon (1973). Since this involves some biomechanics, it is described in the following section, and since the relationship applies to plants as well as animals, the botanical examples are included here as well.

Biomechanics of Body Size

Physical laws come into play when anything is formed, whether animate or inanimate. Often, it is difficult to ascertain the reasons that a certain object is the way it is, but this does not by itself deny that there are physical mechanisms influencing the form and shape of the object. And so it is with the size and shape of organisms. D'Arcy Thompson (1917) wrote extensively on the science of growth and form. An animal's legs must support the weight of its body, the mass of a tree must be supported by the trunk, and tree limbs must be capable of supporting the bending forces created by their weight. McMahon (1973) wrote a simple analysis concerning the biomechanics of these relationships and came up with a reason for the ¾-power function on the weight-specific metabolic law. The following material is taken directly from his paper; the notation used is his also.

Tree trunks, tree limbs, and animal appendages can be represented for mechanical and heat-transfer purposes as cylinders, and the mechanics of cylinders under stress gives revealing insight into the behavior of tree limbs and trunks and the sizes and shapes of animals. A tall, vertical cylinder of length ℓ and diameter d is loaded such that the total weight of the column is acting through its center of mass. This cylinder will fail in compression if the applied stress (force per unit cross-sectional area) exceeds some maximum comprehensive stress σ_{max}. If the cylinder or column is slender enough, it may fail not only by compression but also by elastic buckling, in which lateral displacement causes a toppling movement by exceeding the elastic limit of the material. (Here, "slender enough" means that $\ell/d > 25$. This ratio is exceeded by nearly all tree trunks or limbs.)

According to mechanics (see Timoshenko, 1962), the critical length for buckling is given by

$$\ell_{cr} = 0.851 \, (E/\rho)^{1/3} \, d^{2/3}, \qquad (12.18)$$

where ρ is the weight per unit volume (numerically equal to the density) and E is the elastic modulus of the material. It has been shown that if the weight is distributed over the length of the column or if the cylinder is hollow (providing wall thickness is proportional to diameter), exactly the same functional relation holds, except for a change in the numerical coefficient. It is interesting that the same is true for a cone or a paraboloid of revolution with a change in the numerical constant. A tapered column such as a slender cone can support 2.034 times more height than a cylindrical column of the same volume and material. The two-thirds law prevails in all circumstances, however.

McMahon plotted log height against the log diameter for 576 trees having very slender to very stout trunks. The data are accurately represented by the two-thirds law. He also reports that an analysis of tree limbs shows their lengths to be regulated by a two-thirds law and that the numerical coefficient is a function of the angle at which the limbs leave the main trunk. McMahon used for the tree-height calculations $E = 1.05 \times 10^5 \, \text{kg m}^{-2}$ and $\rho = 6.18 \times 10^2 \, \text{kg m}^{-3}$. He concluded that the proportions of trees are limited by elastic criteria.

The same principles and analysis can be applied to animals. A quadruped standing at rest has its legs subjected to buckling loads, whereas its vertebral column and musculature are receiving bending loads. When the animal runs, the situation is reversed: the legs are supporting bending loads, and the vertebral column is receiving a buckling load through an end thrust. The maximum deflection a structure can sustain under a sudden impulse is just twice that which it can sustain under a steady stress. Each nearly cylindrical element of the animal, whether it be legs or backbone, is subject to a combination of buckling, bending, and torsional loads. As McMahon (1973) writes, "Fortunately the elastic criteria predict the same result independently of the type of gravitational selfloading, namely that every ℓ should be proportional to the two-thirds power of the equivalent d."

The weight W of an animal is proportional to the product of the cross-sectional area of the cylinder ($\pi d^2/4$), and the length ℓ. But from the two-thirds law, $\ell = d^{2/3}$, or $d = \ell^{3/2}$. Hence

$$W \times \ell d^2 = \ell^4 \qquad (12.19)$$

and

$$\ell \propto W^{1/4}; \qquad d \propto W^{3/8}. \tag{12.20}$$

Now suppose a muscle of cross-sectional area A shortens a length $\Delta\ell$ against a force σA in time Δt. Then the power this muscle expends is equal to the work (force times displacement) done per unit time. The power expended by muscle contraction is

$$P = \sigma A \ \Delta\ell/\Delta t. \tag{12.21}$$

According to McMahon, both σ and $\Delta\ell/\Delta t$ are constant. This says that the power output of a particular muscle, and hence all metabolic variables involved in maintaining the flow of energy to that muscle, depend only on the cross-sectional area of the muscle, which, in turn, is proportional to d^2. Therefore, maximal power output is given by

$$P_{\max} \propto d^2 = (W^{3/8})^2 = W^{3/4}. \tag{12.22}$$

This is the weight-specific law of Kleiber, which states that power is proportional to work to the ¾ power. Energy expended by muscle contraction or exercise generally appears as heat within the animal tissue and must pass out to the environment through the surface area of the animal. The surface area of a cylinder is given by

$$A = \pi\ell d + \pi d^2/4. \tag{12.23}$$

For most animals, $\ell/d = 10$, hence $\pi d^2/4 \ll \ell d$, and therefore

$$A = \pi\ell d \propto W^{1/4} W^{3/8} = W^{5/8}. \tag{12.24}$$

This relation suggests that the surface area of animals is proportional to the ⅝ power of their body weight. Observational evidence concerning vertebrates seems to support this power-function relation between surface area and body weight according to McMahon (1973). Earlier workers had shown the exponent to be about 0.67. Then, since the metabolic rate was proportional to body weight to the ¾ power, these workers concluded that metabolic rate was proportional to surface area by ignoring the difference between the powers 0.67 and 0.75. However, to be exact, they should have concluded that $M \propto A^{1.12}$. By solving Eq. (12.24) for W and substituting into Eq. (12.22), one gets

$$M \propto P_{\max} \propto A^{6/5}. \tag{12.25}$$

This suggests that an animal loses heat at a rate which exceeds the linear function of the surface area. This is consistent with the fact that heat is lost from the body surface by radiation, convection, and evaporation and from the respiratory cavity by evaporation, all acting simultaneously. As a crude approximation, if all terms transferring heat from the animal are linearized to a difference in temperature between the mean surface temperature and an "environmental" temperature, it is easy to show that the proportionality factor exceeds the total surface area of the animal.

Body Temperatures

An animal's body temperature is an indicator of its overall energy level and is the result of a balance between heat gained and heat lost. Homeothermic animals generally have relatively steady body temperatures and are usually fairly warm. Poikilothermic animals have variable body temperatures that are closely coupled to the "environmental" temperature. Most poikilothermic animals have relatively little internal heat production, and their body temperatures are the result of external factors such as radiation, convection and conduction; these animals are described as ectothermic. In contrast to these are the animals (including most homeotherms) whose body temperature is substantially determined by their own oxidative metabolism; these animals are described as endothermic. Just as homeothermy is not an absolute invariant of all mammals or birds, so neither is endothermy. The term heterothermy is used to describe those birds and mammals that are not continuously homeothermic. Although the temperatures of ectotherms follow the "environmental" temperature, it is not true that their body temperatures equal the environmental temperature. As described elsewhere in this book, body temperatures of these animals may be much above, and sometimes below, the environmental temperature. Aquatic ectotherms are closer to water temperature than terrestrial ectotherms to air temperature because water has a conductivity approximately 24 times that of air.

Animals acclimate to seasonal changes of temperature over fairly extended periods of time. Many ectotherms become cold-hardened as winter approaches and heat-resistant during the summer. Insects in particular are able to extend their low-temperature limits by increasing the glycol content of their body plasma and also dehydrating during the cooling process. This process is discussed in detail in the chapter concerning temperatures (Chapter 14).

Insects, being small in size, undergo rapid temperature changes. It is easy to estimate these rates by considering an insect body to be a mass m at a specific heat c by means of standard thermal theory. If dq/dt is the rate of heat input to or from an object, then the rate of temperature change dT/dt is given by

$$\frac{dq}{dt} = mc\,\frac{dT}{dt}. \tag{12.26}$$

If metabolic heat production is 0.279 W for a 1.5-g moth (see Heinrich, 1974), and the specific heat is approximately that of water ($4.2\ \mathrm{J\,g^{-1}{}^{\circ}C^{-1}}$), then the rate of warm-up is $0.0445°C\ s^{-1}$. If the specific heat is less than $1.0\ J\ g^{-1}°C^{-1}$, say $0.7\ J\ g^{-1}°C^{-1}$, then the rate of warm-up is $0.036°C\ s^{-1}$. If the body mass is less than 1.5 g, then the rates of warm-up will be even greater. Heinrich (1974) gives values from 0.0167 to $0.150°C\ s^{-1}$ for the rates of increase of thoracic temperatures of large moths (0.5 to 4.0 g). Using Eq. (12.26), one can estimate the ratio of temperature change of very small insects. For example, if the heat input is $0.349\ Wm^{-2}$, which is a reasonable value for heating by sunlight of an insect 2 mm in diameter, then the rate of warm-up is approximately $0.333°C\ s^{-1}$ for a body mass of 4×10^{-3} g and specific heat $2.5\ J\ g^{-1}°C^{-1}$. A small insect responds very quickly to heat input or output, as even this very approximate calculation shows.

Birds tend to be warmer than mammals, their normal body temperatures ranging from about 39° to 43°C, with some a degree or so above or below this range. Mammals generally have body temperatures from 36° to 39°C. With an increase in environmental temperature or air temperature from 10° to 30°C, there is a gradual increase in body temperature of about 2°C. At ambient air temperature or chamber temperature of about 30°C, there is a sudden increase in body temperature of 3° or 4°C until thermal death ensues. Most mammals cannot withstand body temperatures above about 41° or 40°C. When calculating energy budgets of animals, one must use as nearly as possible the appropriate body temperature for the ambient conditions. For a homeotherm, one does not use a constant body temperature for all conditions but a slightly lower than normal body temperature for cool ambient conditions and a higher body temperature for warm ambient conditions. The lowest body temperature which a homeotherm can withstand is used in calculating the lower limit of its climate-space. Each animal species must be analyzed thoroughly in accordance with available physiological information, and the worker must be acutely aware of the conditions under which the data were taken. Excellent comprehensive summaries of temperature information concerning animals are found in the books by Precht et al. (1973) and Rose (1967).

Evaporative Water Loss

An animal is cooled by the evaporation of water, whether by respiratory water loss or by sweating or loss through a permeable skin. In order to evaporate water, the liquid water present must be changed to the vapor state. This requires a certain amount of energy, the latent heat of vaporization, and a diffusion gradient. The quantitative value of latent heat varies with temperature, being $2.26 \times 10^6\ J\ kg^{-1}$ at 100°C, $2.43 \times 10^6\ J\ kg^{-1}$ at 30°C, and $2.51 \times 10^6\ J\ kg^{-1}$ at 0°C. If the air surrounding an animal is very dry, then there may exist a steep gradient of moisture across the air boundary layer from the moist skin or nasal passage to the free air beyond, and evaporative cooling may occur. If the air around an animal is very humid, then there may occur little or no evaporative cooling even though energy may be available since there is little or no moisture gradient across the boundary layer. In every case, the rate of water loss is given by an expression similar to

$$E = \frac{{}_sd_r(T_r) - d_a(T_a)}{r},$$
(12.27)

where ${}_sd_r(T_r)$ is the vapor density of saturated air at the animal surface as a function of the surface temperature T_r and $d_a(T_a)$ is the vapor density of the free air beyond the boundary layer at the air temperature T_a. The quantity $d_a(T_a)$ is equal to $h\ {}_sd_a(T_a)$, where h is the relative humidity of the air and ${}_sd_a(T_a)$ is the vapor pressure of saturated air at the air temperature T_a. The resistance r of the boundary layer is in seconds per meter. The water vapor densities are in kilograms per cubic meter, and therefore E, the evaporation rate, is in kilograms per square meter per second. If the animal surface is not wet by sweat, then vapor is originating inside the pelage or skin and an additional resistance to vapor flow exists. However, one way or another, Eq. (12.27) represents evaporative water loss from an animal surface. One sees that the rate of water loss decreases with increasing humidity at constant air temperature and decreases with increasing air temperature at constant relative humidity. For fixed environmental conditions, water loss increases monotonically with increasing surface temperature of the wetted surface. Wind affects the boundary-layer resistance according to the inverse square root of the wind speed as discussed in Chapter 9. Some organisms seek wind in order to increase evaporative cooling and some avoid wind in order to conserve water. Isopods with highly permeable cuticles sometimes move to wind-exposed sites in order to increase evaporation rate.

Poikilotherms

Animals take great precautions not to lose too much water during periods of heat stress. However, evaporative cooling is an effective way to reduce body temperature for limited periods of time. Most insects, being relatively small, cannot afford to lose a great deal of moisture. Most arthropods are only able to tolerate high temperatures when in air of high humidity, which greatly holds down their water loss. Invertebrate inhabitants of deserts are usually very economical with water use, and many desert insects have highly impervious cuticles. However, as discussed in Chapter 15, these waxy cuticles may suddenly break down at warm temperatures and the permeability to water increase enormously.

The moist skin of amphibians enables them to achieve substantial evaporative cooling, and they sometimes maintain a body temperature as much as 13°C below the environmental temperature. Amphibians often protect themselves against too great a rate of water loss by seeking moist microhabitats in the soil, under leaf litter, etc. When exposed to dry air, amphibians lose water fairly freely. The main control amphibians have over water loss is through their kidneys and urinary loss. Amphibians seem to be unable to take up water from saturated atmospheres.

Reptiles generally do not regulate their body temperatures by means of evaporative cooling, although at high environmental temperatures, vigorous panting keeps body temperatures down. Some lizards are apparently able to lower their body temperature several degrees by evaporative cooling, although others cannot do so. Lizards generally control body temperatures by behavioral means, through regulation of radiation absorbed, convection, and conduction, and use very little evaporative cooling except at high temperatures.

Homeotherms

Birds and mammals have the ability to affect their body temperatures by means of evaporative cooling. Evaporative water loss is generally achieved by means of panting, but for some animals, such as humans, sweating is an important mechanism. A few birds have the means for cutaneous loss of water. The relative importance of sweating and panting among some animals is shown in Fig. 12.4. Animals which have a high capacity for sweating normally have a low capacity for panting. However, there is generally a great deal of variation in these factors among the species of a given animal group.

In birds, evaporative water loss, on a weight-specific basis, varies inversely with body weight as given by the following relations from a review by Dawson and Hudson (1970). For birds generally,

$$E = 0.432W^{0.585}, \tag{12.28}$$

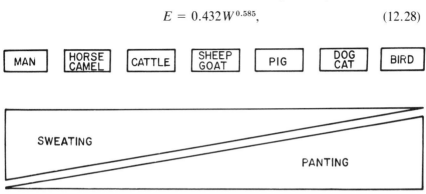

Figure 12.4. Relative sweating and panting by various animals. (Redrawn from Hafez, 1968.)

and for nonpasserines,

$$E = 0.351W^{0.613}, \tag{12.29}$$

where E is evaporative water loss in grams per day and W is weight in grams.

At moderate temperatures (20° to 30°C), evaporation accounts for between one-third and one-tenth of the total heat loss from resting birds. Generally, evaporative water loss increases monotonically with environmental temperatures from a relatively low rate at low temperatures to steeply increased rates at high temperatures, as shown in Fig. 12.5 for the sparrow. Most birds have a similar relationship between evaporative heat loss and ambient temperatures. At environmental temperatures of 43° to 45°C, evaporation will take away more heat than that produced by metabolism. Under these circumstances, a bird will augment panting by rapid gular fluttering. The poorwill and Australian spotted nightjar can dissipate more than three times their metabolic heat production at high ambient temperatures. Generally, gular flutter sets in approximately when the ambient temperature is equal to the body temperature. Rates of gular flutter are often 20 times the normal panting rates and require relatively little additional metabolic heat production, thereby effectively increasing the cooling rate. It is interesting to observe birds in nature when very hot conditions (temperatures above 43°C) occur. Two things are most striking. They always have their beaks wide open when resting and they know where water can be found.

Evaporative heat loss from mammals as percent of total heat loss is shown in Fig. 12.5. It is clear that at low ambient temperatures, evaporative rates are low and increase logarithmically with temperature until they become 30% or more of the total. For humans, the ability to turn on the evaporative sweating mechanism is very effective at high ambient temperatures and may exceed 200% of total metabolic heat production. In this way, a human can survive heat stress for limited periods of time. Clearly, any animal can sustain very high water loss rates for relatively limited times; however, it is doubtful that larger animals reach a steady-state situation under these circumstances. Some mammals lick their fur and add water to its surface, thereby producing increased evaporative cooling.

Morhardt and Gates (1974), in their study of the Belding ground squirrel, obtained the evaporative values shown in Fig. 12.6, given as a percent of total metabolic heat production. The actual values of evaporative energy loss ranged from about 7.0 to 20.9 Wm^{-2}, with minima of about 2.8 Wm^{-2} and maxima of 41.9 Wm^{-2}. For most mammals, the increase of evaporative water-loss rate with increasing ambient temperature is a direct consequence of increased respiratory rate, which follows a logarithmic curve very similar to that shown in Fig. 12.6. It is interesting that the maximum panting rate achieved by some animals, such as dogs, is just the resonant frequency of the thoracic cavity.

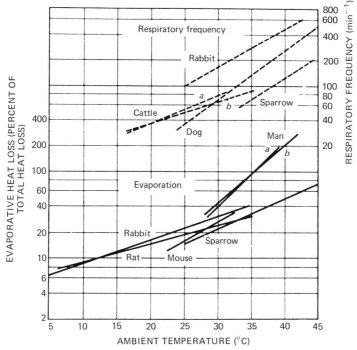

Figure 12.5. Evaporative heat loss from birds and mammals in percent of total heat loss as a function of the ambient temperature. (From Precht et al., 1973.)

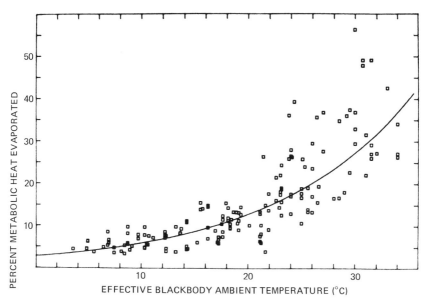

Figure 12.6. Percentage of metabolic heat lost through evaporation as a function of effective temperature for the Belding ground squirrel, *Citellus beldingi beldingi*. (From Morhardt and Gates, 1974.)

To a first approximation, one can estimate the respiratory evaporative water loss from an animal in terms of its breathing rate BR, tidal volume TV, and body temperature T_b (assuming the lung cavity to be at body temperature), the temperature T_a of the free air into which the animal is respiring at a relative humidity h, and by assuming some fraction f of saturation for the air exhaled from the lung cavity. Therefore, the amount of moisture expelled by breathing is given by

$$E = \text{BR} \times \text{TV}[f \, _sd_\ell(T_b) - h \, _sd_a(T_a)], \qquad (12.30)$$

where $_sd_\ell(T_b)$ is the saturation water-vapor density in the lung at the lung or body temperature and $_sd_a(T_a)$ is the saturation water-vapor density of the free air as a function of the air temperature. If $f = 1.0$, it is assumed that the air taken into the lungs is saturated at the temperature of the body. If the air in the lungs is less than saturated, f may take values less than 1.0. It is clear that in extremely humid, warm air, such as encountered in wet tropical regions, respiratory evaporative cooling is very small.

Some animals that have long, narrow nasal passages can recover a portion of the moisture lost by breathing. This is possible because the nasal passages are cooler than the lung with respect to the external environment and water condenses in them. This effect can reduce water loss by 20 or 30% and could be taken into account through the f factor in Eq. (12.30).

Welch and Tracy (1977) have elaborated the theory of respiratory water loss. Information concerning the anatomy and physiology of respiration can be found in Schmidt-Nielsen (1972).

Insulation by Fur and Feathers

Homeotherms, in order to maintain a nearly constant body temperature in cold environments, may either increase metabolic heat production or increase their insulation to reduce heat loss. In terms of energy expenditure, it is more economical to provide increased insulation than increased heat production for protection against cold. Body insulation is provided by fat, pelage, or the boundary layer of air adhering to an animal surface. However, disadvantages are that the boundary layer of air is destroyed by wind and fat is heavy. The insulation quality of stationary air is much greater than that of fat. Mammals generally have a hair coat or fur, whereas birds have evolved feathers for thermal insulation. Fur or feathers are superbly designed for trapping air into small, isolated compartments and provide excellent insulation. Nearly 95% of an animal's coat is occupied by air. Wind and rain disturb fur and tend to reduce its thermal-insulation value somewhat. Mammals can reduce cutaneous and subcutaneous blood flow and thus reduce the temperature of these tissues, the temperature gradient from skin to environment, and therefore heat loss. The gradient from body core to skin is increased, but total heat loss is reduced if cutaneous or subcutaneous conductance is lowered by reduced blood flow. According to Bianca, in Hafez (1968), "the change from maximal peripheral vasodilation to maximal vasoconstriction can augment skin insulation three-fold in cattle and sheep, and six-fold in man." However, a far more effective means of protection against cold is the use of pelage. Feathers are remarkable evolutionary structures and beautifully designed to minimize the effect of wind on their insulative quality. Riegel (1975) has discussed the evolutionary origin of feathers. Mammals and birds, through piloerection, can cause the fur or feathers to stand up and thereby increase coat thickness and insulative quality.

Definition of Insulation, Conductivity, and Conductance

It is well to remind ourselves of the basic definition of insulation I or conductance K in terms of a quantity of heat q passing through a surface area A of thickness d in time t as follows. The transfer of heat through a coat of fur or feathers or a layer of fat is, in principle, the same as that through the wall of a house. This is not strictly true since a coat of fur or feathers is a complex structure into which moving air can penetrate and energy can be exchanged by radiation from various depths. However, for the moment, a pelage can be considered as a simple insulating barrier. The shape of an animal is generally cylindrical or spherical, but as a first approximation, the body covering is considered as a flat slab. The amount of heat q transferred through a wall is proportional to the temperature difference between the hot side T_h and the cold side T_c and the cross-sectional area of the wall A, inversely proportional to the thickness of the wall d, and proportional to the time duration such that

$$q = k \frac{A(T_h - T_c)}{d} t, \qquad (12.31)$$

where k is the proportionality constant known as the thermal conductivity of the wall material. Hence

$$k = \frac{qd}{A(T_h - T_c)t}. \qquad (12.32)$$

Therefore, the units for thermal conductivity are watts per meter per degree Centigrade. Values of conductivity vary from about 421 $Wm^{-1}°C$ for silver to 0.013 $Wm^{-1}°C^{-1}$ for cotton wool or eiderdown. It is conductivity

which is listed for materials in various handbooks. The conductivity of fat
is 0.205 $Wm^{-1}°C^{-1}$, and that of skin is 0.502 $Wm^{-1}°C^{-1}$. Often with animals, one is concerned with the thermal conductance of the whole animal
or the thermal conductivity of fat, fur, or feathers of a certain thickness d.
Hence the conductance K is equal to the conductivity k divided by the
thickness of the material d as follows:

$$K = \frac{k}{d} = \frac{q}{A(T_h - T_c)t}. \tag{12.33}$$

The conductivity of any material is a very specific quantity, whereas the
conductance varies with the thickness. For an animal as a whole, conductance is the average thickness weighed according to heat transfer through
various parts of the animal's surface, which gives the total thermal conductance of the fat, fur, or feathers. Thermal conductance is essentially
the heat flow rate per unit area per unit temperature difference, or watts
per square meter per degree Centigrade.

The insulation I of an animal pelage is the reciprocal of the thermal
conductance. Hence

$$I = \frac{d}{k} = \frac{A(T_h - T_c)t}{q}. \tag{12.34}$$

The units for insulation are square meters per degree Centigrade per
watt. Insulation is also referred to as specific resistance. As with conductance, the term insulation is not strictly an intrinsic property of the material but includes the extrinsic characteristics which involve thickness of
pelage and shape of animal, among other things. Insulation is a proper
description of the resistance to the flow of heat to or from an animal as a
whole.

Now to write the heat-transfer relations for an animal in terms of both
conductance and insulation so that there is no doubt as to how they are
used. Repeating Eq. (12.9), one has, in terms of conductance,

$$M - \lambda E = K(T_b - T_r) \tag{12.35}$$

and in terms of insulation

$$M = \lambda E = \frac{1}{I}(T_b - T_r). \tag{12.36}$$

All previous considerations of heat flow through the body of an animal
are formulated as if the flow were through a flat slab of material. This is
not strictly correct for objects of other shapes. Since some animal parts,
such as heads, may be approximated by spheres or hemispheres and some
parts, such as the body and appendages, by cylinders, it is important to
distinguish the character of flow through these geometrical forms. Heat
flow by conduction through cylinders is given by Eq. (9.12) and that
through spheres by Eq. (9.16).

Heat flow from inside an animal outward may pass through a series of different thermal materials such as fat and fur or feathers. The thermal properties of each material needs to be considered when figuring the total insulation or conductance. If the materials have insulation I_1 and I_2, then the two together have an equivalent insulation of $I = I_1 + I_2$. If the thermal properties are expressed as conductances K_1 and K_2, then the combined conductance K is $1/K = 1/K_1 + 1/K_2$.

Conductivity of Fur

The insulation quality of an animal pelage increases with the thickness of the fur. Figure 12.7 shows the values of insulation for various animals with a winter coat as a function of fur thickness. The units are square meters-seconds-degrees Centigrade per joule. In dividing the fur thickness by the insulation value for each animal using the higher values for squirrel, marten, caribou, reindeer, and white fox, one finds that conductivities are the same, approximately 0.0419 $Wm^{-1}{}^{\circ}C^{-1}$. Clearly, the lemming, beaver, dog, red fox, polar bear, grizzly bear, wolf, and Dall sheep have somewhat higher values of conductivity (lower insulation), about 0.0628 $Wm^{-1}{}^{\circ}C^{-1}$. Davis and Birkebak (1975) list the thermal con-

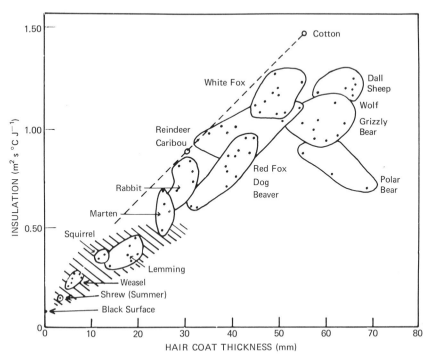

Figure 12.7. Maximum insulation of fur of arctic animals as a function of thickness. (Redrawn from Hafez, 1968.)

ductivity of fur for many animals. Values range from 0.0229 $Wm^{-1}°C^{-1}$ for a shrew to 0.0628 $Wm^{-1}°C^{-1}$ for a polar bear. Other characteristic values are weasel, 0.0272 $Wm^{-1}°C^{-1}$; red fox and beaver, 0.0523 $Wm^{-1}°C^{-1}$; dog and wolf, 0.054 $Wm^{-1}°C^{-1}$; deer, 0.0356 $Wm^{-1}°C^{-1}$; snowshoe hare, 0.0391 $Wm^{-1}°C^{-1}$; and raccoon, 0.0482 $Wm^{-1}°C^{-1}$. However, for each animal species there is a wide variation. A worker using these values is advised to go back to the original literature in order to understand the properties of the particular pelt measured.

The conductivity of a particular hair material does not change a great deal summer to winter, but the insulation value varies with coat thickness, which increases considerably from summer to winter. Birkebak (1966) analyzed the thermal conductivities of various animals as given above and found that the winter values were generally lower than the summer values. Wind and rain affect both conductivity and insulation quality, since moisture increases the conduction of hairs and interstitial air spaces, and both elements tend to compact and reduce the thickness of the coat. The effect of wind and rain on insulation decreases with increasing hair density and coat thickness.

Several years ago, I witnessed the death of many cattle in a late-spring blizzard in New Mexico. The cattle were in open range. Temperatures dropped from about 8°C to about −9°C and wind speeds were 40 to 50 mph. The coats of the cattle became wet, and as the storm grew in intensity, the cattle wandered until caught in deep snow in barrow pits along the highway and railroad. There they thrashed about, generating an increase of metabolic heat. At the same time, their coats were wet from the initial warm conditions. Increased metabolic heat caused the driving snow to continue to wet the coat, and insulation quality was reduced. With the driving winds, heat loss became intolerable, and after many hours metabolic heat production dropped precipitously and sudden death ensued. The cattle were dead standing upright in the vicious storm.

Temperature Gradients

The temperature gradient across a coat of fur varies nonlinearly with distance from the skin as shown in Fig. 12.8. The greatest drop in temperature occurs close to the skin, as one would expect. Measurements by Tregear (1965) show that for a hair spacing of 0.3 mm (1000 hairs/cm^2), a 20-mph wind does not penetrate the fur more than a few millimeters. All the air above the penetration level is well-stirred by wind, and below this level, shown approximately by the arrow in Fig. 12.8, the conductivity of the fur is fairly invariant with depth. The level to which wind penetrates fur depends on the hair density. Rabbit pelt is only slightly penetrated by an 18 mph wind, whereas the sparse hair of a pig is almost completely penetrated by an 8-mph wind. According to Tregear (1965), the conductivity of rabbit fur was 0.037 $Wm^{-1}°C^{-1}$ and did not change with wind

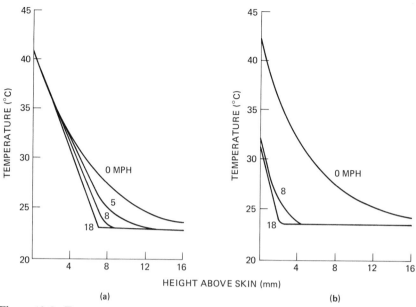

Figure 12.8. Temperature gradients across fur as a function of the height above the skin for various wind speeds for (a) a rabbit and (b) a pig. (Redrawn from Tregear, 1965.)

speed. However, the conductance did change some and was 3.35, 5.44, and 7.12 $Wm^{-2}°C^{-1}$ at wind speeds of 0, 8, and 18 mph, respectively. Conductivity values for a pig were 0.029, 0.063, and 0.105 $Wm^{-1}°C^{-1}$ and conductances were 4.19, 29.73, and 48.15 $Wm^{-2}°C^{-1}$ at wind speeds of 0, 8, and 18 mph, respectively.

Heat-Transfer Mechanisms

Heat transfer through dense fur is primarily conductive and convective, whereas heat transfer through sparse hair is radiative as well as conductive and convective. The dividing line between the two appears to be at a hair density of about 1000 hairs cm^{-2}. Most ungulates and primates have hair density at or below this level, but some have thick skins, such as the pachyderms, or subdermal fat and connective tissue with low blood flow that provides some insulation, such as the pigs and horses. Primates have thin skins and are quite vulnerable to wind cooling. On the other hand, most carnivores and rodents have dense fur which offers good protection against wind. Rabbit fur density ranges from 4,100 to about 11,000 hairs cm^{-2}. Husky dogs can sleep through an arctic gale without being seriously chilled. The appendages of most animals have low hair density, and heat loss from these appendages is serious unless other mechanisms are provided to reduce heat input to legs, tails, and ears. Countercurrent exchange mechanisms exist which can greatly limit the vascular heat flow and convective loss.

The feathers of birds behave in the manner of dense fur, and the heat exchange across a feather coat is largely convective and conductive. However, because of pigmentation, radiative transfer of heat plays an interesting role, as described in the following paragraphs.

According to Davis and Birkebak (1975), free convection in most dense fur is negligible near the skin. However, these authors show that forced convection can make a significant difference in the conductivity of fur. They show that below a certain penetration speed, the effective thermal conductivity is constant, and that above this speed, conductivity increases with increasing wind speed. In fact, these authors find that conductivity increases according to the square root of the wind speed, a relation which agrees with the square-root law for convective heat transfer over cylinders for a broad range of Reynolds numbers. Skuldt et al. (1975) report experiments which show that radiation exchange can play a very significant role in effective fur conductivity. In general, fur conductivity increases as fur density decreases owing to an increase in radiative exchange. Radiative exchange takes place near the outermost parts of the fur unless the hair density is low, and then it can occur down to the skin layer itself.

Pigmented Pelage

A paper which seems to have been overlooked in the literature by many writers is a provocative one by Kovarik (1964) which deals with irradiation of fur by sunlight. This author examines pigmentation of fur or feathers as a protection against high heat load for animals under conditions of simultaneous exposure to intense solar radiation and high ambient temperature. Kovarik quotes Buxton (1923) as follows: "There is only one important exception to the generalization that desert animals resemble in color the soil on which they live, but it is a most remarkable one: in many deserts a number of indigenous animals are black." Kovarik also refers to measurements of cattle body temperatures made in tropical parts of Queensland, Australia, which showed that the only significant correlation between color and temperature is negative. Darker-colored animals were shown to have lower body temperatures than lighter ones.

Kovarik shows that increasing the absorptivity of an animal's pelt not only causes a greater amount of heating in the outermost layers of the coat, but reduces the mean depth at which the heating occurs. The resulting heat flow is divided between that going inward to the animal's skin and that going outward to the environment such that the total thermal load on an animal reaches a maximum for some intermediate value of absorptance rather than for the darkest coloration. When a pelage is of very light color, it reflects a high proportion of the incident solar radiation. When a pelage is of intermediate or relatively high absorptance, such that corresponding to gray or brown coloration, incident sunlight penetrates the fur to considerable depth before being absorbed, and heat is then transferred as readily inward to the skin as outward to the air. But for a black pelage of fur or feathers, solar radiation is largely absorbed near the surface, and the heat generated there is more readily transferred back out to the environment than inward to the skin. These situations only prevail when there is a sufficiently dense pelage of high insulation quality. Kovarik does not give a quantitative determination of the hair density necessary for such phenomena, but one can surmise that it must be about 2000 hairs cm^{-2} or greater. These observations should apply to bird feathers as well.

Thermal Radiation Exchange

There is considerable interest in methods of measuring radiative heat exchange between animals and their environments. Some investigators use net radiometers to measure the difference between incoming and outgoing fluxes of radiation. Other investigators use infrared thermometers or directional radiometers to measure the radiative temperature of an animal's coat. Cena and Clark (1973) describe the mechanism of thermal radiation transfer through an animal pelage and show that, depending on the angle of view, a radiative temperature difference as great as 20°C may be observed between measurements taken normal and tangential to the surface. They give experimental data for a sheep's fleece to verify their theory. These temperature differences are largely the consequence of fur not coming straight out from the skin surface in the direction of the normal to the surface, but lying flat and emerging at an angle. Therefore, the depth into the fur from which a radiometer will receive radiation depends very much on the angle of view, both in terms of azimuth and elevation angles. Straight hairs lying at an angle to the skin surface behave differently from crimped hairs standing straight out from the surface.

Resistance of Pelage

Sometimes it is useful to express the exchange of heat across an animal pelage in terms of resistance, rather than as conductance or insulation. Monteith (1973) and Clark et al. (1973) prefer to use resistance since it is consistent with resistance for convective exchange of vapor. A pelage may have a quantity of heat G entering it from the skin, a net radiative exchange $\mathscr{R}_n$ in the upper layers of pelage, and a convective exchange C with the air beyond the boundary layer. The heat flow G from the skin is transferred across the pelage at a rate determined by the temperature difference between skin temperature T_s and radiant surface temperature of pelage T_r and by the insulation I. Hence one can write

$$G = (T_s - T_r)/I. \tag{12.37}$$

If the numerator and denominator are each multiplied by the quantity ρc_p, where ρ is the density of air in kilograms per cubic meter and c_p is the specific heat of air in joules per kilogram per degree Centigrade, then one gets

$$G = \frac{\rho c_p (T_s - T_r)}{\rho c_p I} = \frac{\rho c_p (T_s - T_r)}{r_h}, \tag{12.38}$$

where thermal resistance r_h $(= \rho c_p I)$ is in seconds per meter. The value of ρ for air is approximately 1.204 kg m^{-3}, and the value of c_p is 1013 J kg^{-1}°C^{-1}, so that the product is 1220 J m^{-3}°C^{-1}.

The energy balance of the pelage is given by

$$G + C + \mathcal{R}_n = 0. \tag{12.39}$$

The convective heat transfer from the pelage across the air boundary layer to the free air at temperature T_a is

$$C = \rho c_p (T_a - T_r)/r_{ha}, \tag{12.40}$$

where r_{ha} is the thermal resistance of the boundary layer.

For any given value of T_a, there is a condition when $T_s = T_r$ for which $G = 0$, according to Eq. (12.38), and $\mathcal{R}_n = -C$, according to Eq. (12.39).

Clark et al. (1973) describe a clever graphical procedure for evaluating the conductance, insulation, and resistance of an animal pelage. They measure skin temperature T_s and radiant surface temperature T_r as a function of net radiation exchange by an animal exposed to a range of incident radiation from a source such as a lamp or from surrounding walls at different temperatures. The results of their measurements with a Dorset Down sheep are shown in Fig. 12.9. The sheep was maintained in a constant-temperature chamber with air temperature at 0°C and mean ventilation rate 0.5 m s^{-1} and irradiated by two 750-W quartz-halogen photoflood lamps. Net radiation was measured with a miniature net radiometer. Skin and air temperatures were measured with thermocouples, and radiant surface temperature was measured with a radiometer or infrared thermometer.

In Fig. 12.9, $\mathcal{R}_n = -C = 307$ Wm^{-2} when the lines representing the T_s values and T_r values intersect at $T_s = T_r = 32.5$°C. Using Eq. (12.40), Clark et al. calculate that $r_{ha} = 142$ s m^{-1}. When the animal is in forced convection, as it was for these measurements, r_{ha} is independent of the temperature difference $T_r - T_a$.

The way in which C changes with $\mathcal{R}_n$ is displayed in Fig. 12.9 by drawing the straight dashed line through the points $(T = T_r = T_s, \mathcal{R}_n = -C)$ and $(T = T_a, \mathcal{R}_n = C = 0)$. The equation of this line is

$$T = T_a - Cr_a/\rho c_p, \tag{12.41}$$

and the convective rate of exchange is read from the abscissa for each point along this line. The slope of the line is proportional to r_{ha}.

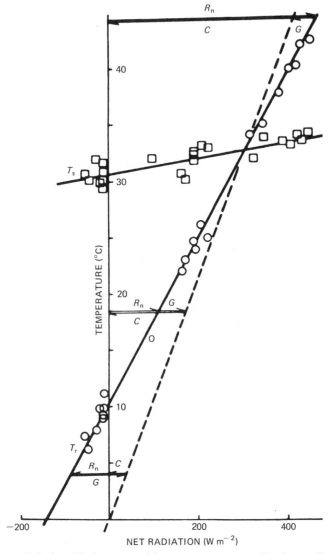

Figure 12.9. Relationship between skin temperature T_s, effective radiative coat temperature T_r, and net radiation $\mathscr{R}_n$ for a sheep with 4.5 cm of fleece at air temperature $T_a = 0°C$. This figure illustrates a graphical method for estimating the partition of sensible heat fluxes between net radiation $\mathscr{R}_n$, convective flux C, and the exchange at the skin surface G. (From Clark et al., 1973b.)

The relation between T_r and $\mathscr{R}_n$ is somewhat more complex and is found by substituting the terms G from Eq. (12.38) and C from Eq. (12.40) into the heat-balance expression, Eq. (12.39), and solving for T_r to give

$$T_r = (\mathscr{R}_n/\rho c_p + T_a/r_{ha} + T_s/r_h)(r_{ha}^{-1} + r_h^{-1})^{-1}, \qquad (12.42)$$

where T_a is held constant. When T_s is either a constant or a linear function of $\mathscr{R}_n$, T_r is linearly related to $\mathscr{R}_n$. The locus of T_r plotted against $\mathscr{R}_n$ is a straight line whose slope is a function of r_h, r_a, and $\partial T_s/\partial \mathscr{R}_n$.

According to the heat-balance expression, Eq. (12.39), $G = -C - \mathscr{R}_n$. This condition is represented on the graph of Fig. 12.9 as the horizontal arrows corresponding to a temperature of 18°C. Arrows indicate the sign convention. The thermal resistance of the animal's pelage is found by noting that when $T_r = T_a$, $C = 0$ and $\mathscr{R}_n = -G = 140$ Wm^{-2}. When $T_s = 30$°C, $r_h = \rho c_p (T_s - T_r)/G = 280$ s m^{-1}. The depth of the sheep's fur was 0.045 m, and the thermal conductivity of the fleece is calculated to be 0.21 Wm^{-1}°C^{-1}.

Radiation Absorbed

The amount of radiation absorbed by an animal depends upon the incident fluxes of radiation, the mean absorptance of the animal surface to the incident fluxes, and the geometrical characteristics of the animal. In many respects, the quantity of radiation absorbed Q_a is the most difficult of all the terms in the heat-budget expression of an animal to evaluate. If an animal is analyzed using a lumped-parameter model for the whole animal, then the mean values of absorptance, emittance, and conductance are used. The absorptance is the fraction of incident radiation of a certain quality, e.g., direct sunlight, scattered skylight, reflected sunlight, or thermal radiation, absorbed over the animal's surface taken as a whole. By contrast, the absorptivity is a localized property of the animal's surface material and represents the fraction of incident radiation absorbed by any specific sample of surface material. In principle, one must know the localized absorptivity of each and every part of an animal's surface in order to determine the total quantity of radiation absorbed. First, an exact description is given of the absorption of radiation by an animal. Then, an approximate lumped parameter description is given for practical applications.

The quantity of direct sunlight absorbed by the ith incremental element of animal surface of projected area $_iA_S$ to a flux of incident direct sunlight S if the surface absorptivity is $_ia_S$ is given by the product $_ia_S \, _iA_S \, S$. Similarly, the amount of incident skylight s absorbed by the ith incremental element of surface area is $_ia_s \, _iA_s \, s$, the amount of reflected direct sunlight is $_ia_{rs} \, _iA_{rs} \, r_sS$, etc. The total amount of radiation absorbed by an animal as a whole is the total sum of all the incremental amounts and is given by the product of the animal's total surface area A and the amount absorbed per unit surface area Q_a. Hence

$$A Q_a = \Sigma \, _ia_S \, _iA_S \, S + \Sigma \, _ia_s \, _iA_s \, s + \Sigma \, _ia_{rs} \, _iA_{rs} \, r_sS$$
$$+ \Sigma \, _ia_{rs} \, _iA_{rs} \, r_s s + \Sigma \, _ia_t \, _iA_g \mathscr{R}_g + \Sigma_i a_t \, _iA_a \, \mathscr{R}_a. \quad (12.43)$$

Clearly, there is far too much effort involved if, in order to evaluate the energy budget of each animal, one must evaluate the radiant energy flux absorbed by each incremental area and add up the total over the entire surface area of the animal. There are a number of simplifying procedures available. The first approximation is to work with mean absorptances for the entire animal. The mean absorptance to direct sunlight is given by

$$a_S = \frac{\Sigma \,_i a_S \,_i A_S \, S}{\Sigma \,_i A_S \, S} = \frac{\Sigma \,_i a_S \,_i A_S}{\Sigma \,_i A_S}. \tag{12.44}$$

The sun is approximately a point source in the sky, and the projected incremental areas $_i A_S$ to direct sunlight change with animal orientation to the sun. The value of the mean absorptance will not usually vary significantly with exposure and orientation to the sun, except for specific animals which have greatly contrasting coloration on dorsal and side surfaces. The total projected area of an animal to direct sunlight will vary considerably with orientation.

Skylight reaches an animal from the entire hemisphere if the animal has a completely unobstructed horizon. In this case, the surface area of the animal which is exposed to the sky is about one half, more or less, of its entire surface area. It is less than half because of occluded surfaces, such as appendages which may be curled underneath the animal. It can be more than half if the animal is oblate and has a flat area such as the belly against the ground. In many instances, one can make good estimates of the exposed surface areas. The mean absorptance to skylight is usually somewhat greater than that to direct sunlight because skylight is relatively rich in ultraviolet and blue wavelengths. The mean absorptance to skylight is given by

$$a_s = \frac{\Sigma \,_i a_s \,_i A_s}{\Sigma \,_i A_s}. \tag{12.45}$$

Both direct sunlight and skylight are reflected from the ground surface. Although each can be treated separately, it is simpler, and does not create much error, to treat them together as a total flux $r(S + s)$ of diffusely reflected sunlight and skylight. The mean absorptance to reflected light is usually given, as an approximation, as the same as the mean absorptance to sunlight. Hence $a_{rS} = a_{rs} = a_s$. The error involved here is also small, as a general rule.

By using mean absorptances, the total amount of radiation absorbed by an animal is written

$$A \, Q_a = a_S \, A_S \, S + a_s \, A_s \, s + a_s \, A_{rS} \, r(S + s) + a_t \, A_g \, \mathcal{R}_g \\ + a_t \, A_a \, \mathcal{R}_a. \tag{12.46}$$

If reflected light is diffusely reflected from the ground hemisphere, using the approximation $A_{rs} = A_g$ further simplifies Eq. (12.46). Since thermal atmospheric radiation emanates from the same hemisphere as skylight, it is possible to use $A_a = A_s$. Therefore, Eq. (12.46) becomes

$$A \, Q_a = a_s \, A_S \, S + a_s A_s \, s + a_s \, A_g \, r(S + s) + a_t(A_g \mathcal{R}_g + A_s \mathcal{R}_a). \quad (12.47)$$

Radiation Emitted

The amount of radiation emitted by the surface of an animal is a product of the emissivity ϵ, the effective emitting area A_r, and $\sigma(T_r + 273)^4$, where σ is the Stefan-Boltzmann constant and T_r is the radiant surface temperature. Therefore,

$$A \, Q_r = \epsilon \, A_r \, \sigma(T_r + 273)^4. \quad (12.48)$$

The emissivity of most animal surfaces is greater than 0.97 according to Monteith (1973).

Surface Area

In order to determine energy exchange for a whole animal, it is necessary to have a measure of the total surface area and its effective areas A_S, A_s, A_g, A_a, and A_r. It is not easy to determine these values, and for many animals one can only make reasonable estimates based on general geometrical considerations such as those given in Chapter 7. For large animals, various mensuration techniques have been developed by animal scientists. Underwood and Ward (1966) worked out the projected areas for human subjects. For small animals, Bartlett and Gates (1967) described a technique which they used successfully with lizards.

Total Surface Area

It is difficult to measure the total surface area of an irregularly shaped object such as an animal. However, if one can make a silver casting of such an animal, then the surface area of the cast is easy to measure by means of an electrical method. Tibbals et al. (1964) first demonstrated the feasibility of this technique for plant leaves. The silver casting was suspended as an electrode in an electrolytic solution contained in a silver-lined cylindrical tank. The resistance from the electrode to the tank wall was measured. For a fixed voltage across the system, from suspended casting (the center electrode) to the walls (as the other electrode), the current is strictly proportional to the surface area of the casting. To determine the amount of surface area exposed to a hemisphere, one can simply coat one part of the casting with an insulator and repeat the electrical measurement. Using this method, Bartlett and Gates (1967) determined the total surface area of a lizard, *Sceloporus occidentalis,* to be 75.8 cm². They were predicting the behavior of this lizard on tree trunks from its energy budget and needed to know the amount of surface area not in contact with the tree. By coating this part with insulation, they found that the lizard exposed 64.5 cm² of surface area to the surrounding environment.

Emittance Area

In order to determine the effective area for the emission of thermal radiation, a silver casting of a small animal is used in the following manner. The casting is suspended in a vacuum chamber with black walls that are maintained at a constant temperature by means of water circulating through coils in the walls. The casting is heated to a temperature above the wall temperature and allowed to cool in the chamber. Using the following relationship, and measuring the temperature of the casting as a function of time, one can calculate the effective emitting area A_r provided all other quantities are known:

$$A_r \, \epsilon\sigma(T_c^4 - T_w^4) = mc \, \frac{dT_c}{dt}. \tag{12.49}$$

The mass of the casting is determined by weighing. The specific heat of silver is known. The emissivity of the silver casting was measured by means of an infrared spectrophotometer. The silver casting used by Bartlett and Gates (1967) had an emissivity of 0.36. The lizard, *Sceloporus occidentalis*, has an effective emitting area of 67.2 cm², or 89% of the actual total surface area of 75.8 cm². Since the lizard had part of its area in contact with the tree, the area from which it emitted radiation while it was on the tree was 0.89 × 64.5 cm, or 57.4 cm². The effective area for the emission of radiation is also the effective area for the absorptance of radiation from extended sources. One can take the same proportion of half the surface area to use for the absorption of radiation from the atmosphere and from the ground. Hence, in the case of the lizard, $A_g = A_a = 33.6$ cm², or 44.3% of the total surface area if the lizard is exposed equally to sky and ground. In this case, the lizard is suspended horizontally above the ground.

Solar Radiation Area

The effective area for the absorption of solar radiation is best determined by drawing the shadow cast on a wall by an animal exposed to a point source of light. This was the technique used by Underwood and Ward (1966) for determining the projected areas of human subjects to solar radiation. Bartlett and Gates (1967) also used this method with the lizard *Sceloporus occidentalis* to obtain the effective area for the absorption of solar radiation as a function of solar altitude and azimuth. These measurements are shown in Fig. 12.10 for a lizard on a tree trunk or fencepost. The lizard

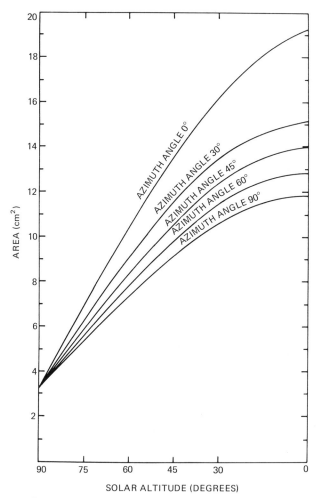

Figure 12.10. Effective area for the absorption of direct solar radiation for a lizard (*Sceloporus occidentalis*) in a vertical position as a function of solar altitude and azimuth angle. (From Bartlett and Gates, 1967.)

is oriented vertically. When the sun is in the zenith (solar altitude 90°), the lizard presents a minimum cross section to solar rays. One could apply the same results to a horizontal lizard by reversing the solar altitude scale. *Sceloporus occidentalis* had a minimum area of 3.3 cm² and a maximum area of 19.2 cm² for absorption of solar radiation. These values are 4.4 and 25.3%, respectively, of the total surface area.

For 50 human subjects standing upright wearing bathing suits or tight-fitting gowns, Underwood and Ward (1966) found that when the areas projected at three azimuth angles of 0, 45, and 90° were averaged, the results were approximated well by a cylinder of height 165 cm and radius 11.7 cm.

Convection Coefficients

Most animals have body shapes which resemble cylinders. One might expect the character of air flow over the surface of an animal to be similar to that over a cylinder. Many tests by various investigators show this to be true; cf. Bartlett and Gates (1967), Porter and Gates (1969), and Porter et al. (1973). In the earlier work of Porter and Gates, it was believed that a good approximation to convective heat transfer for an animal was given by the relation $V^{0.333}/D^{0.666}$, where V is the wind speed and D is the characteristic dimension of the animal. A discussion of convection in Chapter 9 shows that for cylinders, the exact power function of V/D to be used depends upon the Reynolds number of the particular situation of an animal and its environment. The Reynolds number is VD/ν, where ν is the kinematic viscosity of the fluid. Reynolds numbers are often between 300 and 6600. Small animals are usually near the ground surface and do not allow themselves to be subjected to high wind speeds. The wind profile over moderately rough terrain is such that wind speeds reduce rapidly as the ground surface is approached. The range of Reynolds numbers from 300 to 6000 suggests that a power function of about $(V/D)^{0.5}$ for the convective heat-transfer coefficient is a generally good approximation and that other power functions should be used for special conditions of very still air or very high wind speeds and large dimensions. Animals such as horses, cattle, deer, etc., stand erect and are often subjected to greater wind speeds as well as having larger characteristic dimensions, than small animals. Hence large animals would be expected to have Reynolds numbers from 6,600 up to 25,000 or greater. The relation used by Porter and Gates is true for Reynolds numbers less than 4, which implies either very little or no air movement or very small objects. However, determinations by Porter et al. (1973) and data by Weathers (1970) show the following dimensionless relation to hold for cross-flow:

$$\text{Nu} = 0.35 \text{Re}^{0.6}. \qquad (12.50)$$

Porter et al. define their Nusselt and Reynolds numbers in terms of the snout-vent length L of the lizard. Normally, one would use a characteristic dimension in the direction of air flow. However, since the width or diameter of a lizard is proportional to the length, it is possible to use length in the expression for convective heat transfer. Expressing Eq. (12.50) in terms of the convection coefficient, one gets

$$h_c = 1.11 \, V^{0.6} L^{-0.4}. \qquad (12.51)$$

If one compares this equation with Eqs. (9.46) through (9.49), it is clear that it falls among those for large Reynolds numbers. Porter et al. state the following: "Measurements of heat transfer coefficients in outdoor environments are 50 to 100% higher than corresponding values obtained in wind tunnels (Pearman, et al. 1971; and some unpublished results we have obtained in the desert). For all field predictions, the wind tunnel relations were increased by 50% to simulate field conditions."

The plausability arguments for this are two, as follows: that the rough and irregularly shaped surface of an animal may produce more turbulence than a smooth cylinder and result in an increased rate of heat transfer and that air flow is more turbulent in nature than in a wind tunnel because of the roughness of the ground surface and the interference by vegetation with the flow. If one calculates a specific example using Eq. (12.51) and compares with results obtained from Eqs. (9.46) through (9.49), it is clear that Eq. (12.51) gives values which are nearly 100% greater.

Because of the ambiguity involved among various experiments reported in the scientific literature, it seems reasonable, as a general rule, to use the relation

$$h_c = 3.49 \ V^{0.5} \ D^{-0.5}, \tag{12.52}$$

where V is in meters per second and D is in meters.

Mitchell (1976) has found, from a series of experiments, that convective heat transfer for animals is best determined using the convective relationship for a sphere and treating an animal as an equivalent sphere of similar total mass. The procedure recommended by Mitchell is described in Chapter 9.

Appendages

The appendages of animals exchange energy with the environment primarily through radiation and convection. Convective heat exchange may play a particularly significant role in the heat balance of animals since the characteristic dimension of an appendage is usually very much smaller than that for the body itself. Appendages can often be flushed with blood and have a surface temperature several degrees higher than that of the body surface and in this manner facilitate the loss of heat. Or an animal may reduce the blood flow to an appendage by vasoconstriction, thereby keeping the appendage cooler than the body and head.

The antlers of deer, when in velvet, are flushed with blood and are substantially warmer than the rest of the animal's surface. For a red deer in New Zealand when the air temperature was 17°C, Gates (1969b) found the body surface to be 19°C, the head surface 25°C, and the antler surface 32°C. Similar temperature differences were found for a female reindeer with antlers in velvet at the St. Louis Zoo.

Animal appendages have a variety of sizes, shapes, and orientations. Many appendages, such as the legs of most animals and antlers of some animals, have cylindrical shapes. The ears of many animals, such as rabbits, resemble open cylinders with partial closure.

Wathen et al. (1971) investigated the radiative and convective heat exchange from jackrabbit ears and cylindrically shaped appendages. They used the modeling techniques established by Bartlett and Gates (1967) and Tibbals et al. (1964) with metal castings and improved the accuracy of wind-tunnel measurements of convective heat transfer by gold-plating the casting and reducing net radiation exchange. Wathen et al. (1971) found that angles of pitch and yaw between the cylinder axis and wind direction did not give significant systematic variations and that differences were within about 10% of mean values. They obtained, for the convection coefficient for six cylinder models and the casting of a domestic rabbit ear (Belgian hare), the relation

$$h_c = 17.4 \ V^{0.5}, \tag{12.53}$$

where V is in meters per second and h_c is in watts per square meter per degree Centigrade. Here the characteristic dimension is included in the numerical coefficient. The dimension of their cylinders were from 2.8 to 5.2 cm. If these values are put into the usual expression, such as Eq. (12.52), one gets a numerical coefficient between 15.3 and 20.8 for Eq. (12.53).

Schmidt-Nielsen (1964) suggested that the ears of the jackrabbit can radiate to the cold sky and thereby dump heat and also lose heat by convection. If the air temperature is equal to or greater than the animal's surface temperature, however, no heat loss will occur by convection, and in fact heat may be gained. Schmidt-Nielsen et al. (1965) report that jackrabbits can, by vasoconstriction, reduce blood flow through their ears in this situation and cause a reduction in the amount of heat transported into the body. Hence the rabbit's ears can be used to increase its active surface area in order to dump heat or inactivated as heat radiators either by vasoconstriction or by laying the ears against the head. Figure 12.11 is an infrared thermogram of a domesticated rabbit showing the warm ears in contrast to the cooler head and body surfaces. During warm periods, the rabbit positions itself in the shade of a tree or shrub and orients its ears toward the direction of the cold sky. Schmidt-Nielsen (1962) showed that under some circumstances a rabbit may radiate to the cold sky an amount of heat equal to about one-third of its metabolic heat production.

Wathen et al. (1971) calculated the dissipation of metabolic heat by a jackrabbit's ear of total surface area 0.032 m². The animal was in the shade and was exchanging radiation with the sky (view factor, ⅓), ground (view factor, ⅓), and vegetation (view factor, ⅓). The relation for net exchange of radiation by a rabbit ear is

$$Q = \frac{\epsilon \sigma A}{3} \ F[(T_s^4 - T_r^4) + (T_v^4 - T_r^4) + (T_g^4 - T_r^4)]$$
$$= \epsilon \sigma \ AF[(T_s^4 + T_v^4)/3 - T_r^4]. \tag{12.54}$$

The symbols T_s, T_v, T_g, and T_r refer to the temperature of the sky, vegeta-

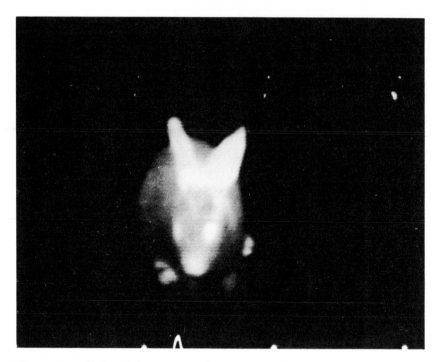

Figure 12.11. Infrared thermogram of a domestic rabbit showing the warmer ears, eyes, nose, and feet and the generally cooler body surface.

tion, ground, and radiant surface of the ear, respectively. The factor F is a view factor for the ear as a whole and has the value 0.8 since the ear is represented by a half-cylinder and each part of the concave inside surface area "sees" another part of the same inside surface. Only 0.8 of the total surface area of the ear is effective as a radiator to the surrounding environment. For example, if one has a thin-walled hollow cylinder and considers the total surface area of the inside as well as the outside surface, the view factor would be 0.5.

For their calculations, Wathen et al. assumed air temperatures of 30°, 35°, and 40°C, a sky temperature of 24°C, a ground-surface temperature of 55° to 60°C, and vegetation temperature of 40°C as typical of midday desert conditions. The rabbit was assumed to hold its ear temperature at body temperature, 40°C. Calculation showed radiant heat loss to the sky to be 0.89 W, net exchange with the vegetation to be zero, and heat gain from the ground to be between 0.94 and 1.28 W, depending on whether the ground surface was at 55° or 60°C. Hence the ears had a net gain of radiant heat from 0.05 to 0.39 W depending on the ground-surface conditions. Air speeds varied from 0.22 to 2.2 m s^{-1}. Wind-tunnel convection coefficients ranged from 7.0 to 27.9 Wm^{-2}°C^{-1}. These authors claim that convection coefficients in the field are 1.4 times greater and therefore used values from 9.8 to 39.1 $Wm^{-2\circ}C^{-1}$. For a 2.5-kg rabbit, body temperatures of 39°, 39.5°, and 40°C and metabolic rates of 8.2, 8.5, and 9.4 W are expected for air temperatures of 30°, 35°, and 40°C for a 2.5-kg animal. For maximum heat loss, the ear temperature is maintained at body temperature. Calculations showed convection loss by the ears to be 2.8 to 11.2 W, 1.4 to 5.7 W, and 0.15 to 0.63 W, respectively, at the three air temperatures. These convective losses amounted to from 34 to 136% of metabolic heat production at 30°C, 16 to 166% at 35°C, and 1.6 to 6.6% at 40°C. Hence, at the lower air temperature, convective heat loss from the ears could amount to nearly all of the metabolic heat production and greatly conserve water loss, but at temperatures of 40°C or above, convective heat loss from the ears is not adequate to more than offset the gains from radiation.

It is straightforward to calculate the rate of blood flow to the ears necessary to carry away metabolic body heat by convection. The quantity of heat q which is convected away must equal the loss in heat from the blood which enters the ear at body temperature T_b and leaves the ear at air temperature T_a. The specific heat of blood c_b is assumed to be the same as that of water, i.e., 4187 J $kg^{-1\circ}C^{-1}$. Hence

$$q = (T_b - T_a)\, c_b. \tag{12.55}$$

Taking the maximum convective heat loss which occurs, i.e., 11.2 W, when the temperature differential is 9°C, one calculates the necessary flow rate to be 0.22 $cm^3\, s^{-1}$. If one assumes that blood contains 4 cm^3 of O_2 per 100 cm^3 and that 1 cm^3 of O_2 can produce 20.9 J of heat, then a resting jackrabbit with a total metabolic heat production of 8.4 W has a total cardiac output of 10 $cm^3\, s^{-1}$. This is more than adequate to supply the maximum necessary blood flow to the ears and other parts as well. If convective heat loss is reduced because of net radiation gain by the ears, then less blood needs to be supplied to the ears. If the blood returning to the body from the ears is not at air temperature, but perhaps more realistically at a temperature $(T_b + T_a)/2$, then the quantity of blood flow required is doubled, to 0.44 $cm^3\, s^{-1}$, and the cardiac output can readily handle this. This example demonstrates the procedure to use for calculating heat exchange and blood flow rates for appendages of any animal.

Behavioral and Special Anatomical Features

Poikilotherms manipulate their body temperatures very largely by means of behavior in terms of environmental posture and exposure. However, for many poikilotherms, metabolic heat production is important to body temperature, and for some, evaporative water loss is extremely important. Although poikilotherms or ectothermic organisms generally have body temperatures near ambient air temperature, a poikilotherm is often substantially warmer or cooler than ambient air.

Many animals use their body heat capacity as a means of temperature adjustment. A classic case is that of the camel, which uses its large body mass (500 kg) to store heat by allowing its core temperature to vary from about 35° to 41°C when deprived of water. Since the specific heat of animals is about 3350 J kg^{-1}°C^{-1}, the amount of stored heat is 10 kJ. If this heat input were dissipated through evaporation, it would require 4.1 liters of water. Hence the ability of the camel to store heat is very essential during hot, dry conditions.

The antelope ground squirrel, *Citellus leucurus,* has a high tolerance for hypothermia, being able to survive body temperatures of 43°C, and stores heat in its body for short intervals when foraging during hot conditions. Unlike the camel, the antelope ground squirrel cannot accumulate heat all day but must frequently dump the stored heat to a cold sink. The animals find such a cold sink in their burrows and rush inside to spread-eagle themselves against the soil and lose heat by condition to the soil and by radiation to the burrow walls. Then, when their body temperature is sufficiently reduced, they can go outside and forage again. Dogs or other animals dumping heat through conduction by lying spread-eagle against a cold floor or the ground are a common sight. The curling up of animals to reduce surface area and conserve heat during cold conditions is well known. Not only do birds ball up into a rounded form, but their feathers undergo pilar erection and increase the quality of insulation. A bird will stand on one leg and tuck the other leg under its body in order to conserve heat. As one would expect from energy-budget considerations, small birds probably utilize sunlight to reduce energy expenditure, as shown by Hamilton and Heppner (1967) and by Lustick (1969) using artificial sunlight as a source. Observations by Ohmart and Lasiewski (1971) showed that the roadrunner, *Geococcyx californianus,* could significantly reduce its energy expenditure during cold conditions (9°C) by becoming hypothermic and reducing its body temperature to about 34.3°C from a normothermic resting temperature of 38.4°C. When hypothermic birds were exposed to radiation, their body temperature increased to its normothermic value, but the temperature of birds at normothermic conditions did not increase.

It is well known that many ectothermic animals such as locusts, grasshoppers, iguanas, and turtles bask in the sun to achieve a body temperature suitable for activity. Large reptiles such as alligators and crocodiles position themselves partially in water and partially in the sun in order to achieve the proper body temperature. Bees aggregate in the hive, penguins cluster in large colonies during intense cold, camels lie down close together during the heat of the day, and other animals group together in order to reduce the effective surface area exposed to the environment. In fact, one way or another, almost all animals use behavior to position themselves appropriately in their environment in order to achieve a proper heat exchange. These matters of position, shape, etc., must all be taken into account when calculating the energy budgets of animals under specific circumstances.

Lizards

Climate-Space

Lizards, which are poikilothermic ectothermic animals, are convenient subjects for energy-exchange studies and have been much used by various investigators. Porter and Gates (1969) determined the climate-space for the white desert iguana, *Dipsosaurus dorsalis*, which will serve as the first example. The animal had a weight of 0.067 kg and a surface area of 1.80×10^{-2} m². The diameter was about 0.015 m. At thermal maximum, body temperature was 45°C, metabolic rate 10.5 Wm⁻², and evaporative water-loss rate 6.3 Wm⁻². At thermal minimum, body temperature was 3°C, metabolic rate 1.4 Wm⁻², and evaporative water-loss rate 1.4 Wm⁻². The average surface absorptance to solar radiation was 0.8 for the dark-colored form and 0.6 for the light-colored form. The fat thickness was taken to be the skin thickness, approximately 10^{-3} m. The conductivity of the skin was 0.502 Wm⁻¹°C⁻¹. Therefore, the conductance K is k/d, or 502 Wm⁻²°C⁻¹.

Since this relationship is expressed by a straight line, it is only necessary to calculate two points on it and plot them on a climate-space diagram in which the ordinate is the temperature T_a and the abscissa is radiation absorbed Q_a. The result is the thermal maximum limit for *Dipsosaurus dorsalis* at a wind speed of 0.1 m s⁻¹. At $T_a = 40°C$, $Q_a = 621$ Wm⁻², and one at $T_a = 50°C$, $Q_a = 531$ Wm⁻².

According to Eq. (12.4), the steady-state energy budget for a lizard is given by

$$Q_a + M - \lambda E_b = \epsilon \sigma \left(T_b - \frac{M - \lambda E_b}{K} + 273 \right)^4$$

$$+ H \left(T_b - T_a - \frac{M - \lambda E_b}{K} \right), \quad (12.56)$$

where $H = 3.49(V/D)^{0.5}$ from Eq. (12.52). The emissivity of the lizard is $\epsilon = 1.0$, and $\sigma = 5.67 \times 10^{-8}$ Wm⁻²K⁻⁴. The diameter D is 1.5×10^{-2} m.

Thermal Maximum Limit

At the thermal maximum, the values used are $M = 10.5$, $\lambda E_b = 6.3$ Wm⁻², $T_b = 45°C$, and $K = 502$ Wm⁻²°C⁻¹. The steady-state energy budget is at $V = 0.1$ m s⁻¹.

$$Q_a + 10.5 - 6.3 = 5.67 \times 10^{-8} \left[45 - \frac{10.5 - 6.3}{502} + 273 \right]^4$$

$$+ 3.49 \left[\frac{0.1}{1.5 \times 10^{-2}} \right]^{0.5} \left[45 - T_a - \frac{10.5 - 6.3}{502} \right]$$

This gives

$$Q_a = 981 - 9.00\ T_a. \tag{12.57}$$

For a wind speed of 2 m s^{-1}, the thermal-maximum climate-space limit is

$$Q_a = 2390 - 40.3\ T_a. \tag{12.58}$$

At $T_a = 40°C$, $Q_a = 778$ Wm^{-2}, and at $T_a = 50°C$, $Q_a = 374$ Wm^{-2}.

Thermal Minimum Limit

Calculation of the climate-space limit for *Dipsosaurus dorsalis* is simpler at the thermal minimum than at the thermal maximum since $M - \lambda E = 0$. At the thermal minimum, $T_b = 3°C$, and for a wind speed of 0.1 m s^{-1}, one gets

$$Q_a = 357 - 9.00\ T_a. \tag{12.59}$$

At $T_a = 0°C$, $Q_a = 357$ Wm^{-2}, and at $T_a = -10°C$, $Q_a = 447$ Wm^{-2}.

For a wind speed of 2 m s^{-1}, the thermal-minimum climate-space limit is

$$Q_a = 450 - 40.3\ T_a. \tag{12.60}$$

At $T_a = 0°C$, $Q_a = 450$ Wm^{-2}, and at $T_a = -10°C$, $Q_a = 853$ Wm^{-2}.

If these straight lines are plotted on the climate-space diagram with the limits to the physical world included, i.e., the lower limit at night when the animal is between ground and clear sky and the upper limit in the daytime when all radiation is included and the shortwave absorptance is 0.6 or 0.8 and the longwave absorptance 1.0, one gets the diagram shown in Fig. 12.12.

The example given above uses all energy quantities as the amount of energy flowing per unit surface area per unit time and the units are watts per square meter. Some authors prefer to give the total quantity of energy transferred to or from the animal as a whole in watts. This is particularly convenient since for most animals, only the total metabolic heat production or the total evaporative water loss is known. It is really six of one, a half-dozen of the other whether per unit surface area or per total animal is used. Porter et al. (1973) used the total animal in their excellent article concerning *Dipsosaurus dorsalis*. Since much of the following information concerning *Dipsosaurus dorsalis* and its environmental conditions is taken from their paper, it is necessary to comment further concerning the notation used therein.

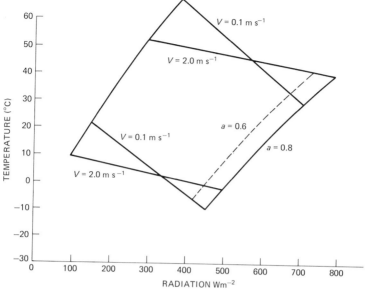

Figure 12.12. Climate-space diagram for the desert iguana, *Dipsosaurus dorsalis*, showing the upper and lower physiological limits, for wind speeds of 0.1 and 2.0 m s^{-1}.

The main problem in going from the one system of units to the other is with the use of symbols. The authors have a distinct ambiguity in notation. This is mentioned here so that readers of that paper may be alerted to the problem. The symbol Q is used with various subscripts to describe both energy flow per unit area per unit time and energy flow per total animal per unit time. One must take note of this in the paper.

In their study, Porter et al. (1973) use an electrical-circuit analog (similar to those used later in this chapter) to describe the total heat flow between a lizard and its environment. Values for heat flow are given here in terms of Q, which is always per unit surface area. They use view factors in their thermal radiation formulas and specific areas for the shortwave-radiation interactions. Their paper is useful in that it gives real values for environmental conditions in the desert.

Metabolic and Evaporation Rates

Metabolic and evaporative-water-loss rates for animals vary with body temperature. When one wants to determine the effect on an animal of a given set of environmental conditions, it is necessary to put into the analysis the temperature dependence of metabolic and evaporation rates. Porter et al. (1973) used the following empirical relations to represent the metabolic and evaporation rates of *Dipsosaurus dorsalis* as a function of body temperature. The metabolic rate was

$$M = 0.0258\ e^{T_b/10}. \tag{12.61}$$

and the evaporation rates were:
For $T_b \leq 20°C$,

$$\lambda E = 0.27, \tag{12.62}$$

for $20°C < T_b < 36°C$,

$$\lambda E = 0.08\ e^{0.586}\ T_b, \tag{12.63}$$

and for $T_b \geq 36°C$,

$$\lambda E = 2.97 \times 10^{-3}\ e^{0.1516}\ T_b, \tag{12.64}$$

with λE in units of watts per kilogram.

It is preferable to express the energy exchange per unit of surface area rather than per unit of mass. Multiplying by the mass of the lizard, 0.067 kg, dividing by the surface area, 0.018 m², and putting the two terms together to form $M - \lambda E$, one gets the following relationships. For $T_b \leq 20°C$,

$$M - \lambda E = 0.096\ e^{T_b/10} - 1.005, \tag{12.65}$$

for $20°C < T_b < 36°C$,

$$M - \lambda E = 0.096\ e^{T_b/10} - 0.298\ e^{0.0586T_b}, \tag{12.66}$$

and for $T_b \geq 36°C$,

$$M - \lambda E = 0.096\ e^{T_b/10} - 0.011e^{0.1516T_b}, \tag{12.67}$$

with $M - \lambda E$ in units of watts per square meter.

Environmental Conditions

Typical desert environmental conditions during a clear midday in April are approximately as follows, according to Porter et al. At 2.5 cm above soil surface, $T_a = 37°C$, and at the soil surface, $T_g = 50°C$. The radiation components are $S + s = 977$, $r(S + S) = 322$, $\mathcal{R}_a = 388$ Wm^{-2}, and $\mathcal{R}_g = 618$ Wm^{-2}. The wind speed near the soil surface is often about 1.0 m s^{-1}.

Bartlett and Gates (1967) determined the effective radiation areas for a lizard. Although they used *Sceloporus occidentalis*, the same ratios can be applied to *Dipsosaurus dorsalis*. The various surface areas are as follows: $A = 180 \times 10^{-4}$ m², $A_S = 45 \times 10^{-4}$ m², and $A_g = A_a = A_4 = 80.1 \times 10^{-4}$ m². The desert iguana has a habit of changing color and having a higher reflectance at body temperatures above about 38°C. Below 38°C, its absorptance is 0.8, and decreases monotonically with temperature; above 43°C, its absorptance is 0.6. For the following calculation, a value of $a_s = 0.6$ is used. The longwave absorptance is $a_t = 1.0$.

It is possible to calculate Q_a for these environmental conditions as follows:

$$
\begin{aligned}
AQ_a &= a_s A_s(S + s) + a_s A_r\, r(S + s) + a_t A_a\, \mathcal{R}_a + a_t A_g\, \mathcal{R}_g \\
&= (0.6 \times 45 \times 977 + 0.6 \times 80.1 \times 322 + 1.0 \times 80.1 \times 388 \\
&\quad + 1.0 \times 80.1 \times 618) \times 10^{-4} \\
&= 2.64 + 1.55 + 3.10 + 4.95 = 12.24.
\end{aligned}
$$

Hence $Q_a = 680$ Wm^{-2}.

Predicted Lizard Temperature

It is now possible to use the environmental data to predict the expected body temperature of the desert iguana. This can be done by putting the environmental data into the energy-budget equation and solving for T_b. This procedure is complicated when done by hand since M and E are functions of the body temperature. An iterative procedure must be used. These calculations are best programmed on a computer which does the iterative successive approximations until a final body temperature which balances the energy budget and the $M - \lambda E$ equations is obtained. A numerical example is given here to illustrate the procedure.

Equations (12.65) to (12.67) are used to calculate $M - \lambda E$ at a given body temperature. The values will not be the same as those used earlier for climate-space determination. One must then determine whether this value of $M - \lambda E$ balances the energy-budget equation [Eq. (12.56)] for the given environmental conditions. If not, then a new T_b is assumed, a new $M - \lambda E$ calculated, and the energy balance tested again.

Assuming a T_b of 45°C, one calculates $M - \lambda E$ to be -1.456 Wm^{-2} from Eq. (12.67). Putting this value and using $K = 502$ Wm^{-2}°C^{-1}, $T_a = 37$°C, $Q_a = 680$ Wm^{-2}, $V = 1.0$ m s^{-1}, and $D = 0.015$ m^2 Eq. (12.56), one gets the following:

$$
\begin{aligned}
680 - 1.45 &= 5.67 \times 10^{-8}\,[45 + 2.88 \times 10^{-3} + 273]^4 \\
&\quad + 3.49\,[1.0]^{0.5}\,[1.5 \times 10^{-2}]^{-0.5}\,[45 - 37 + 2.88 \times 10^{-3}] \\
678.55 &\neq 807
\end{aligned}
$$

The right-hand side is too high, and a new calculation must be made. Try $T_b = 42$°C. Now $M - \lambda E = 0$. Therefore, the right-hand side of the equation gives 680 Wm^{-2} and the left-hand side 700 Wm^{-2}. This is nearly in balance. Only one more iteration should be needed. Try $T_b = 41.4$°C. Now $M - \lambda E = 0.18$ Wm^{-2}, therefore

$$
\begin{aligned}
680 - 0.18 &= 5.67 \times 10^{-8}[41.4 + 273]^4 \\
&\quad + 3.49[1.0]^{0.5}[1.5 \times 10^{-2}]^{-0.5}[41.4 - 37]
\end{aligned}
$$

give 680 Wm^{-2}.

Therefore, 41.4°C is the body temperature which the lizard will assume for these environmental conditions. This temperature is below the thermal maximum, which is about 45°C, and the lizard would tolerate these conditions quite well.

Thermal-Environment Activity Periods

Using essentially the same method as illustrated above, one can calculate for any conditions, below- or aboveground, the body temperature of the desert iguana. Rather than laboriously grinding out iterative calculations by hand, one can program these equations on a computer. Porter et al. did this for the desert iguana for all environments likely to be encountered in the desert. For example, for Palm Springs on 15 July, they showed that daily surface temperatures would reach about 68°C and that an iguana would remain in its burrow until about 0730, emerge into the sun until about 0815, move into the shade and remain there until about 0845, and then retreat to the burrow. Even the shade temperature near the ground surface became too warm for the iguana to remain there after 0845. The animal could remain outside the burrow longer if it climbed a bush where the air temperature was lower than near the ground surface and where there was more air movement. Calculations were done for the 15th day of

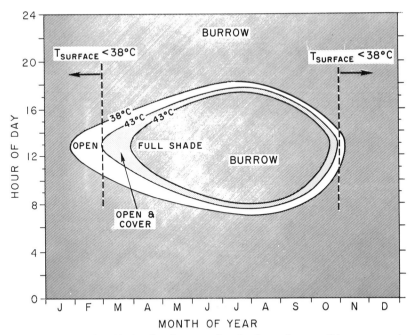

Figure 12.13. Predicted behavioral pattern for the desert iguana, *Dipsosaurus dorsalis*, as a function of time of day and time of year when the animal is either in the burrow or on the surface of the ground. Temperatures give body temperature. (From Porter et al., 1973.)

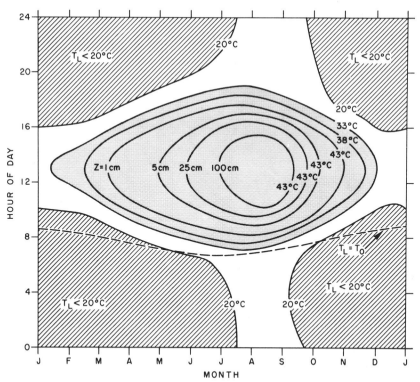

Figure 12.14. Predicted behavioral pattern for the desert iguana, *Dipsosaurus dorsalis,* as a function of time of day and time of year when the animal is able to climb to various heights in bushes. Heights are indicated by values of Z. Temperatures give the lizard body temperature. (From Porter et al., 1973.)

each month of the year. The patterns drawn in Fig. 12.13 are for an animal on the ground surface; those in Fig. 12.14 are for an animal that can climb bushes. It is seen by comparison that the presence of bushes greatly extends the iguana's activity period, particularly if it can climb higher as the season progresses into midsummer.

If able to climb bushes 2.0 m high and remain in deep shade, the iguana could remain outside all day, even in August. While on a bush, the animal is shaded from the direct sun and, to a considerable degree, from the radiant heat of the hot desert sand. Its body "sees" the radiant temperature of the bush, which is close to air temperature.

Porter et al. had some uncertainty concerning the lower active temperature, which they assumed to be about 38°C. They predicted that the soil surface temperature will not reach 38°C for the average year until about 1 March and will stay below this temperature after 1 November. The animal is expected to remain in the burrow most of the time between 1 November and 1 March except during unseasonably warm days. If capable of being active at a body temperature as low as 20°C, the iguana could be nocturnal in the middle of the summer and be outside often in the winter, but this does not seem to be the case. Porter et al. used unpublished data by Mayhew, who gave behavioral sequences in time collected during 10 years of observations of the desert iguana at Palm Springs. His data confirm the predictions of Porter et al. Mayhew indicated a lack of afternoon activity in the summer, although the calculations do predict activity. It is possible that water loss is too great if the animal is outside in both the morning and afternoon during warm summer days.

Biotron Simulation

A particularly nice feature of the investigations made by Porter et al. were the simulations of the desert environment made in a room in the Biotron of the University of Wisconsin. Porter et al. controlled soil temperatures independent of the aboveground environment, where air temperature, humidity, wind speed, and radiation were manipulated to resemble desert conditions. The desert iguana had a burrow in the soil, and there were bushes in the room for shade, and upon which the animal could crawl. A movie camera recorded the location of the iguana as it moved about during the day. It was observed that the animal went through the same behavioral patterns each day as were observed for the lizard on the desert. "On any given day, the darkened lizard emerges slowly and lies on the substrate. As the day progresses, the lizard becomes lighter in color, moves faster, and carries its body higher. Finally, it seeks shade at the base of its bush and also climbs up into the bush. It also uses its burrow to cool off and then re-emerges, the length of stay in the burrow before re-emergence increasing as the day becomes hotter" (Porter et al., 1973, p. 34).

By controlling the aboveground environment separately from soil temperature, the Wisconsin studies showed that soil temperature or animal temperature may act as the emergence cue for the desert iguana. There is some uncertainty about this from their experiments. Kavanau and Norris (1961) have found evidence that circadian rhythms for the burrowing shovel-nosed sand snake, *Chionactis occipitalis,* determine positioning near the surface in anticipation of emergence and that soil surface temperature determines times of actual emergence. The extent to which circadian rhythms play a role in the emergence of the desert iguana is not known.

Water Loss and Food Supply

The thermal environment determines the periods each day when the desert iguana can remain outside the burrow and have its body temperature between 38° and 43°C. When the lizard is outside of the burrow, there is additional energy expenditure due to the heat load. Information concerning the food value, water content, and leaf density of the various desert plant species makes it possible to estimate the time required to obtain food and water. Porter et al. integrated the total metabolism and evaporative water loss over characteristic days for various times of the year. We will not show all the results of their work here since they are available in the original publication. Most of the calculations were done for a 100-g lizard. The very largest of *Dipsosaurus dorsalis* are about 100 g. Most are about 50 g. All numbers can be scaled down for whatever body weight is considered typical. Since Porter et al. used 100 g, this figure is used here also.

Porter et al. calculated the total cumulative daily metabolic requirements for a lizard on the surface at rest and active with a metabolic rate two and five times that of resting metabolism. The value for resting metabolism is probably more realistic since the desert iguana has anaerobic problems and draws down its oxygen supply quickly when active at the higher metabolic rates according to Bennett and Dawson (1972) and Bennett and Licht (1972). Porter et al. also calculated the cumulative daily metabolism for a lizard remaining in the deep soil and one seeking the warmest soil temperature not exceeding 38°C. The metabolic rate at five times resting metabolism cannot be maintained all day, but represents an upper bound for computations of food requirements. The water loss rates calculated represent up to 5 and 24% loss per day of an animal's total body weight and must be restored daily.

The energy content of plants for a carbohydrate–fat mixture is about 19.7×10^6 J kg^{-1} (4705 cal g^{-1}). The digestive efficiency was taken to be from 30 to 50%. For all activity levels, the amount of water (in grams) required is five times greater than the amount of food.

Eating Time Limit

Porter et al. define several kinds of time limits for the animal outside the burrow. The maximum daily activity period is determined by the *thermal time limit* as demonstrated above. The *eating time limit* is the time the animal can remain on the surface using one stomachful of food. The *metabolic time limit* is the time available for surface activity using the calories available from a stomachful of food. The *water time limit* is the time the animal can remain on the surface using the water available from one stomachful of food. All calculations were done on the basis of one stomachful of food. Information concerning how often the desert iguana fills its stomach per day is unavailable. Therefore, rather than assuming an unrealistic value, it is better to make estimates on the basis of one stomachful. Stomach volume and leaf density determine the number of leaves ingested. Eating rate, leaf density, and stomach volume determine how long it takes to fill the stomach. Leaf caloric value and digestive efficiency determine how many joules or calories are available for a given amount of food intake. Leaf water content and digestive efficiency determine the free and metabolic water available from the food. The total water available from the food in a full stomach determines the amount of time available for surface activity. The energy available for surface activity represents that energy available from a full stomach of food which has not been used while the animal is in the burrow. Here we will use centimeters and grams for units since Porter et al. used these units and they involve fewer powers of ten. The calculations are done as follows:

Eating Time Limit (min) = Stomach Volume (cm³)
$\qquad$ × Stomach Fullness (%/100) × Leaf Density (g cm⁻³)
$\qquad$ ÷ Leaf Weight (g/leaf) × Eating Rate (leaves min⁻¹).   (12.68)

Values used for the determination of eating time limit are as follows: stomach volume for a 100-g lizard, 5.3 cm³; leaf density (for *Ambrosia dumosa*), 0.57 g cm⁻³; leaf dry weight, 1.17 to 4.36 mg; leaf water content, 0.43 to 0.78 g H_2O/100 g wet weight; leaf caloric value, 17.2×10^3 J g⁻¹ dry weight; and digestive efficiency, 50 to 70%. For *Larrea divericata*, leaf density is 0.90 g cm⁻³; leaf dry weight, 1.19 to 3.54 mg; leaf water content, 0.47 to 0.53 g H_2O/100 g wet weight; leaf caloric value, 19.1×10^3 J g⁻¹ dry weight; and digestive efficiency, 30 to 50%. The wet weight of *Ambrosia dumosa* is about 18.0 mg/leaf for a large leaf and about 5.47 mg/leaf for a small leaf; the wet weight of *Larrea divericata* is about 7.64 mg/leaf for a large leaf and 2.54 mg/leaf for a small leaf.

Porter et al. assume that one leaf is consumed per bite. An animal smaller than 100 g consumes a leaf per bite but takes less time to fill its stomach. The eating rate may vary considerably from one bite in 4 min to six or eight bites per minute. If a 100-g desert iguana is eating large *Larrea* leaves at a rate of four bites per minute, one can compute the *eating time limit* (ETL) as

$$\text{ETL} = \frac{5.3 \text{ cm}^3 \times 1.00 \times 0.90 \text{ g cm}^{-3}}{7.64 \times 10^{-3} \text{ g/leaf} \times 4 \text{ leaves min}^{-1}} = 156 \text{ min.}$$

Therefore, it will take a 100-g desert iguana 2.6 hr to fill its stomach on *Larrea* leaves. Our animal for the earlier calculations has a body weight of 67 g. This animal will fill its stomach in 105 min.

If a 100-g desert iguana is eating small *Ambrosia* leaves (the preferred diet) at a rate of 4 bites per minute, the ETL is

$$\text{ETL} = \frac{5.3 \text{ cm}^3 \times 1.00 \times 0.57}{5.47 \times 10^{-3} \text{ g/leaf} \times 4 \text{ leaves/min}^{-1}} = 138 \text{ min.}$$

A 67-g desert iguana eating small *Ambrosia* leaves at 4 bites per minute will fill its stomach in 92 min. A very small animal, say a hatchling a few grams in weight, should spend very little time feeding. Porter et al. observed the feeding habits of iguanas on the Mojave Desert in late August during the rainy season. They found that hatchlings spent little time feeding, whereas larger animals, up to 50 g, were taking 75 leaves in a 6-hr day. Some animals were taking as few as 20 and some as many as 100 leaves (bites) per 6-hr day. Calculating the ETL as given above for large *Ambrosia* leaves at 1 bite in 4 min, one gets 336 min, or 5.6 hr.

Metabolic Time Limit

In order to compute the metabolic time limit, one needs to know the energy available for surface activity, which is the total energy available from a stomachful of food minus the energy expenditure in the burrow:

Total Energy Available (J) = Stomach Volume (cm³)
  × Leaf Density (g wet wt cm⁻³) × (1 − Plant Water Content)(%/100)
  × Food Energy (J g⁻¹ dry wt) × Digestive Efficiency (%/100).   (12.69)

A lizard with a digestive efficiency of 80% and a stomach of volume 5.3 cm³ and half full of *Ambrosia* leaves of density 0.57 g cm⁻³ and caloric value 17.2 × 10³ J g⁻¹ dry wt, with a water content of 60%, has, as the total amount of energy available, 8374 J. This value pertains to a 100-g lizard. If the lizard is 67 g, then the total available energy is 5611 J. If the digestion efficiency were 50% rather than 80%, the available energy would be 3509 J for a 67-g lizard and 5234 J for a 100-g lizard. Porter et al. probably should have used a lower digestive efficiency than 80% since they list values about 50% in their Table 2.

A desert iguana in a burrow has a body temperature less than 38°C. Porter et al. computed the energy expenditure of the lizard in the burrow, assuming that the lizard was not within 7.5 cm of the surface.

Porter et al. assumed that inside the times defined by the 38°C contour of their integrated metabolism diagram (not shown here), the desert iguana never exceeds 43°C when on the surface and is at a constant temperature of 38°C when underground. The activity energy for a given day is figured as

Activity Energy (J) = Metabolic Rate at 43°C (J s⁻¹)
            × Activity Time on or above Surface (s)
   + Metabolic Rate at 38°C (J s⁻¹) × Time in Burrow (s).   (12.70)

Porter et al. then calculate the metabolic time limit for various amounts of total energy availability calculated according to Eq. (12.69). If on 1 July, the total energy available is 8374 J, then there are 1.5 hr of activity time available to the 100-g lizard on or above the surface of the ground. If the lizard had a weight of 50 g, then it could be active for about 3 hr. If, however, 10,470 J of total energy were available on 1 July, the 100-g lizard could have a very long active period of about 8 hr. If more energy than this were available, the lizard could not use it all because it would reach a limit for activity imposed by the thermal environment of about 10 hr on 1 July. Earlier in the year, say on 1 April, the thermal time limit occurs at 6 hr, and even less energy (about 6280 J) is used in activity; any additional energy gained would go into fat production.

Fat Production

Porter et al. evaluated the amount of fat production required for a 100-g desert iguana to hibernate from 1 November to 1 March. They integrated the metabolism of a lizard in a burrow at deep soil temperature to obtain a total metabolic requirement of 138 kJ. A lizard seeking the warmest burrow temperatures not exceeding 38°C will need 191 kJ. If only fat is consumed during hibernation and each gram of fat yields 37.7 kJ, then the animal needs 3.5 to 5.1 g of fat to survive the winter. Our 67-g animal will require 92.5 to 127 kJ, or 2.3 to 3.4 g of fat, for hibernation. The number of stomachfuls of food needed, if only used to provide fat, can now be estimated as

No. of Full Stomachs = Energy from Fat Utilized (J)
$\div$ [Conversion Efficiency of Carbohydrate to Fat
$\times$ Digestive Efficiency $\times$ (1 $-$ Plant Water Content)
$\times$ Food Energy (J g^{-1} dry wt) $\times$ Leaf Density (g cm^{-3})
$\times$ Stomach Volume (cm^{-3})]   (12.71)

The values used are conversion efficiency of carbohydrate to fat, 50%; digestive efficiency, 60 and 50%; plant water content, 60 and 50%; food energy, or leaf caloric value, 17.2×10^3 and 19.1×10^3 J g^{-1} dry wt; leaf density, 0.57 and 0.90 g cm^{-3}. Values are for *Ambrosia* and *Larrea*, respectively. For a 100-g lizard with a stomach volume 5.3 cm^3, necessary number of full stomachs of *Ambrosia* $= 37.7 \times 10^3 \times (3.5$ to $5.1) \div$ $[0.50 \times 0.60 \times 0.40 \times 17.2 \times 10^3 \times 0.57 \times 5.3]$ is 21 to 31; number of full stomachs of *Larrea* $= 37.7 \times 10^3 (3.5$ to $5.1) \div [0.50 \times 0.50 \times 0.50 \times$ $19.1 \times 10^3 \times 0.90 \times 5.3] = 12$ to 17.

These are the number of stomachfuls of food required to develop sufficient fat for hibernation. A 67-g lizard would require correspondingly less. If one stomachful per day went for fat production, then from 2 weeks to a month of feeding is needed to acquire sufficient fat for a 100-g lizard to hibernate.

Annual Food Consumption

One can estimate the total number of leaves consumed by a lizard over an entire year. Integrating the resting metabolism curve from 1 March to 1 November gives a requirement of 2.09×10^6 J for a 100-g lizard. Using the same calculation as above, an animal eating *Ambrosia* would need 160 stomachfuls of food, and one eating *Larrea* would need 90 stomachfuls. Adding to this the number of stomachfuls required for fat production, i.e., about 30 and 15, the annual number of stomachfuls would be 190 and 105 for a 100-g animal eating *Ambrosia* and *Larrea,* respectively.

These values are equivalent to 574 g wet weight of *Ambrosia,* and 501 g wet weight of *Larrea.* A 100-g desert iguana would be expected to consume a minimum of 32,000 to 105,000 leaves of *Ambrosia* or 66,000 to 197,000 leaves of *Larrea* depending on the leaf size. These numbers assume that the wet weight of large and small leaves is 18.0 and 5.47 mg for *Ambrosia* and of 7.64 and 2.54 mg for *Larrea.*

Water Time Limit

The *water time limit* is the time an animal can remain on the surface using the water available from one stomachful of food. Assuming steady state, i.e., that water reserves are neither being built up nor depleted, the amount of water available for evaporative loss is

Available Evaporative Water = Plant Water In
$$+ \text{ Water from Metabolism } - \text{ Feces Water Out,} \quad (12.72)$$

where

Plant Water In (g) = Stomach Volume (cm^3) × Leaf Density (g cm^{-3})
$$\times \text{ Plant Water Content } (\%), \quad (12.73)$$

Water From Metabolism (g) = Stomach Volume (cm^3)
$$\times \text{ Leaf Density (g } cm^{-3}) \times (1 - \text{ Plant Water Content})$$
$$\times \text{ Food Energy (J g}^{-1} \text{ dry wt)} \times \text{ Metabolic Water (g } H_2O \text{ cal}^{-1})$$
$$\times \text{ Digestive Efficiency,} \quad (12.74)$$

and

Feces Water Out (g) = Stomach Volume (cm^3)
$$\times \text{ Leaf Density (gm } cm^{-3}) \times (1 - \text{ Plant Water Content})$$
$$\times (1 - \text{ Digestive Efficiency}) \times \text{ Feces Water Content } (\%/100). \quad (12.75)$$

Earlier, the *metabolic time limit* was calculated for a 100-g desert iguana with a stomach half full of *Ambrosia* leaves. The metabolic water production is 3.58×10^{-5} g H_2O per calorie of food (glucose) utilized. Assuming that the leaves have a water content of 60%, as before, and that the lizard has a digestive efficiency of 60% and its feces have a water content of 60%, one gets:

$$\text{plant water in } = 5.3 \times 0.57 \times 0.60 = 1.81 \text{ g;}$$

$$\text{plant water from metabolism in } = 5.3 \times 0.57 \times 0.40$$
$$\times 17.2 \times 10^3 \times 3.58 \times 10^{-5} \times 0.60 = 0.447 \text{ g;}$$

$$\text{feces water out } = 5.3 \times 0.57 \times 0.40 \times 0.40 \times 0.60 = 0.290 \text{ g;}$$

$$\text{available evaporative water } = 1.81 + 0.447 - 0.290 = 1.97 \text{ g.}$$

Porter et al. estimated that an activity time of about 5 hr would be available with this amount of evaporative water for a 100-g desert iguana consuming *Ambrosia*. A 50-g animal would have about 10 hr of activity time if it had 2 g of evaporative water available. No water loss is assumed in the burrow, and the time limit is seen to be largely independent of season. Even if water loss in the burrow was assumed to occur, it would be relatively low because of high humidity in the burrow, and the time limits would be nearly unaffected.

Predator–Prey Interaction

If one can work out the behavioral patterns for several animals in the same manner as shown for *Dipsosaurus dorsalis*, then it is possible to do so for two animals which have a predator–prey relationship. Energy-balance considerations for animals in various microclimates can be used to predict activity times, food requirements, and potential predator–prey interactions.

Porter et al. (1973) developed a hypothetical predator–prey analysis by

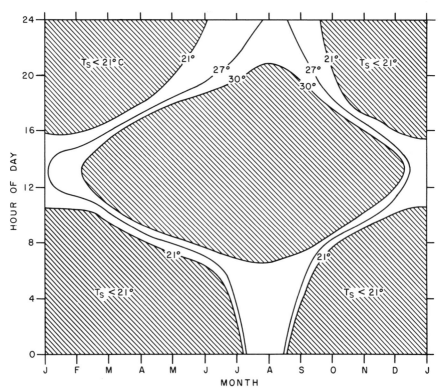

Figure 12.15. Predicted seasonal behavior pattern for a hypothetical desert ant. (From Porter et al., 1973.)

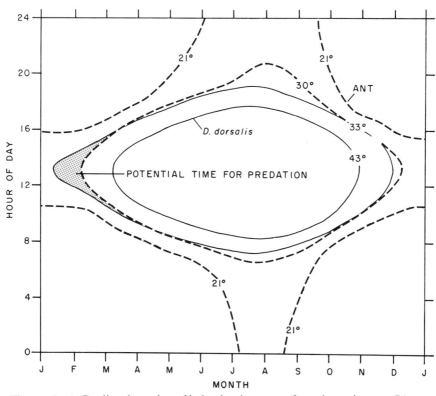

Figure 12.16. Predicted overlap of behavioral patterns for a desert iguana, *Dipso-saurus dorsalis,* and a hypothetical desert ant. The coincidence in time and space indicates the potential time for predation. (From Porter et al., 1973.)

overlapping their analysis of the real-world behavior of the desert iguana with the activity pattern of a hypothetical low-temperature ant. They assumed that the activity temperature range for the ant was 27° to 31°C and predicted the annual activity pattern for the ant on the surface of the ground at Palm Springs, Calif., shown in Fig. 12.15. It is seen that the ant will have a single activity pattern, in the middle of the day, in the winter, a biomodal pattern in spring and fall, and a nocturnal activity period in summer.

In order to determine whether or not the desert iguana can prey on the ant, it is necessary to overlay the activity pattern of the iguana on that of the ant. This is shown in Fig. 12.16; the small cross-hatched area reveals the very small period of the year when the two animals can occupy the same physical environment. Therefore, predation by the desert iguana of the hypothetical low-temperature-preference ant will only be possible at midday between 15 January and about 15 February to 1 March. Since, as a general rule, the iguana does not emerge from its burrow until about 1 March, there will be little or no predation on the ant. Of course, if one considered an ant with a higher temperature preference, there might be a substantial amount of iguana predation on ants at some times of year.

Tropical Lizard Behavior

So far, the thermal behavior of lizards in dry or desert environments only has been examined. Huey (1974) reported detailed observations of the Puerto Rican lizard, *Anolis cristatellus*, showing that this lizard behaviorally regulates its body temperature in an open habitat but passively tolerates lower and more variable temperatures in an adjacent forest where basking sites are few and far between. He suggests that this tropical lizard finds thermoregulation to optimal temperature levels impractical when costs resulting from losses of time and energy are high. *Anolis cristatellus* apparently regulates body temperature behaviorally in habitats where it can shuttle readily between sun and shade but accepts a lower and more variable body temperature in the forest. Thermoregulation under high environmental temperatures is adaptive, but within certain limits: active thermoregulation yields greater benefits than passivity only when costs of regulation are low. *Anolis cristatellus* is a lizard which has evolved to be active within a broad range of body temperatures. Huey suggests that this accounts for its success as a colonist in the Caribbean and that its behavior is in distinct contrast to lizards found in desert or open habitats. He also suggests that the reason for the lack of lizards in temperate-zone forests may be that thermoregulation there is difficult and passivity would not yield sufficient benefits. In other words, it may be too cool in the shade of temperate forests for the lizard's metabolism to function. The next step would be a solid thermal-exchange analysis for *Anolis cristatellus* similar to that of Porter et al. (1973) since Huey's conclusions are somewhat conjectural and it would be good to have a quantitative analysis as evidence. However, this requires detailed physiological data concerning *Anolis* that are probably unavailable.

Lizard on a Tree Trunk

The first detailed energy-exchange analysis of a lizard was presented by Bartlett and Gates (1967) using measurements and observations of the fence lizard, *Sceloporus occidentalis*. It was in this paper that many of the techniques for determining surface areas, radiative areas, convection coefficients, and other physical properties used by subsequent investigators were worked out. The purpose of the energy-budget analysis for *Sceloporus occidentalis* was to predict the position of this lizard on tree trunks as a function of time for a summer day and to compare the predictions with observations made in the field.

Sceloporus occidentalis had the following properties: body weight, 18.4 g; surface area, 75.8 cm²; area in contact with tree trunk, 11.3 cm²; effective area for the absorption of solar radiation, variable, from 3.3 to 19.2

cm^2 (as shown in Fig. 12.10); total emitting area, 67.2 cm^2, or 89% of the actual total surface area; effective emitting area of the lizard not in contact with the tree, 57.4 cm^2; absorptance to direct solar radiation, 0.885; absorptivity to skylight, 0.90; and absorptance to thermal radiation, 0.965. The metabolic rate was 5.51 Wm^{-2}, and the evaporative water loss was 1.81 Wm^{-2}. The mean body temperature was 34.1°C as determined by field measurements. The known range of body temperatures for *Sceloporus occidentalis* was 32.9° to 36.9°C.

The problem which Bartlett and Gates wished to solve was to determine the heat load on the lizard as a function of time of day when the lizard was situated vertically on the trunk of a valley oak, *Quercus lobata,* and to predict the lizard's position during the day. They chose as a study site Chew's Ridge in the Los Padres National Forest, 6 mi SSE of Jamesburg, Calif., since many observations of *Sceloporus occidentalis* had been made there by Bartlett and others. They chose, as those of a typical summer day, the conditions of 21 June. The environmental conditions are described in Bartlett and Gates (1967). Tree surface temperatures were estimated using the technique described by Derby and Gates (1966). The following calculation is for the conditions which prevailed at 1300 on 21 June, a clear day. The solar azimuth was 50°, and the altitude was 72°. For these angles, the lizard surface area available for absorbing direct solar radiation was 6.2 cm^2. The solar intensity perpendicular to the solar rays was 1103 Wm^{-2}, and scattered skylight was 27.9 Wm^{-2}. The thermal radiation emitted by the ground was 223 Wm^{-2}, and that emitted by the atmosphere was 126 Wm^{-2}. The reflectance of the ground surface to sunlight was 0.05. The air temperature was variable, but a value of 23°C is used. The bark temperature on the south side of the tree at 1300 was 41°C. The conductivity of tree bark was 8.37×10^{-2} Wm^{-1}°C^{-1}, and the thickness of the bark across which heat could flow to the lizard was 10^{-3} m.

For the following calculation, used purely as an example, the lizard was assumed to be on the south side of the tree at 1300 on 21 June.

i. *Energy gain.*

Direct solar radiation absorbed $= (a_S \, S \, A_S)$
$$= 0.88 \times 1103 \times 6.2 \times 10^{-4} = 0.61 \text{ W.}$$

Skylight absorbed $= (a_s \, s \, A) = 0.90 \times 27.9 \times 5714 \times 10^{-4} = 0.15$ W.

Reflected sunlight absorbed $= [a_s \, r(S + s) \, A_e]$
$$= 0.88 \times 0.05 \, (1103 + 27.9)57.4 \times 10^{-4}$$
$$= 0.29 \text{ W.}$$

Thermal radiation absorbed $= [a_t(R_a + R_g) \, A_e]$
$$= 0.96(126 + 223) \, 57.4 \times 10^{-4} = 1.93 \text{ W.}$$

Metabolic heat production $= 0.04$ W.

The lizard gains heat from the tree by conduction. The maximum rate of heat exchange, which occurred when the lizard surface temperature was 29.3°C and the temperature difference was 12.7°C, was 1.20 W. The minimum rate was when the lizard was not in direct contact with the tree bark and the surface temperature of the lizard was 38.9°C. In this case, energy was exchanged between lizard and tree by radiation and conduction through air to give a total gain by the lizard of 0.10 W. This calculation is given in Bartlett and Gates (1967). Total energy gain was 4.21 W maximum or 3.12 W minimum.

ii. *Energy loss.* Energy loss occurred by radiation, convection, and evaporation. Three situations are considered.

Maximum rate of heat loss occurred when the surface temperature was 28.9°C and a wind of 2.24 m s^{-1} was blowing across the lizard. The difference in temperature between the lizard's surface and the air was 15.9°C. The convection coefficient used by Bartlett and Gates (1967) was 25.8 Wm^{-2}. Their treatment of convection was somewhat different and more empirical than that used in this book, but the numbers they used are used here for ease of comparison. The evaporative rate of cooling by water loss was 1.4×10^{-2} Wm^{-2}.

$$\begin{aligned} \text{Maximum heat loss} &= 0.96 \times 5.67 \times 10^{-8} \times 57.4 \times 10^{-4} (38.9 + 273)^4 \\ &\quad + 25.8 \times 64.5 \times 10^{-4} \times 15.9 + 1.4 \times 10^{-2} \\ &= 2.94 + 2.64 + 0.01 = 5.59 \text{ W.} \end{aligned}$$

Minimum heat loss was calculated with a lizard surface temperature of 29.3°C and no wind. The convection coefficient in still air was taken as 34.2 Wm^{-2}°C^{-1}.

$$\begin{aligned} \text{Minimum heat loss} &= 0.96 \times 5.67 \times 10^{-8} \times 57.4 \times 10^{-4}(29.3 + 273)^4 \\ &\quad + 34.2 \times 64.5 \times 10^{-4} \times 6.3 + 1.4 \times 10^{-2} \\ &= 2.61 + 0.14 + 0.01 = 2.76 \text{ W.} \end{aligned}$$

Finally, an intermediate situation with body temperature of 34.1°C and a reduced wind speed of 0.447 m s^{-1} gave a heat loss of 4.09 W.

In this laborious fashion, Bartlett and Gates calculated the position of the lizard on a tree trunk as a function of the time of day. The wind speed was one of the most difficult factors to know with precision. The numerical calculations shown above are given as examples as to how the estimates were made. The lizard would be found in positions where energy loss fell between maximum and minimum energy gained. The intermediate value of energy loss of 4.09 W would put the lizard between the maximum and minimum energy gains of 4.21 and 3.12 W and, in fact, quite close to the maximum energy-gain situation. The actual positions of lizards on the tree trunks were observed in the field by Bartlett for 21 June at Chew's Ridge. The agreement between predicted positions and observed positions was very satisfactory. These results are shown in the

paper. Use of a computer would have greatly facilitated these very laborious calculations.

The main point of this example is to show that observational field work
can be much more meaningful when coupled with good theory and quantitative analysis. If analysis is done before going into the field, much
better insight is achieved into the phenomena being observed and one
knows the particular kinds of measurements which must be made.

Alligators

In the recent past, the American alligator, *Alligator mississippiensis,* was
widespread throughout the Gulf and Atlantic Coastal Plain of the southeastern United States. Its territory today is greatly restricted compared to
its original range because of human activities. The alligator is an important animal in the wetland ecosystems and plays a dominant role in
shaping plant communities and maintaining animal life. In order to explore the climatic limits imposed on the alligator, Spotila et al. (1972) investigated the energy budget of this animal. The following example is
taken from that paper; detailed justification of various values used is also
found therein.

Water Loss and Metabolic Rates

The rates of water loss were in part measured and in part inferred from
other work. The Q_{10} for water loss from 10.5° to 19.5°C was 2.4. Metabolic
rates were for a 0.0750-kg alligator but were converted to heat production
per unit surface area. Oxygen consumption was converted to joules by
using 19,260 J $(\ell\, O_2)^{-1}$. The Q_{10} for metabolic rate from 25° to 30°C was
4.0. This figure was used to estimate the metabolic rate at a body tempera-

Table 12.1. Water-Loss and Metabolic Rates for a 0.750-kg Alligator on a Unit-
Surface-Area Basis[a]

Body temperature (°C)	M	λE	$M - \lambda E$	λE_r, cutaneous	λE_b, respiratory
3	0.28	1.82	−1.54	1.26	0.56
5	0.28	2.23	−1.95	1.53	0.70
10	0.35	3.35	−3.00	2.38	0.97
20	0.77	7.89	−7.12	5.51	2.38
30	3.07	17.93	−14.86	13.47	4.46
38	9.28	39.08	−29.8	30.36	10.12

[a] Units are watts per square meter.

ture of 38°C. The metabolic rate at 3°C was assumed to equal that at 5°C. Table 12.1 gives the water-loss and metabolic rates (λE and M) as a function of body temperature. It should be noted that the rates of respiratory water loss exceed the metabolic rate at all body temperatures, i.e., evaporative cooling is always critical to the thermoregulation of an alligator and an adequate water supply is necessary for its survival.

Conduction and Insulation

Heat is exchanged by conduction when the ventral surface of the alligator is in intimate contact with the ground. The conductivity of the animal's insulating layer of fat is the limiting factor in heat flow since values for the thermal conductivity of soils and rocks are very much greater. The conductivity of fat is $k_f = 0.205$ Wm^{-1}°C^{-1}. The conduction of heat between the animal and the ground is given by

$$G = \frac{k_f A_g}{d_f A} (T_b - T_g). \tag{12.76}$$

The thickness of fat is $d_f = 0.007$ m for an alligator of diameter $D = 0.15$ m. The ratio of surface area A_g in contact with the ground to total surface area A is 0.23. Hence

$$G = 6.76(T_b - T_g). \tag{12.77}$$

The maximum temperature difference between alligator and ground is thought to be about 3°C. Hence the greatest rate of heat loss or gain by conduction is 20.3 Wm^{-2}.

The same fat thickness was taken for the insulation of the parts of the alligator body not in contact with the ground surface. The insulation $I = d_f/k_f$, or 0.034 m^2°C W^{-1}. When the climate-spaces were determined for alligators of various sizes, a constant ratio of d_f to D was used, i.e., 0.0467.

Body Temperature

Upper lethal temperature for an alligator is about 38°C, and the lower limit is 3°C. The preferred body temperature is 32° to 35°C.

Climate-Space

The steady-state energy budget for an alligator in contact with the ground is given by

$$Q_a + M = 5.67 \times 10^{-8}[T_b - 0.034 (M - \lambda E_b) + 273]^4$$
$$+ 3.49 V^{0.5} D^{-0.5}[T_b - 0.034 (M - \lambda E_b) - T_a]$$
$$+ \lambda E_b + \lambda E_r + 6.76(T_b - T_g). \tag{12.78}$$

Using the values of M, λE_b, and λE_r given in Table 12.1 for thermal maximum and minimum, one can determine the equation of the lines for the upper and lower climate-space limits.

At the thermal maximum, $T_b = 38°C$, $M = 9.28$ Wm^{-2}, $\lambda E_b = 10.11$ Wm^{-2}, and $\lambda E_r = 30.36$ Wm^{-2}. If $V = 0.10$ m s^{-1} and $D = 0.15$ m, then

$$Q_a = 678 - 2.85\ T_a + 6.76\ (38 - T_g). \tag{12.79}$$

At the thermal minimum, $T_b = 3°C$, $M = 0.28$ Wm^{-2}, $\lambda E_b = 0.56$ Wm^{-2}, $\lambda E_r = 1.26$ Wm^{-2}, and

$$Q_a = 339 - 2.85\ T_a + 6.76\ (3 - T_g). \tag{12.80}$$

Calculation of the climate-space using these equations will yield slightly different limits than those given by Spotila et al. (1972) because a somewhat different convection coefficient is used here. The ground surface temperature may be taken to be approximately 38°C at thermal maximum and approximately 3°C at thermal minimum.

Shown in Fig. 12.17 is the climate-space diagram for an alligator 0.15 m in diameter as given by Eqs. (12.86) and (12.87). The normal physiological climate-space limits at thermal maximum and minimum are the lines A_0 and B_0 when the ground surface temperature is equal to the alligator body temperature. If the ground surface is 3°C warmer or cooler than the animal's body temperature, the physiological limits shift to A_1 and A_2 or B_1 and B_2, respectively, at thermal maximum and minimum. The horizontal line is probably a realistic lower limit for the air temperature of the natural environment in which the alligator lives. The reason for this is that since, under blackbody conditions, the alligator has a minimum air temperature limit of close to 3°C, and this is the air temperature which would be encountered at night, it is unlikely that daytime air temperatures will be less than this value. The alligator clearly could withstand colder air temperatures when basking in full sunshine, but unless it went into warm water at night, or went underground, it could not withstand the cold nighttime air temperatures associated with these cold daytime temperatures. Heat transfer by an alligator in water or underground is largely by conduction and would demand water or soil temperatures well above 3°C.

Spotila et al. computed the climate-spaces of alligators of different diameter. These diagrams are shown in Fig. 12.18. The influence of body size is most pronounced in small animals. For small animals, wind limits the maximum air temperatures that the animal can withstand when in steady state in full sunshine. Once the body size is greater than about 0.20 m, further change in diameter has very little influence.

An alligator can shuttle between water and the terrestrial environment, between sun and shade, etc. Steady-state energy-budget calculations do not provide the limits to such behavior. An alligator can exceed its climate-space limits for periods of time that are small compared to its time constant. If an alligator were to bask in sunshine at air temperatures well above the limiting line, then it would have to be well within the climate-space quadralateral before and after basking so that its body

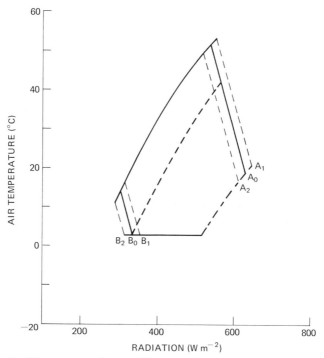

Figure 12.17. Climate-space for an alligator, *Alligator mississippiensis,* 0.15 m in diameter for a wind speed of 0.10 m s^{-1}. The alligator cannot survive in any habitat where the temperature of available blackbody cavities falls below 3°C, the lower horizontal line. The upper and lower physiological limits shift to A_1 and A_2 or B_1 and B_2 from A_0 and B_0 when the ground surface is 3°C warmer or 3°C cooler than the animal body temperature. (Diagram from Spotila et al., 1972.)

would be cool to begin with and able to cool off afterward. The thermal time constant of reptiles is approximately given by

$$\tau = 48.9 \, D^{1.38}. \tag{12.81}$$

If D is in meters then τ is in hours according to analysis by Spotila et al. (1973). This is discussed in Chapter 13 in the consideration of time-dependent energy budgets. A reptile of diameter 0.20 m has a time constant of 5.3 hr; if the diameter is 1.0 m, the time constant is 49 hr. If, however, the reptile's diameter is 0.05 m, the time constant is 0.78 hr, or about 47 min. Most adult alligators are of sufficient size to have considerable thermal lag and can spend fairly long periods outside their steady-state thermal limits.

Alligators are aquatic animals whose existence is dependent upon the availability of standing water. At all stages in their life history, from hatchling to adult, they rely on water as their primary medium for locomotion and main source of food. If the water supply dries up, the animals' behavior patterns are seriously disrupted and their thermoregulatory abil-

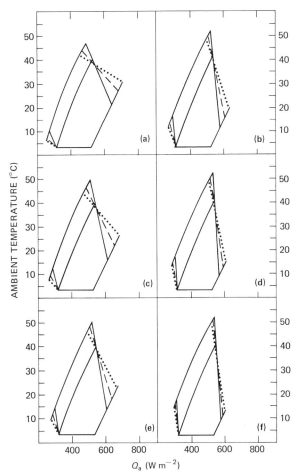

Figure 12.18. Climate-space diagrams for alligators (*Alligator mississippiensis*) of different diameter *D* for wind speeds of 0.1, 1.0, and 5.0 m s^{-1}: (a) *D*, 2 cm; (b) *D*, 20 cm; (c) *D*, 5 cm; (d) *D*, 50 cm; (e) *D*, 10 cm; (f) *D*, 4.0 cm.

ity is destroyed. Alligators enter water to avoid thermal stress. During the summer, they avoid the sun during the heat of the day and restrict most of their terrestrial activities to the hours of darkness. During the winter, they bask on land during sunny days but remain in water when the air temperature is below the water temperature. All of this behavior is completely consistent with the climate-space diagram as calculated from steady-state energy-budget theory.

Chipmunks

The chipmunk is a fairly ubiquitous animal throughout the cooler regions of North America. In the Sierra Nevada mountains of California, four species of chipmunk are found altitudinally zoned and contiguously allopatric along the east slope of the cordillera as shown in Fig. 12.19. Dr. H. Craig Heller studied these chipmunks for his Ph.D. thesis at Yale University. The thesis was designed to answer a very fundamental question as to which factors determine and maintain the lines of contact between these different populations. If the lines of contact represent physiological limitations within each species, then the fundamental niches of the species do not overlap and various physiological, and perhaps behavioral, interspecific differences exist. But if the physiological limitations are very broad, with each species maintaining a limited distribution, it may be that competitive exclusion occurs. Heller's thesis was undertaken to find out which explanation is correct. Since the study involved careful physiological studies, and therefore energy-balance studies, in addition to behavioral investigations, Heller worked with D. M. Gates for part of the analysis. Together, they worked out the energy-balance studies reported by Heller and Gates (1971). The following is a summary of those results.

It is noted from Fig. 12.19 that these chipmunk species array themselves altitudinally from high to low elevation in the order *Eutamias alpinus*, *E. speciosus*, *E. amoenus*, and *E. minimus*. The aggressively dominant species *E. alpinus* and *E. amoenus* appear to limit the altitudinal ranges of *E. speciosus* and *E. minimus* through interference competition (see Heller, 1971). Comparison of the species' adaptation to aridity and high temperatures indicated that *E. amoenus*, *E. speciosus*, and *E. alpinus* are physiologically unable to occupy the habitat of *E. minimus*. That the habitats of the four species extend from the alpine to the desert, suggests that energy-budget considerations may be an important aspect of the adaptation and distribution of these species.

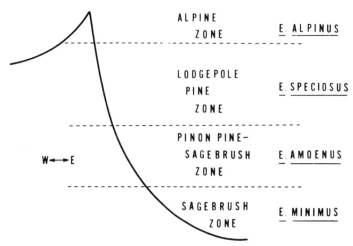

Figure 12.19. Altitudinal zonation of four *Eutamias* species that inhabit the east slope of the Sierra Nevada mountains in California. (From Heller and Gates, 1971.)

Animal Properties

Metabolic and water-loss rates were measured as a function of ambient temperature within a metabolic chamber. The metabolic rates, expressed as heat production per surface area, are shown in Fig. 12.20. Table 12.2 lists most of the pertinent physical and physiological data for the four species of *Eutamias*. The water-loss rate was not partitioned into respiratory and skin evaporation. Experiments conducted in a small wind tunnel gave an empirical determination of the convection coefficients over a limited range of wind speeds. The values obtained are given as a linear function of the wind speed V. This seemed to Heller and Gates to be a reasonable approximation at the time, and since these expressions were used in their article, they are also used here.

The evaporative water-loss rates require some explanation. Heller and Gates used the evaporation rate at 35°C to compute the maximum limit of

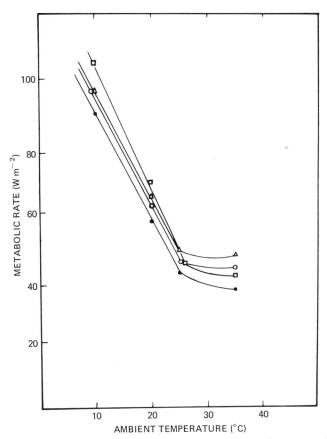

Figure 12.20. Surface-area-specific metabolic rates as a function of ambient temperature for four *Eutamias* species: (●) *E. alpinus;* (△) *E. speciosus;* (○) *E. amoenus;* (□) *E. minimus.* (From Heller and Gates, 1971.)

Table 12.2. Properties of Four Species of the Chipmunk Genus *Eutamias* in the Sierra Nevada Mountains

Species	W (g)	A (cm^2)	M (Wm^{-2})	E (Wm^{-2})	$M - \lambda E$ (Wm^{-2})	I (m^2 °C W^{-1})	h_c (Wm^{-2} °C^{-1})	$I(M - \lambda E)$
Eutamias alpinus	39.1	87	36.3	18.8	17.5	0.171	$0.049V + 2.25$	2.985
Eutamias speciosus	70.7	121	47.5	20.2	27.3	0.118	$0.037V + 1.65$	3.221
Eutamias amoenus	42.4	94	43.3	18.1	25.2	0.126	$0.053V + 2.46$	3.161
Eutamias minimus	35.2	77	40.5	14.0	26.5	0.128	$0.044V + 3.25$	3.401

the climate-space even though the thermal-maximum body temperature was 40°C. The reason for doing so was that they did not consider the chipmunk capable of sustaining the maximum rate of evaporative water loss long enough to be in a steady state. It is likely as a result, that Gates and Heller underestimated the thermal maximum limit to the chipmunk's climate-space. Metabolic rates were not measured at 40°C, but only up to 35°C. The other reason for using an evaporative water-loss rate measured at 35°C is that this gave a consistent set of values for $M - \lambda E$. It is likely that $M - \lambda E$ would not have been too different as a body temperature of 40°C was approached. Whether right or wrong, this was the basis for computing the maximum limit to the climate-space.

The pelage of the chipmunk had a mean absorptance to solar radiation of about 0.73. The maximum body temperature was considered to be 40°C and the minimum 4°C during hibernation.

Climate-Space

The chipmunk is a diurnal animal and will not be exposed to the night sky. Using the numbers given in Table 12.2, one gets the climate-spaces for the four species of *Eutamias* shown in Fig. 12.21. The climate-spaces of *E. alpinus*, *E. amoenus*, and *E. minimus* are very similar, but the upper limit of *E. speciosus* in still air is much lower at high values of Q_a. The ambient temperature at which *E. speciosus* can tolerate its maximum Q_a in still air is about 12°C, which is lower than the T_a at which the other species can tolerate the same Q_a. The dotted lines indicate the T_a in still air at which each species can tolerate the level of Q_a which is maximal for *E. speciosus* superimposed on the other climate-spaces. The prediction that *E. speciosus* would suffer heat death before the other species when exposed to a dry heat was confirmed by laboratory tests. The reasons for the greater heat stress in *E. speciosus* are seen in Table 12.2. This species has a larger M, λE, and $M - \lambda E$, smaller I, and a substantially greater body size than the other three species. Convective cooling is lowest for this species. The quantity $I(M - \lambda E)$ is of intermediate value. The energy budget is written

$$Q_a + M - \lambda E = 5.67 \times 10^{-8} [T_b - I(M - \lambda E) + 273]^4$$
$$+ h_c[T_b - T_a - I(M - \lambda E)]. \quad (12.82)$$

It is clear that if the left-hand side is maximal (as it is for *E. speciosus*), then, at fixed body temperature of 40°C, the right-hand side can balance only if the air-temperature limit goes down substantially. The animals could raise their upper limits by increasing evaporative water loss, but in an arid environment, the animal must also remain within the limits of its water budget. The upper limits could be raised by increasing T_b. *Eutamias minimus*, for example, responds to dry heat more through hyperthermia and less through evaporative water loss than the other three species. The

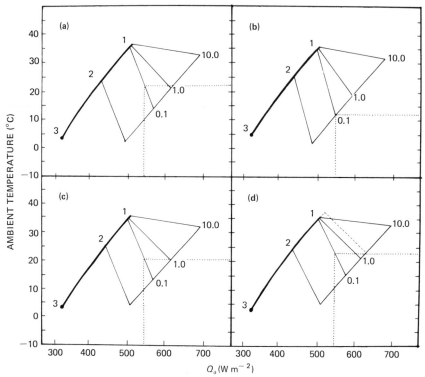

Figure 12.21. Climate-spaces for four *Eutamias* species: (a) *E. alpinus;* (b) *E. speciosus;* (c) *E. amoenus;* (d) *E. minimus.* (From Heller and Gates, 1971.)

dashed lines on the climate-space diagram for *E. minimus* indicate the extension which results from an increase of T_b to 42°C with a wind velocity of 1.0 m s^{-1}. At any Q_a, the animal could withstand a 3.6°C increase in T_a. If the same body temperature increase occurred in still air, the air-temperature increase would be even greater, probably 5° or 6°C.

Before completing our discussion of the four mountain species of chipmunk, let us look at a climate-space diagram computed for a chipmunk of the Great Lakes region, *Tamias striatus*, from some data taken in my laboratory several years ago by Dr. A. Morhardt. The climate-space is shown in Fig. 12.22. Morhardt used a higher thermal maximum for T_b, i.e., about 42.5°C, than Heller and Gates. It is my feeling that the upper bounds in this diagram are somewhat more realistic than the much more conservative upper bounds for the four species of *Eutamias*. However, there is a question of limiting water supply at the upper limit. If the chipmunk cannot maintain a fairly high water-loss rate, then it cannot endure at as high a temperature as shown here. Figure 12.22 also shows the limitations to the left of the blackbody line. A chipmunk will encounter such a situation in the daytime when shaded from the sun and exposed to the clear sky.

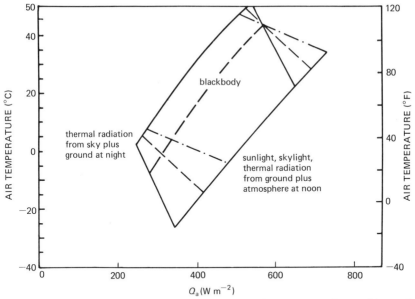

Figure 12.22. Numbers on the right represent climate-space for a chipmunk, *Tamias striatus,* at wind speeds of (—) 0.1 m s⁻¹, (– –) 1.0 m s⁻¹, and (–·–) 10.0 m s⁻¹. The animal is 0.037 m in diameter.

It is possible for *Tamias striatus* to exist at the lower thermal limits of its climate-space for limited periods of time. The chipmunk has the fur insulation and metabolic capacity to endure temperatures below freezing. However, it does not usually do so, since it is a hibernating animal and remains in its burrow during the winter. Heller and Gates elected to draw a lower limit to the climate-space diagram for *Eutamias* based on maximum sustainable metabolism rather than maximum attainable metabolism. The maximum sustainable metabolism is limited by food supply during cold climatic conditions.

Chipmunks are hibernators. They forage for food during active times and store it in their burrows for consumption during the 4 or 5 months of cold conditions aboveground. It is the amount of food available for winter consumption that determines the average maximal metabolic rate the animal can sustain during the inactive time.

White-Tailed Deer

The white-tailed deer, *Odocoileus virginianus,* is widely distributed in North America and adapts to a wide variety of ecological conditions. Moen (1968a,b) has reported on energy-exchange studies of the white-tailed deer in Minnesota during severe winter conditions. Moen quotes a number of investigators who report that deer urgently need cover during severe winter situations and that as temperatures drop and snow depths increase, deer tend to congregate in areas providing good protection. However, Moen questioned some of these assertions because of the fact that deer exist in regions of western Minnesota and in North Dakota and South Dakota in habitats which contain no coniferous cover and only scattered areas of deciduous vegetation.

Moen was interested in the amount of metabolic heat production needed by a deer to survive certain environmental conditions. He measured the nocturnal-radiation environment of the deer in an open field under clear skies for various air temperatures and then solved the energy-budget equation in order to determine the metabolic heat required for the animal to be in energy balance. He did calculations for 30-, 50-, and 70-kg deer of diameter 0.305 m. He obtained the following expression between the weight W in kilograms and the surface area A in square meters:

$$A = 1.375 + 9.1 \times 10^{-3} W. \tag{12.83}$$

Moen used a standard evaporative water loss of 12.00 Wm^{-2} for each deer. Rather than using an equation similar to Eq. (12.90) and calculating the heat loss for a given environmental condition, he made an animal simulator from which he could measure the rate of heat loss by radiation. This consisted of a 1-gallon metal can filled with warm water over which was stretched the hide of a deer. With an infrared radiometer, he measured the radiant heat loss from the surface of the fur for various environmental conditions. He used a standard heat-transfer formula for convective heat loss from the animal. Rather than attempting to reproduce all of his numbers here, a calculation is given to show the metabolic heat production for a given set of conditions. Moen did not give an explicit number for the insulation of the fur. He did indicate that the fur was about 0.02 m thick. Using a conductivity of 4.2×10^{-2} $Wm^{-1}{}^{\circ}C^{-1}$, one gets $I = 0.476$ $m^{2}{}^{\circ}C$ W^{-1}. The body temperature was 39°C. In still air, with $V = 0.10$ m s^{-1}, one gets, for the energy budget of a deer,

$$Q_a + M - 12.00 = 5.67 \times 10^{-8} [212 - 0.476(M - 12.00) + 273]^4$$
$$+ 2.00[39 - T_a - 0.476(M - 12.00)]. \tag{12.84}$$

Since Moen measured the radiation environment for various air temperatures, it is possible to use approximately his values as given in Table 2 of Moen (1968a). For example, at $T_a = 0$°C, $Q_a = 182.8$ Wm^{-2}, and at $T_a = -20$°C, $Q_a = 134.6$ Wm^{-2}. Using the first of these values, one can solve the above equation for M. Because M enters to the fourth power as well as linearly, it is necessary to use an iterative procedure to solve for M. The value of M which will balance the equation when $T_a = 0$°C and $Q_a = 182.8$ Wm^{-2} is approximately 123.5 Wm^{-2}. This is a high metabolic rate and is about 3.7 times the basal metabolic rate, which Moen gives as 33.2 Wm^{-2} (20.9 kcal hr^{-1}). The student can try these calculations for various temperatures, radiation amounts, and wind speeds and then compare the results with those given in Moen's paper. One can also calculate the amount of feed required for a deer to maintain such a metabolic rate by assuming a caloric value for the food of about 16.7×10^6 J kg^{-1} (4 kcal g^{-1}).

According to Cook and Hamilton (1942), white-tailed deer tend to bed down under any convenient cover such as a tree or shrub or lie in the lee of a rock. Deer often will bed down near the base of single trees. Moen, however, found that the animals in his study area did not seek deciduous cover during periods of extreme cold and high winds, but were found in the open near standing corn. Tracks indicated that the deer were active in the open during these cold nights:

> Deer were also observed bedding in the open on north exposures during sub-zero air temperatures and north winds up to 17 m.p.h. On another occasion, 4 deer were bedded on a hill in a picked corn field when the air temperature was $-20°C$ and falling, the wind velocity was 5 m.p.h., and snow was beginning to fall. Four more deer came from a marsh, bedded near the first group, and all eight began feeding in the field as darkness approached.

Moen shows in his paper the importance of adequate diet, particularly critical for smaller animals during very severe winter conditions. During the daytime, a deer will take advantage of solar radiation and when browsing during cold conditions, will remain in the sunshine.

Distributed-Parameter Model

The first attempt to use a "distributed parameter" model for an animal was made by Birkebak (1966). He criticized earlier models as being unrealistic and said that their oversimplified form obscured certain fundamental relationships. Nevertheless, as has been shown in this chapter and elsewhere, a lumped-parameter model can take into account all of the pertinent heat-transfer mechanisms and allows many valid conclusions to be drawn. Certainly, it is desirable to use a "distributed parameter" model wherever feasible, but often the results using it are not significantly more accurate, owing to other imprecisions concerning variables and parameters.

Birkebak (1966) suggests that a bird can be represented as a sphere when in the usual night configuration and by a combination of spherical and cylindrical forms when in the normal day configuration. Two bird species were studied by Birkebak in detail, the cardinal, *Richmondena cardinalis,* and the English sparrow, *Passer domesticus.* Two mammals were studied, the long-tailed shrew, *Sorex cinereus,* and the red squirrel, *Tamascirus hudsonicus preblei.*

Cardinal

Birkebak (1966) gives a table showing the detailed measurements on the head, body, and tail of a cardinal. Birkebak wrote the heat loss from the cardinal in the following form, which is based on heat transfer by conduction,

$$q = \frac{\Sigma \, 2\pi \, K\ell(T_b - T_r)}{\ell n(r_r/r_b)} + \frac{\Sigma \, 2\pi \, r_r \, r_b \, K(T_b - T_r)}{r_r - r_b}$$
$$+ \, KA(T_b - T_g) + \lambda EA. \quad (12.85)$$

The symbols T_r and r_r are the temperature and radius of the outer surface of the bird or animal, T_b and r_b are the body temperature and inside radius, K is conductance, which varies over the animal, λEA is the evaporative water loss, and ℓ is the length of the cylindrical midportion of the body. The first term is heat transfer by conduction from a cylinder. This term was applied to the central body portion. The second term represents heat transfer by conduction from a hemisphere. This term was applied to the head, breast, and tail section; hence the summation is used to indicate that more than one body portion is involved. Birkebak assumed the outside surface temperature of the feathers to be at the ambient air temperature. Birkebak's numbers converted to watts are as follows: head as a hemisphere, $q_1 = 4.62 \times 10^{-3} \, \Delta T$; breast as a hemisphere, $q_2 = 7.55 \times 10^{-3} \, \Delta T$; tail section as a hemisphere, $q_3 = 7.55 \times 10^{-3} \, \Delta T$; and body as a cylinder, $q_4 = 2.62 \times 10^{-3} \, \Delta T$. The rate of total heat loss is

$$q_T = q_1 + q_2 + q_3 + q_4 = 22.34 \times 10^{-3} \, \Delta T, \quad (12.86)$$

where ΔT is the temperature difference between the body temperature at 40°C and the environmental air temperature. The outer temperature of the cardinal was taken to be at the air temperature. The sums given above ignore evaporative water loss. When water loss is accounted for, Birkebak gets good agreement between his calculations and Dawson's (1958) observations concerning heat production of the cardinal versus environmental air temperature. Equation (12.86) applies only to the temperature range below thermoneutrality. These calculations assumed an average integument thickness of 0.02 m. When an average integument thickness Δr of 0.01 m was used, the total body heat loss in watts was

$$q_T = 27.44 \, (T_b - T_r). \quad (12.87)$$

English Sparrow

Following the same procedure for the English sparrow as for the cardinal, Birkebak determined the total body heat loss without evaporative cooling to be

$$q_T = 20.93 \, (T_b - T_r). \quad (12.88)$$

Birkebak gets very good agreement with Kendeigh's (1944) results when evaporative cooling is included, but he does not give the values he uses. According to Steen (1958), the English sparrow during cold nighttime conditions balled and fluffed up, and this configuration corresponds to a spherical form. Birkebak gets a heat loss about 15% higher than that measured by Steen for these conditions. For daytime conditions, the birds were not balled up, and the heat loss was greater than during the night for any given environmental conditions. The effective feather thickness during the night was assumed to be about twice that during the day.

Red Squirrel

The red squirrel was chosen for study by Birkebak (1966) because good physiological information had been obtained for this species by Irving et al. (1955). The squirrel was described in terms of a spherical head, a hemispherical tail region or rear end, and a cylindrical body. Actually, Birkebak divided the body into three cylindrical sections. The legs were also considered as cylinders. For every part of the body (head, tail section, and legs), Birkebak measured the average fur thickness and obtained an inner and outer radius for use in the calculations. The heat loss calculated here has been converted to our units. Radii are given in meters and heat loss in watts. Thermal conductivity is in watts per meter per degree Centigrade.

Heat loss from the head as a sphere is

$$q_1 = \frac{4 \pi r_r r_b k_1 \Delta T}{r_r - r_b}, \tag{12.89}$$

or

$$q_1 = \frac{4 \pi (0.0148)(0.0258)(0.0301)}{0.0110} = 0.0131 \ \Delta T.$$

Heat loss from the forward body section as a cylinder is

$$q_2 = \frac{2 \pi L k_2 \Delta T}{\ell n(r_r/r_b)}, \tag{12.90}$$

$$q_2 = \frac{2 \pi (0.0456)(0.0325) \ \Delta T}{\ell n \ (0.0299/0.0139)}$$

$$\text{or } 12.2 \times 10^{-3} \ \Delta T.$$

Heat loss from the midsection of body as a cylinder is

$$q = \frac{2 \pi (0.0456)(0.0325) \ \Delta T}{\ell n(0.0303/0.0143)} = 12.4 \times 10^{-3} \ \Delta T.$$

Heat loss from the back body section as a cylinder is

$$q_4 = \frac{2\,\pi(0.0456)(0.0334)\,\Delta T}{\ell n\,(0.0312/0.0132)} = 11.2 \times 10^{-3}\,\Delta T.$$

Heat loss from the tail area as a hemisphere is

$$q_5 = \frac{2\,\pi\,r_r\,r_b\,k_5\,\Delta T}{r_r - r_b}, \tag{12.91}$$

or

$$q_5 = \frac{2\,\pi\,(0.0132)(0.0312)(0.0334)\,\Delta T}{0.0180} = 4.9 \times 10^{-3}\,\Delta T.$$

Heat loss from the hind legs (as far as can be determined this is the equation Birkebak must have used) is

$$q_6 = \frac{2\,\pi\,L\,k_6\,\Delta T}{\ell n(r_r/r_b)}, \tag{12.92}$$

or

$$q_6 = \frac{2\,\pi\,(0.10)(0.0634)\,\Delta T}{\ell n\,(0.0169/0.0059)} = 37.9 \times 10^{-3}\,\Delta T.$$

Heat loss from the front legs is

$$q_7 = \frac{2\,\pi\,(0.07)(0.0651)\,\Delta T}{\ell n(0.0158/0.0048)} = 24.1 \times 10^{-3}\,\Delta T.$$

Apparently, Birkebak used a thermal conductivity for the fur on the legs about twice that of the fur on the body.

The total heat loss is given by

$$q_T = q_1 + q_2 + \cdots q_7 = 115.8 \times 10^{-3}\,\Delta T. \tag{12.93}$$

Often, however, part of a leg is in contact with the body and does not lose heat. Sometimes, a countercurrent exchange mechanism is used in the extremities to reduce peripheral heat loss and results in a steep temperature gradient along the leg. Taking these things into account, one can assume that only about half the leg's surface loses heat. If, on this basis, the heat loss from the appendages is taken to be $30.9 \times 10^{-3}\,\Delta T$ rather than $62.0 \times 10^{-3}\,\Delta T$, then the total heat loss is $84.9 \times 10^{-3}\,\Delta T$ rather than $115.8 \times 10^{-3}\,\Delta T$. With these two estimates of the heat loss, the body contributes between 46 and 73%, whereas the appendages contribute from 54 to 27%, respectively. Note that the appendages may contribute from a quarter to a half of the total heat loss and are therefore extremely significant heat exchangers. It is possible that they contribute even less when the animal is hard pressed to conserve heat. But when the animal is under heat stress and needs to dump heat, the appendages can help considerably. Some rodents lick their appendages when overheated, causing further heat loss through evaporative cooling. Birkebak obtained good agreement between calculated heat loss and observed values. In fact, observations fall between the lines 84.9×10^{-3} and 115.8×10^{-3} ΔT for a body temperature of 40°C.

Additional Information

I had hoped to include in this chapter descriptions of all animal orders, but time and space have precluded doing so. The field of animal energetics is advancing rapidly. Many papers are appearing, and it is not possible to refer to all of them. Only a few are mentioned here.

Literature concerning the energetics of insects is considerable. Very few papers, however, actually consider the energy budget of an insect. A few useful papers are the following: Bartholomew and Casey (1977), Bartholomew and Epting (1975), Cena and Clark (1972), Clark et al. (1973b), Heinrich (1974), and West-Eberhard (1978). An arachnid is reported on in a paper by Riechert and Tracy (1975) concerning the energy balance of a spider.

Callahan (1977a,b) describes evidence that the antennal sensilla of some moths are dielectric waveguides or resonators for infrared radiation and that insect pheromones and host plants emit chemiluminescence at these same far-infrared wavelengths, in the 17-μm region.

The thermal energetics of fish are treated by the papers of Carey and Teal (1969a, 1969b) on tuna and sharks and Erskine and Spotila (1977) on black bass.

There has been much interest concerning the mechanisms of heat exchange in large reptiles and a debate as to the amount of ectothermy or endothermy in these animals. See the following papers: Bakker (1975), Spotila et al. (1973), and Tracy (1975). Large reptiles have large heat capacity and large time constants. A paper by Farlow et al. (1976) shows convincingly that the plates along the arched back and tail of *Stegosauraus* served an important thermoregulatory function as convection fins. All reptiles thermoregulate to some extent, through body orientation to sunlight or wind.

An important application of biophysical ecology is to the energetics of eggs. See Bakken et al. (1978) and Packard et al. (1977).

Problems

1. If the Q_{10} for an animal is 3.0, what is the increase in metabolic rate if the body temperature increases 5°C?

2. Determine the buckling length for a tapered tree trunk if the elastic modulus is $E = 10.5 \times 10^5$ kg m^{-2} and the density is $\rho = 6.18 \times 10^2$ kg m^{-3} for trunk diameters of 0.25, 0.5, and 1.0 m. Remember that a tapered trunk can support a height 2.034 greater than a cylindrical column.

3. The fur of a shrew has a thermal conductivity of $0.0229 \ Wm^{-1}{}^{\circ}C^{-1}$ and thickness of 0.003 m, and that of a polar bear has a conductivity of $0.0628 \ Wm^{-1}{}^{\circ}C^{-1}$ and thickness of 0.04 m. Determine the thermal conductance for each animal. Which animal has the most effective insulation against heat loss?

4. Determine thermal resistance of the pelage for the shrew and polar bear of Problem 3.

5. A sheep has a dry metabolic rate $(M - \lambda E)$ of $70 \ Wm^{-2}$, and its fleece has a thermal conductivity of $0.06 \ Wm^{-1}{}^{\circ}C^{-1}$. The animal's body temperature must remain at 37°C. The fleece is 0.04 m thick, and the diameter of the sheep's body is 0.25 m. The air temperature is $-20°C$, and the wind speed is $1.0 \ m \ s^{-1}$. What is the minimum value of Q_a for which the sheep can survive in this environment?

6. The desert iguana, *Dipsosaurus dorsalis*, has a body diameter of 0.015 m. Its thermal conductance is $K = 502 \ Wm^{-2}{}^{\circ}C^{-1}$. At the thermal maximum, body temperature is 45°C, metabolic rate $10.5 \ Wm^{-2}$, and evaporative heat loss $6.3 \ Wm^{-2}$. For a wind speed of $1.0 \ m \ s^{-1}$, determine the relationship between Q_a and T_a at the thermal maximum.

7. Imagine a summer day with the following sequence of conditions:

Time	T_a	$S + s$	$r(S + s)$	T_q
0600	25	100	20	23
0800	30	500	100	32
1000	35	900	180	38
1200	40	980	196	45
1400	42	900	180	47
1600	38	500	100	43
1800	32	100	20	34

The absorptivity of a lizard to shortwave radiation is 0.8, and that to long-wave radiation is 1.0. Using the information from Problem 6 and computing R_a and R_g, indicate which times of day the lizard can be fully exposed to sunshine and when it must remain in the shade or in the burrow. The wind speed is $1.0 \ m \ s^{-1}$.

Chapter 13

Time-Dependent Energetics of Animals

Introduction

In the previous chapters, the simplest form of steady-state energy-budget analysis for plants and animals was used, i.e., a "lumped parameter" model. This very basic approach is now extended to include other variables, and a rearrangement of the variables that will yield insight to the difficult problems of field measurements is made. Further, it is necessary to include time-dependent analysis so that the transients that animals experience can be understood. Much of this analysis was developed by Dr. George Bakken while working in my laboratory and has appeared in papers by Bakken and Gates (1975) and Bakken (1976).

The refined analysis given in this chapter yields considerably greater insight into the various strategies (including ectothermy and endothermy) used by different animals to cope with a vast array of environmental conditions. From this analysis, one can gain an understanding of the efficiency of evaporative cooling by respiratory water loss, sweating through the skin, and sweating or wetting of pelage. The results of this analysis have implications concerning the evolution of endothermy. Although some of the relationships developed here are algebraically complex, they are mathematically straightforward and relatively easy to follow and understand.

Electrical analogs are used for the description of heat flow between an animal and its environment and within the animal itself. Heat flow is regarded as equivalent to current flow in an electrical circuit, temperature difference as equivalent to voltage difference, and thermal resistance or conductance as equivalent to electrical resistance or conductance. A necessary assumption of the model is that heat flow is a linear process. We

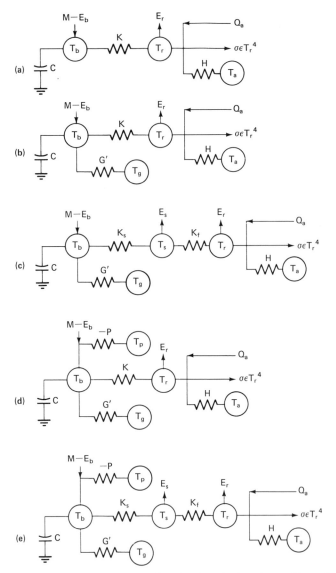

Figure 13.1. Thermal circuit diagrams for the electrical analogs of heat flow to or from the environment for various classes of animals: (A) animal without pelage and with no ground conduction; (B) animal without pelage and with ground conductance; (C) animal with pelage and with ground conduction; (D and E) animal with metabolism and water loss linearly dependent on body temperature but without pelage and with pelage, respectively. All quantities are defined in the text. (From Bakken and Gates, 1975.)

have seen that this is not, in general, true, since radiation exchange is proportional to the fourth power of the absolute surface temperature of the animal or any surface radiating to the animal. However, it was also shown, in the discussion of Newton's law in Chapter 12, that two surfaces whose temperatures are not very different are effectively exchanging energy proportional to the temperature difference. If radiative exchange can be approximated as a linear process, then, since conduction and convection are linearly related to the temperature difference between an animal or object and the environment, it is possible to treat the total heat-transfer description in linear form.

The electrical-circuit analog to heat flow between an animal and its environment is shown in Fig. 13.1. Kirchoff's law states that the total current flowing into any juncture in an electric circuit is equal to the total current flowing out from that juncture. In terms of heat flow, this law simply states that at any temperature node T_x, the heat flowing in is equal to the heat flowing out. Nodes correspond to isothermal surfaces of the body such as the interface between core and skin, skin and pelage, and pelage and environment. In a distributed-parameter model, the surface elements are considered as infinitesimal, and the heat flow changes element by element over the animal's surface. When a lumped-parameter model is used, the entire surface of the animal is considered as being some mean temperature. This may be only a crude approximation to the real situation. The series of circuit analogs shown in Fig. 13.1 represents animals of various internal complexities and undergoing heat exchange with the environment by means of various mechanisms.

Animals without Pelage

Time-Dependent Equation

The simplest model is shown in Fig. 13.1a, which represents an animal without fur or pelage undergoing external heat exchange by radiation, convection, and evaporation only. In this example, heat exchange by conduction, such as to a warm or cold substrate, is not considered. Also, metabolic heat production and evaporative water loss are not considered as a function of body temperature. This model is solved first, and then the more complex cases are considered. Notation used here is essentially the same as in other chapters.

Kirchoff's law applied to the surface node T_r shown in Fig. 13.1a yields

$$H(T_r - T_a) + \epsilon\sigma(T_r + 273)^4 + \lambda E_r = K(T_b - T_r) + Q_a, \quad (13.1)$$

where H is the convection coefficient and K is the thermal conductance between the body core and the skin surface. All terms in Eq. (13.1) are in watts per square meter. As previously defined, heat flow from the animal to the environment is regarded as positive, as is heat production within the animal.

Applying Kirchoff's law to the isothermal core of the animal at temperature node T_b shown in Fig. 13.1a yields

$$C \frac{dT_b}{dt} + K(T_b - T_r) = M - \lambda E_b, \qquad (13.2)$$

where C is the heat capacity of the body and the term $C \, dT_b/dt$ represents heat lost or gained by the body as the body temperature decreases or increases over time. Simply for analytical convenience, the quantity $M - \lambda E_b$ is assumed to be constant and independent of T_b and T_a. Later in this chapter, the dependence of this term on temperature is considered.

Although it is possible to solve Eqs. (13.1) and (13.2) simultaneously as they stand, the term T_r^4 introduces a serious complexity. To facilitate solution, it is convenient to linearize the radiation term by expanding it about a mean surface temperature $\overline{T}_r$ in the following manner:

$$\begin{aligned}(T_r + 273)^4 &= [\overline{T}_r + 273 + T_r + 273 - (\overline{T}_r + 273)]^4 \\ &= [\overline{T}_r + 273]^4 + 4(\overline{T}_r + 273)^3 [T_r + 273 - (\overline{T}_r + 273)].\end{aligned} \quad (13.3)$$

It is convenient to denote a linearized radiation coefficient $\mathcal{R} = 4\epsilon\sigma(\overline{T}_r + 273)^3$ and, when Eq. (13.3) is introduced into Eq. (13.1), to form a term $Q_n = Q_a - 3\epsilon\sigma(\overline{T}_r + 273)^4 - 273 \, \mathcal{R}$.

The error involved in the linearization process is $6\epsilon\sigma(\overline{T}_r + 273)^2 \, (\overline{T}_r - T_r)$. Hence, when $\overline{T}_r + 273$ is about $300°K$, the error is 0.6% when $\overline{T}_r - T_r$ is equal to $10°C$ and only proportionately larger when $\overline{T}_r - T_r$ is greater than $10°C$.

Equation (13.1) is now written

$$H(T_r - T_a) + \mathcal{R}(T_r + 273) + \lambda E_r = K(T_b - T_r) + Q_n. \quad (13.4)$$

Before proceeding to the time-dependent solution to Eqs. (13.2) and (13.4), it is simpler to solve the steady-state equation in a manner analogous to our treatment in the previous chapter.

Steady-State Solution

In a steady state situation, $dT_b/dt = 0$, and Eq. (13.2) becomes

$$K(T_b - T_r) = M - \lambda E_b. \qquad (13.5)$$

Solving Eq. (13.5) simultaneously with Eq. (13.4) by eliminating T_r and obtaining T_b, one gets

$$T_b = (M - \lambda E_b) \frac{(K + H + \mathcal{R})}{K \, (H + \mathcal{R})} - \lambda E_r \left(\frac{1}{H + \mathcal{R}} \right)$$

$$+ \frac{H \, T_a + \mathcal{R} \, \overline{T}_r + Q_n}{H + \mathcal{R}} \qquad (13.6)$$

The following is recognized as being equivalent to an insulation term,

$$I = \frac{K + H + \mathcal{R}}{K(H + \mathcal{R})}.$$ (13.7)

Inspection of Eq. (13.6) shows that the first two terms contain the physiological quantities M, E_b, E_r, and I. Consider for a moment an inanimate object of the same size, shape, color, etc., as the animal. The physiological quantities in this case are zero, and the temperature of the inanimate object, according to Eq. (13.6), is

$$T_e = \frac{HT_a + \mathcal{R}\,\overline{T}_r + Q_n}{H + \mathcal{R}}.$$ (13.8)

This is called the *operative environmental temperature,* which is the temperature of an inanimate object of zero heat capacity, or in steady state, with the same size, shape, and radiative properties as the animal and exposed to the same microclimate. It can also be considered to be the temperature of a blackbody cavity producing the same thermal load on the animal as the actual nonblackbody microclimate, and is therefore the true environmental temperature. The inanimate object representing the animal has a temperature which integrates all factors affecting its energy status, as for example radiation and convection.

The first two terms in Eq. (13.6) represent the *physiological offset temperature* T_Δ by which an animal may offset T_b from T_e by means of metabolism, evaporation, and insulation:

$$T_\Delta = T_b - T_e = (M - \lambda E_b)I - \lambda E_r \left(\frac{1}{H + \mathcal{R}}\right).$$ (13.9)

If $\lambda E_r = 0$, one gets "Newton's law of cooling." And, of course, one can write

$$T_b = T_e + T_\Delta.$$ (13.10)

An animal without pelage may have contact with the ground so that heat is exchanged by conduction. The circuit diagram for such a case is shown in Fig. 13.1b.

Animals with Pelage

Time-Dependent Equation

For an animal with pelage, heat is conducted from the body core to the skin surface and from the skin surface to the pelage surface. Hence the thermal conductance K must be separated into two components, as illustrated in Fig. 13.1c: the conductance of the skin K_s and of the pelage K_f. Evaporation may take place from both the skin and the pelage surface and is denoted by λE_s and λE_r, respectively. Conduction to the ground at temperature T_g is denoted by a conductance G. By appling Kirchoff's law to the three nodes at core, skin–fur interface, and fur surface, one gets the following three equations:

$$C\frac{dT_b}{dt} + K_s(T_b - T_s) + G(T_b - T_g) = M - \lambda E_b \qquad (13.11)$$

$$K_f(T_s - T_r) + \lambda E_s = K_s(T_b - T_s) \qquad (13.12)$$

$$H(T_r - T_a) + \epsilon\sigma[T_r + 273]^4 + \lambda E_r = K_f(T_s - T_r) + Q_a. \qquad (13.13)$$

If one linearizes Eq. (13.13) as before, it becomes

$$K_f(T_s - T_r) + Q_n - \mathcal{R}(T_r - \overline{T}_r) - H(T_r - T_a) - \lambda E_r = 0. \qquad (13.14)$$

Solving Eq. (13.14) for T_r, one gets

$$T_r = \frac{K_f T_s + \mathcal{R}\overline{T}_r + HT_a + Q_n - \lambda E_r}{K_f + \mathcal{R} + H}. \qquad (13.15)$$

Substituting this expression for T_r in Eq. (13.12), one gets

$$T_s = \frac{(K_s T_b - \lambda E_s)(K_f + \mathcal{R} + H) + K_f(\mathcal{R}\overline{T}_r + HT_a + Q_n - \lambda E_r)}{(K_s + K_f)(H + \mathcal{R}) + K_s K_f}. \qquad (13.16)$$

Steady-State Solution

For a steady-state situation, $C\, dT_b/dt = 0$. Equations (13.11), (13.12), and (13.13) are solved simultaneously by eliminating T_s between the first two equations and T_r between this result and the third equation. One then identifies the following quantities:

$$K_{sf} = \frac{K_s K_f}{K_s + K_f} \qquad (13.17)$$

$$I = \frac{K_{sf} + H + \mathcal{R}}{(K_{sf} + G)(H + \mathcal{R}) + K_{sf}G} \qquad (13.18)$$

$$T_e = \frac{K_{sf}(HT_a + Q_n) + G\,T_g(K_{sf} + H + \mathcal{R})}{(K_{sf} + G)(H + \mathcal{R}) + K_{sf}G} \qquad (13.19)$$

$$T_\Delta = (M - \lambda E_b)I - \lambda E_s\left[\frac{K_{sf} + \left(\dfrac{K_s}{K_s + K_f}\right)(H + \mathcal{R})}{(K_{sf} + G)(H + \mathcal{R}) + K_{sf}G}\right]$$
$$- \lambda E_r\left[\frac{K_{sf}}{(K_{sf} + G)(H + \mathcal{R}) + K_{sf}G}\right]. \qquad (13.20)$$

It is of interest to note that the insulation term contains quantities that involve both the animal and the environment. The combined skin and fur conductances are primarily a property of the animal, although fur conductance is affected by wind and moisture. The convection coefficient H is influenced by the wind, but also by the shape, size, and orientation of the animal. Conductance to the ground depends on the amount of the animal's surface in contact with the ground and the conductivity of the fat, skin, and compressed fur. The insulation also contains the radiation coefficient $\mathcal{R}$, which depends on the radiant temperature of the pelage surface and its emissivity. Therefore, we have the interesting situation in which the insulation I, often assumed to be a property of the animal, is determined by both the animal and its environment.

Consideration of the operative environmental temperature T_e as expressed by Eq. (13.19) shows it to be a combination of animal properties and environmental characteristics.

General Time-Dependent Relations

We now return to the time-dependent form of the heat-transfer equation. It is the goal, after T_s from Eq. (13.16) is substituted into Eq. (13.11), to reduce Eq. (13.11) to the form

$$C \frac{1}{K_o} \frac{dT_b}{dt} + T_b - T_e - \frac{M^*}{K_o} = 0, \tag{13.21}$$

where K_o is the overall heat conductance for the animal and M^* is effective net metabolic heat production, often called the dry metabolic heat production.

After substitution of Eq. (13.16) into Eq. (13.11), one first divides through by the coefficient of T_b. Then one groups all purely environmental heat-flow terms ($\mathcal{R}\,\overline{T}_r$, $G\,T_g$, $H\,T_a$, Q_n) into one term and all the physiological heat-flow terms (M, λE_b, λE_s, λE_r) into another term. Then, Eq. (13.11) takes the form

$$C \left[\frac{(K_s + K_f)(H + \mathcal{R}) + K_s K_f}{[K_s K_f + G(K_s + K_f)](H + \mathcal{R}) + GK_s K_f} \right] \frac{dT_b}{dt} + T_b$$

$$- \left[\frac{K_s K_f(HT_a + \mathcal{R}\overline{T}_r + GT_g + Q_n) + GT_g(K_s + K_f)(H + \mathcal{R})}{[K_s K_f + G(K_s + K_f)](H + \mathcal{R}) + GK_s K_f} \right.$$

$$- \left. \frac{\begin{bmatrix} (M - \lambda E_b)[(K_s + K_f)(H + \mathcal{R}) + K_s K_f] \\ - \lambda E_s K_s(K_f + \mathcal{R} + H) - \lambda E_r K_s K_f \end{bmatrix}}{[K_s K_f + G(K_s + K_f)](H + \mathcal{R}) + GK_s K_f} \right] = 0. \tag{13.22}$$

By identifying the quantity K_{sf} as the combined series conductances of K_s and K_f, one gets

$$K_{sf} = K_s K_f/(K_s + K_f). \tag{13.23}$$

If one takes the bracketed part of the first term of Eq. (13.22), divides through by $K_s + K_f$ and substitutes K_{sf}, according to Eq. (13.23), for the term $K_s K_f/(K_s + K_f)$, then the part in brackets becomes $1/K_o$ where K_o is

$$K_o = \frac{(K_{sf} + G)(H + \mathscr{R}) + G\,K_{sf}}{K_{sf} + H + \mathscr{R}}. \qquad (13.24)$$

If both the numerator and denominator of the second bracketed term of Eq. (13.22) are divided by $K_s K_f$ and Eq. (13.23) is used, one gets, for the operative environmental temperature,

$$T_e = \frac{K_{sf}\,(HT_a + \mathscr{R}\overline{T}_r + Q_n + GT_g) + GT_g(H + \mathscr{R})}{(K_{sf} + G)(H + \mathscr{R}) + GK_{sf}}. \qquad (13.25)$$

This expression represents the temperature of the environment as seen by the animal, or the temperature a blackbody enclosure would have if it completely surrounded the animal and delivered to it an amount of energy equal to that received from the environment.

The last bracketed term in Eq. (13.22), when rearranged, becomes

$$(M - \lambda E_b)\frac{1}{K_o} - \lambda E_s \left[\frac{(K_{sf}/K_f)(K_f + H + \mathscr{R})}{(K_{sf} + G)(H + \mathscr{R}) + GK_{sf}}\right]$$
$$- \lambda E_r \left[\frac{K_{sf}}{(K_{sf} + G)(H + \mathscr{R}) + GK_{sf}}\right]. \qquad (13.26)$$

Multiplying and dividing by $(K_{sf} + H + \mathscr{R})$, one gets

$$\frac{M^*}{K_o} = \frac{1}{K_o}\left[M - \lambda E_b - \lambda E_s \left(\frac{K_{sf}(K_f + H + \mathscr{R})}{K_f(K_{sf} + H + \mathscr{R})}\right)\right.$$
$$\left. - \lambda E_r \left(\frac{K_{sf}}{(K_{sf} + H + \mathscr{R})}\right)\right]. \qquad (13.27)$$

This equation is put in the following form by rearranging the quantities in the bracket of the λE_s term and dividing the numerator and denominator in the bracket of the λE_r term by K_{sf}:

$$M^* = M - \lambda E_b - \lambda E_s \frac{1 + \dfrac{H + \mathscr{R}}{K_f}}{1 + \dfrac{H + \mathscr{R}}{K_{sf}}} - \lambda E_r \left[\frac{1}{1 + \dfrac{H + \mathscr{R}}{K_{sf}}}\right]. \qquad (13.28)$$

The quantity M^*, the *effective net metabolic heat production*, is essentially the dry metabolic heat production with correction factors applied to λE_r and λE_s to account for the degree of thermal coupling to the body core.

After completing all these algebraic manipulations of Eq. (12.22) and substituting the transformation given in Eqs. (13.24), (13.25), and (13.28), one arrives at an equation equivalent in form to Eq. (13.21). Actually, all the other cases involving an animal without pelage or without conduction to a substrate are derivable from Eqs. (13.22) through (13.27) by taking the appropriate limits. An animal without pelage has $K_f \rightarrow \infty$ and $K_{sf} \rightarrow K_s$. The coefficient of the λE_s term in Eq. (13.27) is identical with the coefficient of the λE_r term; i.e., λE_s and λE_r are equivalent in the absence of pelage. An animal without conduction to the substrate has $G = 0$ in all of the above equations.

Evaporative Cooling

The heat required to vaporize water on the skin or pelage surface of an animal comes from both the body core and the environment. This is seen from the expression in Eq. (13.24), where λE_b, λE_s, and λE_r are multiplied by different coefficients involving parameters related to environmental conditions. Following Bakken (1976), the coefficients may be regarded as giving the efficiency of evaporation at a particular site defined as follows: efficiency of evaporation = heat of vaporization supplied by body core ÷ total heat of vaporization supplied by both core and environment. The efficiency is unity if all of the heat of vaporization is used to cool the core. The efficiency is less than unity if some of the heat of vaporization cools the environment close to the animal's surface.

In Eq. (13.28), the coefficient of E_b is unity since the action of λE_b is directly on the body core. The coefficient of λE_s is

$$\mathscr{E}_s = \frac{1 + (H + \mathscr{R})/K_f}{1 + (H + \mathscr{R})/K_{sf}}. \tag{13.29}$$

Since $K_{sf} < K_f$, the denominator of $\mathscr{E}_s$ is larger than the numerator, and the efficiency of evaporation from the skin surface is less than unity. For an animal with considerable pelage, $K_f \ll K_s$ and $K_{sf} \simeq K_f$. Therefore, the efficiency of cutaneous evaporative cooling from the skin surface $\mathscr{E}_s$ is nearly unity and essentially independent of $H + \mathscr{R}$. Effective pelage restricts the transport of water vapor, and the cooling rate is small. Strong sweating will wet the fur, and then λE_r becomes large. In this case, the water-vapor gradient across the fur becomes small and λE_s is negligible.

The efficiency of evaporation at the outer surface of the animal is the coefficient of $\mathscr{E}_r$ in Eq. (13.28). Hence

$$\mathscr{E}_r = \frac{1}{1 + (H + \mathscr{R})/K_{sf}}. \tag{13.30}$$

The efficiency of evaporation from the fur $\mathscr{E}_r$ is always less than the efficiency of evaporation from the skin $\mathscr{E}_s$ since the denominators are the same but the numerator of $\mathscr{E}_s$ is always greater than unity. Furthermore, the efficiency $\mathscr{E}_r$ is always poor since K_{sf} is almost always less than half $H + \mathscr{R}$. The efficiency $\mathscr{E}_r$ is usually less than 0.67, or 67%. The poor coupling of evaporative cooling to the body core contributes to the poor effectiveness of water use in sweating and saliva spreading. This is shown by the fact that sheep with heavy wool pelage sweat very much less than thinly furred cattle. Desert rodents rarely resort to saliva spreading, in contrast to laboratory mice and other nondesert rodents.

An animal without pelage has a surface evaporative efficiency of

$$\mathscr{E}_r = \frac{1}{1 + (H + \mathscr{R})/K_s} \tag{13.31}$$

since $K_{sf} = K_s$ when there is no pelage. Many animals lacking pelage have large values of K_s, in which case the term $(H + \mathscr{R})/K_s$ becomes small and $\mathscr{E}_s$ becomes greater than 0.8. Bakken (1976) gives a table showing estimated values for $\mathscr{E}_r$ for furred animals and those lacking fur under various conditions and also for $\mathscr{E}_s$ for animals with dry fur. He shows that as the animal size (diameter) increases, either $\mathscr{E}_r$ or $\mathscr{E}_s$ decreases. These factors decrease with an increase in wind speed.

It becomes clear from the above discussion that respiratory water loss from the body cavity, λE_b, is much more effective for cooling than cutaneous water loss λE_s, which in turn is more effective than evaporative cooling λE_r from the pelage surface. Profuse sweating is uneconomical for an animal with effective pelage insulation. Desert rodents with closely balanced water budgets should avoid fur wetting. Wetting does increase the rate of evaporation and, although not of great efficiency, is practiced by some mesic rodents during thermal stress. Small mammals living at the ground surface experience low wind speeds, so that their convection coefficient is small. Under these conditions, saliva spreading, though inefficient, can give significant cooling. Birds, on the other hand, have large convection coefficients because of high wind speeds. For birds, $K_f \ll K_s$ and $K_{sf} \simeq K_f$. Hence wetting of feathers is ineffective for cooling.

Relation to "Newton's Law"

Equation (13.21) is in the form of "Newton's law" of cooling. Animals that rely primarily on internal metabolic heat production or evaporative cooling (endotherms) have a nearly constant body temperature. The net heat flow to or from the environment is then equal to the rate of internal production or less, and

$$M - \lambda E = K(T_b - T_e), \tag{13.32}$$

where $\lambda E = \lambda E_b + \lambda E_s + \lambda E_r$. This is equivalent to the steady-state form of Eq. (13.21) in which $dT_b/dt = 0$,

$$M^* = K_0(T_b - T_e). \tag{13.33}$$

Animals that rely primarily on external sources and sinks for thermoregulation (ectotherms) are assumed to have zero metabolic heat production or evaporation. Heat flow comes entirely from the environment. From Eq. (13.21), one has

$$\frac{dT_b}{dt} = -\frac{K_o}{C}(T_b - T_e).$$ (13.34)

It is clear, however, that an animal rarely meets either of the sets of conditions necessary to create a "Newton's law" type of heat-exchange situation. In fact, the only proper distinction of heat flow is given by Eq. (13.21) and the full elaboration of the terms involved. Bakken and Gates (1975) state that "this should be a sufficient warning against the use of oversimplified analysis, and should clearly indicate that heat transfer in animals shows the strong interactions between the animal and its environment so characteristic of all ecological problems."

Time-Dependent Solution

Animals are usually in transient states a significant fraction of the time during the day. As a first approximation, an animal is considered to have a single body temperature T_b which is a function only of time. It is assumed that the internal thermal conductance is large compared to the external thermal conductance. The animal or object is surrounded by a medium at temperature T_∞. If the object or animal is at temperature T_0 at time $t = 0$ and cools to temperature T_∞, then the change in internal energy must equal the loss of energy externally at any instant,

$$dT = -\frac{K_o}{C}(T - T_\infty)\,dt,$$ (13.35)

where the minus sign indicates that internal energy decreases when $T > T_\infty$. This equation is equivalent to

$$\frac{dT}{T - T_\infty} = -\frac{K_o}{C}\,dt.$$ (13.36)

Integration gives

$$\frac{T - T_\infty}{T_0 - T_\infty} = e^{(K_o/C)t},$$ (13.37)

where the quantity C/K_o has the dimensions of time and is known as the time constant τ since the exponent must be dimensionless, C is the heat capacitance, and K_o is the overall heat-transfer coefficient. The time constant is the time required for the body temperature to change by 63% or the temperature difference $(T_0 - T)$ to reach 63% of the initial temperature difference $(T_0 - T_\infty)$. This is seen by putting $t = \tau$ in Eq. (13.37) and writing

$$T - T_\infty = (T_0 - T_\infty)e^{-1}$$ (13.38)

$$T = (1 - e^{-1})\,T_\infty + T_0\,e^{-1}.$$ (13.39)

Hence

$$T_0 - T = (1 - e^{-1})(T_0 - T_\infty). \qquad (13.40)$$

It is well known that precisely the same analytical expressions represent the charging or discharging process in an electrical circuit in which the time constant is RC, the product of the circuit resistance R and the capacitance C.

Going back to Eq. (13.35) and rewriting it in terms of $\tau = C/K_0$, one gets

$$\frac{dT}{dt} + \frac{T}{\tau} = \frac{T_\infty}{\tau}. \qquad (13.41)$$

This suggests that Eq. (13.21) can be written

$$\frac{dT_b}{dt} + \frac{T_b}{\tau} = \frac{T_e}{\tau} + \frac{M^*/K_0}{\tau}. \qquad (13.42)$$

Animal Properties and Time Constants

The physical and physiological characteristics of six animals are listed in Tables 13.1 and 13.2. The animals are listed in order of the weight-to-surface-area ratio. The smallest animal, the masked shrew, has the smallest value of W/A, or the largest surface area per unit weight, whereas the pig has the largest value of W/A, or the smallest surface area per unit weight. The heat capacity per unit surface area (C) increases systematically from the shrew, desert iguana, cardinal, jackrabbit, and domestic sheep to the pig.

Also listed in Tables 13.1 and 13.2 are the heat-transfer characteristics of the six animals. The desert iguana is the only ectothermic animal; the other five are endothermic animals. The convection coefficient H decreases as the size of the animals increases. The values of H and I are calculated for still air conditions; H and I will vary with the square root of the wind speed if the animal is in wind. These numbers are based on the values in Porter and Gates (1969) but have been converted to SI units. The conductance of fur is determined using the conductivity of air, 0.025 $Wm^{-1}{}^\circ C^{-1}$, divided by the fur thickness, and that of fat is determined using the conductivity of fat, 0.205 $Wm^{-1}{}^\circ C^{-1}$, divided by the fat thickness. From the values of the heat-transfer coefficients, one can determine T_e for any air temperature and amount of radiation absorbed or amount of net radiation.

The time-constant characteristic of an animal is an interesting number. The time constant $\tau = C/K_0 = IC$ and depends on the heat capacity of the animal and overall conductance to the environment, or overall insulation. Animals with feathers or fur have very large time constants. The specific heat of protoplasm is 3430 $J\ kg^{-1}{}^\circ C^{-1}$. The heat capacity as used

Table 13.1. Values of Physical Dimensions and Thermal Conductances for Various Animals[a]

Animal	W (kg)	A (m²)	W/A (kg m⁻²)	C (J m⁻² °C⁻¹)	D (m)	d_f (m)	d_s (m)	K_f (Wm⁻² °C⁻¹)	K_s (Wm⁻² °C⁻¹)	K_{sf} (Wm⁻² °C⁻¹)
Masked shrew	0.0045	0.0021	2.14	7,340	0.018	0.002	0.001	12.5	205	11.8
						0.003	0.001	8.13	205	7.82
Desert iguana	0.0508	0.0175	2.90	9,947	0.015	0	0.001	–	205	–
Cardinal	0.045	0.011	4.09	14,029	0.050	0.005	0.001	5.00	205	4.88
						0.010	0.002	2.51	102	2.45
Jackrabbit	2.30	0.174	13.2	45,276	0.100	0.008	0.001	3.13	205	3.08
						0.015	0.002	1.67	102	1.64
Sheep	69.6	1.06	65.7	225,350	0.250	0.128	0.001	0.195	205	0.195
						0.100	0.010	0.250	20.5	0.247
Pig	102.8	1.36	75.6	259,308	0.360	0.001	0.001	25.0	205	22.3
						0.003	0.035	8.33	5.86	3.44

[a] Heat capacity C is expressed per unit of surface area and uses a specific heat for protoplasm of 3430 J kg⁻¹ °C⁻¹. Upper and lower values correspond to thermal maximum and minimum. Numbers used are from Porter and Gates (1969).

Table 13.2. Values of Physiological Properties, Heat-Transfer Parameters, and Time Constants for Various Animals[a]

Animal	M (Wm^{-2})	λE_b (Wm^{-2})	λE_r (°C)	T_b (°C)	$\bar{T}_r$ (Wm^{-2}°C^{-1})	$\mathscr{R}$ (Wm^{-2}°C^{-1})	H (Wm^{-2}°C^{-1})	I (W^{-1}m^2°C)	τ (hr)
Masked shrew	139	26.0	26	41.0	33.5	6.53	8.22	0.152	0.31
	346	0	0	37.5	−5.5	4.34	8.22	0.208	0.42
Desert iguana	10.5	6.28	0	45.0	45.0	7.29	8.38	0.069	0.19
	0.140	0.140	0	3.0	3.0	4.77	8.38	0.081	0.22
Cardinal	76.8	76.8	0	42.5	42.5	7.12	4.94	0.288	1.12
	107	3.5	0	38.5	−16.0	3.85	4.94	0.522	2.03
Jackrabbit	62.8	62.8	0	43.7	43.7	7.20	3.49	0.423	5.32
	77.5	9.07	0	37.5	−4.0	4.41	3.49	0.736	9.25
Sheep	91.4	74.0	20.9	41.7	52.3	7.81	2.21	5.23	327
	69.8	7.0	5.6	39.5	−99.4	1.19	2.21	4.34	272
Pig	57.9	69.8	75.4	41.7	41.8	7.08	1.10	0.167	12.0
	124	7.0	7.0	36.0	0.3	4.63	1.10	0.465	33.5

[a] In each row, the upper number listed is for thermal maximum and the lower number is for thermal minimum. Numbers used are from Porter and Gates (1969).

here is the heat capacity per unit of surface area since all other quantities are expressed per unit of surface area. If the weight (or mass) of an animal is W and its surface area is A, then $C = 3430\ W/A$. Clearly, the product IC must be in seconds. The time constants for six animals are given in Table 13.2. The values are given in hours since the values in seconds are very large numbers. The time constants for the animals change in an interesting manner. Although values generally increase with increasing mass of the animals, they also depend on the effective insulation against heat loss possessed by the animal. The desert iguana and the cardinal are of comparable body weight but have very different time constants. The lack of thermal insulation on the iguana gives it a very short time constant, about 12 min, whereas the cardinal has a time constant of 1 to 2 h because of the superb insulation quality of the feathers. The combination of heavy fleece and massive body gives the sheep a time constant of about 300 h, or 12 days, which is a surprisingly large number. The pig, of comparable weight per unit surface area, has a time constant of 12 to 33 h, which seems reasonable. It is the incredibly good insulating quality of its wool pelage that gives the sheep such a low cooling or heating rate.

Body-Temperature-Dependent M and λE_b

In the preceding analysis, the metabolic and evaporative water-loss rates were taken as constant, or at least given discrete values. However, metabolic and evaporative water-loss rates normally depend on body temperature. So far in our analysis, M, λE_b, and λE_r affect the final temperature and rate of change of body temperature but not the thermal insulation I or the time constant τ. However, a purely iterative type of analysis applied to large reptiles by Spotila et al. (1973) showed that I and τ are affected by body-temperature-dependent metabolic and evaporative water-loss rates, which in turn are dependent on body temperature. For example, as T_b increases, $M - \lambda E_b$ decreases and eventually becomes negative, tending to decrease T_b. The net effect is that equilibrium temperature is reached more quickly if there is no feedback and hence a smaller τ.

General Equations

It is relatively straightforward to allow the net metabolic heat production $(M - \lambda E_b)$ to be linearly dependent on T_b while λE_s and λE_r are kept constant. Various functional relationships might be considered between $(M - \lambda E_b)$ and T_b, but the following is a simple approximation with which to begin. A body temperature $T_b = T_p$ is taken as a linearization point, and the dependence of $(M - \lambda E_b)$ on T_b is given as

$$M - \lambda E_b = (M - \lambda E_b)_p + P(T_b - T_p), \tag{13.43}$$

where $(M - \lambda E_b)_p$ is the value of $(M - \lambda E_b)$ at $T_b = T_p$ and $P = d(M - \lambda E_b)/dT_b$.

The term $P(T_b - T_p)$ has exactly the same form as the conductive and convective terms $K_s(T_b - T_s)$, $K_f(T_s - T_r)$, $G(T_b - T_g)$, and $H(T_r - T_a)$ in Eqs. (13.11) to (13.13) and earlier equations. The variable metabolism term may be regarded as putting a fictitious negative conductance $-P$ connecting T_b with a heat source of temperature as shown in Fig. 13.1d and e. A real conductance of this sort would change the insulation and the time constant, and the mathematically equivalent body-temperature-dependent metabolic heat production does the same.

If the circuit analysis of Fig. 13.1e is carried out using Kirchoff's law as previously, but including the body-temperature-dependent net metabolic heat production, the result is

$$I^a = \frac{K_{sf} + H + \mathcal{R}}{(K_{sf} + G - P)(H + \mathcal{R}) + K_{sf}(G - P)} = \frac{I}{1 - IP}, \quad (13.44)$$

where I is given by Eq. (13.18). The symbol I^a is used to indicate an apparent value of the insulation, as contrasted with the insulation obtained for the animal when $(M - \lambda E_b)$ is independent of body temperature. As done earlier, an operative environmental temperature T_e^a and an offset temperature T_Δ^a are derived. Therefore,

$$T_e^a = \frac{K_{sf}(HT_a + Q_n) + GT_g(K_{sf} + H + \mathcal{R})}{(K_{sf} + G - P)(H + \mathcal{R}) + K_{sf}(G - P)} = \frac{T_e}{1 - IP} \quad (13.45)$$

$$T_\Delta^a = [(M - \lambda E_b)_p - PT_p]I^a$$

$$- \lambda E_s \frac{K_{sf} + [K_s(K_s + K_f)](H + \mathcal{R})}{(K_{sf} + G - P)(H + \mathcal{R}) + K_{sf}(G - P)}$$

$$- \lambda E_r \frac{K_{sf}}{(K_{sf} + G - P)(H + \mathcal{R}) + K_{sf}(G - P)}$$

$$= \frac{(T_\Delta)_p}{1 - IP} - PT_pI^a. \quad (13.46)$$

Here $(T_\Delta)_p$ is T_Δ evaluated at $T_b = T_p$.

If P is positive, i.e., $(M - \lambda E_b)$ increases with an increase of T_b, then I^a increases and $\tau^a = I^aC$ increases. If P is negative, i.e., $(M - \lambda E_b)$ decreases with an increase of T_b, then I^a decreases and τ^a decreases. An explosive rise of body temperature and heat death occur when I^a and τ^a become negative so that the exponent in Eq. (13.37) becomes positive. This occurs when

$$P \geq \frac{(K_{sf} + G)(H + \mathcal{R}) + K_{sf}G}{K_{sf} + H + \mathcal{R}} = \frac{1}{I}. \quad (13.47)$$

That this must be true is readily seen from Eq. (13.45) since for I^a to be negative, IP must be greater than one and therefore P must be equal to or greater than $1/I$.

It is interesting to compare a nonfurred animal with a furred animal with respect to the potential for an explosive rise in body temperature. For the naked animal, $K_{sf} = K_s$. Since K_s is much greater than K_f, $K_{sf} \to K_f$ in the case of a furred animal. For the nonfurred, naked animal, I is very much less than for the furred animal, and a much larger positive P is required to produce explosive heat death than for a furred animal. An animal insulated with pelage is much more sensitive to the effects of a metabolic rate dependent on body temperature than an animal lacking pelage. This fact may have been of importance during the evolution of endothermy.

Time-Dependent Solution

The differential equation which represents the time-dependent heat exchange for any animal is of the form given by Eq. (13.42), and earlier Eq. (13.21). These equations may be rewritten as follows:

$$\frac{dT_b(t)}{dt} + \frac{T_b(t)}{\tau} = \frac{T_e + T_\Delta}{\tau}. \tag{13.48}$$

Both T_e and T_Δ are time-dependent functions describing environmental conditions and an animal's physiological response.

The simplest type of solution to Eq. (13.49) is when there is a step function or sudden change in conditions. This sort of situation occurs when an animal shuttles between two microclimates. An animal in a microclimate T_e has a body temperature $T_b = T_e + T_\Delta$ and moves into a new microclimate for which T_e changes to T_e' and its physiological state changes from T_Δ to T_Δ'. The animal may allow its body temperature to vary with time $T_b(t)$ until a new equilibrium, $T_b' = T_e' + T_\Delta'$, is reached as $t \to \infty$. Applying these boundary conditions to the solution of Eq. (13.48), one gets a solution of the same form as Eq. (13.37),

$$T_b(t) = (T_e' + T_\Delta') + [(T_e + T_\Delta) - (T_e' + T_\Delta')]e^{-t/\tau}, \tag{13.49}$$

where the time constant $\tau = IC$.

Shuttling

An animal shuttling between hot and cold climates has equilibrium body temperatures two microclimates of T_b^{1h} and T_b^{1c}. However, the animal has a range of preferred body temperatures with upper and lower set-point body temperatures, not as high as nor as low as the equilibrium values, and of T_b^h and T_b^c, respectively. In the hotter microclimate, the animal is considered to have an equilibrium body temperature $T_b^{1h} = T_e^{1h} + T_\Delta^{1h}$, and in the colder microclimate, $T_b^{1c} = T_e^{1c} + T_\Delta^{1c}$. As a first approximation, it is assumed that behavior is purely thermoregulatory with minimum activity. The animal's initial state is in the colder microclimate, with $T_b(t) = T_b^c$; it then moves into the warmer microclimate. When its body temperature has risen to $T_b(t) = T_b^h$, the animal will return to the cooler microclimate, thereby completing its cycle. This behavior is shown in graphical form in Fig. 13.2.

The shuttling behavior shown in Fig. 13.2 is described mathematically by Eq. (13.48), which can be used to compute the length of time that the animal can spend in each microclimate as a function of the set points and microclimate conditions. For an animal in the warmer microclimate, one can write, from Eq. (13.48),

$$T_b(t) = T_b^{1h} + (T_b^c - T_b^{1h})e^{-t/\tau_h}. \tag{13.50}$$

The elapsed time after the microclimate has abruptly changed from cold to hot, which is the time the animal spends in the warmer conditions, is given by

$$\Delta t_h = -\tau_h \, \ell n \, \frac{T_b^h - T_b^{1h}}{T_b^c - T_b^{1h}}. \tag{13.51}$$

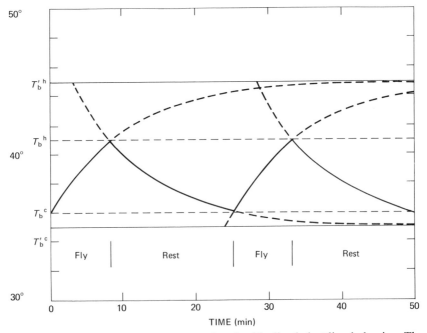

Figure 13.2. Body temperature versus time for idealized shuttling behavior. The upper and lower set points for body temperature are T_b^h and T_b^c. The equilibrium body temperature in the hot environment is T_b^{1h} and in the cold environment T_b^{1c}. The legend "fly" and "rest" indicates the activity state for a hummingbird shuttling. (From Bakken and Gates, 1975.)

Similarly, when the animal's temperature has reached the upper set-point temperature T_b^h, it shuttles to the cold microclimate, and its temperature begins to drop. The elapsed time it spends in the colder conditions is given by

$$\Delta t_c = -\tau_c \, \ell n \, \frac{T_b^c - T_b^{1c}}{T_b^h - T_b^{1c}}. \tag{13.52}$$

These elapsed times are indicated in Fig. 13.2. Of course, in nature, behavior is not purely thermoregulatory, and other considerations make behavior less regular. Cabanac and Hammel (1971) and Hammel et al. (1967) show laboratory shuttling behavior for the lizard *Tiliqua scincoides*.

The kinds of other considerations that may cause an animal to interrupt a regular thermoregulatory behavior are food gathering or territorial defense. The animal may wish to maximize the time spent in whichever microclimate contains the food resource. Bakken and Gates (1975) defined a duty cycle D which gives the fraction of total active time spent in the food-containing microclimate as a function of the microclimate and physiological conditions. For the case with the food resource in the warm environment,

$$D_h = \frac{\Delta t_h}{\Delta t_h + \Delta t_c}, \tag{13.53}$$

and for the food resource in the cool environment,

$$D_c = \frac{\Delta t_c}{\Delta t_h + \Delta t_c}. \tag{13.54}$$

The amount of time spent in each part of the duty cycle depends on the choice of T_b^h and T_b^c. The duty cycle for time spent in the hotter environment depends on the proximity of T_b^h to T_b^{1h}. Because food-gathering efficiency often increases with the length of each excursion, optimizing strategies exist for maximum foraging per unit time under any given set of T_b^h, T_b^c, T_b^{1h}, and T_b^{1c}. These strategies very probably are employed by lizards, turtles, snakes, and other animals.

If, for example, an animal is feeding in a warm environment, it may maximize the time spent in this microclimate by making short excursions, which is effectively equivalent to increasing T_b^c. The slope of the warming curve is minimized, and the slope of the cooling curve maximized, as T_b^c approaches T_b^h, whereas the duration of each excursion is reduced. The duty cycle when an animal can be in a warm microclimate is about one-third of the total cycling time when T_b^c and T_b^h are far apart, but when they are close together, the warm duty cycle becomes about one-half of the total cycle, although the total cycle is now very short.

This result can be studied analytically. The absolute maximum limit on the duty cycle, D^{max}, is

$$D_{\text{h}}^{\text{max}} = \lim_{T_{\text{b}}^{\text{c}} \to T_{\text{b}}^{\text{h}}} D_{\text{h}} = \frac{\tau_{\text{h}}(T_{\text{b}}^{\text{h}} - T_{\text{c}})}{\tau_{\text{h}}(T_{\text{b}}^{\text{h}} - T_{\text{c}}) + \tau_{\text{c}}(T_{\text{h}} - T_{\text{b}}^{\text{h}})} \qquad (13.55)$$

for infinitely short excursions. A similar analysis involves maximizing the time spent in the cold state. In this case,

$$D_{\text{c}}^{\text{max}} = \lim_{T_{\text{b}}^{\text{h}} \to T_{\text{b}}^{\text{c}}} D_{\text{c}} = \frac{\tau_{\text{c}}(T_{\text{h}} - T_{\text{b}}^{\text{c}})}{\tau_{\text{c}}(T_{\text{h}} - T_{\text{b}}^{\text{c}}) + \tau_{\text{h}}(T_{\text{b}}^{\text{c}} - T_{\text{c}})}. \qquad (13.56)$$

As long as $T_{\text{b}}^{\text{c}} \leq T_{\text{b}} \leq T_{\text{b}}^{\text{h}}$, $D_{\text{h}}^{\text{max}}$ sets the upper bound and $D_{\text{h}}^{\text{min}} = 1 - D_{\text{c}}^{\text{max}}$ the lower bound on the duty cycle in the hot climate. Equation (13.53) sets the duty cycle for maximum length excursions. Similarly, $D_{\text{c}}^{\text{max}}$ sets the upper and $D_{\text{c}}^{\text{min}} = 1 - D_{\text{h}}^{\text{max}}$ the lower boundary on excursions into the cold microclimate. The duty cycle $D_{\text{c}}^{\text{max}}$ will be reduced and $D_{\text{h}}^{\text{max}}$ increased if T_{b} is allowed to exceed T_{b}^{h}, and $D_{\text{c}}^{\text{max}}$ will be increased and $D_{\text{h}}^{\text{max}}$ decreased if T_{b} falls below T_{b}^{c}.

If foraging efficiency and exposure to predation is known as a function of duration of the excursion, the optimum strategy may be computed. Unfortunately, studies of shuttling fail to record information sufficient to make accurate checks of predictions. Once again, this illustrates the danger of field observations being done without a sufficient theoretical analysis in advance, with the result that neither correct nor complete data is obtained. This is a difficult point for many field biologists to understand. However, many field biologists will readily use analysis when it is developed. I am sure that everyone involved with the following examples would have used analysis if the state of the art had been more advanced at the time of the observations.

Heath (1965) studied the shuttling of horned toads, *Phrynosoma*, and made excellent observations of behavior, determining T_{b}^{c} and T_{b}^{h} but providing only the most meager microclimate data. For example, in one series of outdoor experiments, D, Δt_{h}, and Δt_{c} were determined, but only T_{a} and the temperature of the substrate in the shade were recorded. In another series of experiments, a laboratory set-up was used which would, in principle, allow the microclimate to be determined by duplicating the field experimental conditions. However, no quantitative data was given on D, Δt_{h}, or Δt_{c}, only T_{b}^{h} and T_{b}^{c}.

Shuttling behavior is not restricted to ectotherms; many endotherms exhibit this behavior. Ground squirrels shuttle in and out of their burrows during hot conditions, and hummingbirds shuttle from one part of their habitat to another. Camels even shuttle on a diurnal basis, letting their body temperature increase as much as 9°C from dawn to late afternoon. Observations made by Stiles (1971) of the time budgets in Anna's hummingbird, *Calypte anna*, is another example of such frustration with field data and analysis. Stiles noted that percent time spent flying was a decreasing function of air temperature T_a and "blackbulb" temperature T_{bb}. The fraction was almost exactly the same when T_a and T_{bb} were the same, regardless of whether the bird was breeding or not, whereas the percentage of time spent feeding was different. This suggests that the hummingbird was essentially shuttling since the combined environmental and metabolic heat load while flying and hovering eventually resulted in overheating. However, these findings could also reflect decreasing requirements for metabolic heat production since day and night temperatures are usually covariant and food resources are more abundant during breeding season, when the percent time spent feeding is reduced relative to percent time spent in all flight activities. Expected shuttling behavior and daily food requirements must be computed from microclimate and physiological data and the results compared with observation before a choice between the two hypotheses is possible.

The agreement in total flight time strongly suggests the occurrance of shuttling. When $T_a = 29°C$ and $T_{bb} = 40°C$, a nonbreeding male hummingbird spent 6.5% of total time flying, and when $T_a = 28°C$ and $T_{bb} = 41°C$, a breeding hummingbird flew 6.25% of the time. Since T_{bb} is probably closer to T_e, the 0.25% difference is in the correct direction. However, trying to interpret these results with more precision is out of the question because of the lack of adequate physiological and microclimate data.

It is very difficult to use field data that do not include a proper measure of T_e to make an accurate interpretation of observations. The only practical way to obtain T_e for small animals is by the use of casts, as described by Bakken and Gates (1975).

General Time-Varying Environment

A more general case for an animal is for the operative environmental temperature T_e to vary in a relatively smooth fashion. This applies particularly to large animals in pure forest or grassland habitats when T_e varies smoothly within a 24-hr fundamental period. Any smoothly varying periodic function may be represented by a Fourier series. The function need only go through one cycle, so that in practice any function may be so represented. Consequently, we may represent T_e by the Fourier series

$$T_e = A_0 + \sum_{j=1}^{\infty} A_j \cos\left(\frac{j\pi}{F} t\right) + B_j \sin\left(\frac{j\pi}{F} t\right), \qquad (13.57)$$

where the duration of the cycle is $2F$, from $-F$ to F. The normal time base of hours, minutes, and seconds is quite unsatisfactory since it results in a large number of conversion terms in the result for T_b. A more useful time base, commonly used in meteorology, is the hour angle h of the sun. The hour angle, expressed in radians, has the value $-\pi$ at $t = 0000$, $-\pi/2$ at 0600, 0 at 1200, $\pi/2$ at 1800, and π at 2400, that is, is equal to the angle between the local longitude and the sun. The conversion factors are

$$t_h \times \frac{\pi}{12} = h$$

$$t_{min} \times \frac{\pi}{720} = h \qquad (13.58)$$

$$t_s \times \frac{\pi}{43,200} = h.$$

These conversions must be applied to all parameters containing time units so that K, G, H, and P are in Joules per square meter per degree Centigrade per radian, M, E, and Q_a are in joules per square meter per radian, and $t = h$ in radians. Then, Eq. (13.57) becomes

$$T_e = A_0 + \sum_{j=1}^{\infty} A_j \cos(jh) + B_j \sin (jh), \qquad (13.59)$$

where

$$A_0 = \frac{1}{2\pi} \int_{-\pi}^{\pi} T_e (h) \, dh$$

$$A_j = \frac{1}{\pi} \int_{-\pi}^{\pi} T_e (h) \cos (jh) \, dh$$

$$B_j = \frac{1}{\pi} \int_{-\pi}^{\pi} T_e (h) \sin (jh) \, dh.$$

Then, by assuming that T_A is constant or $(M - E_b)$ is a linear function of T_b as in Eq. (13.43), Eq. (13.48) becomes

$$\frac{dT_b}{dh} + \frac{T_b}{\tau} = \frac{T_A}{\tau} + \frac{1}{\tau} \left[A_0 + \sum_{j=1}^{\infty} A_j \cos (jh) + B_j \sin (jh) \right]. \qquad (13.60)$$

The boundary condition for the solution of Eq. (13.60) is $T_b(h = 0) = T_b^0$, with the result

$$T_b(h) = (A_0 + T_A) + \left[(T_b^0 - (A_0 + T_A)) - \left(\sum_{j=1}^{\infty} (A_j - j\tau B_j) \cos^2 \phi \right) \right] e^{-h/\tau}$$

$$+ \sum_{j=1}^{\infty} \cos \phi \, [A_j \cos (jh - \phi) + B_j \sin (jh - \phi)], \qquad (13.61)$$

where the phase angle ϕ is

$$\phi = \cos^{-1} \left[\left(\frac{1}{1 + (j\tau)^2} \right)^{1/2} \right] = \sin^{-1} \left[\left(\frac{(j\tau)^2}{1 + (j\tau)^2} \right)^{1/2} \right]. \qquad (13.62)$$

The exponential term diminishes for large values of h ($h \gg \tau$). Since A_0 is just the average value of T_e, $A_0 + T_A$ is the average body temperature for $h \gg \tau$, and the sine and cosine terms give the fluctuations of T_b about the time-average body temperature. Each Fourier component (j) of the fluctuation of T_b with time follows the corresponding jth Fourier component of T_e with a phase lag ϕ that varies from $\phi = 0$ for $j\tau \ll 1$ to $\phi = \pi/2$ for $j\tau \gg 1$ and an amplitude cos ϕ that varies from 1 for $j\tau \ll 1$ to 0 for $j\tau \gg 1$. The significance of the high-frequency (large j) components T_e for T_b thus drops off rapidly as j increases. For large animals with large τ, only the low frequency (small j) components of T_e need be considered. Small animals with correspondingly small values of τ will respond to much higher frequency components of T_e.

The Fourier-series solution may be used to study the general influence of diurnal and seasonal variations of T_e on animals of various time constants. This, for example, permits analysis, such as that by Spotila et al. (1972) of the response of body temperatures of large reptiles to daily and seasonal fluctuations in climate, to be done much more simply and clearly.

The predominant feature of the earth's climate, averaged over several years, is a cyclic daily fluctuation of T_a and Q_a superposed on a very slow cyclic seasonal fluctuation. In practice, the daily cycle is subject to variation due to changes in cloud cover and movement of warm and cold air masses. However, the essential features of the climate experienced by an animal may be represented by the average conditions for the study of generalized responses. This is equivalent to taking only the A_0 and A_1 terms of the Fourier-series solution, Eq. (13.61). The A_0 term represents the average daily operative environmental temperature and thus corresponds to the seasonal conditions. This implies a steady-state solution for body temperature in response to seasonal fluctuations. Since the time constant of even a large dinosaur would never have exceeded a few days, this is a valid procedure. The A_1 term represents the amplitude of the daily cycle of T_e and may be allowed to vary seasonally along with A_0. The daily cycle has a maximum near solar noon ($h = 0$) and a minimum near midnight ($h = \pm \pi$), so that the cosine term only will give a good representation. In practice, peak values of T_e will be reached after solar noon since air temperature lags behind solar radiation by 1 to 2 hr, so that a nonzero B_1 term is required to phase T_e properly with h. This problem may be circumvented by keeping hour angle units but shifting the calculation hour angle h' ahead of the true solar hour angle h so that B_1 becomes zero. This shift may be corrected easily when presenting the results of the calculation. The body temperature for a short segment of the average seasonal conditions represented by A_0 and A_1 is then

$$T_b(h') = (T_\Delta A_0) + \cos \phi A_1 \cos (h' - \phi). \qquad (13.63)$$

This solution assumes that the animal is exposed to the effects of air temperature, wind, and radiation continuously in a stable part of the habitat, which generally implies that the animal does not retreat to a burrow or other shelter. Also, it is assumed that T_Δ is constant or determined by T_b, so that the result is most immediately applicable to reptiles and amphibians. However, the approach may be modified to describe some of the essential properties of birds and mammals as well.

Evolution of Thermoregulation

There have been many hypotheses concerning the evolution of thermoregulation, and in particular the evolution of endothermy. Homeothermy is thought to have originated in order to prolong the time that an animal spends at its preferred body temperature for activity or digestion. An animal that can control its rate of heat gain or loss is able to warm quickly and cool slowly, thereby increasing the time spent at its preferred body temperature. If τ_o is large and τ_h is small, the time spent in the cold environment is increased.

During evolution, a slow elevation of the basal metabolic rate resulted in an increased capacity to control the rate of change of body temperature. Vasoconstrictive control of I, and therefore τ, makes this possible. There is evidence of this in altricial animals, such as juvenile birds without adequate insulation, where some thermoregulation is possible. In the absence of significant metabolic heat production, insulation would be a handicap, since it would slow the rate of warming of animals with high body temperatures. This would be especially true in a cold environment, where insulation would exclude heating by solar radiation.

The response of birds and mammals to cold stress includes muscular thermogenesis by shivering. The most spectacular example of endothermy in the absence of metabolic heat production is the intermittent endothermy of the flight muscles of sphingid moths as described by Heinrich (1971).

According to certain experts (Cowles and Bogert, 1944; Heath, 1964, 1965; Brattstrom, 1965), heliothermic lizards maintain relatively high body temperatures by absorbing substantial amounts of solar radiation and for this reason have never developed fur. Operative temperatures in the preferred range of body temperatures for basking lizards, 35°–42°C, exist only during daylight hours when Q_a is large enough to produce an operative environmental temperature substantially greater than the air temperature. A short time constant is necessary to allow T_b to reach the preferred value while such operative temperatures are available. However, an animal with a lower preferred body temperature, in the 20°–30°C range, would find exposure to the operative temperatures in existence during full sun highly stressful.

Cowles (1940, 1946) assumed that the evolution of insulation preceded endothermy as a defense against heat stress caused by strong solar radiation. A large, well-insulated animal with limited net metabolic heat production would have a large time constant, and therefore a mean daily body temperature near the daily average operative environmental temperature. This was shown by Spotila et al. (1973) in their analysis of the heat budgets of large reptiles. By means of a large time constant, these animals could avoid the stress of large daily body-temperature fluctuations. This was of particular value to the large reptiles living in exposed environments, where there would have been a tendency to overheat in the direct sunlight.

Bakken and Gates (1975) describe the reverse situation, in which large land tortoises control their heating and cooling rates by adjusting their circulation. The Galapagos marine iguana, *Amblyrhynchus cristatus,* has a remarkable capacity for altering the rate of heat transfer and heats twice as fast as it cools. The insulation provided by the shell of turtles is effective in rejecting solar radiation since the shell can be heated to a temperature which normally would be destructive to living tissue. The shell not only adds insulation but allows vasoconstriction to reduce thermal conductance to values that result in surface temperatures above thermal injury.

Bakken and Gates (1975) suggest that a protomammal evolved some form of hair as a protection against solar radiation and that the presence of hair allowed a minor increase of metabolic heat production to result in an adaptive offset temperature. T_Δ. Selection for increased metabolic heat production would assist animals living in temperate climates of the middle latitudes, where considerable seasonal variation occurs. However, there would have been a simultaneous problem with heat stress under warm conditions, so that rather than elevated basal metabolism for coping with cold stress, there might have been increased muscular activity. Bakken and Gates (1975) state the following:

> One possible objection is that, under warm conditions, the metabolic heat generated by normal locomotor activity of an insulated animal would also elevate T_b, producing heat stress. If an initially reptilian metabolism with a low overall rate and dependence on anaerobic metabolism for bursts of activity is assumed, the average rate of heat production will be low enough that evaporative cooling by panting or saliva spreading could reduce the net heat production to an essentially zero or even negative (net evaporative cooling) value. These mechanisms are well developed in many reptiles, including crocodilians, lizards, and turtles, for dealing with an external heat load. Evaporative cooling mechanisms would presumably be a necessary preadaptation to allow the animal to enter hot, high-radiation environments that would provide subsequent selection for insulation. As an example, the wool insulation in sheep, which use panting as the primary mechanism for evaporative cooling, decreases the heat load significantly in the presence of strong solar radiation (Yeats, 1967).

The effect of body temperature on metabolic rate must be considered in connection with the evolution of endothermy. Assume that an animal has evolved some sort of pelage protection against strong solar radiation and has at the same time a low aerobic metabolism and depends on an-aerobic metabolism for activity. The animal's insulation would retard the rate of cooling under cool conditions but would also slow the rate of warming, so that the animal would not become active until late in the day. Foraging and prey pursuit would result in the production of metabolic heat. The body temperature would generally be low, and evaporative cooling would not be much used to offset metabolic heat production. The animal would probably cease foraging soon after sundown but then would be faced with the problem of maintaining an adequate stomach tempera-ture for digestion. However, since this hypothetical animal is insulated, it would have a longer time constant $\tau = IC$, which would increase the digestion time. The metabolic temperature factor P would be positive, and it is likely that P would be greater than the thermal conductance $1/I$, since the animal is somewhat insulated. As a result, the time constant would become $\tau^a = IC/(1 - IP)$, decidedly greater than the value $\tau = IC$, and would result in an increased digestion time. The amplifying influ-ence of P and I together is extremely important in maintaining a tempera-ture difference T_Δ from the environment so that digestion will continue. Some physical activity during digestion will also assist with maintenance of a temperature difference.

Again, Bakken and Gates (1975) describe these arguments in some de-tail:

> Other scenarios of this sort can also be used to illustrate the amplifying effect of the temperature coefficient on insulation, which could lead to selection for increased insulation, increased aerobic metabolism, and ultimately endothermy as we know it. The scenario described is particu-larly attractive since it illustrates how selective forces resulting from fall-ing seasonal temperatures would tend toward nocturnality that has fre-quently been postulated for therapsid proto-mammals and early mammals (Jerison, 1971). The capacity for nocturnal foraging would open new niches closed to animals with a relatively high body tempera-ture requirement and without a usable heat-production and storage capacity.

The key factor with regard to the evolution of endothermy is the selec-tion pressure for aerobic metabolism. A hypothetical predator, while chasing prey, would utilize anaerobic metabolism and draw down its ox-ygen supply. Repayment of the oxygen loss after the chase would be com-plicated by the aerobic metabolism requirement at the resulting elevated body temperature and the necessity for digestion. All this would result in an increased tendency for aerobic capacity. The animal would take ad-vantage of its improved insulation and aerobic capacity to forage noctur-nally. Insulation would allow it to retain heat generated from increased muscular activity, or possibly from muscular contractions produced by a form of shivering. Sphinx moths use this strategy, as described by Bartholomew and Epting (1975). Eventually, the increase in aerobic capacity would allow more activity and result in an evolutionary advantage when foraging, chasing prey, and escaping from predators. According to Bakken and Gates (1975): "An increase in basal metabolic rate to the level associated with contemporary mammals would result if it is assumed that the scope of aerobic metabolism is limited so that an increase in active metabolism results in an increase in basal aerobic metabolism. The increase in active aerobic metabolic capacity would thus be the key to endothermy and the elevated basal metabolic rate seen in mammals and birds."

Problems

1. If the masked shrew were without fur on its body, what would be its time constant? Use the numbers given in Tables 13.1 and 13.2.

2. For a sheep shorn of its fleece, determine the time constant using the numbers given in Tables 13.1 and 13.2.

3. If a cardinal dies, how long will it take for its body temperature to drop from 41° to 35°C if the air temperature is 25°C? Use the cardinal's time constant as evaluated from thermal maximum.

4. At the thermal maximum, determine the physiological offset temperature for the masked shrew using the numbers given in Tables 13.1 and 13.2. Then determine the operative environmental temperature at thermal maximum in still air ($V = 0.1$ m s^{-1}) if the air temperature is 30°C and the net radiation is 359 Wm^{-2}.

5. What is the physiological offset temperature for the cardinal at thermal minimum using the numbers given in Tables 13.1 and 13.2? Compare this value with that for the desert iguana at thermal minimum.

Chapter 14

Photosynthesis

Introduction

In order to improve our understanding of the physiological ecology of plants, we wish to develop an analytical model of whole-leaf photosynthesis that will relate the metabolic processes within plant leaves to conditions of light, temperature, humidity, wind, and concentration of carbon dioxide and oxygen near the plant. In order to begin our formulation, we must first understand some of the general characteristics of photosynthesis.

Primary productivity, or photosynthesis, is fundamental to all life on earth. Photosynthesis fixes approximately 6×10^{13} kg of carbon into organic compounds each year. The atmosphere contains 70×10^{13} kg of carbon, which is supplied to plants in the form of CO_2 at an atmospheric concentration of 0.03% (vol/vol). The carbon source for photosynthesis in lakes, rivers, and oceans is in the form of CO_2 and HCO_3^- dissolved in the water. It is estimated that 90×10^{13} kg of carbon is exchanged each year between the oceans and the atmosphere, mostly by physical and chemical processes. The terrestrial biosphere contains a total carbon reservoir of 176×10^{13} kg, and the oceans contain $3,900 \times 10^{13}$ kg. In addition to exchanging carbon dioxide, plants also exchange water vapor and oxygen with the atmosphere.

Photosynthesis

Biochemistry of Photosynthesis

Photosynthesis is an extremely complex series of reactions which uses the energy of visible light to reduce CO_2 to sugar. The overall process consists of those reactions which convert light energy into chemical energy (light reactions) and those which fix CO_2 into carbohydrate (dark reactions). The former require the presence of light, whereas the latter can occur in the dark if the proper energy-rich compounds, ATP and $NADPH_2$, are present.

The light reactions begin with the absorption of light quanta by chlorophyll molecules and excitation (increased energy levels) of electrons in the pigment. Several hundred chlorophyll molecules are associated in an array much like an antenna, with one special chlorophyll molecule that acts as a reaction center gathering excited electrons and passing them on to the electron-transport chain. Excited electrons pass on to the reaction center and then through the electron-transport chain, which captures the energy as ATP and $NADPH_2$. The electrons lost from the chlorophyll are replaced by electrons obtained from the splitting of water, with molecular oxygen being produced as a by product. The ATP and $NADPH_2$ are used in the dark, or Calvin cycle, reactions, in which CO_2 is converted to sugar. Initially, CO_2 and ribulose diphosphate (RuDP) are combined in the presence of RuDP carboxylase to give two molecules of phosphoglycerate. ATP and $NADPH_2$ are used in the Calvin cycle to regenerate RuDP and produce glucose. A more detailed description of the biochemistry of photosynthesis is given in Nobel (1974) and Salisbury and Ross (1978).

Photosynthesis is not a particularly efficient process. Of the total amount of light energy incident on a plant, only 1 to 7% is converted to useful products. Photons of red light of 660 nm have an energy content of 180 kJ mole^{-1}, only about one-third of the energy necessary for the reduction of 1 mole of CO_2. Blue light is more energetic, but not sufficiently so to supply the necessary energy. Although it might seem that three photons would supply sufficient energy for the reduction of one CO_2 molecule, experiments show that about eight photons must be supplied simultaneously. This number is known as the quantum requirement for photosynthesis. Most of the radiant energy absorbed by a leaf goes into heat. Although physiologists generally consider this to be wasted energy, it is nevertheless important in maintaining a plant at a proper thermodynamic level so that the chemical reactions proceed at a suitable rate. Radiant energy in the form of heat is also used to evaporate water and help drive water transport in the xylem.

Plants not only capture CO_2 to produce organic compounds, releasing oxygen in the process, but also respire and use oxygen to oxidize sugars and other compounds to CO_2 and water. All plants undergo dark respiration, the major portion of which occurs in the mitochondria and does not require light, and dark respiration is a major source of chemical energy for plant cells. Another form of respiration known as photorespiration occurs only in light and is controlled by RuDP carboxylase-oxygenase. Photorespiration appears to be counterproductive since the energy released is not used directly by plant cells. However, it would be surprising if evolution had created a system that did not have some benefits. Photorespiration may serve as a source of the amino acids glycine and serine or play a protective role by draining off excess energy-rich compounds formed in the chloroplasts when light intensity is great and CO_2 concentrations are low. It is generally agreed that glycolic acid is the primary source of CO_2 released during photorespiration.

Photosynthetic Pathways

In the context of our present knowledge, there appear to be three types of photosynthetic pathways in plants, referred to as the C_3, C_4, and CAM photosynthetic pathways. The main distinction between the C_3 and C_4 pathways is the first product of CO_2 assimilation. In C_3 plants, the first product of CO_2 assimilation by the Calvin cycle is the three-carbon compound 3-phosphoglyceric acid (PGA). Glucose, usually considered to be the final product of photosynthesis, is then formed from PGA and RuDP, which are regenerated by a complex chain of reactions. In C_4 plants, the first products of CO_2 assimilation are four-carbon acids such as oxaloacetic, malic, and aspartic acid. During assimilation, CO_2 is enzymatically attached to phosphoenolpyruvic acid (PEP) to produce oxaloacetic acid, which is then converted to the other acids.

Differences between C_3- and C_4-pathway plants are considerable. In addition to mesophyll cells, C_4 plants have bundle-sheath cells which surround the vascular bundles of the leaves. The internal structures of the chloroplasts in the mesophyll and bundle-sheath cells of C_4 plants differ, reflecting the different function of the two tissues. Carbon dioxide is assimilated in the mesophyll cells by the C_4 cycle. The four-carbon acids produced in the mesophyll are then transported to the bundle-sheath cells, where they are decarboxylated and the CO_2 is refixed by RuDP carboxylase. The C_4 plants are known as high-efficiency plants since their photosynthetic capacities under ambient conditions are often two or three times greater than those of the C_3 plants. The C_4 plants exhibit little or no photorespiration, whereas photorespiration may be as much as 50% of the rate of photosynthesis in C_3 plants. Photosynthesis has a higher temperature optimum for C_4 plants, often between 35° and 47°C, than for C_3 plants, where it is often 30°C or less. The C_4 plant achieves maximum photosynthesis at higher light intensities than C_3 plants. The C_4 plants have low compensation points (CO_2 concentration at which net photosynthesis is zero), whereas C_3 plants have compensation points ranging upward from about 40 ppm. Photosynthesis by C_4 plants is not inhibited by oxygen, except at very high concentrations well above normal amounts.

Typical C_4-pathway plants include maize, sugarcane, sorghum, pigweed, and Bermuda grass, whereas spinach, tobacco, wheat, rice, and beans are typical C_3 plants. There are vast numbers of each type in nature, and some plant families have evolved a mixture of C_3 and C_4 species.

Welkie and Caldwell (1970) have reported on the apparent dichotomy of C_3 and C_4 species occurring in various plant families from a study of herabarium specimens. Presence or absence of bundle-sheath cells was the discriminating anatomical feature used. It appears that C_4 plants are better adapted to dry, hot habitats than C_3 plants, but it is interesting to note that desert vegetation contains a mixture of C_3, C_4, and CAM plants. Johnson (1975) has reported on gas-exchange studies of desert plants. He points out that desert shrubs with C_4 anatomy seem to be restricted to the genus *Atriplex* of the family Chenopodiaceae. The desert plant *Tidestromia oblongifolia*, which has a high temperature optimum for photosynthesis (47°C), is a C_4-pathway plant, whereas ubiquitous desert shrubs such as *Larrea divericata* and *Ambrosia dumosa* are C_3 plants.

Stowe and Teeri (1978) report that the percentage of C_4 dicot species of plants in a geographic region is strongly correlated with summer pan evaporation and dryness ratio. However, they also showed that the distribution of C_3 species in the C_4 dicot families (nine families native to North America) correlates strongly with summer pan evaporation. These plant families appear to have adaptive characteristics for dryness in addition to the photosynthetic pathway. Teeri and Stowe (1976) showed that the number of C_4 grasses present correlates strongly with minimum daily July temperature. Hence the C_4 photosynthetic pathway may confer different ecological properties to plants in different phylogenetic groups. Downes (1969) showed, for a number of crop plants, weedy species, and xerophytes, that the C_4 species use only about half as much water as C_3 species for the production of an equal amount of dry matter. The North American dicot plant families containing the most C_4 plants include the Amaranthaceae, Chenopodiaceae, and Euphorbiaceae, but the percentage of C_4 species does not exceed 4.5% of the total in any family. The proportion of C_4 species among all dicot species is 4.4% in the Sonoran Desert, 3.4% in Arizona, 2.8% in Texas, 2.0% in Kansas, 0.8% in Missouri, 0.4% in North and South Carolina, and 0.1% in the Great Smoky Mountains.

The desert succulents are all perennials with the CAM (crassulacean acid metabolism) photosynthetic pathway. The important families of CAM-type plants are the Cactaceae, Agavaceae, and Liliaceae. These plants are characterized by either succulent or thickened fibrous photosynthetic organs. Plants with the CAM pathway have the option of fixing CO_2 at night by a mechanism similar to that used by C_4 plants or during the day by a mechanism similar to that of C_3 plants. The CAM plants open their stomates at night, take in CO_2 by diffusion, combine the CO_2 with PEP (provided from stored carbohydrates), and form malate. In the light the next day, with stomates closed, the malate is decarboxylated to form pyruvate and CO_2, which is then available for photosynthesis. By closing their stomates during the day, CAM plants greatly reduce transpiration. Johnson (1975) reports on measurements of assimilation efficiency and transpiration ratios for desert CAM plants, as well as C_3 and C_4 desert plants. The assimilation efficiency is defined as $e = P/P_M$, and the transpiration ratio as TR $= E/P$, where P is the photosynthetic rate, P_M is the photosynthetic rate at saturating CO_2, and E is the transpiration rate. He reported e values for C_4 summer annuals and C_3 winter-spring annuals of 0.89 and 0.31, respectively, and TR values of 560 and 225. For a CAM plant, *Opuntia basilaris,* he found e and TR to be 0.92 and 45 in the winter and 0.85 and 153 in the summer. It appears that the CAM plants are more efficient than C_3 or C_4 plants with respect to water use under both winter and summer conditions. The CAM plants have much more favorable TR values during the summer than the C_4 plants, as a direct result of nighttime gas exchange and the CO_2 storage system. Johnson (1975) writes as follows: "Thus the inverted pattern of gas-exchange characteristic of CAM's must be considered a major strategy for water-use efficiency. The evolution of this strategy is a form of environmental selection coupled with the development of an appropriate physiological system that could store CO_2 in the dark and yet use it for photosynthetic energy accumulation in the light."

Analytical Models

The simplest photosynthetic model for a whole leaf includes three processes: energy exchange, diffusion of carbon dioxide and oxygen, and the biochemical reactions related to photosynthesis.

Rates of photosynthesis and respiration within a leaf are dependent on temperature, light, carbon dioxide and oxygen concentrations, and other factors such as enzyme availability. Leaf temperature is determined by exchange of radiation (including light) between the leaf and the environment, loss of heat by convection, and evaporation of moisture by the outward diffusion of water. The rate at which gases flow to or from a leaf depends on the mechanism of gas diffusion. The diffusion rate of a gas is proportional to the difference in concentration outside and inside the leaf and inversely proportional to the resistance of the pathway through the boundary layer, stomates, substomatal cavity, cell walls, and cells to the chloroplasts, mitochondria, or peroxisomes. The rate at which carbon dioxide flows into or out of the leaf must equal the rate at which the biochemical reactions of photosynthesis and respiration occur within the leaf. The biochemical reactions at the metabolic centers are complex and depend upon the conditions existing within the leaf. All these processes, energy exchange, gas exchange, and chemical kinetics, are interconnected and interdependent. In order to describe analytically the manner by which a whole leaf carries on photosynthesis, it is necessary to account for all these interrelationships. This process was begun in Chapter

3, where the simplest photosynthetic model for a whole nonrespiring leaf was described. Here we shall extend the analysis to include all primary processes. It is not our intention here to present the most detailed models available, but to present one approach to the analysis which will illustrate the methodology involved. Recent reviews of all models are given by Tenhunen et al. (1980a) and Tenhunen et al. (1980b).

There are several approaches to modeling whole-leaf photosynthesis, but all give similar results. Some approaches begin with the biochemical processes of photosynthesis and attempt to make these the cornerstone of the model, whereas other approaches model from a biophysical basis and bring in the biochemical events where necessary. The problem is that, when working with the whole leaf (a necessity for understanding the ecology of a plant), one is seeing the sum total of all chemical reactions and gas diffusions going on simultaneously and integrated over time. For whole-leaf photosynthesis, processes of gas exchange and chemical kinetics must be considered along with the thermodynamics of the leaf. We shall begin with a biophysical approach and then, later in the chapter, give an elaboration of the biochemical kinetics.

Analytical models of whole-leaf photosynthesis have been given by Chartier (1966, 1970), Chartier and Prioul (1976), Charles-Edwards (1971), Hall (1971), Lommen et al. (1971), Peisker (1974, 1976), Prioul and Chartier (1977), Reed et al. (1976), and Tenhunen et al. (1976, 1977).

Photosynthesis Model without Respiration

Gas Exchange

A model of photosynthesis for a nonrespiring leaf was developed in Chapter 3. The diffusion of CO_2 from the atmosphere to the sites of fixation in the chloroplasts is described by Fick's law,

$$P = \frac{C_a - C_c}{R}, \tag{14.1}$$

where P is the photosynthetic rate (mmole per square meter per second), C_a is the concentration of CO_2 in the air (mmole per cubic meter), C_c is the concentration of CO_2 in the chloroplasts (mmole per cubic meter), and R is the resistance to CO_2 diffusion from the air through the leaf boundary layer, stomata, intercellular air spaces, cell walls, and cytoplasm into the chloroplasts (in seconds per meter) as shown in Fig. 14.1a.

The rate of CO_2 fixation is a function of CO_2 concentration in the chloroplasts and the light level. Both of these functional dependencies have the same general form as shown in Fig. 14.2. The response of photosynthesis to CO_2 concentration is easier to understand than that to light. Various empirical descriptions of these curves have been given. The

A. Without Respiration   B. With Respiration

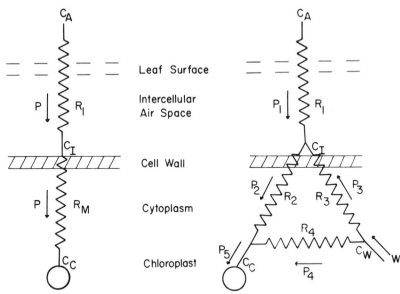

Figure 14.1. Electrical analogs for CO_2 exchange in leaves: (A) simplified resistance network without a respiratory source of CO_2; (B) resistance network with respiration. Fluxes are positive in the direction of the arrows. Symbols: R, resistance; P, CO_2 flux; C, CO_2 concentration; W, flux of respiration.

curves are known as Blackman curves (first described by Blackman and Smith, 1911) if characterized by two linear portions connected by a sharp elbow and Michaelis–Menten curves if more rounded and will be distinguished analytically later, following the mathematical description.

Dependence on Carbon Dixoide Concentration

The shape of the curve relating photosynthesis to CO_2 concentration may be described by a Michaelis–Menten type of equation, even though it is not a true Michaelis–Menten situation since a series of biochemical reactions are involved rather than a single reaction. The description is

$$P = \frac{P_{\mathrm{M}}}{1 + \dfrac{K}{C_{\mathrm{c}}}}, \tag{14.2}$$

where P_{M} is the rate of photosynthesis in mmole per square meter per second at saturating C_{c} and a specific leaf temperature and K is a coefficient equal to the chloroplast concentration of CO_2 (in mmole per cubic meter) at which $P = P_{\mathrm{M}}/2$.

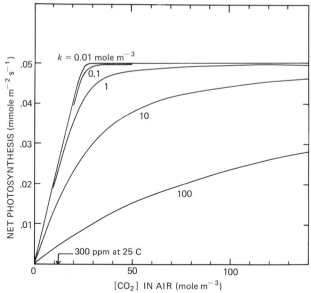

Figure 14.2. Photosynthesis as a function of CO_2 concentration in air for several values of the affinity coefficient K. This example was calculated with $R = 500$ s m^{-1}, $T = 30°C$ (optimum), and $L = 400$ Wm^{-2} (equivalent to full sunlight).

Dependence on Light

A typical response of photosynthesis to light is shown in Fig. 14.3. The general shape is similar to that for CO_2 response shown in Fig. 14.2. The rate of photosynthesis depends on the concentrations of photoproducts such as ATP or $NADPH_2$. At low light intensities, ATP or $NADPH_2$ is produced approximately as a linear function of light intensity, whereas at higher light intensities, the rate of production levels off to some maximum value. It is possible, therefore, to distinguish two components of the light-response curves for photosynthesis, a linear region and a plateau region in which enzymes in the Calvin cycle may regulate the response. A Michaelis–Menten equation such as Eq. (3.12) of Chapter 3 is thus not an adequate description of the response of photosynthesis to light (it was used there for simplicity).

The light response of photosynthesis is described by the following empirical expression, which was first worked out by Smith (1937, 1938):

$$P_M = \frac{c_1 P_{ML} L}{(1 + c_1^2 L^2)^{1/2}}.$$ (14.3)

The quantity P_{ML} is the rate of photosynthesis at saturating light and carbon dioxide concentration for a specific leaf temperature, c_1 is an empirically determined constant, and L is the photosynthetically active radi-

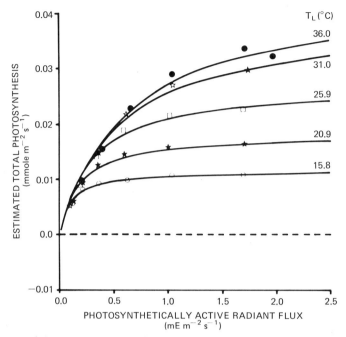

Figure 14.3. Light response curves for photosynthesis at several leaf temperatures for *Phaseolus vulgaris*. The abscissa is in photosynthetically active radiant flux.

ant flux between 400 and 700 nm (in einsteins per square meter per second).

It is useful to rewrite Eq. (14.3) in terms of the initial slope of the light-response curve. The initial slope α is readily determined by measurement and represents the maximum efficiency of incident light-energy conversion. By differentiating Eq. (14.3) and letting L approach zero, the following relationships are obtained:

$$\frac{dP_M}{dL} = \frac{c_1 P_{ML}}{(1 + c_1^2 L^2)^{1/2}} - \frac{c_1^3 L^2 P_{ML}}{(1 + c_1^2 L^2)^{3/2}} \tag{14.4}$$

$$\lim_{L \to 0} \frac{dP_M}{dL} = \alpha = c_1 \cdot P_{ML}. \tag{14.5}$$

Therefore,

$$c_1 = \frac{\alpha}{P_{ML}}, \tag{14.6}$$

and substitution into Eq. (14.3) gives

$$P_M = \frac{\alpha L}{\left(1 + \dfrac{\alpha^2 L^2}{P_{ML}^2}\right)^{1/2}}. \tag{14.7}$$

Equation (14.7) contains two unknown properties of the plant, the maximum efficiency to light α, in moles per einstein, and the maximum capacity for carboxylation P_{ML}, in mmoles per square meter per second. The factor P_{ML} is strongly temperature dependent, as described in the next section. A term known as the quantum yield defines the maximum efficiency of absorbed light conversion and is equivalent to α as used here. Emerson and Lewis (1941) showed that quantum yield is independent of temperature at zero oxygen concentration. Therefore, whenever we work with C_3 plants at the low oxygen concentrations necessary to reduce photorespiration and define the photosynthetic surface properly, we shall consider α to be temperature independent. A typical value of α is about 50 mmole E^{-1}.

At normal oxygen concentrations in air, Ehleringer and Bjorkman (1977) observed α to be temperature dependent in C_3 plants and temperature independent in C_4 plants. They also showed that α in *Encelia californica,* a C_3 plant, exceeded that in *Atriplex rosea,* a C_4 plant, at low temperatures but fell below that of *A. rosea* at temperatures above 30°C.

Temperature Dependency of the Light Maximum

Rate of photosynthesis depends on carbon dioxide concentration, light intensity, and leaf temperature. By plotting photosynthesis on a vertical axis, with two horizontal axes—temperature and light intensity—for conditions of saturating carbon dioxide and low oxygen, one gets a three-dimensional surface. Varying carbon dioxide or oxygen concentration will move the surface down. Our purpose is to arrive at an analytical description of these surfaces. The maximum photosynthetic rate under these conditions (without respiration and at optimal temperature, saturating carbon dioxide concentration, low oxygen concentration, and saturating light) is designated as P_{MLT}. Any point corresponding to these conditions, but at a leaf temperature other than optimum, is designated as P_{ML}. Any point on the P_{ML} surface at any light level is a maximum rate for low oxygen (i.e., no photorespiration) and saturating carbon dioxide and is designated as P_M. The temperature dependence of P_{ML}/P_{MLT}, described by $G(T_K)$, is shown in Fig. 14.4. It is now necessary to give an analytical expression of this temperature function.

The light and carbon dioxide maxima of photosynthesis depend on (a) the temperature of activation of the rate-limiting enzymatic step of the process and (b) the temperature of activation for denaturation of the rate-limiting enzyme at high temperatures. Johnson et al. (1954) have discussed the temperature dependence of these types of enzymatic processes in detail and show that they are expressed by the function

$$G(T_K)\ 100 = \frac{P_{ML}}{P_{MLT}}\ 100 = \frac{c_2\ T_K\ e^{-\Delta H^{\neq}/R\ T_K}}{1 + e^{-\Delta H_1/R\ T_K}\ e^{\Delta S/R}}, \tag{14.8}$$

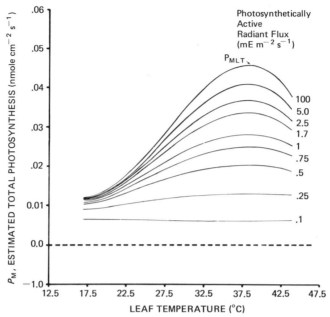

Figure 14.4. Temperature response curves for photosynthesis at several light intensities for *Phaseolus vulgaris*.

where c_2 is a constant, T_K is the leaf temperature (in degrees Kelvin), R is the gas constant (8.31 J mole^{-1} °K^{-1}), $\Delta H{\neq}$ is the energy of activation for the enzyme-catalyzed photosynthetic reaction (in joules per mole), ΔH_1 is the energy of activation for the denaturation equilibrium (in joules per mole), ΔS is the entropy of the denaturation equilibrium (in entropy units) and e is base of natural log. Although this temperature function was worked out for single-system enzyme reactions, its validity for a complex process such as photosynthesis has been established.

The quantity $\Delta H{\neq}$ can be estimated from the slope of a regression obtained for $\ln(P_{ML}/P_{MLT} \times 100)$ as a function of inverse absolute leaf temperature when leaf temperature is well below the optimum. The quantity ΔH_1 is obtained from a similar regression when high leaf temperatures (above the optimum) have reduced the photosynthetic rate by denaturation to 20 to 40% of the maximum rate. For *Phaseolus vulgaris,* values of $\Delta H{\neq} = 48{,}465$ J mole^{-1} and $\Delta H_1 = 175{,}682$ J mole^{-1} were reported by Tenhunen et al. (1976). The details of the procedures involved are described in the reference. After $\Delta H{\neq}$ and ΔH_1 are determined, values for c_2 and ΔS are obtained from the slope and the photosynthetic rate at the temperature optimum $_0T_K$ using the expressions

$$\frac{\Delta H{\neq} + RT_K}{\Delta H_1 - \Delta H{\neq} - RT_K} = e^{-\Delta H_1/RT_K}\, e^{\Delta S/R}, \qquad (14.9)$$

where ΔS is the only unknown, and

$$100 = \frac{c_2\ _oT_K\ e^{-\Delta H \neq /RT_K}}{1 + E^{-\Delta H_1/RT_K}\ e^{\Delta S/R}}, \tag{14.10}$$

where c_2 is the only unknown. Best-fit values for *Phaseolus vulgaris* were $\Delta S = 557.16$ J mole^{-1} °C^{-1}, $c_2 = e^{17.925}$, and $P_{MLT} = 0.03697$ mmole m^{-2} s^{-1}.

Photosynthesis Model

By substitution from Eqs. (14.7) and (14.8), Eq. (14.2) becomes

$$P = \frac{\alpha L\ P_{MLT}\ G(T_K)}{\left[1 + \dfrac{K}{C_c}\right][P_{MLT}^2 G^2(T_K) + \propto\ ^2L^2]^{1/2}}. \tag{14.11}$$

By eliminating C_c between Eqs. (14.1) and (14.11), one gets a quadratic expression for the rate of photosynthesis by a whole leaf as a function of carbon dioxide concentration, light intensity, and temperature. However, it is cumbersome to carry all of the details of Eq. (14.11) into the simple algebra of solving the quadratic for P. Therefore, we introduce P_M into Eq. (14.11), where

$$P_M = \frac{P_{MLT}\ \alpha\ L\ G(T_K)}{[P_{MLT}^2\ G^2(T_K) + \alpha\ ^2L^2]^{1/2}}. \tag{14.12}$$

Solving Eq. (14.2) for C_c, substituting into Eq. (14.1), and solving the resulting quadratic for P, as in Chapter 3 [see Eq. (3.10)], one gets

$$P = \frac{(C_a + K + RP_M) - [(C_a + K + RP_M)^2 - 4C_a RP_M]^{1/2}}{2R}. \tag{14.13}$$

The quadratic has the form

$$ax^2 + bx + c = 0, \tag{14.14}$$

with solutions

$$x = \frac{-b \pm \sqrt{b^2 - 4ac}}{2a}. \tag{14.15}$$

Every quadratic equation has two roots, as indicated in Eq. (14.15) by the plus and minus signs. It is necessary to determine which root is a valid solution in the context of our current problem. The procedure is algebraically complicated and is written out fully in Appendix B of Lommen et al. (1971). By taking the positive sign for the limit as $C_a \to \infty$, $P \to C_a/R$ or infinity, and by taking the negative sign, $P \to 1.0$. It is clear from Fig. 3.12 that P does not go to infinity with increasing CO_2 concentration but to a relative value of 1.0. Therefore, the negative root is the proper solution.

Note also, from Eq. (14.13), that

$$\lim_{R \to \infty} P = \frac{C_a}{R}.$$
(14.16)

This equation indicates that at high values of resistance, CO_2 molecules reaching the chloroplasts are assimilated so quickly that C_c is essentially zero. Thus rate of CO_2 diffusion determines photosynthetic rate for large R. Also, it can be shown from Eq. (14.13) that

$$\lim_{R \to 0} P = \frac{P_M}{1 + \dfrac{K}{C_a}}.$$
(14.17)

This equation indicates that at very low resistance, rate of photosynthesis is determined by the rate of CO_2 fixation.

Equation (14.13), then, represents the simplest situation of CO_2 fixation by a whole leaf which has no respiration, including dark respiration. This equation cannot be applied to plants with any confidence, but it is a useful step toward a more complete analysis.

Leaf Resistance

For the nonrespiring leaf, the diffusion pathway leading from the air outside the leaf to the intercellular air space and from the cytoplasm to the chloroplast is relatively simple, as shown in Fig. 14.1a. The total resistance R of this diffusion pathway is divided into two major components, R_1 and R_m. The component R_1 consists of the boundary-layer resistance and the stomatal resistances; the component R_m consists of all liquid phase resistances between the intercellular air spaces and the chloroplasts. Total resistance is affected by any factor that changes any of its components. Boundary-layer resistance changes with wind speed, leaf size, and leaf orientation. For CO_2, the boundary-layer resistance is taken to be the boundary-layer resistance for water vapor multiplied by 1.56 (the ratio of the diffusion coefficients for H_2O and CO_2 in air at 0°C).

Stomatal Resistance

Stomatal resistance is known to be a function of light, leaf water potential, the phase of endogenous rhythm, leaf temperature, CO_2 concentration and relative humidity of the air, and perhaps other factors. It is possible to include each of these responses in the analytical description of the leaf resistance to carbon dioxide and water-vapor diffusion. The stomatal resistance of a well-watered leaf changes with light intensity, as shown in Fig. 14.5. This relationship, described by an inverse Michaelis–Menten equa-

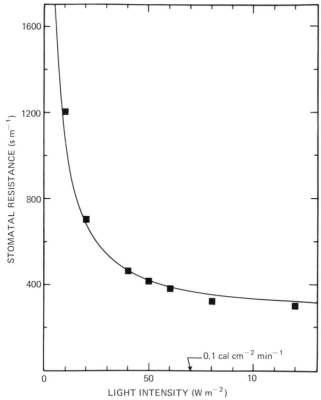

Figure 14.5. Stomatal resistance as a function of the light intensity: (■) data of Gaastra (1959); (–) calculated from Eq. (14.18). Data points are for turnip, as reported by Gaastra (1959).

tion, is

$$R_s = \frac{L + K_{RL}}{L} \, _{min}R_s, \tag{14.18}$$

where R_s is the stomatal resistance to CO_2 diffusion (in seconds per meter), $_{min}R_s$ is the minimum value of R_s reached at high light intensities, and K_{RL} is the value of L such that $R_s = 2(_{min}R_s)$. If L is in watts per square meter, then K_{RL} is in the same units. The solid line shown in Fig. 14.5 is calculated from Eq. (14.18) for $K_{RL} = 350$ Wm^{-2} and $_{min}R_s = 250$ s m^{-1}. The data given in Fig. 14.5 are from Gaastra (1959) for a turnip leaf in 0.03% CO_2 and at an air temperature of 20.3°C.

Leaf water potential is a function of several environmental factors. As sufficient information becomes available concerning these effects, it will be possible to develop a functional relationship for the dependence of stomatal resistance on leaf water potential. The endogenous rhythm is usually such as to produce stomatal opening during daylight and closure at night. The influence of temperature on stomatal opening is described later in the chapter. In the analysis described by Lommen et al. (1971), no temperature dependencies were assumed. This procedure is presented here before temperature effects are examined.

Properties of the Model

Now that the simplest photosynthesis model (for a nonrespiring leaf) has been developed, it is important to describe some of its properties before proceeding with the more complex model that includes respiration. Photosynthesis as a function of leaf temperature, as derived from the model for a nonrespiring leaf, was shown in Fig. 3.14. It is evident from this illustration that when resistance is high, photosynthesis depends almost entirely on rate of CO_2 diffusion and is nearly independent of temperature. When resistance is relatively low, gas diffusion is not limiting, and the effect of temperature on the biochemical events is significant. The values of the coefficients used in the calculation of Fig. 3.14 are $C_a = 12.5$ mmole m^{-3}, $L = 400$ Wm^{-2}, $K = 10$ mmole m^{-3}, and $T = 30°C$ (optimum).

The effect of changing values of K on photosynthetic rate as a function of CO_2 concentration in the air is shown in Fig. 14.2. The values used for this plot were $R = 500$ s m^{-1}, $L = 400$ Wm^{-2}, and $T = 30°C$ (optimum). Another diagram of photosynthesis versus CO_2 concentration with variable resistance R and K equal to 10 mmole m^{-3} is shown in Fig. 14.6. The values of C_a, L, and T are the same as given above. In both Figs. 14.2 and 14.6, when $RP_M/K \gg 1$, a Blackman-type curve is approached and when

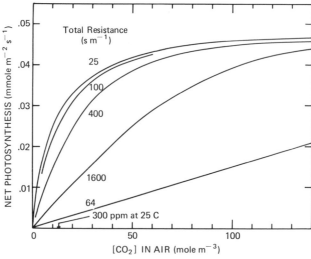

Figure 14.6. Photosynthesis as a function of CO_2 concentration in air for several values of total CO_2 diffusion resistance. Here $K = 10$ mole m^{-3}, $T = 30°C$ (optimum), and $L = 400$ Wm^{-2} (equivalent to full sunlight).

$RP_M/K \ll 1$, a Michaelis–Menten-type curve is approached. Blackman curves are characterized by two linear portions connected by a sharp elbow and result when resistance to CO_2 uptake controls the rate of photosynthesis. Michaelis–Menten curves are more rounded and result when biochemical reactions control the rate of photosynthesis, i.e., when K is large or R is very small.

Respiration

Carbon dioxide is produced in the cells of leaves at various sites. Evolution of carbon dioxide in the dark is largely associated with the mitochondria and is often referred to as mitochondrial respiration, designated W_m. However, since it is not clear that carbon dioxide production is restricted to mitochondria, we shall refer to this type of respiration as "dark" respiration. Light-induced respiration, the bulk of which involves glycolate metabolism, involves both the mitochondria and the peroxisomes and is therefore designated W_p. This type of respiration is referred to as photorespiration.

Measurements made with C_3 plants at an oxygen concentration of 1.5% show that there is little observable photorespiration, and any residual respiration at low oxygen concentrations will be considered dark respiration. Our understanding of the respiration process is not as adequate as is desirable for construction of a complete model of the respiration components. Some of these inadequacies will be mentioned later in the chapter. For the moment, however, we shall assume that photorespiration is suppressed at low oxygen concentrations.

Residual respiration may be determined by means of a linear extrapolation of the initial portion of either the curve of net-photosynthesis-versus light intensity or that of net photosynthesis versus carbon dioxide concentration to the intercept below zero net photosynthesis. The initial portion of a light-response curve determined at saturating carbon dioxide concentration is linear as the result of limitations by the photoreactions that generate ATP and $NADPH_2$. The behavior of the light response is discussed by Tenhunen et al. (1976). The initial portion of a carbon dioxide — response curve is linear owing to mesophyll resistance to carbon dioxide transfer, as shown by Jones and Slatyer (1972). One problem with the assumption that extrapolation of carbon dioxide response to zero gives an estimate of the total residual respiration is that there may be recycling of carbon dioxide within the leaf in the light. This problem is discussed later in the chapter.

Dark Respiration

Dark respiration may be suppressed in light except at low light intensities. There is evidence for dark respiration in the light however; see Chapman and Graham (1974), Raven (1972), and Marsh et al. (1965). Irvine (1970) measured the rate of respiration at various light intensities in sugarcane, a species with a low CO_2 compensation point. He used an isotope method which labeled and traced dark respiration but not peroxisomal respiration. Dark respiration declined sharply with increasing light intensity, reaching a very low rate at light intensities greater than one-fourth of full sunlight. This functional behavior is represented by the expression

$$W_m = W_{ML} \left\{ \exp \left[(-\ln 2) \left(\frac{L}{L_H} \right) \right] \right\}, \qquad (14.19)$$

where W_{ML} (in mmoles per square meter per second) is the value of W_m at zero light intensity and optimum temperature, L is the light intensity over the wavelength range 400 to 700 nm, measured in watts per square meter, or einsteins per square meter, and L_H is a constant such that when $L = L_H$, the quantity $\exp[(-\ln 2)(L/L_H)] = \frac{1}{2}$. It is, in fact, this need to describe the shape of the monotonically decreasing curve relating respiration to light intensity that forces one to write the exponential function in this form. Chapman and Graham (1974) disagree and say that there is no light effect.

Dark respiration is independent of both CO_2 and O_2 concentration in the air according to Forrester et al. (1966). If this is true, then one is justified in omitting any such dependence in the W_m term. If a dependence on these concentrations is discovered, it can readily be included.

Dark respiration is temperature dependent and increases monotonically with increasing temperature. This functional dependence may be written in analytical form by means of the Arrhenius equation,

$$W_{ML} = J \, e^{[-(E/RT_K)+A]}, \qquad (14.20)$$

where J is a proportionality constant with units of mmoles per square meter per second, E is the apparent activation energy in joules per mole, R is the gas constant, T_K is the absolute temperature of the leaf, e^{-E/RT_K} is the Boltzmann factor, or the fraction of molecules possessing sufficient energy to react (in mmole^{-1}), and A is an empirically determined constant.

If light affects the rate of dark respiration, the complete equation is given by

$$W_m = J \, e^{[-(E/RT_K)+A]} \, e^{-[\ln 2 \, (L/L_H)]}. \qquad (14.21)$$

If there is no effect of light, this expression reduces to Eq. (14.20).

Photorespiration

Photorespiration is considerably more complex than dark respiration. Photorespiration is the uptake of O_2 and release of CO_2 in the light by photosynthesizing tissue. It involves formation of glycolate in the chloroplasts, possibly from Calvin-cycle intermediates. Glycolate is transported to the peroxisomes, where it is oxidized to glyoxylate and glyoxylate is metabolized to glycine. Glycine is metabolized to serine in the mitochondria, with the release of varying amounts of CO_2. These various metabolic products cycle among the chloroplasts, peroxisomes, and mitochondria. Photorespiration is absent or not detectable in C_4-type plants. All C_3-type plants have photorespiration.

It has been shown by Decker (1959), Holmgren and Jarvis (1967), Hew et al. (1969), and Jackson and Volk (1970) that photorespiration increases with increasing light intensity. Hofstra and Hesketh (1969) indicate that the temperature dependence of photorespiration probably follows the temperature response of photosynthesis, although according to Jackson and Volk (1970), this may not be true at temperatures above the photosynthetic optimum. Photorespiration increases with increasing oxygen concentration according to Forrester et al. (1966).

Internal CO_2 Concentration

If one measures net photosynthetic rate as a function of carbon dioxide concentration in the air outside a leaf, conversion of this function to that relating photosynthesis to carbon dioxide concentration in the intercellular air spaces inside the leaf is straightforward. The procedure involves measuring the rate of loss of water vapor, or the transpiration rate E, and deriving the water-vapor resistance r_ℓ as given by

$$r_\ell = \frac{_sd_\ell(T_\ell) - h\ _sd_a(T_a)}{E} \tag{14.22}$$

when the leaf temperature T_ℓ, air temperature T_a, and relative humidity h are measured. The quantity $_sd_\ell(T_\ell)$ is the water-vapor density at saturation as a function of leaf temperature, and $_sd_a(T_a)$ is the water-vapor density of the air at saturation as a function of air temperature. Since all quantities except r_ℓ are known, one can solve for r_ℓ. (This analysis depends on the air movement across the leaf surface being vigorous enough to reduce the boundary-layer resistance to a small value.) When r_ℓ for water vapor is determined, this value then gives the resistance R_1 to carbon dioxide since $R_1 = 1.56r_\ell$. The quantity 1.56 is the ratio of the diffusion coefficients of H_2O and CO_2.

Knowing the value of R_1, one can determine the intercellular CO_2 concentration C_i from the net photosynthetic rate P_1 and the CO_2 concentration in the air outside of the leaf by means of Eq. (14.25).

Competitive Inhibition between O_2 and CO_2

It is extremely difficult by means of laboratory measurements to distinguish between the photosynthetic and photorespiratory processes going on within a leaf. Measurements of gas exchange outside the leaf fail to take into account photorespired CO_2 recycled internally, and it remains impossible to measure CO_2 concentrations at the fixation sites. Bowes et al. (1971) and Bowes and Ogren (1972) have shown that O_2 competitively inhibits carboxylase activity of RuDP carboxylase-oxygenase and that CO_2 competitively inhibits the oxygenase activity of the same enzyme. This competitive inhibition between O_2 and CO_2 is known as the Bowes–Ogren hypothesis. From classical equations describing the rate of an enzymatically regulated reaction in the presence of a competitive inhibitor, the rate of carboxylation in the presence of oxygen is given as a modification of Eq. (14.2) in the form

$$P = \frac{P_M}{1 + \dfrac{K(1 + [O_2]/K_O)}{C_c}}, \tag{14.23}$$

where $[O_2]$ is the concentration of oxygen at the fixation site and K_O is a constant equal to the concentration of oxygen that results in a doubling of K. An increase in the oxygen concentration will increase the denominator and produce a reduction in the net photosynthetic rate.

Photorespiration, or rate of oxygenation at the fixation site, may be treated in a similar manner, in terms of the CO_2 efflux:

$$W_P = \frac{W_M}{1 + \dfrac{K_O\,[1 + C_c/K]}{[O_2]}}. \tag{14.24}$$

The quantity W_M is the maximum rate of photorespiration at saturating oxygen, and K_O is the oxygen concentration at which $W_p = W_M/2$ with $C_c = 0$. Oxygen concentration at the reaction site is essentially equal to that in the air. However, C_c may be very low because of diffusion resistances. An increase in the internal concentration of carbon dioxide results in an increase in the denominator and produces a decrease in the photorespiration rate. Just as an increase in internal carbon dioxide concentration increases the photosynthetic rate, so also an increase in oxygen concentration increases the photorespiration rate.

Resistances

The sites of fixation (chloroplasts) and respiration (mitochondria and peroxisomes) are distributed throughout the leaf cellular tissue. Carbon dioxide may flow from outside the leaf through the intercellular air space across the cell walls and through the cytoplasm to the chloroplasts, it may originate at the sites of respiration and flow directly through the cytoplasm to the chloroplasts, or it may flow from respiration sites out through the cell walls, intercellular air spaces, stomates, and boundary layer to the free air. Although it is possible to describe a three-dimensional network of carbon dioxide sources and sinks and flow pathways, there is little need to do so at this stage of our analysis. We would, in addition, be unable to measure some of these components. Instead, we shall describe a two-dimensional resistance network with a single fixation site and a single respiration site.

Leaf Resistance

A two-dimensional electrical analog of carbon dioxide exchange in leaves is shown in Fig. 14.1b. The carbon dioxide concentration at the chloroplasts is C_c, that at the respiration sites is C_w, and that outside the leaf is C_a. The quantity R_1 is the resistance between the outside air and the intercellular air space (IAS) and comprises the boundary-layer and stomatal resistances. The CO_2 concentration is assumed to be uniform throughout the IAS. Resistances R_2 and R_3 are the direct resistances between the IAS and the sites of photosynthesis and respiration, respectively. Both contain cell-wall and cytoplasm resistance as components. Resistance R_4 is the direct resistance between the sites of respiration and the chloroplasts. Indirect diffusion pathways from the IAS to the chloroplasts are represented by $R_3 + R_4$. Therefore, the total diffusive resistance to a CO_2 molecule between the IAS and the chloroplasts is R_2 in parallel with $R_3 + R_4$. Total resistances between the IAS and the sites of respiration and between the sites of respiration and the chloroplasts can be analyzed similarly.

Estimates of these resistances can be made. A reasonable assumption is that $R_2 = R_3$. Since the sites of both respiration and photosynthesis are located in a relatively thin layer of cytoplasm near the cell wall, the diffusion paths represented by R_2 and R_3 are of similar length, and $R_4 < R_2$ since R_4 does not contain a cell-wall component of resistance and the chloroplasts, mitochondria, and peroxisomes are often in close proximity.

Circuit Theory for Diffusion Pathways

The triangular circuit diagram shown in Fig. 14.1b is completely analogous to the electrical circuit diagram familiar to engineers and physicists. Simple methods have been worked out for solving such circuit diagrams. Kirchoff's rule states that there must be mass, or current, balance at any junction point in a circuit. The concentrations of CO_2 at these junctions are C_a, C_i, C_w, and C_c. The fluxes of CO_2 flowing between any two junctions are P_1, P_2, P_3, P_4, and P_5. The flux of respiration from the sites of respiration is shown in Fig. 14.1b as W. These various currents or fluxes, as given by Fick's diffusion law, are

$$P_1 = \frac{C_a - C_i}{R_1} \tag{14.25}$$

$$P_2 = \frac{C_i - C_c}{R_2} \tag{14.26}$$

$$P_3 = \frac{C_w - C_i}{R_3} \tag{14.27}$$

$$P_4 = \frac{C_w - C_c}{R_4}. \tag{14.28}$$

Applying Kirchoff's rule at the junctions, one gets

$$P_2 = P_1 + P_3 \tag{14.29}$$

$$W = P_3 + P_4 \tag{14.30}$$

$$P_5 = P_2 + P_4. \tag{14.31}$$

The quantity P_1 is the flux of CO_2 entering or leaving the leaf and is readily measurable. The quantities P_2, P_3, P_4, and P_5 are internal fluxes of CO_2 and are impossible to measure directly. The quantities W, R_1, R_2, R_3, and R_4 can be either measured or estimated. Therefore, the fluxes P_2 . . . P_5, if expressed in terms of P_1, W, and R_1 . . . R_4, can be determined by means of measured or estimated quantities. In order to do this, the following algebraic procedure is necessary.

Rearranging Eqs. (14.27) and (14.28) gives

$$C_w = C_i + P_3R_3 \tag{14.32}$$

and

$$C_w = C_c + P_4R_4. \tag{14.33}$$

These two expressions are then equated as follows:

$$C_i + P_3R_3 = C_c + P_4R_4 \tag{14.34}$$

Solving Eq. (14.26) for C_c and Eq. (14.30) for P_4 and substituting into Eq. (14.34) gives

$$C_i + P_3R_3 = (C_i - P_2R_2) + (W - P_3)R_4. \tag{14.35}$$

Solving Eq. (14.29) for P_3 and substituting into Eq. (14.26) gives

$$C_i + (P_2 - P_1)R_3 = C_i - P_2R_2 + (W - P_2 + P_1)R_4, \tag{14.36}$$

and solving for P_2 gives

$$P_2 = \frac{P_1(R_3 + R_4) + WR_4}{R_2 + R_3 + R_4}. \tag{14.37}$$

Then, since $P_2 = P_1 + P_3$,

$$P_3 = \frac{WR_4 - P_1R_2}{R_2 + R_3 + R_4}, \tag{14.38}$$

and since $P_4 = W - P_3$,

$$P_4 = \frac{W(R_2 + R_3) + P_1 R_2}{R_2 + R_3 + R_4}. \tag{14.39}$$

And finally, since $P_5 = P_2 + P_4$,

$$P_5 = P_1 + W. \tag{14.40}$$

Therefore, if P_1, W, R_2, R_3, and R_4 are known, one can determine P_2, P_3, P_4, and P_5 using Eqs. (14.37) through (14.40). The concentrations of CO_2 at the junctions C_i, C_c, and C_w can now be determined in terms of known quantities by rearranging Eqs. (14.25), (14.26), and (14.27) as follows:

$$C_i = C_a - P_1 R_1, \tag{14.41}$$

$$\begin{aligned} C_w &= C_i + P_3 R_3 \\ &= C_a - P_1 R_1 + P_3 R_3, \end{aligned} \tag{14.42}$$

$$\begin{aligned} C_c &= C_i - P_2 R_2 \\ &= C_a - P_1 R_1 - P_2 R_2. \end{aligned} \tag{14.43}$$

The CO_2 flux P_5 to the chloroplast must be equal to the rate of the biochemical reaction in the chloroplast, which depends upon the CO_2 concentration according to the Michaelis–Menten expression Eq. (14.23), which contains the competitive inhibition factor β,

$$P_5 = \frac{P_M}{1 + \dfrac{K\beta}{C_c}}, \tag{14.44}$$

where

$$\beta = 1 + \frac{[O_2]}{K_O}. \tag{14.45}$$

By substituting P_2 from Eq. (14.37) into Eq. (14.43), one gets C_c in terms of the known quantities P_1, W, R_1, R_2, R_3, and R_4 as follows:

$$\begin{aligned} C_c &= C_a - P_1 R_1 - \frac{P_1 R_2(R_3 + R_4) + W R_2 R_4}{R_2 + R_3 + R_4} \\ &= C_a - P_1 \left[R_1 + \frac{R_2(R_3 + R_4)}{R_2 + R_3 + R_4} \right] + W\left(\frac{R_2 R_4}{R_2 + R_3 + R_4} \right). \end{aligned} \tag{14.46}$$

Equation (14.46) may now be written in the simplified form

$$C_c = C_a - P_1 S_1 - W S_2, \tag{14.47}$$

where S_1 and S_2, the equivalent network resistances, are defined as follows:

$$S_1 \equiv R_1 + \frac{R_2(R_3 + R_4)}{R_2 + R_3 + R_4} \tag{14.48}$$

$$S_2 \equiv \frac{R_2 R_4}{R_2 + R_3 + R_4}. \tag{14.49}$$

It is convenient to make some simplifying assumptions which will be useful later in our model formulation. If $R_2 \simeq R_3$ and R_4 is a small fraction of R_2, e.g., R_2/M, then

$$S_1 \simeq R_1 + R_2 \frac{(M + 1)}{(2M + 1)} \tag{14.50}$$

$$S_2 \simeq \frac{R_2}{(2M + 1)}. \tag{14.51}$$

If M is a large number, then $S_1 \to R_1 + R_2/2$ and $S_2 \to 0$. This approximation is convenient for certain situations. Requiring M to be very large places the site of CO_2 evolution in respiration close to the site of CO_2 fixation.

Photosynthesis Model with Respiration

All the components required for the full analytical model describing photosynthesis for a whole leaf, with respiration taken into account, have been derived in the above paragraphs. Now it is necessary express Eq. (14.44) in quantities that can be measured, i.e., P_1, W, C_a, S_1, and S_2. This is done by substitution from Eqs. (14.40) and (14.47) to give

$$P_1 + W = \frac{P_M}{1 + \dfrac{K\beta}{C_a - P_1 S_1 - W S_2}}. \tag{14.52}$$

Quadratic Expression for Photosynthesis

By cross-multiplying and clearing Eq. (14.52) of fractions, one gets

$$S_1 P_1^2 - [C_a + K + S_1(P_M - W) - W S_2]P_1 \\ + (C_a - W S_2)(P_M - W) - W K = 0. \tag{14.53}$$

Again, this is a quadratic equation of exactly the same form as Eq. (3.10), and the solution is undertaken in precisely the same manner as given in Appendix B of the Lommen et al. (1971) paper. The only difference is that with respiration W included, one gets

$$\lim_{C_a \to \infty} P_1 = P_M - W, \tag{14.54}$$

$$P_1 = \frac{[C_a + K\beta + S_1(P_M - W) - W S_2]}{2S_1} \\ - \frac{\left\{ [C_a + K\beta + S_1(P_M - W) - W S_2]^2 \atop - 4S_1[(C_a - W S_2)(P_M - W K\beta)] \right\}^{1/2}}{2S_1}, \tag{14.55}$$

and

$$W = W_p + W_m. \tag{14.56}$$

We wish to derive all model parameters from measurements of P_1 versus C_a, oxygen, and light for a whole leaf. In order to do this, it is convenient to make a few approximations. The first step is to treat stomatal resistance as a separate problem. It was shown above how a measurement of the transpiration rate permits determination of R_1. Then, from this value of R_1, C_i can be determined from C_a and P_1 by means of Fick's law. It is this procedure which allows one to modify Eq. (14.55) to represent the photosynthetic rate as responding to C_i directly and eliminate R_1 from S_1. Matters are somewhat simplified if one considers the sites of respiration to be close to the sites of fixation. This makes M very large in Eqs. (14.50) and (14.51), and $S_1 \simeq R_2/2 = R_m$, where R_m is the mesophyll resistance and $S_2 = S_1/(1 + M) = R_m/(1 + M)$. The reason R_m is half of R_2 is that R_m represents the resistance of CO_2 flux pathways through R_2 and R_3 of Fig. 14.1b considered in parallel. Hence $1/R_m = 1/R_2 + 1/R_3$. But since $R_2 = R_3$, $1/R_m = 2/R_2$ and $R_m = R_2/2$.

If one substitutes these approximations into Eq. (14.55), then the useful form of the quadratic equation for expressing the photosynthetic response to the internal carbon dioxide concentration becomes

$$P_1 \simeq \frac{a - (a^2 - b)^{1/2}}{2R_m}, \tag{14.57}$$

where

$$a = \left[C_i + K\beta + R_m(P_M - W) - W\frac{R_m}{1 + M} \right]$$

and

$$b = 4R_m \left[\left(C_i - W\frac{R_m}{1 + M} \right)(P_M - W) - W K \beta \right].$$

Objectives of the Model

The objectives of this analysis are twofold: to determine the values of all necessary parameters that describe the photosynthetic response of a leaf to environmental conditions of light intensity, carbon dioxide concentration, oxygen concentration, and leaf temperature and to predict the photosynthetic response of a leaf to any combination of environmental conditions that may occur.

An analytical model which is to give an adequate description of the photosynthetic response of C_3 plants to environmental conditions must be consistent with all observations. The analysis used here separates the effects of light and temperature from those of carbon dioxide and oxygen and provides possible insight into the photosynthetic and photorespiration processes of C_3 plants. A successful analytical description must account for the following behavior:

1. The saturation of net photosynthesis and suppression of oxygen inhibition at high CO_2 concentration.
2. The decrease in photosynthesis and increase in photorespiration as oxygen concentration is increased above 1%.
3. The decrease in the slope of the carbon dioxide–response curves with increasing oxygen concentration.
4. The linear increase of the CO_2 compensation curve with increasing oxygen concentration.
5. Rate of O_2 evolution into CO_2-free air.
6. The relationship of net photosynthesis rate to carbon dioxide and oxygen concentrations.

Evaluation of the Parameters

It is clear, from Eqs. (14.57) and (14.24), that in order to describe a set of observational data, one must determine the values of five parameters: R_m, the mesophyll diffusion resistance from the intercellular air space to the fixation site; K, the leaf affinity coefficient for fixation; K_O, the leaf affinity constant for oxygenation; P_M, the photosynthesis rate at saturating carbon dioxide concentration; and W_M, the maximum rate of oxygenation at saturating oxygen concentration.

Measurements of net photosynthesis are made first with the leaf in air at a low oxygen concentration (1%) in order to suppress photorespiration but not restrict dark respiration. Dark respiration is determined by an extrapolation of the initial slope of the P_1 versus C_i or L curves. The amount of residual respiration is essentially independent of carbon dioxide concentration but is a function of leaf temperature; it may or may not be independent of light intensity.

Values for K, P_M, and R_m are obtained initially from analysis of measured net photosynthesis P_1 versus internal carbon dioxide concentration C_i. Actually, measurements are obtained of P_1 versus C_a (the carbon dioxide concentration of the air in the leaf chamber), and these curves are then replotted in terms of C_i as determined by the process described previously (see "Internal CO_2 Concentration").

The procedure used for analyzing the net-photosynthesis curves makes use of a nonlinear least-squares analysis which requires derivatives of P with respect to the parameters, as described by Tenhunen et al. (1976, 1977). These procedures are too elaborate to describe here in detail. A general computational program involving three subroutines is used.

Subroutine A solves the quadratic equation, Eq. (14.57), for net photosynthesis when photorespiration is essentially zero. The partial derivatives of the quadratic equation with respect to K, P_M, and R_m are determined.

Subroutine B performs an iterative procedure to determine the chloroplast carbon dioxide concentration C_c that competitively interacts with oxygen in the chloroplast and thus the relative velocities of photosynthesis and photorespiration as given by Eqs. (14.57) and (14.24). Subroutine B gives estimates of the oxygen-dependent parameters K_O and W_M at constant values of K, P_M, and R_m as determined using subroutine A. Partial derivatives of P_1, determined by means of Eq. (14.57), are estimated with respect to K_O and W_m.

Subroutine C leads to a final solution for the fit of net-photosynthesis data using least squares. Reasonably good estimates of K, P_M, R_m, K_O, and W_M are obtained with subroutines A and B. Subroutine C is a generalized version of subroutine B. Iteration is performed with respect to all five parameters. The photosynthesis function is solved in the same manner as in subroutine B. The only addition to the subroutine is to determine the partial derivatives of the function P_1 of Eq. (14.57) with respect to the parameters K, P_M, and R_m.

It is useful to assume that W_M equals P_M since both W_P and P_5 are regulated at RuDP carboxylase-oxygenase and are limited by the production of RuDP in the Calvin cycle and ATP or $NADPH_2$ in the light reactions. It appears, from the analysis of data of Joliffe and Tregunna (1973), that the activation energies of W_M and P_M with respect to temperature are equal. Laing et al. (1974) suggest that the carboxylation and oxygenation reactions of RuDP carboxylase-oxygenase have nearly identical activation energies.

Parameter Values

There is very little information available concerning the values of K and K_O because of a lack of the proper set of measurements of photosynthesis response to carbon dioxide concentration at various concentrations of oxygen, especially 1%, and various leaf temperatures. In the paper by Lommen et al. (1975), a K value of 10 mmole m^{-3} was used. Tenhunen et al. (1977) derived values at 25°C for sunflower of $K = 0.0183$ mmole m^{-3}, $K_O = 0.001$ as a partial function vol/vol, $P_M = 0.028$ mmole $m^{-2} s^{-1}$, $W_M = 0.01$ mmole $m^{-2} s^{-1}$, and $R_m = 184$ s m^{-1}. Tenhunen et al. (1979) obtained, at 25°C for wheat, values of $K = 0.493$ mmole m^{-3}, $K_O = 0.0823$, $W_M = P_M = 0.0346$ mmole $m^{-2} s^{-1}$, and $R_m = 227$ s m^{-1}. Jones and Slatyer (1972) report a value for whole cotton leaves of $K = 0.23$ mmole m^{-3}.

The apparent affinity constants for carbon dioxide and oxygen, K and K_O, of whole plant leaves differ considerably from values reported for pure enzymes and must not be considered equivalent to the latter. The values of K and K_O may reflect the properties of the enzyme RuDP carboxylase-oxygenase in the leaf system, but the reactions are so complex that the values represent "system" constants rather than simple enzyme response. The particular values of K and K_O obtained depend on the resistance network assumed and the values of R_m.

There may be a temperature dependency to the affinity constants, but very little is known concerning this. Tenhunen et al. (1979) report temperature behavior of K and K_O for wheat as shown, but the results are highly uncertain. The K values are 0.493, 0.201, and 1.61 mmole m^{-3} at 25°, 30°, and 35°C, respectively, and the K_O values are 0.0823, 0.0294, and 0.273 at the same temperatures. Values of R_m show a logical progression from 227 to 128 s m^{-1} with increasing temperature.

Model Demonstration

With a simplified set of values being used for convenience, it is now possible to demonstrate the properties of the analytical model. The parameter values chosen for wheat at 25°C are R_m = 200 s m^{-1}, M = 100, K = 0.5 mmole m^{-3}, K_O = 0.1, P_{ML} = 0.035 mmole m^{-2} s^{-1}, and α = 50 mmole E^{-1}.

The carbon dioxide response of photosynthesis is calculated for wheat at a leaf temperature of 25°C for oxygen concentrations of 1 and 21% and a light intensity of 15 × 10^{-4} E m^{-2} s^{-1}. It is assumed that $W_M = P_M$. Comparison with the data obtained by Ku and Edwards (1977; see Table 14.1) is shown in Fig. 14.7. Agreement is very good, as it is also (not shown) for data at leaf temperatures of 30° and 35°C. Net photosynthetic rate increases with diminishing oxygen concentration as is expected.

The calculated response of net photosynthesis to light at various levels of carbon dioxide and oxygen is given in Fig. 14.8. It is clear from this sequence of curves that either high CO_2 or low O_2 will suppress photorespiration and increase net photosynthesis. The lowest curve is at 300 ppm CO_2 and 21% O_2. The highest curve is at 1000 ppm CO_2 and 1% O_2. The two intermediate curves are at 1000 ppm CO_2, 21% O_2 and 300 ppm CO_2, 1% O_2. The initial slopes of curves A, B, and C are identical, which indicates equal light-utilization efficiency. This is a result of the large value of K_O and C_c making the apparent value of K_O, which is K_O[1 +

Table 14.1. Values of the Parameters in the Analytical Model Derived to Fit the Observational Data for Wheat Obtained by Ku and Edwards (1977)

Parameter	Value			Units
	Leaf temperature			
	25°C	30°C	35°C	
$W_M = P_M$	0.035	0.039	0.050	mmole m^{-2} s^{-1}
K	0.493	0.201	1.61	mmole m^{-3}
K_O	0.082	0.029	0.273	decimal fraction (vol/vol)
R_m	227	194	128	s m^{-1}

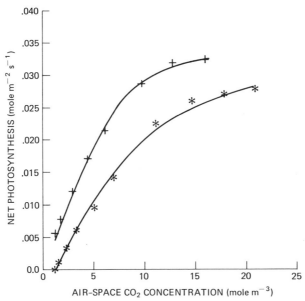

Figure 14.7. Net photosynthesis as a function of internal concentration of CO_2 for oxygen concentrations of (+) 1% and (*) 21% for wheat at 25°C. Data points are from Ku and Edwards (1977). K, 0.493 mole m^{-3}; K_O, 0.082; R_m, 227 s m^{-1}; P_M ($= W_M$), 0.0346 mmole m^{-2} s^{-1}; L, 0.0015 E m^{-2} s^{-1}.

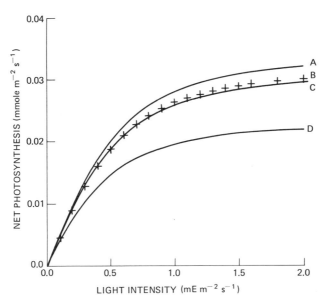

Figure 14.8. Net photosynthesis as a function of light intensity: (A) 1000 ppm CO_2 and 1% O_2; (B) 1000 ppm CO_2 and 21% O_2; (C) 300 ppm CO_2 and 1% O_2; (D) 300 ppm and 21% O_2. K, 0.493 mole m^{-3}; K_O, 0.082; R_m, 227 s m^{-1}; $P_M(= W_M)$, 0.0346 mmole m^{-2} s^{-1}; L, 0.0015 E m^{-2} s^{-1}.

(C_c/K)], even higher. At 300 ppm CO_2 and 21% oxygen, the initial slope of the light curve is lower, in a manner consistent with the observations of Ehleringer and Bjorkman (1977) for C_3 plants.

It is expected that photorespiration rates will increase with increasing light intensity. The predicted relationship between photorespiration rate and light intensity is shown in Fig. 14.9; the upper line is at a carbon dioxide concentration of 300 ppm and an oxygen concentration of 21%. The increase in photorespiration with increasing light intensity results from an increase in P_M and a decrease in the apparent oxygen affinity constant for photorespiration $K_0[1 + (C_c/K)]$ as C_c decreases with increased fixation. At low oxygen concentration or high carbon dioxide concentration, photorespiration is nearly zero.

We now explore the behavior of the analytical photosynthesis model with a change in the parameter values. Let K change from 0.5 to 0.05 mmole m^{-3} and K_0 from 0.1 to 0.01 mmole m^{-3}. Net photosynthesis as a function of the air-space concentration of carbon dioxide at oxygen concentrations of 1 and 21% is shown in Fig. 14.10. As the K value becomes smaller, the photosynthetic response to carbon dioxide concentration becomes greater. In other words, photosynthesis reaches its maximum response at a lower CO_2 concentration. As the value of K_0 decreases, the influence of increasing oxygen concentration becomes greater in reducing net photosynthesis.

The analysis allows calculation of the predicted magnitude of photores-

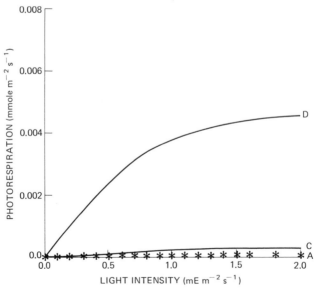

Figure 14.9. Photorespiration as a function of light intensity: (A) 1000 ppm CO_2 and 1% O_2; (C) 300 ppm CO_2 and 1% O_2; (D) 300 ppm CO_2 and 21% O_2. K, 0.493 mole m^{-3}; K_0, 0.082; R_m, 227 s m^{-1}; $P_M(=W_M)$, 0.0346 mole m^{-2} s^{-1}; T, 25°C.

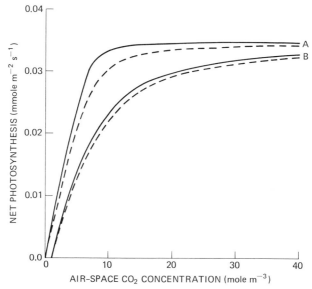

Figure 14.10. Calculated net photosynthesis as a function of air-space CO_2 concentration for two values of K and K_0 such that $K/K_0 = 5$ at 25°C and oxygen concentrations of (A) 1% and (B) 21%: (—) $K = 0.5$ mole m^{-3}, $K_0 = 0.1$; (--) $K = 0.05$ mole m^{-3}, $K_0 = 0.01$.

piration, with an allowance for an assumed amount of dark respiration as a function of air-space concentration of carbon dioxide and oxygen concentration. The results are illustrated in Fig. 14.11. The calculations are for a temperature of 25°C. There is a small but definite amount of residual photorespiration at low oxygen concentration. Mitochondrial, or dark, respiration simply biases the curves upward.

Calculation of net and total photosynthesis at 25°C and 21% oxygen gives the results shown in Fig. 14.12. It is of particular interest that the competitive interaction between photosynthesis and respiration results in enhancement of total photosynthesis (net photosynthesis plus photorespiration) rather than diminution. At low carbon dioxide concentrations, the photorespiratory flux of CO_2 helps to maintain an effectively high CO_2 concentration at the chloroplast, which contributes to a favorable competitive situation for total CO_2 fixation. The high photorespiratory flux also results in lower net rates of photosynthesis owing to simultaneous diffusion of CO_2 out of the leaf. Since a leaf with photorespiration uses photoproducts more effectively, even at low concentrations of carbon dioxide in the intercellular air space, it is possible that photorespiration protects the leaf from damaging effects of photoproducts during periods of water stress and high irradiance. The rate of total photosynthesis (rate of carboxylation, or Calvin-cycle cycling) at zero carbon dioxide concentration in the intercellular air space may be a measure of this protective ef-

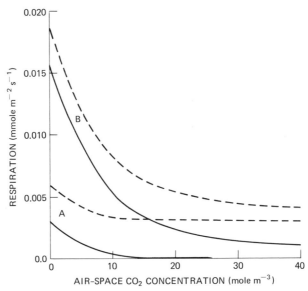

Figure 14.11. Respiration as a function of air-space CO_2 concentration for $W_M = 0$ and 0.003 mmole $m^{-2}\,s^{-1}$ and O_2 concentrations of (A) 1% and (B) 21% at 25°C: (—) $W_M = 0$ mmole $m^{-2}\,s^{-1}$; (--) $W_M = 0.003$ mmole $m^{-2}\,s^{-1}$. K, 0.5 mole m^{-3}, K_{OK}, 0.1; R_m, 200 s m^{-1}; and M, 100.

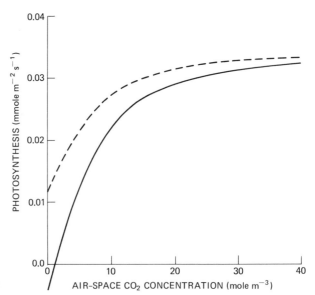

Figure 14.12. Photosynthesis as a function of the air-space CO_2 concentration for an oxygen concentration of 21% and at 25°C: (—) net photosynthetic rate P_1; (--) total photosynthetic rate $P_5 = P_1 + W$. K, 0.5 mole m^{-3}; K_0, 0.1; R_m, 200 s m^{-1}; M, 100.

fect of photorespiration and is clearly a function of oxygen concentration. The oxygen dependency is shown in Fig. 14.11. At low oxygen concentrations, there is little protective ability, but at normal ambient concentrations, protection is considerable.

The temperature dependency of photosynthesis in plants is poorly understood. How temperature affects the carbon dioxide affinity, oxygen inhibition of photosynthesis, maximum capacity for photosynthesis, and solubility of O_2 and CO_2 is not well known. However, one can demonstrate some temperature responses of net photosynthesis for wheat in air of 300 ppm CO_2 and 21% O_2. If the total diffusion resistance of the leaf is 1000 s m^{-1} and the parameter values for wheat are used, one calculates a temperature response shown by the solid line in Fig. 14.13. If the total leaf resistance is 1500 s m^{-1}, the apparent temperature optimum shifts from about 24° to 21°C, and if the resistance decreases to 500 s m^{-1}, the temperature shift is from 24° to 33°C. Calculations show that a change in total leaf resistance from 100 to 2000 s m^{-1} will change the apparent temperature optimum from 40° to 20°C. An increase in the value of P_{MLT} will shift the optimum to lower temperatures.

As shown in Chapter 15, plants that are acclimated to cold regions have lower temperature optima than plants acclimated to warm regions but at the same time have higher total resistance, more photorespiratory products, higher sucrose and starch content, and higher levels of RuDP car-

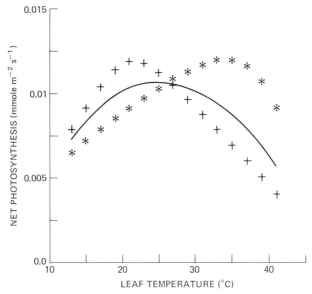

Figure 14.13. Temperature responses of net photosynthesis at 300 ppm external CO_2 concentration and 21% oxygen. Solid-line response is for a total diffusion resistance R of 100 s m^{-1}. The temperature optimum shifts to a lower temperature for (+) R = 1500 s m^{-1} and to a higher temperature for (*) R = 500 s m^{-1}.

boxylase. All of these changes could contribute to the lowering of the temperature optimum. Both K and K_0 may be temperature dependent and may significantly affect the temperature optimum. At the moment, it is difficult to separate cause and effect, since one is dealing with a complex system with many influencing factors.

Biochemical Model of Photosynthesis

It is useful to derive a model of whole-leaf photosynthesis based on principles of chemical kinetics. Many of these principles were outlined by Rabinowitch (1951) and used by Chartier (1966, 1970) and Chartier and Prioul (1976) to formulate a leaf model of photosynthesis.

If A is the CO_2 acceptor molecule, ribulose 1,5-diphosphate, in the Calvin–Benson pathway, K_1 is the rate constant associated with the enzyme rubulose-diphosphate carboxylase, and ACO_2 represents the Calvin–Benson pathway intermediates, or the carboxylated acceptor, then

$$[A] + C_c \xrightarrow{K_1} [ACO_2]. \tag{14.57a}$$

If H is the reducing power (NADPH) necessary to regenerate ribulose diphosphate, K_2 is a rate constant for reducing the Calvin–Benson pathway intermediates, and π is the carboxylation product resulting, then

$$[ACO_2] + [H] \xrightarrow{K_2} [\pi] + [A]. \tag{14.58}$$

If the amount of reducing power H depends directly on the light intensity L and the rate constant is K_3, then

$$[H] = K_3 L \tag{14.59}$$

and

$$[A_0] = [ACO_2] + [A], \tag{14.60}$$

where $[A_0]$ represents the concentration of the total carbohydrate pool and is assumed to be constant in all chemical models of this type for the normal range of CO_2, O_2, and light intensity. If P_5 is the gross photosynthetic, or fixation, rate at the chloroplast, then at steady state,

$$P_5 = K_2[ACO_2][H] = K_1[A]C_c. \tag{14.61}$$

Solving from Eq. (14.60) for [A], using $K_3 L$ for [H] from Eq. (14.59), substituting into the second quality of Eq. (14.61), and solving for $[ACO_2]$, one gets

$$[ACO_2] = \frac{K_1[A_0]C_c}{K_1 C_c + K_2 K_3 L}. \tag{14.62}$$

Then, substituting this value into the first equality of Eq. (14.61), one gets

$$P_5 = \frac{K_1 K_2 K_3 \, L[A_o] C_c}{K_1 C_c + K_2 K_3 L}. \tag{14.63}$$

The initial slope α of the light-response curve is an estimate of the maximum efficiency of light utilization. By taking the limit as L goes to 0 of the derivative of Eq. (14.63) with respect to L, α can be determined:

$$\lim_{L \to 0} \frac{dP_5}{dL} = K_2 K_3 [A_o] = \alpha. \tag{14.64}$$

The initial slope of the P_5 versus C_c curve is the reciprocal of the carboxylation resistance as shown by taking the limit of the derivative of Eq. (14.63) with respect to C_c:

$$\lim_{C_c \to 0} \frac{dP_5}{dC_c} = K_1 [A_o] = \frac{1}{R_c}. \tag{14.65}$$

Then, by substituting from Eqs. (14.64) and (14.65) into Eq. (14.63), one gets

$$P_5 = \frac{C_c}{R_c + \dfrac{C_c}{\alpha L}}. \tag{14.66}$$

This expression has the form of the Michaelis–Menten equation we have used earlier.

The gross photosynthetic rate is equal to the sum of the net photosynthetic rate P_1 plus the dark respiration and photorespiration rates W_m and W_p:

$$P_5 = P_1 + W_m + W_p. \tag{14.67}$$

From Eq. (14.1),

$$C_c = C_a - P_1 R. \tag{14.68}$$

Substituting from Eqs. (14.67) and (14.68) into Eq. (14.66), one gets

$$P_5 = \frac{C_a - P_1 R}{R_c + \dfrac{C_a - P_1 R}{\alpha L}}. \tag{14.69}$$

At low light intensity,

$$\frac{C_a - P_1 R}{\alpha L} \gg R_c \tag{14.70}$$

and

$$P_5 \cong \alpha L. \tag{14.71}$$

Flux of carbon dioxide is determined by the light utilization efficiency, and carboxylation resistance is negligible.

At high light intensity,

$$\alpha L \gg C_a - P_1 R \qquad (14.72)$$

$$P_5 \cong \frac{C_a - P_1 R}{R_c}. \qquad (14.73)$$

Substituting Eq. (14.67) into Eq. (14.73) and solving for P_1, one gets

$$P_1 \cong \frac{C_a - R_c(W_m + W_p)}{R + R_c}, \qquad (14.74)$$

which is the maximum photosynthetic rate under saturating light intensity.

These results depend upon the following set of assumptions:

1. H_2O and CO_2 follow the same diffusion pathway and CO_2 resistance can be derived from the resistance to H_2O diffusion.
2. First-order, irreversible kinetics are valid for a homogeneous system in which each reaction site experiences the same average light intensity and internal carbon dioxide concentration.
3. The amount of CO_2 acceptor pool is constant. This assumption may not be valid, and a change in A_o will affect the carboxylation resistance R_c and the shape of the photosynthetic-response curve.
4. The carboxylation reaction is regulated by a single valued rate function K_1. However, K_1 may depend on the concentration of oxygen and carbon dioxide at the fixation site.

Chartier and Prioul (1976) modified the analysis given above to include the influence of photorespiration and competitive inhibition of RuDP carboxylase-oxygenase. They rewrote Eq. (14.60) to include the intermediates B formed in the regeneration of RuDP as follows:

$$[A_o] = [ACO_2] + [B] + [A]. \qquad (14.75)$$

Then they replaced Eq. (14.58) by

$$[ACO_2] + [H] \xrightarrow{K_2} [\pi] + [B] \qquad (14.76)$$

and

$$[B] + [X] \xrightarrow{K_4} [A] + [X], \qquad (14.77)$$

where $[X]$ is a regulating-enzyme cofactor and K_4 is a rate constant.

Combining Eqs. (14.76) and (14.77) with Eq. (14.57), one gets

$$P_5 = K_1 [A]C_c = K_2[ACO_2][H] = K_4[B][X], \qquad (14.78)$$

and combining with Eq. (14.75) leads to

$$[A_o] = \frac{P_5}{K_1 C_c} + [ACO_2] + \frac{K_2[ACO_2][H]}{K_4[X]} \qquad (14.79)$$

or

$$\frac{1}{P_5} = \frac{1}{[A_o]K_1C_c} + \frac{1}{K_2[H][A_o]} + \frac{1}{[A_o]K_4[X]}. \qquad (14.80)$$

Now, making use of Eq. (14.59) and the definitions for α and R_C, one gets

$$\frac{1}{P_5} = \frac{1}{C_c/R_c} + \frac{1}{\alpha L} + \frac{1}{P_{ML}}, \qquad (14.81)$$

where P_{ML} is the maximum gross photosynthetic rate at saturating light and carbon dioxide concentration and is equal to $[A_o]K_4[X]$.

Leaf respiration involves oxygen combining with the acceptor molecule at a rate governed by the rate constant K_5,

$$[A] + [O] \xrightarrow{K_5} [AO_2], \qquad (14.82)$$

where $[O]$ is the concentration of oxygen at the reaction site and $[AO_2]$ is the oxygenation product of the photorespiration reaction.

From Eqs. (14.82) and (14.78) and the definition for R_c, one gets

$$W_p = K_5[A][O] = P_5K_5[O]/K_1C_c = P_5[O]/R_o(C_c/R_c), \qquad (14.83)$$

where $R_o = 1/[A_o]K_5$, an oxygenation resistance similar to the carboxylation resistance R_o.

The purpose of carrying the biochemical analysis to this stage is to show the technique necessary to derive photosynthetic response curves. Various complexities arise as one carries the analysis further. They are discussed in detail by Prioul and Chartier (1977) but are not pursued further here since they suggest that the different biochemical models cannot be distinguished on the basis of measurement made with the whole leaf photosynthesis. This is the reason why we have attempted to keep simple the biophysical model used throughout most of this chapter. Nevertheless, it is important for the student to realize that biochemical analysis may add significantly to the achievement of a practical analytical model. The advancement of these various approaches must proceed in parallel. It is important to emphasize once again that a great deal of the response by plants to environmental factors is governed by biochemical reactions at the metabolic level.

Problems

The following problems are to be done in sequence. Each one builds on the previous problem and systematically leads the student through the procedure necessary for determining all values of photosynthesis and respiration.

1. Determine $G(T_K)$ for temperatures of 10°, 15°, 20°, 25°, 30°, 35°, 40°, and 45°C using the following values for the parameters in the temperature

function for photosynthesis, Eq. (14.8): $\Delta H\neq$, 48,465 J mole^{-1}; ΔH_1, 175,682 J mole^{-1}; ΔS, 557.16 J mole^{-1} °C^{-1}; c_2, $e^{17.925}$, or 6.092 × 10^7; and R, 8.317 J mole^{-1} °C^{-1}. Plot $G(T_K)$ versus temperature and draw the curve.

2. Using the parameter values for a plant leaf P_{MLT} = 0.04 mmole m^{-2} s^{-1} and α = 0.05 mole E^{-1}, determine the values of P_M as a function of light intensity when the leaf temperature is 25°C. Use light-intensity values of 2 × 10^{-5}, 5 × 10^{-5}, 2 × 10^{-4}, 5 × 10^{-4}, and 2 × 10^{-3} E m^{-2} s^{-1} for photosynthetically active radiation. Use the value of $G(T_K)$ calculated in Problem 1. Plot P_M versus light intensity at 25°C. Repeat the calculation of P_M for leaf temperatures of 10°, 20°, 30°, and 40°C and plot these on the same graph as a function of light intensity. Also plot P_M versus leaf temperature for each value of light intensity.

3. Assume that there is no light effect on dark respiration and determine dark respiration rates at leaf temperatures of 10°, 20°, 25°, 30°, and 40°C when E = 94,888 J mole^{-1}, A = 41.886, and J = 10^{-8} mole m^{-2} s^{-1}. Use Eq. (14.20). Plot W_m versus T_K and draw the curve.

4. In order to determine the rate of photosynthesis, it is necessary to assume a value for W_M. We shall assume that $W_M = P_M$. These factors are functions of leaf temperature and light intensity. At a light intensity of L = 5 × 10^{-4} E m^{-2} s^{-1} and a leaf temperature of 25°C, $W_M = P_M$ = 0.0169 mmole m^{-2} s^{-1}. Determine the rate of photorespiration W_P using Eq. (14.24) when K_0 = 0.01 (vol/vol), K_C = 0.10 mmole m^{-3}, and $[O_2]$ = 0.20 (vol/vol) for a chloroplast carbon dioxide concentration of C_c = 12.5 mmole m^{-3}, which is the ambient air concentration of carbon dioxide. Now obtain $W = W_p + W_m$.

5. Determine, using Eq. (14.55), the photosynthetic rate for a leaf temperature of 25°C and a light intensity of 5 × 10^{-4} E m^{-2} s^{-1} when S_1 = R_m = 200 s m^{-1} and $S_2[=S_1/(1 + M)]$ = 1.98 s m^{-1}.

6. Now use Fick's law of diffusion [Eq. (14.1)] to obtain a new value of C_c using the value of P from Problem 5 and assuming $R = S_1$ = 200 s m^{-1} and C_a = 12.5 mmole m^{-3}. This value of C_c is not the same as assumed in Problems 4 and 5; therefore, new values for W_P, W, and P must be determined using this new value of C_c. With the new value of P, use Fick's law to determine another value of C_c. If this value does not agree with the last, then recalculate W_P, W, and P and repeat the iteration until agreement is reached. This procedure will then have given the final correct values for C_c, W_P, W, and P for a leaf temperature of 25°C and light intensity of 5 × 10^{-4} E m^{-2} s^{-1}.

7. Now go back to Problem 4 and repeat all the steps of Problems 4, 5, and 6 for the same light intensity and a new leaf temperature, say 20°C. With additional effort, one can determine C_c, W_P, W, and P for all combinations of temperature and light intensity, e.g., T_K = 10°, 20°, 30°, and 40°C and L = 2 × 10^{-5}, 5 × 10^{-5}, 2 × 10^{-4}, 5 × 10^{-4}, and 2 × 10^{-3} E m^{-2} s^{-1}. This is a laborious procedure when done by hand but is easy when programmed on a computer.

Chapter 15

Temperature and Organisms

Introduction

Temperature is an extremely important environmental variable to all life. The temperature of any object, solid or fluid, living or nonliving, is a measure of its energy status, determined by energy exchange with the environment. The distance of the earth from the sun is such that the planet earth is neither too warm nor too cold for higher forms of life. The rate of spin of the earth on its axis makes the length of day and night agreeable to life. If the rate of spin were slower, daytime temperatures would be much warmer and nighttime temperatures much colder. If the rate of spin were more rapid, then day and night would be shorter and temperature extremes would be less great. The molecular composition of the atmosphere greatly affects the climate near the ground and has been an important factor in the evolution of life.

The average global surface temperature of the earth is about 13°C. The coldest standard air temperature ever recorded is −88.3°C, measured at the Soviet Union's Vostok Station Pole of Cold in Antarctica on 24 August 1960. The warmest standard air temperature ever recorded is 57.8°C, measured at two places on earth, Azizia, Tunisia (on 13 September 1922), and San Luis, Mexico (on 11 August 1933). Higher air temperatures can be measured next to the soil surface; in the case of a dark soil in full sun on a hot day, these temperatures can exceed 70°C. Lower air temperatures exist in the upper atmosphere, but it is unlikely that anything much colder than −88.3°C will be found near the ground surface. Further discussion of temperature extremes, and climatic conditions in general, can be found in Gates (1972).

There are several ways by which we can understand the responses of plants and animals to temperature changes. One way is by means of bio-climatology, including the study of biological events associated with past climates, or paleoclimatology. Another way, through physiology, is to study the interaction of microclimate and plants and animals, particularly thermal effects. A third way is to study the behavior of whole communities of organisms in response to climatic change. Many reviews have been written concerning the subject of temperatures and life. These include Uvarov (1931), Andrewartha and Birch (1954), Johnson et al. (1954), Johnson (1957), Miller (1961), Went (1961), Gates (1962, 1972), Johnson and Smith (1965), Rose (1967), Ladurie (1971), and Precht et al. (1973).

Conclusions drawn from observations of climatic change and the effect of temperature on ecosystems or particular plants are broad and general in character, whereas conclusions drawn from physiological observations are highly specific and precise. All types of information are of value in our attempt to assess the influence of temperature on organisms and ecosystems. One can conclude from agricultural observations, both historic and modern, that a shift in the mean growing-season temperature by only a few degrees has major impact on crop productivity. The year-to-year variations of mean growing-season temperatures are of the order of several degrees Centigrade, and these variations show up significantly in crop production. On the other hand, if one considers various physiological events that occur in plants or cold-blooded animals such as insects, one finds great sensitivity to certain temperature changes occurring at critical stages of the life cycle. Sometimes only a degree or two Centigrade will mean a considerable difference in the success or failure of a particular population. Most organisms can withstand fairly large temperature variations, however, and there is usually sufficient plasticity within populations that temperature extremes are survived, particularly if they are not too long-lasting.

Temperature History of the Earth's Climate

Climate is never static. Climates constantly vary and fluctuate. During much of the past 500 million years, temperatures often averaged 10°C warmer than at the present time. Much of the Silurian, Devonian, and Carboniferous periods of geological time, for example, were warm, with intervals of both very dry and very wet conditions. The Carboniferous period ended with an ice age that extended into Permian times. During the Mesozoic era that followed, climates were generally warm. The Cenozoic era (about 66 million years ago) was warm and moist. During the Eocene epoch (about 50 million years ago), palms, laurels, and magnolias grew in southeastern Alaska, swamp forests were found in Tennessee, and alligators lived in Wyoming. The latest glacial epoch, called the Pleistocene, began about 1 million years ago. Major ice sheets have advanced and receded at least four times during this period, the last advance ending approximately 10,000 years ago. During historical times, climates have varied in relatively short periods, often producing alternate times of abundance or famine. Even today, modern society, with its critical dependence upon fossil fuels, is highly vulnerable to severe climatic change. It is not the place here to describe geological and glacial events in detail except to comment that much can be learned concerning plant and animal responses to climatic change through paleontology and paleoclimatology.

During the first part of the nineteenth century, until about 1880, temperatures were decreasing worldwide. From 1880 until approximately 1945, there was a distinct warming trend, as shown in Fig. 15.1. The maximum change of the mean annual global temperature was about 0.9°F (0.5°C) and that of the mean winter global temperature about 1.4°F (0.8°C) between 1880 and 1945. Since 1945, a sharp drop of the mean annual global temperature has occurred, and at the present time, temperature conditions are approaching those which were existent globally about 1900. During the warming period, some locations experienced much greater increases of temperature than others. In the United States generally, there was an increase of nearly 1.1°C from the coldest decade to the warmest. The average temperature change for the Blue Hill observatory outside Boston, Mass., was 3°C. Other stations throughout the Northern Hemisphere experienced less extreme changes during the warming period. Average temperatures changed much less in the Southern Hemisphere than in the Northern Hemisphere, and at some locations no change whatsoever took place. Specifically, the primary amelioration

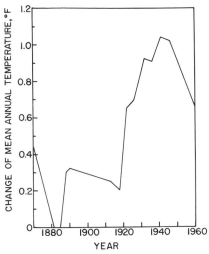

Figure 15.1. Change in the mean annual temperature of the Northern Hemisphere from 1870 to 1960. (From Gates, 1972.)

took place in the Arctic, in temperate regions of the Northern Hemisphere including the United States, Europe, and Siberia, and, to a lesser extent, in the temperate zones of the Southern Hemisphere and the tropics. The regions of the world generally without much amelioration were northeast Canada, a large part of South America, the southwest quarter of Africa, parts of central Asia, Pakistan, the Indian Ocean, southeastern Asia, and Australia.

These changes of seasonal and annual temperatures have had notable effect on the growing season of plants. At Oxford, England, the growing season is defined in terms of temperatures 6°C or above. The average length of the growing season at Oxford during the twentieth century was 2 to 3 weeks longer than during the nineteenth century, but has decreased by 1 to 2 weeks since 1950. Although not always true, during the warming trend (1880–1945), there was generally less precipitation than during the cooling period (1945–present).

It was in the period 1800–50 that the coolest summer (1816) and the coldest winter (1829–30) occured in the long sequence of European temperature records (until the "winter of the century" in 1962–63). It was, however, the decade 1689–98 which represented the coldest overall temperatures and which is referred to as "The Little Ice Age." Grain failed to ripen or else froze, and there was great scarcity during 1693 and 1694 in western Europe. One-third of the population of Finland died in the famine of 1697. Crops were destroyed by cold in 1709, and tens of thousands of people died throughout Europe. The intense cold affected all crops, poultry, cattle, and wildlife. The eighteenth century, however, experienced considerable amelioration, and harvests in Sweden became bountiful in the years 1730–39.

One of the very best methods for assessing the influences of climate and, in particular, temperature on plant and animal life is to study the various phenological records from different parts of the world. Phenology is the study of natural phenomena that recur periodically in response to climatic and seasonal changes. Included in phenological studies are such matters as bud formation, leafing of trees and shrubs, flowering, ripening of fruit, bird and insect migration, animal behavior, leaf senescence, and autumn coloring.

Physiology and Temperature

Introduction

The most precise information concerning the effects of temperature on organisms comes from physiology. The literature on this subject is truly voluminous. Much of it has been summarized in several very important books on the subject, including Belehradek (1935), Johnson et al. (1954), Johnson (1957), Rose (1967), and Precht et al. (1973).

All organisms are sensitive to temperature at all stages of their life histories, whether seed, egg, larva, or adult. Organisms have minimum and maximum temperatures below or above which they cannot survive. Within this temperature span, they have a narrower range within which they are active, and somewhere within the active temperature range, an optimum temperature exists near which the organism functions best physiologically. Sometimes the optimum temperature for growth is not the "best" temperature for longevity or some particular activity. Eggs of an insect, for example, may mature most rapidly at one temperature but have a higher survival rate if development occurs more slowly at a lower temperature.

Organisms maintain a dynamic balance with their changing environment through a complex of metabolic reactions. Metabolic response to the thermal environment is enzyme mediated and both genetically and phenologically influenced. At the cellular level, metabolic response is biochemical and is strongly affected by a complex of enzymes. However, within the genetic capacity of a given organism, many physiological adjustments can be made to compensate for or adjust to environmental change. Physiological response is determined considerably by the environmental history of the organism and the suddenness of the environmental change.

The ability of an organism to survive either very high or very low temperatures depends upon many subtle biochemical cellular changes within the organism and in particular upon certain repair mechanisms. Balance or lack of balance between rates of thermal damage and rates of repair at the cellular level determines the survival of an organism undergoing thermal stress. Since these processes of damage or repair are integrative with time, the length of time an organism undergoes thermal stress is critical. Certain conditioning responses require extensive periods of time for adjustment, as for example the formation of glycerol, which helps to remove water from cells and thereby protects from rupture by freezing.

Usually, the change in environmental thermal conditions, whether natural or man-made, is relatively small. The likelihood of sudden thermal death is not then so great, but the ability to complete the life cycle can be seriously affected. When grain or fruit fails to mature, it is not that the plants were killed, but simply that conditions were not propitious for maturation. Even though thermal death may not ensue, the rate of growth of an organism may be affected sufficiently to influence its productivity, competitive position in the ecosystem, or resistance to other factors. With regard to thermal death, the threat of cooling below the freezing point is likely to be more serious than the chance of getting slightly too warm. For most organisms, the low temperature threshold for survival is probably more abrupt than the high temperature threshold. Many organisms, however, particularly tropical plants, do not need to go below freezing in order to be killed by cold and are severely damaged by temperatures substantially above freezing. The freezing point of water does represent an abrupt environmental transition, as of course does the boiling point of water. The environmental condition of the world in which we live often crosses below the freezing point but only approaches the boiling point in very special circumstances.

Extremes of temperature, either hot or cold, cause damage to cells and if sustained, will cause the death of plants. At high temperatures, the effects are apparently from enzyme inactivation, denaturation, and protein coagulation. Dehydration, membrane rupture, and cell-wall damage result from ice formation at temperatures below freezing. Plants have evolved protective mechanisms in order to mitigate or tolerate such unfavorable temperature circumstances. Cells generate glycerol, which acts as an "antifreeze" against ice formation. Protection against freezing can be achieved if cells can become dehydrated and thereby reduce the formation of needle ice in the cells. Rate of cooling is critical to the freezing process in plants and the degree of resultant damage.

Freezing of Plant Cells

A great deal is known concerning the freezing mechanism in plants, yet a great deal remains to be learned. It is a fascinating subject, worthy of much discussion. Here, however, only a summary is in order. An excellent review is given by Mazur (1969). The following events may occur as cells are cooled:

1. Cells and their external medium may supercool as the temperature falls. When ice forms in the external medium, the cell membrane keeps ice from seeding the cell interior, so the cell remains unfrozen and supercooled.
2. As the temperature falls, dehydration of the cell may occur. As the extracellular solution is converted to ice, the concentration of extracellular solutes rises, and the aqueous vapor pressure falls. Since the cell is supercooled, its aqueous vapor pressure exceeds that of the extracellular water, and water flows out of the cell and freezes extracellularly. Hence dehydration occurs if the cell is sufficiently permeable to water or cooling is sufficiently slow. But if the cell is not sufficiently permeable to water or is cooled too rapidly, it will not remain in equilibrium, but will continue to supercool until it eventually equilibrates by internal freezing and damage results.
3. As more and more water is converted to ice during cooling, the solubility of some electrolytes, whether intra- or extracellular, may be exceeded, and these electrolytes will precipitate as a result.

During warming, the events will essentially be reversed. Progressive melting of the external solution will cause the external aqueous vapor pressure to rise above that of cells that have dehydrated during cooling. As a result, water will flow into the cell at a rate dependent on both the warming velocity and the permeability of the cell to water.

Photosynthesis

Photosynthesis in green plants is a complex series of biochemical events which depend upon light, temperature, and the concentrations of carbon dioxide, oxygen, and water. Certain nutrients such as phosphorus and nitrogen must be available to make possible the very complex molecular arrangements that occur within chloroplasts, including synthesis of the chlorophyll molecule itself. Photosynthesis involves photochemical events and therefore depends upon the availability of light. Photosynthesis is also a thermochemical process, and thus the chemical reactions vary with temperature.

There is much evidence available concerning the temperature dependence of photosynthesis in plants. The effect of temperature on photosynthesis is shown from the many measurements which have been made of net assimilation of carbon dioxide by whole leaves of various plant species as a function of temperature. Net assimilation of carbon dioxide is the difference between gross photosynthesis and respiration. Measurements are made by passing air through a chamber containing a plant leaf in which the chamber temperature is carefully controlled. A measurement is made of the concentration of carbon dioxide entering and leaving the chamber. The difference in carbon dioxide concentration of the air stream flowing across the plant leaf is a measure of net assimilation, also termed the net photosynthesis. Respiration for any plant leaf increases monotonically with temperature as shown in Fig. 15.2. The process of respiration releases carbon dioxide from carbohydrates, whereas photosynthesis

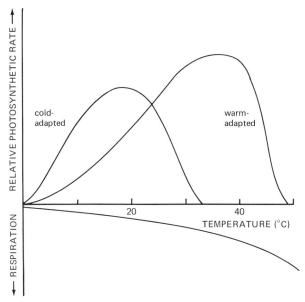

Figure 15.2. Idealized curves of relative net photosynthetic rate and respiration rate as a function of temperature for coldand warm-adapted plants.

fixes carbon dioxide in the process of building carbohydrates. Net photosynthesis, which occurs when carbon dioxide fixation exceeds respiration, results in an uptake of carbon dioxide from the air. The rate of carbon dioxide uptake is plotted as the positive upward values in Fig. 15.2, and the release of carbon dioxide by respiration is shown as the negative values. At very high temperatures, respiration always exceeds photosynthesis; at low and intermediate temperatures, photosynthesis exceeds respiration.

Also shown in Fig. 15.2 are two generalized curves representing net photosynthesis by plants that are cold- and warm-adapted. Entirely different plant species have become genetically adapted to cold and hot climates. In temperate regions, there are species representative of both cold and warm climates, as well as many species associated strictly with temperate conditions. As mentioned elsewhere, temperate conditions are not always particularly "temperate," but have enormous temperature variations. In fact, most warm-tropical climates have far less temperature variation than temperate climates. The plants of temperate regions exhibit a relationship between net photosynthesis and temperature intermediate between that of cold- and warm-climate-adapted plants.

The temperature response of net photosynthesis for many plants is shown in Figs. 15.3 and 15.4. Plants are genetically "locked in" to their photosynthetic response to temperature, as well as to their response to light, carbon dioxide and oxygen concentration, and other factors. Nevertheless, most plants can undergo a certain amount of phenotypic adaptation to habitat conditions. Such phenotypic shifts in the characteristic response of photosynthesis to temperature are exhibited in Figs. 15.3 and 15.4. Relative rather than absolute values are shown here since our attention is focused primarily on the temperature response. Let us now study these figures in detail.

In Fig. 15.3, we see the temperature response of three species adapted to cold and temperate climates, *Pinus cembra, Fagus silvatica,* and *Picea abies,* and two species adapted to warm climates, *Ficus retusa* and *Acacia craspedocarpa.* The most cold-adapted species has an optimum temperature for photosynthesis at 12°C, and the most warm-adapted has an optimum at 35°C. In fact, the pine will not function much above 35°C, and the *Acacia* will not function at 12°C. In Fig. 15.3, we see further comparisons of species from diverse habitats. *Cercidium floridum,* a desert tree, has an unusually low temperature optimum for a desert plant, whereas *Tidestroma oblongifolia,* a desert perennial, has a remarkable high temperature optimum. Sorghum is a tropical grass and has a high temperature optimum. Cotton and sunflower have optima suited to warm climates also. *Zea mays,* corn, also has an optimum near 40°C and is cultivated tropical grass. Warm nights are favorable to the growth of corn, as well as warm days for photosynthesis. Most temperate-zone plants, except for alpine and montane species, have temperature optima in the vi-

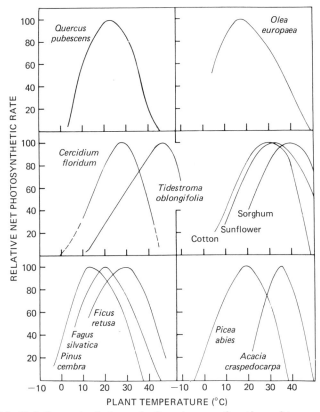

Figure 15.3. Relative net photosynthetic rate as a function of temperature for various plants.

cinity of 20°C, as exhibited by *Quercus pubescens, Olea europaea, Fagus silvatica,* and *Picea abies* in Fig. 15.3. Alpine or arctic plants have low temperature optima as for example 10°C for *Artemesia tridentata* or 15°C for *Oxyria digyna* and *Solidago virgaurea* (shown in Fig. 15.4).

The temperature optimum for a plant shifts with the conditions prevailing during its growth. This is evident in Fig. 15.4, in which the temperature optima for plants in their natural habitats shift slightly toward lower temperatures as the season progresses beyond the summer solstice. Plants grown in the greenhouse can also have a temperature-optima shift when grown in cooler or warmer conditions of a few degrees, as shown for *Mimulus cardinalis* in Fig. 15.4, or as much as 12°C, as shown for *Oxyria digyna.* A shift of temperature optimum also occurs for plants grown in high or low light intensities, as shown by the curves for *Ranunculus glacialis* in Fig. 15.4. The temperature optimum for net photosynthesis is controlled by enzymes located within the metabolic centers of the leaf. The enzyme chemistry of a particular plant species probably is largely genetically determined, with only some adaptive adjustment to changes in

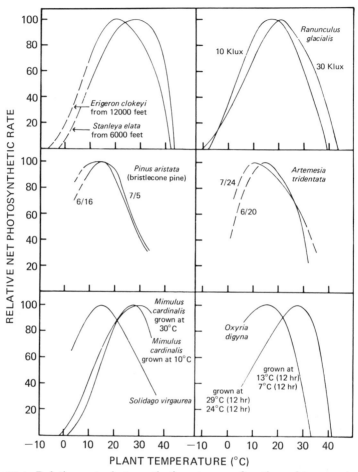

Figure 15.4. Relative net photosynthetic rate as a function of temperature for various plants.

environmental conditions being possible. One must continually keep in mind that other environmental and plant factors influence to some degree the photosynthetic response to temperature.

Clearly, the greatest sensitivity of plants to temperature is not in the vicinity of the temperature optimum but to either side of the optimum. Usually, the slope of the curve is steeper on the high-temperature side than on the low-temperature side. If normal atmospheric conditions are such that plant temperatures are about optimum, and warmer conditions are then imposed on the environment, one would expect a decrease in plant productivity. If, however, environmental conditions are cool, and plant temperatures are usually below optimum, then with a shift to warmer conditions, plant productivity will increase.

The influence of temperature on whole plants is somewhat different from that on individual leaves because of the effect of respiration in stems and roots. The smaller respiratory losses for single leaves, especially at low temperatures, cause them to have higher net assimilation rates and higher optimum temperatures than whole plants.

Wilson (1966) shows that over a fairly broad temperature range around the optimum temperature, plants have less than 10% variation in net assimilation. This temperature range was 17°–29°C for duckweed, 12°–30°C for grape, 18°–32°C for cucumber, 18°–33°C for sunflower, and 23°–36°C for maize. Temperature fluctuations within these ranges in the outdoor habitat cause relatively little variation in net assimilation, and indeed, fluctuations of light will be far more significant. Temperature tends to be important in plants with high optimum temperatures that are grown in cool climates. Maize, which has a high temperature optima, often shows a significant decrease in net productivity when planted in cool-temperate climates. According to Wilson (1966), sunflower and broad bean growing in Scotland in the summer and wheat growing in southern England in the winter and spring exhibit significant temperature influence. Sugar beet, potato, barley, and sunflower growing during the English summer show no effect of temperature on net assimilation.

Higher Plants

The effect of temperature on higher plants must be considered at each stage of the life cycle, i.e., dormancy, germination of the seed, morphogenesis, growth and photosynthesis, flowering, fertilization, and fruit set. We will not consider most of these stages in detail here since they have been well described in other books.

Cold Hardening

Woody plants undergo cold hardening during the autumn, can generally withstand very low temperatures during the winter, and lose their cold hardening during the spring. Plants will undergo cold hardening when temperatures are gradually lowered below 6°C. There is enormous variability among plants in their ability to undergo cold hardening. Dormant willow twigs can withstand the temperature of liquid nitrogen after appropriate hardening. Twigs collected during the winter require much less additional hardening to survive being immersed in liquid nitrogen than twigs collected in the spring or summer. Winter wheat will rapidly lose its resistance to cold when exposed to warm temperatures. All cold-tolerant plants lose their resistance during the summer. Sugars in the plant cells seem to play an important role in the process of cold hardening, just as they play a role in the process of heat resistance.

Leaf moisture of conifer needles generally declines as cold hardening occurs. Clements (1938) reported extensively on the cold resistance of forest trees in North America. Ponderosa pine was a little less cold hardened than fir, which could withstand temperatures of $-32°C$ during the winter but only $-19°C$ in the spring. *Pinus cembra* could withstand temperatures of $-47°C$, whereas *Rhododendron ferrugineum* could not tolerate temperatures lower than $-15°C$. Rhododendrons are susceptible to winter killing in certain climates if not covered by snow. Although seasonal changes in cold resistance occur with most plants of north-temperate regions, changes in cold resistance of Mediterranean evergreens with season are very small, and many cannot withstand temperatures below $-13°C$ according to Larcher (1954).

Conifers from the northern Rocky Mountains undergo cold hardening during the autumn, and all native conifers can withstand temperatures of $-60°C$. Parker (1961) showed that eastern white pine needles were hardened to $-196°C$ in late winter, but that southern pine species hardened only slightly during the autumn and could not withstand temperatures less than $-24°C$ during the winter. Deciduous trees generally undergo the same amount of cold hardening as conifers in any particular region of the world. There appears to be no lower limit to the cold hardiness of plants, some of which can withstand temperatures of $-183°C$ or below. Buds of arctic birches cannot withstand temperatures lower than about $-60°C$.

Recently, Sakai and Weiser (1973) reported on the freezing resistance of 70 tree species in North America. They tested dormant 1-year-old twigs from mature trees in the five major tree regions of North America. All twigs were subjected to an artificial hardening regime by holding samples at $-3°C$ for 2 weeks, $-5°C$ for 1 week, and $-10°C$ for 3 days to induce maximum freezing resistance. After hardening, twigs were cooled at 5°C increments each day to $-30°C$ and then to $-80°C$ in 10°C increments. Some twigs were removed from each test regime and tested for viability by being kept at room temperature, the evergreen twigs in polyethylene bags saturated with water vapor and the deciduous twigs in water. Viability was estimated from the degree of browning or lack of it.

Table 15.1 summarizes the extremely interesting results of Sakai and Weiser. As a general rule, the cold hardiness of tree species is directly related to the average minimum annual temperature at the northern limits of natural ranges and artificial plantings. Sakai and Weiser stated some of their conclusions as follows:

> Extremely hardy conifers and deciduous trees cover boreal and high altitudes in the Rocky Mountains. On the other hand, tender broad-leaved evergreen trees and conifers which can survive freezing to only $-15°C$ are confined to the Pacific and southeastern coast regions. Broad-leaved evergreen genera such as *Cinnamomum, Myrica, Osmanthus,* and *Quercus* which emigrated to the Japanese main islands after the Ice Age, survived freezing to only $-10°C$. The same genera are native to Florida

Table 15.1. Freezing Resistance of Tree Species

Relative-hardiness classification	Representative species	Average minimum temperature at the northern limits of growth (°C)		Observed freezing resistance (°C)
		Natural range	Artificial plantings	
Tender evergreen	Quercus virginiana	−3.9 to −6.7	−9 to −12	−7 to −8
Hardy evergreen	Magnolia grandiflora	−9 to −12	−8 to −20	−5 to −20
Hardy deciduous	Liquidambar styraciflua	−8 to −20	−26 to −29	−25 to −30
Very hardy deciduous	Ulums americana	−37 to −46	−40 to −43	−40 to −50
Extremely hardy deciduous	Betula papyrifera	below −46	below −46	below −80
	Populus deltoides	−32 to −34	−37 to −45	below −80
	Salix nigra	−32 to −34	−37 to −45	below −80

and coastal areas of California and Oregon. Winter minimum temperatures appear to be the principal limitation on the natural ranges of these tender genera.

In general, the same relationship holds in the case of many hardy deciduous and coniferous trees (except for *Populus, Betula,* and *Salix*). *Populus deltoides* and *Salix nigra,* growing in northern Indiana, were found to survive freezing to at least −80°C, but they naturally range to winter minimum of − 28.9°C. This suggests that winter minimum temperature is not a principal factor governing the geographic distribution of these extremely hardy trees.

Many of the Arcto-Tertiary flora became extinct or restricted to the Pacific coast (Sequoia) and southeastern United States (*Taxodium, Liquidambar,* etc.) during their southward movement. Winter twigs of *Taxodium distichum* (swamp bald cypress) growing along the Wabash River of far southwestern Indiana and in flooded valley lands in Mississippi, survived freezing down to at least − 30°C after artificial hardening. *Metasequoia,* which became extinct in North America, has a similar level of hardiness, and *Gingko bilboa,* which also became extinct in North America by the close of the Tertiary period, was found to resist freezing to − 40°C. It seems unlikely that low environmental temperatures were responsible for the extinction of these hardy species. Other factors such as summer drought, as previously noted, might have been the limiting factor. In this connection it is interesting that most of the relic genera of the Arcto-Tertiary flora are now confined to moist valley bottoms and flood plains of the sub-tropics in North and Central America and southern China. *Sequoia sempervivens* (redwood) is restricted to the California and Oregon coastal regions where heavy summer fogs from the ocean provide a humid atmosphere. Humid mild climates seem to be optimal for relic genera such as *Taxodium distichum, Nyssa aquatica,* and *Liquidambar styraciflua.*

The manner of cold hardening makes a substantial difference in the degree of cold resistance, but there is conflicting evidence concerning the effect of fluctuating ambient temperature during the hardening process. Repeated freezing and thawing will give results different from prolonged freezing. Time is always an important element in any thermal phenomena involving life.

Trees in cold regions often have their trunks crack open in winter. Differential heating or cooling of wood can cause it to split. Generally, cracking of trees occurs during cooling, but there is some evidence that cracking may occur during heating. Apple-tree cracks, for example, are observed on the south side of trees. Parker (1963) noted that most cracks on city trees of New Haven, Conn., occur on southeastern or southwestern sides of the trunks. Sunshine is absorbed by the bark and expands the sapwood, which is frozen, and the heartwood cracks since it is still cold and unable to expand. The cracking of trees during very cold days in the northern woods results in a loud noise very much like a rifle shot. Cracking is particularly common after the sap starts running in the spring.

Some authors report that cracks start just before sunrise, an observation that would seem to dispel the notion that a primary cause is the sunshine warming the south side of the tree.

Derby and Gates (1966) developed an analytical model to predict the temperature profiles within tree trunks and compared their analysis to observations made of the trunks of aspen trees. The problem of freezing or thawing of water in the wood is discussed in this paper, and a method of treating phase change analytically is described. The method used by Derby and Gates involved a finite-difference iterative procedure in which the cross section of a tree trunk was segmented into 12 pie-shaped sections, each of which was divided into 5 radial pieces from the center of the tree to the outer surface. This gave a total of 60 sections, each connected to neighboring sections by means of a conductance. The trunk was considered to be heterogeneous and anisotropic. Incident solar and thermal radiation on the outside bark, as well as reradiation and convection, were taken into account. This is another example of a biological problem requiring physics and analytical methods. A comparison between measured and predicted temperatures proved that the model was extremely good.

A range of plants from various habitats were cold hardened in a controlled environment by Parker (1963); the results are given in Table 15.2. It is clear that there is an abrupt dichotomy between the tree species from *Cryptomeria japonica* through *Ilex opaca,* which would be damaged by the coldest winter temperatures in the northern United States, and those

Table 15.2. Cold Resistance in January at New Haven, Conn., of Leaves of Various Woody Plants[a]

	Origin	Temperature withstood (°C)
Abies guatemalensis	Costa Rica	−6
Cupressus lusitanica	Costa Rica	−10
Cryptomeria japonica	Japan	−20
Pinus palustris	Northern Florida	−24
Pinus taeda	North Carolina	−25
Hedera helix	Unknown	−25
Ilex opaca	Native	−25
Chanaecyparis pisifera	Japan	−42
Rhododendron catawbiense	Unknown	−43
Taxus baccata	Unknown	−45
Juniperus virginiana	Connecticut	−52
Kalmia latifolia	Connecticut	−54
Picea excelsa	Europe	−58
Tsuga canadensis	Connecticut	−60
Pinus sylvestris	Europe	−62
Pinus strobus	Connecticut	−70

[a] From Parker (1963).

from *Chamaecyparis pisifera* through *Pinus strobus,* which would be able to withstand the coldest temperatures. Plants from the high mountains of Costa Rica were unable to sustain even moderately cold temperatures. Many tropical plants are injured when cooled to about 6°C and die when cooled to a few degrees above the freezing point. Some tropical plants can withstand light frosts, and some date palms withstand temperatures down to −7°C. Usually, date palms, bromeliads, palmettos, and other such plants will kill-back when air temperatures drop to near freezing but will then recover and grow again.

Table 15.3 shows the summer and winter resistance of some evergreens from work done by Larcher (1954). These plants, when hardened during the winter, can withstand temperatures from 4° to 8°C lower than they can sustain when hardened during the summer months. Physiological changes occur in plants during late summer and early autumn which prepare them for winter. Deciduous forest trees from Connecticut had an increase in hardiness from June to August of 2° to 3°C.

Trees of a given species but of different ecotypes show variations in cold hardiness. White ash of southern origin was winter-killed in Massachusetts, whereas specimens of northern origin were somewhat hardy. As trees age, they seem more capable of becoming dormant and hardy in the autumn. *Pinus excelsa* in the Swiss Alps has been found to consist of different altitudinal ecotypes which have varying hardiness. Red ash, brought to the Harvard Forest from various origins, seemed to have three ecotypes, according to Parker (1963): "one from northern states which was slow-growing, winter-hardy, and lost its leaves in autumn; one from the southern coastal plain down to South Carolina which was fast-

Table 15.3. Summer and Winter Resistance of Evergreens from the Northern Mediterranean Region[a]

	Summer resistance (°C)	Winter resistance (°C)
Nerium oleander	−5	−9
Laurus nobilis	−6	−10
Arbutus unedo	−6	−11
Viburnum tinis	−6	−11
Olea europea	−6	−13
Pinus pinea	−6	−13
Quercus ilex	−6	−13
Chamaerops humilis	−13	−13
Trachycarpus fortunei	−13	−14
Cupressus sempervirens	−8	−16
Cedrus atlantica	−7	−16
Cedrus deodara	−7	−18
Citrus trifoliata	−14	−20

[a] From Larcher (1954).

growing, sensitive to cold, and retained its leaves through two killing frosts; and a third type from New York which was intermediate between the other two."

The roots of a tree are most sensitive to cold during the winter. Many roots are killed by soil temperatures of $-10°$ to $-15°C$. Fortunately, soil temperatures are usually much less extreme than air temperatures and lag behind. Diurnal variations of temperature penetrate only about 30 cm beneath the surface and are substantially attenuated by depth. Changes of surface temperatures of several days duration penetrate to 50 cm or more. Long-term annual changes can be measured down to several meters depth. Therefore, roots generally are not subject to the extremes of cold that the aerial plant parts must endure. Roots of alder were only able to sustain temperatures of $-2°C$, whereas *Pinus ponderosa* roots attained a winter hardiness of $-12°C$. Summer minima were often $-2.5°$ to $-3.5°C$. Apple-tree roots are usually killed at $-9°$ to $-12°C$ in winter, but in summer can only withstand temperatures of $-3°C$. Snow cover gives good insulation to tree roots in northern regions. Young trees, particularly root tips, are more tender and susceptible to winter kill than older plants. Again, northern species and ecotypes generally have more cold resistance than southern ones.

Fruit buds and flowers are usually more sensitive to cold than wood or bark but can become winter hardened with gradual cooling. Apple fruit buds were more sensitive than bark to cold and were found to winter kill at $-1.3°$ to $-1.8°C$. During the month of June, flower buds were sensitive to temperatures of $-4°$ to $-5°C$. Spraying of fruit tree flowers and buds with water during periods of frost affords some protection. As the water freezes, the latent heat released prevents freezing of the flower or buds during periods of moderate or light frost.

Heat Resistance

Many people have studied the resistance to heat and cold by plants and the thermostability of plant cells and proteins. O. L. Lange (1965) has reported the response of whole plants to heat, and Alexandrov (1964) and Alexandrov et al. (1970) give reviews concerning the cytological affects of high and low temperatures. For a time, there appeared to be a contradiction between observations concerning the response of the whole plant to heat and the response of individual cells, but recently this difficulty has been resolved by Lutova and Zavadskaya (1966). Basically, the problem was that Lange observed an increase in heat resistance with a rise in the temperature at which the plant was cultivated, whereas Alexandrov found no change in the thermal stability of plant cells under different conditions of cultivation. An increase in the temperature at which the plant is cultivated activates the repair process to thermal injury; however, the so-called primary thermostability of the plant cells, as observed through protoplasmic streaming, is uninfluenced by conditions of cultivation. The resistance of a plant to damage by heat or cold is time-dependent. The longer a plant is kept at a high or low temperature that is potentially damaging, the greater is the effect of temperature on the cell or plant viability.

The variability of plants and plant leaves with regard to heat damage is enormous. Susceptibility to high temperatures depends upon many factors, including age, season of year, vigor of growth, previous temperature history, light intensity, and genetic, or inherited, tolerance. Figure 15.5, for example, shows the relative heat resistance of leaves of four plant species according to their positions on the stem. The young leaves of *Ilex aquifolium* are the most susceptible to heat damage, and older leaves have greater heat tolerance; similar results are found for frost resistance in this plant. On the other hand, older leaves of *Atrichum undulatum* are more vulnerable to heat damage. The youngest and oldest leaves of *Kalanchoe blossfeldiana* are most heat resistant, with minimum heat tolerance generally exhibited by the third or fourth leaf below the bud or apex. In this species, there is a correlation between the lowest heat resistance and highest photoperiodic sensitivity. *Mnium undulatum*, the youngest and oldest leaves have the least heat resistance, and the leaves of intermediate age have the greatest resistance. These observations are from Lange (1967).

Figure 15.6 shows the temperature at which heat damage occurs for four plant species, as reported by Lange (1965). The plants were held for 30 min at the temperature given. Four weeks after the exposure, the percent of leaf area damaged was determined. In *Ilex,* heat damage began at

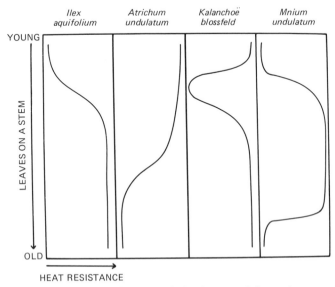

Figure 15.5. Relative heat resistance of the leaves of four plant species as a function of age. (According to Lange, 1967.)

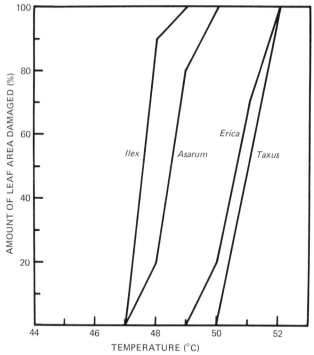

Figure 15.6. Percent of leaf area damaged when exposed for 30 min to the temperatures shown. Assessment of damage was made 4 weeks after exposure, during which time the plants had been returned to normal temperature conditions. (Redrawn from Lange, 1965.)

47°C and became almost total by 48°C. In *Asarum,* damage began at 47°C and reached 80% by 49°C, whereas in *Taxus,* damage began at 50°C and was total by 52°C. It is interesting to remind ourselves that *Tidestroma oblongifolia* had a photosynthetic temperature optimum at 47°C and was still functioning well at temperatures of 52°C and above. The enormous variability of the plant world continues to be evident in temperature data.

Heat damage of plant leaves depends critically upon the length of exposure time. Various cytological and physiological changes occur in plant cells as a function of the degree of heating. Alexandrov (1964) has reviewed these processes in great detail. Some of the processes that he used as criteria of heat damage or heat resistance include respiration, photosynthesis, exit of substances from plant cells into the medium in which they were being observed, suppression of plasmolysis, exit of pigments from the vacuoles, loss of vital staining, luminescence of the chloroplasts, persistance of protoplasmic streaming, changes in the cytoplasm and the nucleus, and change in viscosity. Generally, a decrease in protoplasmic streaming or photosynthesis occurs before plasmolysis or a decrease in vital staining or respiration. In cells of *Campanula persicifolia,* for example, protoplasmic streaming and photosynthesis ceased at 44°C, whereas plasmolysis and suppressed respiration at 60°C. Thus streaming and photosynthesis respond to an earlier stage of damage. After consideration of the many processes affected by high temperatures, Alexandrov decided to use cessation of protoplasmic streaming as the main indicator of heat damage in plant cells. The time that protoplasmic streaming persists at a given temperature is a measure of the heat resistance of cells. Plotting persistence of streaming as the ordinate and temperature of heating as the abscissa, Alexandrov (1964) obtained the relationships shown in Fig. 15.7. The position of the break in the curve and the angle between the two more or less straight-line sections is significant. For the right-hand straight-line section, time of cessation of protoplasmic streaming is determined by the rate of thermal denaturation of certain proteins involved in the streaming movement. The sharp break in the curve in the zone of mild heating indicates that at lower temperatures, or with more prolonged heating, the cells are capable of counteracting the heat injury. At temperatures corresponding to the left-hand portion of the curve, repair mechanisms restore protoplasmic streaming once heating of the cells is stopped. In fact, when a plant cell is heated to a temperature to the left of the "break" in the curve, the repair mechanism is constantly trying to overcome the process of protein denaturation. Table 15.4 is taken from Alexandrov (1964) and shows, for various plants, the temperatures at which protoplasmic streaming stops after 5 min of heating, and also the temperature to which the cells must be raised to cause irreversible heating damage and beyond which streaming cannot be restored. It is

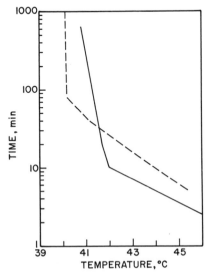

Figure 15.7. Cell damage resulting from heat exposure, as determined by persistence of protoplasmic streaming, as a function of temperature: (—) *Campanuta persicifolia;* (--) *Hepatica nobilis.* (Redrawn from Alexandrov, 1964.)

Table 15.4. Effect of Temperature on Protoplasmic Streaming

	Temperature at which streaming stops after 5 min of heating (°C)	Temperature for irreversibility (°C)
Tradescantia fluminensis	44.5	52.0
Zebrina pendula	46.7	53.0
Campanula persicifolia	44.0	52.0
Ficus radicans	45.8	54.6
Chlorophytum elatum	47.0	56.0
Catabrosa aquatica	44.7	48.5
Oxilatoria tenuis	49.2	54.0
Phormidium autumnale	52.2	56.4

not that these species are of particular importance or interest, only that they illustrate the temperatures at which many plants undergo heat damage. Some plants that are adapted to cold climates might have lower critical temperatures for heat damage, and, of course, some that are adapted to warm climates may have higher critical temperatures.

Heat tolerance of higher plants depends upon many factors. Lange (1965) lists the following:

1. Changes due to aging.
2. Increase in the heat hardiness of leaves with growing water saturation deficit.
3. Influences of preexisting temperature, humidity, and water conditions.
4. Influence of stage of development.
5. Endogenous daily fluctuations.
6. Seasonal variations. (Usually, one peak of hardiness is encountered in summer and one in winter, but some plants have a single peak during the winter months.)

Heat Production by Respiration

As a general rule, there is negligible heat production in plants owing to respiration. For most plants under normal conditions, it is virtually impossible to detect a temperature change produced by metabolic events. On the other hand, the spontaneous combustion of hay is a temperature phenomena well known to most everyone and is the result of a temperature increase produced by bacterial respiration. The temperature optimum for respiration and growth in most bacteria is below 40°C, but in the case of thermophilic bacteria, this temperature optimum is usually above 50°C. When hay is not dried properly, ordinary bacteria will develop in it, raising its temperature to above 40°C. Above this temperature, thermophilic bacteria begin to take over, raising the temperature to above 70°C. Then, autooxidation takes over, and the temperature at the center of the haystack may exceed the ignition temperature; spontaneous combustion is the result. According to Brock and Darland (1970), some thermophilic bacteria can survive temperatures as high as 92° to 100°C in alkaline hot springs, depending upon the altitude, but only 75° to 80°C in acidic hot springs.

It is also known that the temperature in the spadix of certain Araceae (arums) such as *Monstera delicosa* rises as much as 14°C above air temperature just before opening of the flower. This temperature rise is caused directly by respiration. According to Nagy et al. (1972), the inflorescence of *Philodendron selloum* temporarily maintains a core temperature of 38° to 46°C despite air temperatures of 4° to 39°C. The heat is produced primarily by small, sterile male flowers whose rate of oxygen consumption rivals that of flying hummingbirds and sphinx moths.

Growth

Plant growth in most species is at an optimum between 25° and 35°C. Many arctic and alpine species, however, have optima between 12° and 20°C, and some tropical and desert plants have optima between 35° and 40°C. Since growth occurs at the apex, it is interesting to note the effect of temperature on rate of change of apex size. For chrysanthemum, doubling of apex volume took 5.1 days at 5°C, 2.1 days at 17°C, 1.9 days at 20°C, and 2.0 days at 22°C. Temperature does not affect apical growth nearly as much as it affects rate of leaf expansion. Leaf area is generally larger for plants grown at higher temperatures than those grown at low temperature. For many plants, night temperatures have a more pronounced effect on growth than day temperatures.

Plants in their natural habitats are subject to greatly fluctuating soil and air temperatures. Temperatures change diurnally and seasonally as well as from moment to moment. When grown at constant temperature and light, many plants have abnormal growth. In fact, some species lose their ability to grow when maintained too long under constant conditions. More leaf variation in size and shape occurs in peas when grown under constant conditions than under diurnally varying conditions. Tomatoes exposed to fluctuating temperatures had greater stem elongation, dry matter production, and fruit set than plants grown at a constant temperature. For tomatoes, the optimum night temperature was 17°C and the optimum day temperature 26°C. Young tomato seedlings, however, showed best growth when kept close to 25°C. In the case of tomato plants other than seedlings, photosynthesis is operating during the day and has a high temperature optimum, whereas growth occurs at night and has a low temperature optimum. Most plants require average nighttime temperatures, except in the case of *Saintpaulia*, which needs slightly warmer nights than days. Optimum night temperatures for growth decrease with increasing age and decreasing light intensity. Not all plants require fluctuating diurnal temperatures. Cucumbers, french beans, peanuts, sugarcane, and wheat, for example, do best under fairly constant conditions. This is generally true for tropical plants. On the other hand, some plants such as succulents, which are adapted to dry continental climates, require a large temperature change between day and night. It is clear that plants have evolved to fit the conditions of the habitats available to them.

Trees

The growth of trees is determined by many factors, including temperature, moisture, wind, sunshine, nutrients, pathogens, and other organisms. The perennial nature of trees means that growth responds to events occurring many months or even 1 or 2 years beforehand. Tree growth integrates over time the various factors of environmental influence, often in a synergistic fashion. For this reason, it is sometimes difficult to specify precisely how a single variable affects tree growth. However, a great deal of information is available concerning tree growth, and the effect of temperature can be specified reasonably well. The work of Fritts (1966) is particularly notable in this respect.

Trees require a fairly general supply of moisture; a season of prolonged drought is unfavorable to growth. Monsoon forests survive droughts of 4 or 5 months because of adequate storage of water from the rainy season. Climates that are dry or have prolonged drought with short periods of good rain are conducive to grasslands. Continental interiors often have good summer rains and extended winter drought and thus are suitable for grass growth but unsuitable for trees.

As a general rule, the vegetation type of a region (e.g., forest, grassland, or desert) is determined primarily by moisture, whereas the flora (e.g., taiga, savanna, or steppe) is determined by temperature. There are, of course, exceptions to this general rule; for example, the poleward extension of forest growth is limited largely by temperature, and edaphic conditions may overrule climatic factors and cause forest instead of grassland or vice versa.

A tree does not have a single climate or a single temperature about it, but a range of temperatures, from the root zone to the soil surface to the air near the ground to the air in the crown. Air temperature in the crown varies with height and compass direction around the tree. Different parts of a tree respond to these various temperature regimes at any given time in a variety of ways. Without getting too involved in the complexities of the problem, it is possible to describe in general the effect of temperature on tree growth. A good review of the influence of environment on tree growth is given by Gaertner (1964).

Usually, cold temperatures produce a more definitive limit to tree growth than warm temperatures because of the phase transition at the freezing point of water. The usual binomial curve showing the relationship between growth and temperature indicates increasing growth with increasing temperature up to an optimum and then a rapid decrease of growth at high temperatures. The low-temperature limits for growth of various species tend to be much more well defined than the high-temperature limits. A distinction must be made between the effect of temperature on growth and the temperature limits for survival of a species. Temperature affects the ability of tree roots to absorb moisture. In general, very little water movement occurs within roots when the soil temperature is below 0°C. Norway pine and Scots pine in Finland had radial growth affected by temperatures during the growing season, but height growth affected by temperatures of the previous season. In spruce, early-summer temperatures were most important for growth; for pine, late-summer temperatures were more important. The growth of trees varies a great deal from species to species and depends upon the individual physiological requirements of the species. Growth often stops before temperature becomes limiting during the summer and seems to be determined by factors other than temperature.

During the 1930s in Nova Scotia, it was noticed that yellow and white birch stands were losing vigor and considerable dieback was occurring. No organism could be found that was responsible for this loss, and as a result, investigators began a meticulous search for the cause. Because rootlet mortality was reported to be the first symptom of dieback, it was thought that temperature changes in the soil might be the cause. This suspicion was supported by the fact that during a 30-year period prior to 1950, the average summer temperatures in Nova Scotia had risen from 1.0° to 1.4°C. Research reported by Redmond (1955) showed that temperature increases in the soil upset the population balance of fungi-inhabiting roots and the rhizosphere and disturbed the development of mycorrhizae. The presence of mycorrhizae on rootlets is responsible for nitrogen fixation and is therefore essential for the health of the plant. Small increases in soil temperature above those normally occurring in the soil result in greatly increased mortality of rootlets of yellow birch. An increase of 2°C after 100 days resulted in an increase in mortality from 6% to about 60%; an increase of 7°C resulted in 100% mortality. The mortalities for yellow birch and sugar maple are shown in Fig. 15.8 as a function of increase in soil temperature over normal values. Normal soil temperatures were between about 10° and 17°C from June to September at maximum depth for yellow birch rootlets. When these normal soil temperatures were increased by the amounts shown in Fig. 15.8, mortality increased as shown. This demonstrates that sustained increases in the average soil temperature, in this case for 100 days, resulted in a dramatic increase in mortality for yellow birch. In the case of sugar maple, increased mortality was not significant until the average soil temperature was increased by more than 5°C.

The dieback of birch trees with increase in soil temperatures is confirmed by other, less quantitative observations. In northern Michigan during the 1920s, it was observed that when surveyors and others cut back the forest and exposed birch trees to full sunshine, the birch trees

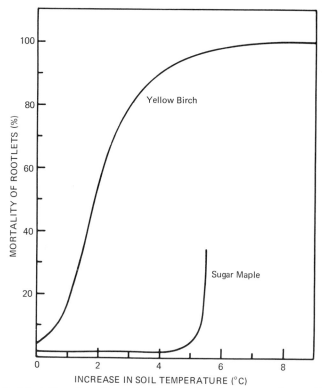

Figure 15.8. Mortality of rootlets of yellow birch and sugar maple as a function of the increase in soil temperature over normal values, which are 10° to 17°C during June to September at maximum depth for rootlets. (Redrawn from Redmond, 1955.)

died within a year or two. This pattern was repeated again and again, but the association between dieback of the birch trees and degree of exposure was not generally recognized. Gates (1930), however, did report his observations of this phenomena.

Some investigators consider the evidence showing that low temperatures limit the northward extension of tree species to be insufficient. On the other hand, it is one thing to plant trees north of their natural range and have them survive and quite another to have them survive and propagate in regions colder than those within their normal range. Usually, tree seedlings will be particularly vulnerable to cold kill. A great mass of data does exist, according to Parker (1963), showing that the range of forest trees is affected directly by cold temperatures. In temperate climates, cold damage to trees occurs most frequently in autumn and spring, but spring frosts are usually most damaging. There are always exceptions in nature because of the vagaries of climate. An extensive warm period during winter may cause buds to burst, leaves to emerge, and other physiological changes to occur in plants. If this is followed by intense cold, then killing of buds, leaves, flowers, and other plant parts may occur. Artificial heat applied to a localized region can produce precisely such effects on plants. Late-spring freezes are a particular hazard to trees. Some investigators have held that late-spring freezes limit the northward extension of many native trees in Europe. Damaging spring frosts are apparently quite common in Europe. Examples exist in all parts of the world of freezing temperatures producing varying degrees of plant damage. Under certain conditions, and for certain parts of particular species, damage may occur when temperatures are only $-2°$ or $-3°C$, but in other cases, damage will not occur until temperatures reach $-10°C$ or below. Much depends upon the degree of cold hardening that has occurred and whether conditions have been such that some of the cold hardening is lost after having been acquired. In Texas, during a cold wave following a warm period in winter, temperatures of $-4°C$ killed various exotic conifers that previously had withstood winter temperatures of $-9°C$. Winter damage to longleaf pine in Virginia is reported and to *Acacia, Grevilla,* and *Eucalyptus* trees in southern California. Exotic palms are particularly sensitive to winter temperatures and are good indicators of the minimum temperatures reached in a region. Date palms cannot withstand temperatures below about $-6°C$ for any length of time and cease growth when daily temperatures are below 9°C for several days.

The phenomena of "red belt" formation in conifers, which is a browning of the needles, seems to relate to very cold temperatures (say, $-31°C$) preceded by warm temperatures (7°C). This is a common phenomena in mountain regions where *Foehn* or chinook winds occur, producing a sudden warming followed by great cold. Intensive sunshine also seems to have much to do with "red belt" formation since conifers seem to develop alot of anthocyanin under these conditions. Often, a belt or zone showing the "red belt" phenomena will develop on the side of a mountain and run at constant elevation for several miles. Often, such phenomena are the result of extreme desiccation in the conifer needles owing to the simultaneous extremes of temperature, dry air, and sunshine. In mountain regions with snow cover and clear skies, the intensity of direct and reflected solar radiation can be very great.

Tree limits are determined by a combination of factors including desiccation, a short growing season and lack of photosynthesis, lack of snow and exposure to very cold winter temperatures, lingering of snow in the summer around small trees, and abrasion and desiccation by strong persistent winds. Tree limits occur not only in mountains but along streams of grassland and desert regions, as well as along the edge of tundra in the far north. The tree limits between watercourses and grasslands are almost certainly a moisture effect, but those along the edge of the arctic tundra are strongly correlated with isotherms. Bryson (1966) has shown that a relationship exists between the northern border of the boreal forest and the

summer position of the Arctic front through northern Canada. Furthermore, Bryson suggests that during the past 4000 years, the northern
boundary of trees has systematically moved northward with movement of
the Arctic front. Tree limits on mountains often correlate well with certain
isotherms, such as the 10°C means isotherm for July.

Grasslands

The grassland ecosystems of the world are more influenced by moisture
than by temperature. The distribution of grasses from warmer to cooler
regions is largely determined by winter-killing of the more temperature-
sensitive species. Temperatures during the growing season are usually
adequate for germination and other physiological processes. Grasslands
of the arctic and alpine tundras will be treated separately since they have
characteristics which make them a special case.

Grasses, like all plants, have certain preferred temperatures for
growth, flowering, seed set, germination, etc. Grasses are specifically
adapted for storage of nutrients and products of photosynthesis in an extensive root system. Grassland soils are characterized by a deep humus
zone. The humus zone results from the fact that the aerial parts of all
grasses die off annually; in addition, the extensive root systems of perennial grasses decay each year, and all parts of the plant must be replaced
each growing season in the case of annual grasses. Chernozem soils are
most characteristic of grassland regions. Leaching in these soils is low
compared to the more intensive leaching in the podzol soils characteristic
of forests. The moisture and temperature conditions of the soil regime are
particularly critical to the growth of grasses.

Enormous temperature fluctuations occur diurnally in the soil and air
near the ground surface. Plant parts, insects, microorganisms, and the
various animals that inhabit this zone must withstand these temperature
fluctuations in order to survive. Exposed dark soil may reach temperatures as high as 60°C when fully sunlit. Maximum and minimum temperatures attenuate very rapidly with depth in the soil. Short time variations of
a few minutes duration at the surface are not noticeable below a few centimeters depth. Diurnal variations penetrate to about 15 cm, whereas only
the annual cycle of variation is felt at depths up to 1 m. This means that
the soil temperatures at depths of 20–40 cm or greater are extremely
steady from day to day. A heat source imposed on this regime at the
ground surface will be felt at these depths if sustained for several days.
Generally, it is not soil temperature that determines the distribution of
grasses, but soil moisture. Shreve (1924), studying soil temperatures and
moisture of different slopes, concluded that vegetational differences on
opposing north and south slopes were determined by soil moisture rather

than soil temperature. Snow cover is an effective insulator over the soil surface. There is often no freezing of soil when the surface is covered with about 1 m of snow, even when air temperatures plunge to $-20°C$ or below. Freezing of the upper 7 to 10 cm of soil occurs at very low temperatures when there is about 1 m of snow, but freezing goes to 0.46 m or more when snow depths are less than 0.3 m. Without snow cover, soils in northern United States where air temperatures are below $-20°C$ may freeze to depths of 1 m during extended cold periods of many weeks.

Germination of various grasses varies greatly with temperature. *Poa pratensis* germinates only with warm days (26°C) and warm nights (20°C). *Poa scabrella* and *Poa bulbosa* require cool days around 10°C and night temperatures of between 3° and 10°C. Sudan grass, *Sorghum vulgare* var. *sudanese,* germinates best when temperatures are 30° to 38°C. Desert summer ephemerals germinate best when temperatures are 27° to 33°C, and desert winter ephemerals require temperatures of 15° to 18°C. In most instances, the right combination of moisture and temperature is required for germination. Went (1949) discusses the temperature influence on germination and growth of grasses. He found that two grama grasses had different optimum germination temperature requirements; *Bouteloua barbata* and *Bouteloua aristidoides* both required 27°C days, but *B. aristoides* needed warmer nights than *B. barbata.*

Various species of grasses have different temperatures at which growth is initiated, reaches a maximum rate, and ceases. Kentucky bluegrass, *Poa pratensis,* grows best during cool periods and does poorly during very warm conditions. With Kentucky bluegrass, top-growth recovery following cutting was slow at 15°C in the absence of soil nitrogen. When nitrogen was applied, rhizomes turned and pushed through the soil surface. At 27°C, without nitrogen, no new roots were produced, but some of the rhizomes appeared aboveground. However, with nitrogen at this temperature, rhizomes and roots died. At 38°C, there was very little top growth, and all rhizomes and roots died irrespective of the amount of nitrogen available.

Laude (1953) showed that the growth of many perennial grasses is correlated with low temperatures rather than the availability of moisture. Laude showed that 7 out of 20 species of grasses tested were dormant during the summer. Some grasses, such as *Oryzopsis milacea, Stipa cernua,* and *Phalaris tuberosa,* which are dormant during summer, resume growth in the autumn when temperatures are cooler. Photoperiod, as well as soil moisture, may also play a role in some cases but temperature is the dominant factor in other cases.

At certain stages in the development of a cereal, a small increase in temperature can result in complete growth arrest, and even death. For example, when the second leaf begins to emerge in Marquis wheat, growth continues up to a temperature of 32.5°C, but at 33°C, chlorophyll production stops and the plants die. For wheat, an increase in temperature over the range 10° to 30°C resulted in narrower and thinner leaves. The length and area of leaves increased with increasing temperature from 10° to 20°C and declined at temperatures of 30°C. The optimal temperature for root growth is about 15°C. The optimum for root growth of oats, a temperate cereal, seems to be about 20°C; in comparison, maize, a tropical grain, has an optimum at 30°C. In wheat, vegetative growth is greatest at about 25°C for a 12-hr day. Growth of oats is optimal over a wider range than wheat, the temperature optima being 20° to 30°C in the day and 25° to 30°C at night. Maize growth is optimal at day temperatures as high as 35°C (in fact, there is some evidence of optimum up to 40°C) and night temperatures of 25°C. Corn does grow very actively, undergoing cell expansion, at night, particularly during very warm nights.

Arctic and Alpine Ecosystems

Arctic regions are characterized by extreme habitats with very cold and long winters, summers generally low in temperature and fairly short in duration, strong winds, long photoperiod, low light intensities, low nitrogen supplies, low precipitation, and relatively simple populations of plants and animals. Those organisms that occupy the Arctic must be physiologically adapted to cold temperatures. Temperate plants and animals must also sustain much winter cold, but during the summer months, arctic plants and animals must metabolize even when the temperature remains quite low. The activity of most animals is reduced in the Arctic, and as a result, pollinating insects are less active and certain plants are not easily cross-pollinated. However, bumblebees, Lepidoptera, and certain Diptera are sufficiently active to cross-pollinate many plants. Cross-pollination seems to occur in some plants that also undergo self-pollination.

An arctic site will have a flora of vascular plants which correlates well with the mean July temperature of the site. Summer temperatures are much more significant in regulating the flora of an arctic site than mean annual temperatures. All parts of arctic plants can survive freezing temperatures at any season of the year. Whereas plants of temperate zones have roots which are sensitive to cold, arctic plants have roots which survive freezing with ease. In temperate regions, there is usually adequate snow cover during the colder winter periods to afford protection to root systems, but adequate snow cover is often absent in arctic regions.

During sunny periods, the ground surface often gets very warm, and the basal rosette, cushion forms, and dense mats of many arctic plants facilitate the capture of this energy, allowing some metabolic activity to occur. Often, the presence of red anthocyanin pigment in these plants further enhances the warming process through the capture of sunlight. Many arctic plants have red anthocyanin mixed in with the green pigments of leaves to become effectively almost black absorbers of sunlight. Savile (1972) reviews work on arctic plants in which a temperature of 3.5°C was recorded in a clump of saxifrage in northern Greenland when the air temperature was −12°C and the temperature in a dark clump of moss was 10°C. Deeply pigmented plants often absorb sufficient sunlight through the snow in the spring to begin assimilation while still essentially buried. The snow becomes macrocrystalline and forms cavities surrounding the plants. Sunlight penetrates the snow, is absorbed, and warms the plant.

Krog (1955) showed that willow catkins with transparent hairs reached notably higher temperatures in the sun than those catkins which had either the hairs removed or painted black. In fact, when the air temperature was about 0°C, the catkins of *Salix polaris* were 15° to 25°C above air temperature in full sunshine. The semitransparent catkins allow sunlight to penetrate the hairs and be absorbed in the almost black scales surrounding the flower buds. Since the heat is absorbed within the catkin, the hairs offer resistance to reradiation and heat loss through convection. This same mechanism for heating may occur in the fur coats of certain animals. In other instances, black-colored animals absorb sunlight near the outermost surface of the fur or (in certain carabid beetles) the elytra and thereby reradiate heat effectively. This could be an important strategy in hot climates. Wilson (1957) reports extensive measurements of arctic plant temperatures. Many arctic plants are active a little below 0°C and often need to function at a few degrees above zero. Sunshine plays a dominant role in leaf temperatures of arctic plants, but temperature differences between leaf and air are about half those in the temperate zone because the sun is low in the sky. Since wind is a characteristic feature of the arctic environment, sheltering of plants from the wind may make a difference in leaf temperature of several degrees, particularly in sunshine.

Arctic plants have their photosynthetic activity geared to low temperatures. Mooney and Billings (1961) compared the photosynthetic activity of arctic and alpine plant populations. They found that arctic plants have a higher photosynthetic rate, with an optimum at lower temperatures, than alpine populations of the midlatitudes. Mooney and Johnson (1965) compared the physiology of an alpine population of *Thalictrum alpinum* from a latitude of 37°31′ in California with that of an arctic population from 68°06′ in Alaska and found the peak photosynthetic rate of the arctic plants to be 5°C lower than that of the alpine plants. Optimum temperatures for photosynthesis in arctic and alpine plants are generally from 12° to 20°C. An example is illustrated in Fig. 15.4.

Alpine environments of the midlatitudes generally have among the greatest diurnal temperature fluctuations of any ecosystems. Furthermore, these environments are characterized by intense amounts of sunshine during the growing season and considerable amounts during the

winter. The energy environment of the alpine tundra and plant temperatures is described by Gates and Janke (1965). During the summer, daytime
temperatures seldom exceed 20°C, but the summer night may have temperatures of −5°C or below. There are frequent transitions across the
freezing point of water. All alpine vegetation is low in height; there are numerous cushion plants, grasses, forbs, and herbs of short stature, and
Krumholtz (the prostrate forms of evergreens) in the vicinity of timberline. Study of the alpine vegetation reveals that the plants are responding
to the wind profile. Persistent winds are generally characteristic of alpine
regions. Calm days are rare. Where there is protection from the wind, in
the lee of rocks or in crevices or depressions, the vegetation grows taller
than elsewhere. Many alpine soils are dark and strongly absorb the incident solar radiation. Soil surface temperatures may often exceed air temperatures by 40°C. In fact, the very high soil surface temperatures cause
high mortality among plant seedlings during warm days with strong solar
irradiation. On the other hand, the cushion plants benefit from the warmth
of solar heating. Temperatures of alpine plants have been further reported
by Salisbury and Spomer (1964) and Spomer and Salisbury (1968). When
skies were cloudy, the temperatures of alpine plants were close to air temperature, but when the sun shone, leaf temperatures were as much as
22°C above air temperature. *Geum turbinatum,* alpine avens, is of great
abundance in alpine habitats but seldom occurs below timberline. Its distribution in the alpine environment may be determined largely by soil temperatures, since the plants seem to grow best where soil temperatures at a
depth of 15 cm do not exceed 15°C. Germination, sprouting, vegetative
development, and dormancy induction are directly regulated by soil temperatures. Impairment of normal functioning owing to supraoptimal soil
temperatures makes other factors such as shading, mineral availability,
etc., more critical to the distribution of these *Geums.* Spomer and Salisbury suggest that the abrupt increase in soil temperatures in subalpine
forests relative to alpine areas, plus shading by the canopy, accounts for
the general lack of *Geums* in the subalpine forest.

The problem of what causes timberlines to exist where they do has
been addressed by many ecologists. Timberlines have an apparent coincidence with a monthly mean air isotherm of 10°C. As one moves up a
mountain slope, very few environmental factors change abruptly with
changes in altitude. Solar radiation does not increase abruptly as one
moves up the slope. Air temperatures do not change abruptly, but rather
gradually. Precipitation does not change suddenly, although snow accumulation may shift with the topography. The one thing which does change
abruptly with position on the slope is the wind speed. Wind speed can
make a substantial difference in the degree to which trees of the high
mountains are coupled to air temperature. Cold air at high elevations will
cool the trees warmed by the intense radiation of the high mountains; thus

degree of cooling is determined by the wind. When the wind changes abruptly with topography, the degree of convective cooling also changes, and much heat is taken from the trees, heat which is not then available for the thermal chemistry of photosynthesis and respiration. Wind also produces moisture stress on trees. I believe, therefore, that wind is the primary determinant of the location of timberline. However, snow accumulation and soil moisture content are determined by the air flow over the mountain ridges and will also influence tree growth. Above timberline, winds blow persistently, and the alpine plants must be prostrate, remaining near the ground surface where the wind strength is much reduced and radiation warms the soil. Microclimate permits growth of grasses, sedges, and herbs adequate for their survival. The alpine environment is harsh, but vegetation can sustain itself even though growth is often extremely slow. The influence of wind on plants is thoroughly discussed by Grace (1977).

Temperature Effects on Insects

Introduction

Generally, the temperature of an insect is close to air temperature or the environmental temperature. Insects, being very small, have relatively large convection coefficients. Therefore, their body temperatures are affected strongly by air temperature and rate of air movement. Insects in sunshine can achieve body temperatures of near air temperature to as much as 20°C above air temperature, depending upon their size, shape, color, and amount of hair. Convective heat loss is reduced substantially by thick coats of hair on some insects. Insects exposed to a radiation intensity of 1047 Wm^{-2} at a wind speed of 0.5 m s^{-1} by Digby (1955) exhibited the body temperatures, relative to air temperature, shown in Fig. 15.9 as a function of time of exposure. The shapes of these curves are exactly what one would expect for the heating curve of any inanimate object, and the rate of temperature increase or the time to reach equilibrium is directly related to the time constant for the insect. A very small insect such as *Drosophila melanogaster* is very close to the air temperature because of strong convective cooling. Larger insects can have body temperatures considerably above air temperature. Extremely pubescent insects such as *Bombus terrestis* have very high body temperatures when exposed to high levels of radiation because convective cooling is extremely ineffective in lowering body temperature. Certainly, body color plays a very important role in the temperatures assumed by insects in sunshine. Black locusts possessed body temperatures 3° to 3.5°C higher than green-colored locusts. Examples of these kinds of temperature effects are numerous.

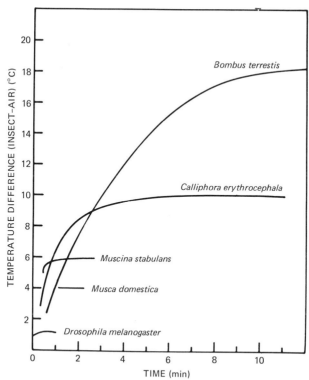

Figure 15.9. Difference between insect body temperature and air temperature as a function of time for insects in air of 20°C moving at 0.5 m s⁻¹ when exposed to radiation of 1047 Wm⁻². (Redrawn from Digby, 1955.)

Temperature, of course, never acts alone. Changes in insect physiology, morphology, reproduction, movement, feeding rate, and so on, depend upon a set of environmental factors including humidity, wind, light intensity, and concentrations of oxygen and carbon dioxide, as well as past history. When dealing with whole ecosystems, we must concern ourselves not only with single insect species but also with insect–insect interactions, insect–other animal interactions, and insect–plant interactions. Physiological requirements, morphological changes, behavior, etc., are different for each species. These factors interact in many subtle and complex ways to modify any single environmental factor such as temperature.

The volume of literature concerning temperature effects on insects is enormous. The work of Uvarov (1931) is well known, and the chapter by Clarke on insects and temperature in Rose (1967) gives most of the pertinent general references since 1931. Much other information is found in Andrewartha and Birch (1954).

Moisture Loss

Insects generally can do relatively little cooling by the mechanism of moisture loss, primarily because they have a limited water supply by virtue of such small size. Insect integuments are of very low permeability to water, most water loss occurring through the tracheae. Observations by Beament (1959) concerning the permeability of arthropod cuticle showed a dramatic increase of permeability at specific temperatures for various species. The results of Beament's studies are illustrated in Fig. 15.10. The sudden increase in permeability is caused by the loss of orientation of the wax molecules in the thin layer of cuticular wax that maintains water within the insect. Beament found that the temperature at which permeability suddenly increases is a function of the habitat conditions in which the insect lives. (A similar sudden breakdown in permeability of leaf cuticle was observed by Gates et al. (1964) for *Mimulus,* monkeyflower, a genus in the Scrophulariaceae.) Certain insects had an increase in cuticle

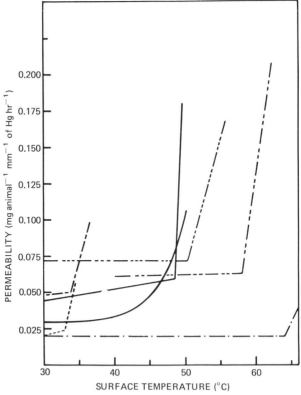

Figure 15.10. Permeability of arthropod cuticle to water as a function of insect surface temperature: (—) *Shistocera gregaria;* (– – –) *Ixodes ricinus;* (–––) *Nemotus larva;* (– – – –) *Rhodnius prolixis* nymph; (–·–) *Ornithodorus* (– – – – –) *Tenebrio molitor.* (Redrawn from Beament, 1959.)

permeability at temperatures as low as 33°C, and others as high as 64°C, but for most it was between 40° and 50°C.

Rapid loss of water, other than momentary, is usually fatal to an insect. This is one of the potential effects of increased heat in an already-warm environment, i.e., pushing of temperatures above the thresholds such as those below which integument permeability is low and stable.

Growth

Growth of insects is affected by temperature such that the rate of growth increases with temperature until an optimum temperature is reached, above which rates of growth fall off rapidly with increasing temperature. The body size achieved by an insect varies with temperature. Usually, larger body size is associated with growth at lower rather than higher temperatures. Sometimes, an insect achieves greater ultimate size when grown in a fluctuating-temperature environment than when grown under constant conditions. For some insects, growth depends upon symbiosis with a microorganism upon which the insect depends for some nutrient. The microorganism may be particularly sensitive to temperature, and its removal by high temperature may inhibit growth of the insect. For example, the cockroach depends upon bacteria which are destroyed when the insect is kept at 37° to 39°C, and growth is impaired. Similar findings have been reported for termites.

Mutation Rates

Mutation rates in the genes of insects may vary considerably with temperature. Larvae of *Drosophila* kept at −6°C for 25 to 40 min showed a tripling of the rate of production of lethal mutants on the X and 2nd chromosomes, according to Birkina (1938). The frequency with which the Y chromosome was lost in *Drosophila* males was considerably greater at 26°C than at 18°C. Many very subtle genetic and cytological changes occur in insects when temperatures are varied. The ratio of nuclear DNA to cytoplasmic RNA may change considerably with temperature, affecting cell growth and cell division. With an increase of temperature above 20°C, there is, for some insects, a decrease in cell number and an increase in cell size.

Activity

Insects generally have two states of activity: a high-energy, or active state, which is open and unsteady, and a low-energy, or inactive, state, which is closed and steady. It is the latter state that the insect uses to survive extreme conditions. The inactive state is known as diapause and is achieved only slowly. Any number of factors in the environment can be a warning of impending adverse conditions and a signal for an insect to switch into diapause. Temperature, photoperiod, moisture, or other factors can induce diapause. To survive, a species of insect must endure the conditions to which it is subject at any stage of development. If an insect species achieves high population densities, then conditions must have been favorable at most stages of its life cycle.

All animals utilize adenosine triphosphate (ATP) as energy currency for energizing muscle tissue. Energy is released from ATP by the enzyme ATPase. The activity of this enzyme is very temperature dependent. The Q_{10} of this enzyme from the cockroach *Periplaneta americana* has a value of 4.0, and values for *Locusta migratoria* are similar. In fact, the response of this enzyme to temperature probably accounts for the observed relationship between muscular activity and temperature.

The oxygen consumption, or metabolic rate, of insects increases with temperature, as shown in Fig. 15.11 for *Pollenia rudis*. This curve is averaged from the data of Argo (1939). Once again, one is impressed with the abrupt increase of rate of oxygen consumption over a narrow range of temperatures. For some insects, a plateau is reached for a particular temperature range, over which metabolic rate is insensitive to changes in temperature. The rate of development of insect eggs usually increases with temperature up to some optimum and drops off very rapidly at higher temperatures, as shown in Fig. 15.12 for eggs of *Noctua pronuba*.

Many insects have the ability to overwinter and undergo supercooling. Insects produce sorbitol or glycerol from glycogen in their blood plasma. Many have concentrations as high as 5 M at very low temperatures, in the vicinity of $-48°C$. The ant *Camponotus pennsylvanicus pennsylvanicus* had no glycerol in its tissue at 20° to 25°C but when slowly cooled to 0°–5°C during a period of 6 days, had a concentration of 10% glycerol. Slow warming of the ant caused the glycerol to disappear.

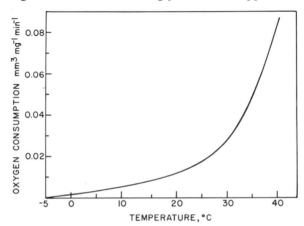

Figure 15.11. Metabolic rate of *Pollenia rudis*, measured by oxygen consumption, as a function of temperature. (Redrawn from Argo, 1939.)

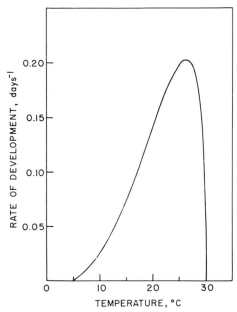

Figure 15.12. Rate of development of the eggs of *Noctua pronuba* as a function of temperature.

Most insects are able to adjust when reared under warmer or cooler conditions than normal, but there are always some species which show relatively little ability to adjust. For example, the mosquito *Anopheles minimus* did not extend its lethal limit when reared at 30° to 35°C. Extension of tolerance to cold always seems to be greater than that of tolerance to heat.

Generally, 50°C seems to be the maximum temperature for survival by most insects, but there is enormous variation, and some insects can withstand substantially higher temperatures. The upper temperature limits for insects are well documented by Uvarov (1931), who presents a superb comprehensive review of the literature up to that time concerning insects and climate. Fleas cannot withstand temperatures above 32°C, and in this respect are one of the least heat resistant of all insects. The tsetse fly, which transmits sleeping sickness to cattle and human, cannot withstand high temperatures and therefore remains among the cool, green leaves of the forest until cattle come nearby. Then it darts out and alights on an animal, feeding and infecting. It then returns to the cool leaves of the forest. The tsetse fly cannot remain long on the hot, sunlit surface of the cattle, but it can bite and do its damage quickly.

The codling moth does serious damage to apples. Accidently introduced into North America from Europe, it has a warm optimum temperature. In New England, the warmest part of the summer is sufficiently short that the codling moth has only one generation per year. However, in the Shenandoah Valley of Virginia, it has two or more generations each year, and further south, in an area that includes the Ozarks, it may have three broods during the warm season and cause great damage to apple orchards.

The temperature below which larvae of the Mediterranean fly remain active is 14°C; that for larvae of the Hawaiian fly is 16°C. The rice-stem borer of Japan is active when temperatures are between 24° and 26°C but also requires a July rainfall between 150 and 250 mm. If it is too dry or too wet, the rice-stem borer will not be active, even at optimum temperatures. The rate at which crickets chirp is a function of the temperature, although it also varies with the species of cricket. As a general rule, if the temperature is between 45° and 80°F, then, by adding 37 to the number of chirps counted in 15 s, you get the approximate air temperature in degrees Fahrenheit.

Edwards (1960) carried out a statistical analysis of records kept in Saskatchewan over a long period of time relating to observations of grasshopper abundance and mean monthly temperatures from July through September in the preceding 3 years. Most grasshopper outbreaks seem to follow periods of hot weather.

The northern limit (in the Northern Hemisphere) of the distribution of an insect species is often determined by the lowest temperature reached in the habitat. *Heliothus armigera,* an eastern North American moth, is a serious pest of corn and cotton in southern states. In New York state, most pupae are killed by cold, and in Canada all pupae are killed every winter. The population of this moth in Canada is maintained only by reinvasion each summer. The cutworm *Euxoa segetum,* a pest in the Lower Volga, has large populations only when there is substantial snowfall for several winters in sequence. The snow blanket keeps temperatures in the soil, where the pupae overwinter, much warmer than they would be otherwise. Unusually cold winters with little snow cut back the populations of many insect species which, although cold hardened, cannot withstand the very coldest temperatures encountered and still have a high percent survival.

European Pine-Shoot Moth

One of the most detailed studies concerning the effect of climate on an insect is the investigation by G. W. Green (1968) of the European pine-shoot moth. A description of the behavior of this species will serve as an example showing the effect of temperature on various stages of insect development.

The European pine-shoot moth was introduced to North America from Europe about 55 years ago. This pest inflicts heavy damage on stands of red and Scots pines in the United States and Canada and has recently spread to stands of pondersoa pine in Washington and Oregon. Adults emerge in early summer and lay their eggs on pine shoots. The eggs require 2 weeks to hatch. The first- and second-instar larvae feed on needle tissue. The third-instar larvae feed on buds in August and by early September have bored in for the winter. The third-instar larvae, after becoming cold hardened over the winter in the buds of the pine shoots, eventually molt, begin feeding, and emerge in the spring. The larvae pupate in late May or early June within hollowed-out shoots or buds, and between mid-June and mid-July the adults emerge to begin the life cycle once again. Let us now review the effects of temperature on each stage of this life cycle.

The flight of the adult insects about and just after the summer solstice is necessary for mating, oviposition, and dispersal. The female adult releases a potent pheromone sex attractant on the day or evening of emergence and mates at this time. The flight activity of adults is a function of light intensity and air temperature, as shown in Fig. 15.13 (lower left). Optimum conditions for flight are temperatures of 22°C and a light intensity of 125 fc. Flight does not occur at temperatures below 12°C or above 28°C. Such a low light intensity of course indicates that flights occur at dawn or dusk, and the optimum temperature is likely to occur in the evening hours rather than during the early morning. The eggs of the European pine-shoot moth are laid on the bark of shoots near the bases of pine needles. The incubation time shown in Fig. 15.13(left center) decreases with increasing temperature up to 30°C and then increases with higher temperature. The success percentage of hatching is nearly 100% in moist air for temperatures from 15° to 30°C and then decreases abruptly with higher temperatures, as shown in Fig. 15.13(left center). In dry air, optimum hatching success was achieved at 18°C, with desiccation causing a falling off at higher temperatures. An abrupt decrease in success also occurs at temperatures above 30°C.

Newly hatched larvae are light and temperature sensitive. Rate of movement increases with increasing temperature, up to about 32°C. With prolonged exposure at this temperature, movement becomes disoriented. Whereas the larvae are strongly attracted to a light source at temperatures below 30°C, they turn away from the light and behave photonegatively at temperatures exceeding 34°C. The young larvae crawl along the shoots and needles of the pine branches until they locate the needle sheath. They then spin a silken web and bore into the sheath, at which time they begin feeding. The time from hatching to feeding is a function of temperature and is shortest when temperatures are 30° or 31°C, as shown in Fig. 15.13 (upper left). If temperatures are about 34°C or higher, the larvae become overheated and quickly go to the shaded underside of the twig or drop on silk to a more shaded site. Then search for a feeding site begins again. Warm temperature very decidedly extends the time between hatching and feeding.

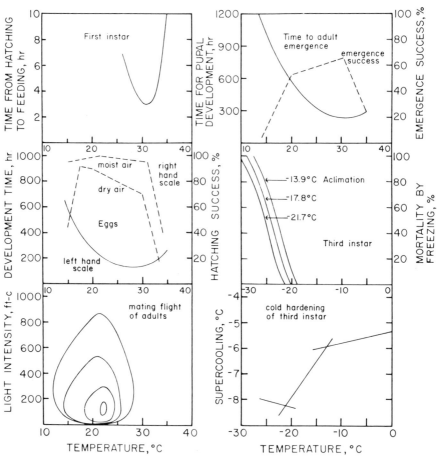

Figure 15.13. Temperature effects on various stages of the life cycle of the European pine-shoot moth. (Data from Green, 1968.)

The feeding larvae destroy the base of needles. They then molt and move on to another sheath to destroy more needle bases. Finally, during the third instar, they bore into the buds and prepare to overwinter. During September, cold hardening of larvae occurs at any temperature above −14°C. At temperatures between −22° and −14°C, however, active cold hardening occurs, during which time the mean supercooling points of larvae are depressed about 2°C and reach temperatures as low as −26°C. The mean supercooling points of larvae are shown in Fig. 15.13 (lower right) as a function of the hardening temperature. The rate at which hardening occurs is much more rapid at low than at high temperatures. Larvae collected in the field were found to be cold hardened to −20°C in September, −21°C in November, −24.4°C in January, and −26.5°C in March. A mortality curve for overwintering larvae in relation to temperature and conditions of hardening is shown in Fig. 15.3(right center).

Larvae spending the winter in buds of the red and Scots pine, in order to survive, must not encounter temperatures below that to which they are cold hardened. As a general rule, the shoot moth does not occur beyond the $-30°C$ minimum winter isotherm, except on small trees which are covered by snow. The insulating effect of snow is very good, and temperatures of branches within the snow are much higher than that of the air above. The temperature difference between pine shoot and air can be as much as $-42°C$ depending on the depth of snow cover.

In the late spring, the larvae emerge from dormancy and begin feeding again. Pupation occurs in late May or early June inside the shoots and buds. In Fig. 15.13(upper right), one sees the developmental time of pupae and the emergence success of adults in relation to constant environmental temperature. Since the environmental temperature is the temperature of the shoot or bud, Green investigated the shoot or bud temperature in relation to air temperature. He found shoot and bud temperatures to be 4° to 6°C above air temperature on clear, sunny days and about 3.5°C above air temperature on cloudy days. The time to complete pupal development is minimal at 30°C, and the emergence success of adults from pupae peaks at about 30°C. The final investigation in Green's life-cycle study was to learn how fecundity of virgin females varies with temperature. He was able to show that egg production increased with temperature between 17° and 28°C and decreased at temperatures above 28°C. It is clear that temperatures which favor rapid pupal development also favor egg production, with a 2°C difference in optimum temperatures.

Clearly, the intensity of infestation by this insect depends upon the seasonal sequence of temperatures at each stage of the life cycle. Persistently undesirable temperatures at any time during the annual cycle can reduce an insect population or limit its growth. Most insects have temperature responses during each period of the life cycle similar to those of the European shoot moth, but of course differ in details such as optimum temperatures and the exact nature of the temperature response.

Development

The rate of development of insects is strongly affected by temperature, as in the cases of *Noctua pronuba* and the European pine-shoot moth. Morphological changes in insects are also known to occur with temperature. These effects are many and varied. Temperature affects the length of time for wing development, wing venation, and in some instances even the shape of wings. In bees, lower temperatures produce shorter wings. The chitinous structures of an insect change with temperature, and the legs, antennae, palpi, head, thorax, and other parts are affected. The size and number of facets in the compound eye of *Drosophila* vary inversely with temperature. The genital organs of insects change development with temperature, and in certain insects, male parts modify in the direction of female parts, or vice versa, under high temperatures. Coloration of insects changes enormously with temperature, particularly in Lepidoptera. Respiration rates change with temperature, as has been discussed above, and other physiological rates are also affected. As in the example of the European pine-shoot moth, the fertility of an insect is affected by temperature. Rate of movement is strongly influenced by the thermal environment. The life-span of insects is longer at lower temperatures than at higher temperatures. Often, the length of life of an insect is nearly doubled by a temperature reduction of 10°C at moderate and intermediate temperatures. The influence of temperature on insect longevity is discussed by Baumberger (1914).

Behavior

Insect behavior is greatly influenced by temperature. A few insects in particular have been studied in detail for temperature effects because of their economic importance. Clark (1949) made observations of the early-stage nymphs of the plague locust, *Chortoicetes terminifera*, whose preferendum temperature is about 42°C. Andrewartha and Birch (1954) describe Clark's observations as follows:

> Clark studied the behavior of early-stage nymphs of this locust in an outbreak center in New South Wales. The first requirement of large numbers was satisfied in this case; at the time when the eggs had just hatched, the first instar nymphs were estimated to occur in the vicinity of the egg beds in numbers up to 2,000 per square yard in some situations. Notwithstanding the fact that their parents had been gregarious, the first instar nymphs were neither markedly gregarious nor excitable. They soon distributed themselves over the area in relation to the occurrence of green plants suitable for food. Most of them spent the night sheltering in the vegetation or in other places. At daybreak, the temperature in these situations was about 2°C, and the hoppers remained motionless. They became more active as the temperature increased, until at 20°C "disability" due to cold could no longer be detected, and their movements were "normal." Between 20 and 45°C the hoppers sought out and crowded into the warmest possible situations at ground level, taking up what the author calls the "basking formation". They crowded closely together, often with their bodies touching and broadside to the sun. On cool days, basking groups persisted until late in the afternoon, usually breaking up as the shadows lengthened and the temperature fell. On hot days, the basking groups would disperse as the temperature approached 45°C, and the hoppers would seek more shady places; ordinarily the groups would form again later in the afternoon. This behavior was governed by the existence of a preferendum at about 42°C. Outbursts of jumping would occur in the basking groups at irregular intervals, and, as time went on, the innate gregariousness and excitability of the hoppers increased. The first occurrence of gregarious mass migration was observed in the second

instar in situations where the numbers were highest; elsewhere in situations where the numbers were lower, it was not observed until the fourth or fifth instar.

Clark (1949) summarizes his observations as follows:

The change from individualistic to gregarious behavior is a result of increasing responsiveness of hoppers to the presence and movements of others. It is affected by a period of crowding during which mutual contact, both visual and mechanical, becomes probably the most common experience of hoppers. External influences, especially temperature, play an important causal role in crowding. A period of crowding is apparently essential for the progeny of swarms of this species to develop gregarious behaviour, i.e., the capacity for mass migration. This is probably true for other locusts. The length of the necessary period of crowding varies in relation to population density and probably temperature conditions.

Kennedy's (1939) account of the behavior of the desert locust *Schistocerca gregaria* in the Sudan was in general agreement with the results described here for *Chortoicetes*. Kennedy concluded that "aggregation" of *Schistocerca* was the result of a "diurnal regime determined largely by temperature differences and changes involving prolonged basking on the small patches of bare sheltered and sunlit ground among the vegetation. Aggregation results because of the limited amount of bare ground available for basking."

Appendices

Appendix 1

Conversion Factors
Système International (SI) Units and cgs Equivalents

Quantity	SI units	cgs units
Length	1 m	$= 10^2$ cm
Area	1 m^2	$= 10^4$ cm^2
Volume	1 m^3	$= 10^6$ cm^3
Mass	1 kg	$= 10^3$ g
Density	1 kg m^{-3}	$= 10^{-3}$ g cm^{-3}
Force	1 kg m s^{-2} = 1 N (newton)	$= 10^5$ dynes
Pressure	1 kg m^{-1} s^{-2} = 1 N m^{-2}	$= 10$ dynes cm^{-2}
		$= 10^{-2}$ mbar
Work, energy	1 kg m^2 s^{-2} = 1 J (joule)	$= 10^7$ ergs
Power	1 kg m^2 s^{-3} = 1 W (watt)	$= 10^7$ ergs s^{-1}
Heat energy	1 J	$= 0.2388$ cal
Heat or radiation flux	1 W	$= 0.2388$ cal s^{-1}
Heat flux density	1 W m^{-2}	$= 2.388 \times 10^5$ cal cm^{-2} s^{-1}
		$= 1.433 \times 10^{-3}$ cal cm^{-2} min^{-1}
		$= 1.433 \times 10^{-3}$ langley min^{-1}
Latent heat	1 J kg^{-1}	$= 2.388 \times 10^{-4}$ cal g^{-1}
Specific heat	1 J kg^{-1} °C^{-1}	$= 2.388 \times 10^{-4}$ cal g^{-1} °C^{-1}
Thermal conductivity	1 W m^{-1} °C^{-1}	$= 2.388 \times 10^{-3}$ cal cm^{-1} °C^{-1} s^{-1}
Thermal resistance	1 s m^{-1}	$= 10^{-2}$ s cm^{-1}

Appendix 2

Physical Constants

Constant	Value
Speed of light in a vacuum	2.9979×10^8 m s^{-1}
Avogadro's number	6.0225×10^{23} mole^{-1}
Planck's constant	6.6256×10^{-34} J s
Gas constant	8.3143 J mole^{-1} $^\circ$K^{-1}
Stefan–Boltzmann constant	5.6697×10^{-8} W m^{-2} $^\circ$K^{-4}

Appendix 3

Properties of Air, Water Vapor, and Carbon Dioxide

Tempera-ture ($^\circ$C)	Density of (air (kg m^{-3})) Dry	Sat.	Latent heat (J kg^{-1})	Thermal conductivity of air (mW m^{-1} $^\circ$K)	Molecular diffusion coefficients (m^2 s^{-1}) D_v	D_c
0	1.292	1.289	2.501×10^6	24.3	21.2×10^{-6}	12.9×10^{-6}
5	1.269	1.265	2.489×10^6	24.6	22.0×10^{-6}	13.3×10^{-6}
10	1.246	1.240	2.477×10^6	25.0	22.7×10^{-6}	13.8×10^{-6}
15	1.225	1.217	2.465×10^6	25.3	23.4×10^{-6}	14.2×10^{-6}
20	1.204	1.194	2.454×10^6	25.7	24.2×10^{-6}	14.7×10^{-6}
25	1.183	1.169	2.442×10^6	26.0	24.9×10^{-6}	15.1×10^{-6}
30	1.164	1.145	2.430×10^6	26.4	25.7×10^{-6}	15.6×10^{-6}
35	1.146	1.121	2.418×10^6	26.7	26.4×10^{-6}	16.0×10^{-6}
40	1.128	1.096	2.406×10^6	27.0	27.2×10^{-6}	16.5×10^{-6}
45	1.110	1.068	2.394×10^6	27.4	28.0×10^{-6}	17.0×10^{-6}

Appendix 4

These charts give the altitude and azimuth of the sun as a function of the true solar time and the declination of the sun. Each chart is for a separate latitude. To find the solar altitude and azimuth, proceed as follows.
1. Select the chart appropriate to the latitude.
2. Find the solar declination δ corresponding to the date from the following.

Declination	Approximate dates
+23°27′	June 22
+20°	May 21, July 24
+15°	May 1, Aug. 12
+10°	April 16, Aug. 28
+5°	April 3, Sept. 10
0°	March 21, Sept. 23
−5°	March 8, Oct. 6
−10°	Feb. 23, Oct. 20
−15°	Feb. 9, Nov. 3
−20°	Jan. 21, Nov. 22
−23°27′	Dec. 22

3. Determine the true solar time as follows.
    (a) Add 4 minutes to the local standard time for each degree of longitude the observer is east of the standard meridian, or subtract 4 minutes for each degree west of the standard meridian, to get the local mean solar time.
    (b) Add algebraically the equation of time from the following table to the local mean solar time to get the true solar time.

Equation of Time Correction in Minutes

Month	Day of month							
	1	5	9	13	17	21	25	29
January	−3	−5	−6	−8	−9	−11	−12	−13
February	−13	−14	−14	−14	−14	−14	−13	—
March	−12	−11	−11	−10	−9	−8	−6	−5
April	−4	−3	−2	−1	0	+1	+2	+3
May	+2	+3	+3	+3	+4	+4	+3	+3
June	+2	+2	+1	0	−1	−1	−2	−3
July	−3	−4	−5	−6	−6	−6	−6	−6
August	−6	−6	−6	−5	−4	−3	−2	−1
September	0	+1	+2	+4	+5	+7	+8	+9
October	+10	+11	+12	+13	+14	+15	+16	+16
November	+16	+16	+16	+16	+15	+14	+13	+12
December	+11	+10	+8	+6	+4	+2	0	−2

4. Read the altitude and azimuth at the point determined by the declination and true solar time.

Solar Altitude and Azimuth[a]

0°

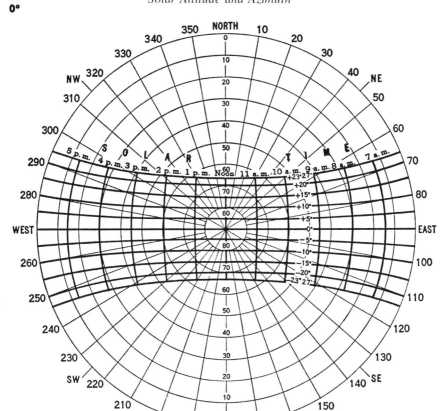

[a] From the Smithsonian meteorological tables, edited by List (1963).

10° N.

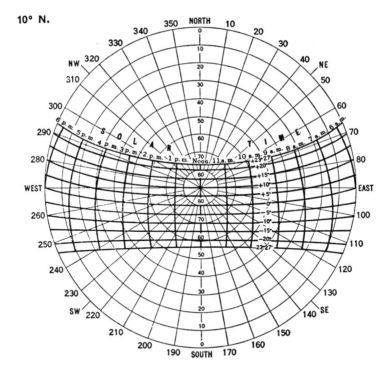

20° N.

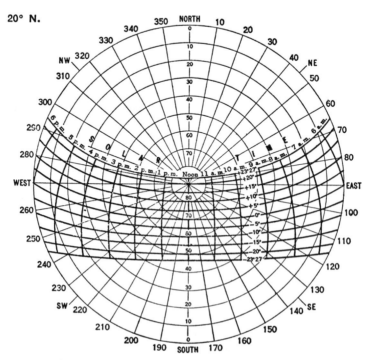

30° N.

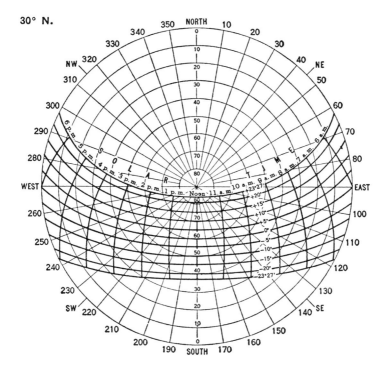

40° N.

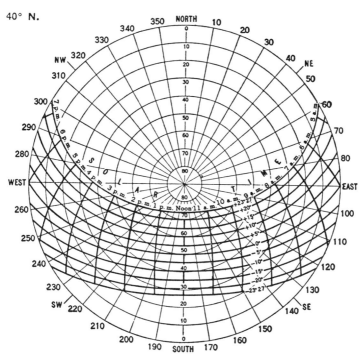

50° N.

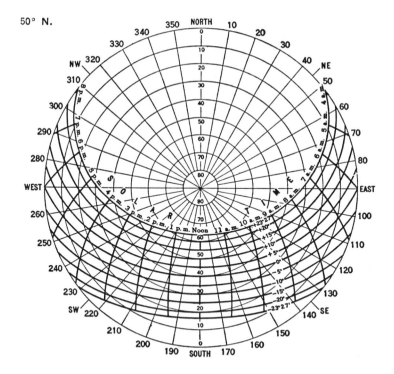

60° N.

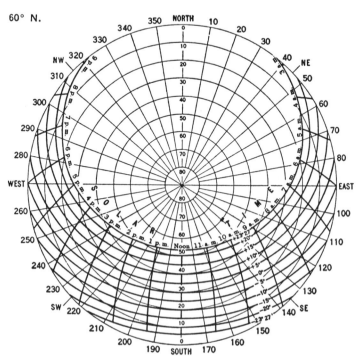

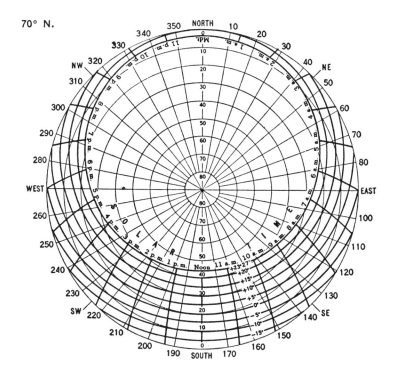

70° N.

References

Ajello, J., J. King, Jr., A. L. Lance, and C. W. Odd (1973). Ground-level ultraviolet solar flux in Pasadena, California. *Bull. Amer. Meteorol. Soc.* 54:114–15.

Alexandrov, V. Y. (1964). Cytophysiological and cytoecological investigations of heat resistance of plant cells toward the action of high and low temperature. *Quart. Rev. Biol.* 39:35–77.

Alexandrov, V. Y., A. G. Lomagin, and N. L. Feldman (1970). The responsive increase in thermostability of plant cells. *Protoplasma* 69:417–58.

Allen, W. A., and A. J. Richardson (1968). Interaction of light with a plant canopy. *J. Opt. Soc. Amer.* 58:1023–31.

Allen, W. A., H. W. Gausman, A. J. Richardson, and J. R. Thomas (1969). Interaction of isotropic light with a compact plant leaf. *J. Opt. Soc. Amer.* 59:1376–79.

Allen, W. A., H. W. Gausman, and A. J. Richardson (1970). Mean effective optical constants of cotton leaves. *J. Opt. Soc. Amer.* 60:542–47.

Anderson, M. C. (1967). Photon flux, chlorophyll content, and photosynthesis under natural conditions. *Ecology* 48:1050–53.

Andrewartha, H. G., and L. C. Birch (1954). *The Distribution and Abundance of Animals*. Chicago: Univ. Chicago Press.

Ångstrom, A. (1915). A study of the radiation of the atmosphere. *Smith. Misc. Coll.* 65:1–159.

Ångstrom, A. (1924). Solar and terrestrial radiation. *Quart. J. Roy. Meteorol. Soc.* 50:121–25.

Ångstrom, A. (1951). Actinometric measurements. In *Compendium of Meteorology*, pp. 50–57. Boston: Amer. Meteorol. Soc.

Ansari, A. Q., and W. E. Loomis (1959). Leaf temperatures. *Amer. J. Bot.* 46:713–17.

Argo, V. N. (1939). The effect of temperature upon the oxygen requirements of certain adult insects and insect eggs. *Entomol. Soc. Amer. Ann.* 32:147–63.

Askenasy, E. (1875). Ueber die Temperature welche Pflanzen in Sonnenlicht annehmen. *Bot. Zeit.* 33:441–44.

Bailey, I. W., and E. W. Sinnott (1916). The climatic distribution of certain types of angiosperm leaves. *Amer. J. Bot.* 3:24–39.

Baes, C. F., H. E. Goeller, J. S. Olson, and R. M. Rotty (1971). Carbon dioxide and climate: The uncontrolled experiment. *Amer. Sci.* 65:310–20.

Baker, D. N., and D. L. Myhre (1969). Effects of leaf shape and boundary layer thickness on photosynthesis in cotton (*Gossypium hirsutum*). *Physiol. Plant.* 22:1043–49.

Bakken, G. S. (1976a). A heat transfer analysis of animals: Unifying concepts and the application of metabolism data to field ecology. *J. Theor. Biol.* 60:337–84.

Bakken, G. S. (1976b). An improved method for determining thermal conductance and equilibrium body temperature with cooling curve experiments. *J. Theor. Biol.* 1:169–75.

Bakken, G. S., and D. M. Gates (1974). Linearized heat transfer relations in biology. *Science* 183:976–77.

Bakken, G. S., and D. M. Gates (1975). Heat transfer analysis of animals: Some implications for field ecology, physiology, and evolution. In D. M. Gates and R. B. Schmerl (eds.), *Perspectives of Biophysical Ecology*. New York: Springer-Verlag, Chap. 6, pp. 225–290.

Bakken, G. S., V. C. Vanderbilt, W. A. Buttemer, and W. R. Dawson (1978). Avian eggs: Thermoregulatory value of very high near-infrared reflectance. *Science* 200:321–23.

Bakker, R. T. (1975). Experimental and fossil evidence for the evolution of tetrapod energetics. In D. M. Gates and R. B. Schmerl (eds.), *Perspectives of Biophysical Ecology*. Chap. 21, pp. 365–99.

Balding, F. R., and G. L. Cunningham (1976). A comparison of heat transfer characteristics of simple and pinnate leaf models. *Bot. Gaz.* 137:65–74.

Bange, G. G. (1953). On the quantitative explanation of stomatal resistance. *Acta Bot. Neerland.* 2:255–97.

Bartholomew, G. A., and T. M. Casey (1977). Endothermy during terrestrial activity in large beetles. *Science* 195:882–83.

Bartholomew, G. A., and R. J. Epting (1975). Rates of post-flight cooling in sphinx moths. In D. M. Gates and R. B. Schmerl (eds.), *Perspectives of Biophysical Ecology*. New York: Springer-Verlag.

Bartlett, P. N., and D. M. Gates (1967). The energy budget of a lizard on a tree trunk. *Ecology* 48:315–22.

Baum, W. A., and L. Dunkelman (1955). Horizontal attenuation of ultraviolet light by the lower atmosphere. *J. Opt. Soc.Amer.* 45:166–75.

Baumberger, J. P. (1914). Studies in the longevity of insects. *Entomol. Soc. Amer. Ann.* 7:323–53.

Beament, J. W. (1959). The waterproofing mechanism of arthropods, I: The effect of temperature on cuticle permeability in terrestrial insects and ticks. *J. Exp. Biol.* 36:391–422.

Belehradek, J. (1935). *Temperature and Living Matter*. Berlin: Gebruder Bornstraeger.

Bener, P. (1963). The diurnal and annual variations of the spectral intensity of ultraviolet sky and global radiation on cloudless days at Davos, 1590 m a.s.l. Physikalisch-Meteorologisches Observatorium Davos, Davos-Platz, Switz., Contract AF 61(052)-618, Tech. Note 2.

Bennet, I. (1965). Monthly maps of mean daily insolation for the United States. *Solar Energy* 9:145–52.

Bennett, A. F., and P. Licht. (1972). Anerobic metabolism during activity in lizards. *J. Comp. Physiol.* 81:277–288.

Bennett, A. F., and W. R. Dawson (1972). Aerobic and anaerobic metabolism during activity in the lizard. *J. Comp. Physiol.* 81:289–299.

Bennett, H. E., J. M. Bennett, and M. R. Nagel (1960). Distribution of infrared radiance over a clear sky. *J. Opt. Soc. Amer.* 50:100–6.

Benson, L., E. A. Phillips, and P. A. Wilder (1967). Evolutionary sorting of characters in a hybrid swarm, I: Direction of slope. *Amer. J. Bot.* 59:1017–26.

Bergen, J. Y. (1904). Transpiration of sun leaves and shade leaves of *Olea europaea* and other broad-leaved evergreens. *Bot. Gaz.* 43:285–96.

Berland, T. G., and V. Y. Danilchenko (1961). *The Continental Distribution of Solar Radiation.* Leningrad: Gidrometeoizdat.

Bierhuizen, J. F., R. O. Slatyer, and C. W. Rose (1965). A Porometer for Laboratory and Field Operation. Tech. Bull. 41, Inst. for Land and Water Management Research, Wageningen, pp. 182–91.

Billings, W. D., and R. J. Morris (1951). Reflection of visible and infrared radiation from leaves of different ecological groups. *Amer. J. Bot.* 38:327–31.

Birkebak, R. C. (1966). Heat transfer in biological systems. *Internat. Rev. Gen. Exp. Zool.* 2:269–344.

Birkina, B. N. (1938). The effect of low temperature on the mutation process in *Drosophila melanogaster*. *Biol. Zhurn.* 7:653–60.

Birth, G. S. (1971). Spectrophotometry of biological materials. Ph.D. thesis. Purdue Univ., W. Lafayette, Ind.

Bjorkman, O. (1966). The effect of oxygen concentration on photosynthesis in higher plants. *Physiol. Plant.* 19:618–33.

Black, J. N. (1956). The distribution of solar radiation over the earth's surface. *Archiv. Met. Geoph. Bioklim. Ser. B.* 7:165.

Blackman, F. F., and A. M. Smith (1911). Experimental researches on vegetable assimilation and respiration, IX: On assimilation in submerged water plants and its relation to the concentration of carbon dioxide and other factors. *Proc. Roy. Soc.* B85:389–412.

Blackman, G. E. (1956). Influence of light and temperature on leaf growth. In F. L. Milthorpe (ed.), *The Growth of Leaves*. London: Butterworths.

Blackman, G. E., J. N. Black, and A. W. Kemp (1955). Physiological and ecological studies in the analysis of plant environment, X: An analysis of the effects of seasonal variation in daylight and temperature on the growth of *Helianthus annuus* in the vegetative phase. *Ann. Bot.* 33:353–60.

Blair, W. F. (1947). Variation in shade of pelage of local populations of the cactus mouse (*Peromyscus eremicus*) in the Tularosa Basin and adjacent areas in southern New Mexico. *Contr. Lab. Vert. Biol.* 37:1–7.

Bliss, L. C. (1956). A comparison of plant development in microenvironments of arctic and alpine tundras. *Ecol. Monogr.* 26:303–37.

Bodenheimer, F. S. (1952). Problems of animal ecology and physiology in deserts. Proc. Intl. Symp. Des. Res., Jerusalem. Pp. 1–23.

Bodenheimer, F. S., A. Halpering, and E. Swirski (1953). Experiments on light transmission through some animal integuments. *Bull. Res. Counc. Israel* 2:436–38.

Botkin, D. B., J. G. Janak, and J. R. Wallis (1972). Some ecological consequences of a computer model of forest growth. *J. Ecol.* 60:849–72.

Bott, T. L., and T. D. Brock (1969). Bacterial growth rates above 90°C in Yellowstone hot springs. *Science* 164:1411–12.

Bowen, I. S. (1926). The ratio of heat losses by conduction and by evaporation from any water surface. *Phys. Rev.* 27:779–87.

Bowes, G., and W. L. Ogren (1970). The effect of light intensity and atmosphere on ribulose diphosphate carboxylase activity. *Plant Physiol.,* vol. 46, suppl. 7.

Bowes, G., and W. Ogren (1972). Oxygen inhibition and other properties of soybean ribulose 1,5-diphosphate carboxylase. *J. Biol. Chem.* 247:2171–2176.

Bowes, G., W. L. Ogren, and R. Hageman (1971). Phosphoglycolate production catalyzed by ribulose diphosphate carboxylase. *Biochem. Biophys. Res. Commun.* 45:716–22.

Boyer, J. S. (1969). Free energy transfer in plants. *Science* 163:1219–20.

Brattstrom, B. H. (1965). Body temperatures of reptiles. *Amer. Midl. Nat.* 73:376–422.

Bregetova, L. G., and A. I. Popova (1963). Leaf temperature regime of plants in Tadzhikistan. *Akad. Nauk. Tadzhikskoi SSR,* Otdel fiziologii i biofiziki rastenii, Temat. sb., no. 3, pp. 29–40, Dushanbe. (Transl. by H. W. Milner for the Crerar Library, Chicago.)

Brock, T. D. (1967). Life at high temperatures. *Science* 158:1012–19.

Brock, T. D. (1969). Microbial growth under extreme conditions. *Symp. Soc. Gen. Microbiol.* 19:15–41.

Brock, T. D., and G. K. Darland (1970). Limits of microbial existence: Temperature and pH. *Science* 169:1316–18.

Broecker, W. S. (1975). Climatic change: Are we on the brink of a pronounced global warming? *Science* 189:460–63.

Brown, H. T., and F. Escombe (1900). Static diffusion of gases and liquids in relation to the assimilation of carbon and translocation in plants. *Phil. Trans. Roy. Soc. Ser. B* 193:223–91.

Brown, H. T., and F. Escombe (1905). Researches on some of the physiological processes of green plants with special reference to the interchange of energy between the leaf and its surroundings. *Roy. Soc. London Proc. Ser. B* 76:29–111.

Brown, H. T., and W. E. Wilson (1905). On the thermal emissivity of a green leaf in still and moving air. *Roy. Soc. London Proc. Ser. B* 76:122–37.

Brown, W. H. (1919). Vegetation of the Philippine Islands. Philippine Bureau of Science, Dept. of Agric. and Natural Res., Manila., no. 13, pp. 1–434.

Brunt, D. (1932). Notes on radiation in the atmosphere. *Quart. J. Roy. Meteorol. Soc.* 58:389–418.

Bryson, R. (1966). Air masses, streamlines, and the boreal forest. *Geog. Bull.* 8:228–69.

Bryson, R. A., and W. M. Wendland (1966). Tentative climatic patterns for some late-glacial and post-glacial episodes in central North America. Paper presented at the Conference on Environmental Studies of the Glacial Lake Aggasiz Region, Winnipeg, Manitoba, Canada, 4–6 November 1966.

Budyko, M. I. (1963). *Atlas Teplovogo Balansa*. Gidrometeorologicheskoe Izdatel'skoe, Leningrad.

Buffo, J., L. J. Fritschen, and J. L. Murphy (1972). Direct solar radiation on various slopes from 0 to 60 degrees north latitude. USDA Forest Service Research Paper PNW-142.

Buxton, P. A. (1923). *Animal Life in Deserts*. London: Edward Arnold and Co.

Byrne, G. F., C. W. Rose, and R. O. Slatyer (1970). An aspirated diffusion porometer. *Agr. Meteorol.* 7:39–44.

Cabanac, H. P., and H. T. Hammel (1971). Peripheral sensitivity and temperature regulation in *Tiliqua scincoides*. *Internat. J. Bioclimatol.* 15:239–43.

Cain, S. A., and J. D. O. Miller (1933). Leaf structure of *Rhododendron catawbiense* Mich. *Amer. Midl. Nat.* 14:68–82.

Cain, S. A., G. M. Castro, J. M. Pires, and N. T. da Silva (1956). Applications of some phytosociological techniques to Brazilian rain forest. *Amer. J. Bot.* 43:911–41.

Callahan, P. S. (1977a). Moth and candle: The candle flame as a sexual mimic of the coded infrared wavelength from a moth sex scent. *Appl. Opt.* 16:3089–93.

Callahan, P. S. (1977b). Tapping modulation of the far infrared (17 μm region) emission from the cabbage looper moth pheromone (sex scent). *Appl. Opt.* 16:3098–3102.

Callendar, G. S. (1938). The artificial production of carbon dioxide and its influence on temperature. *Quart. J. Roy. Meteorol. Soc.* 64:223–37.

Cameron, R. E. (1971). Antarctic soil microbial and ecological investigations. In *Research in the Antarctic*. American Association for the Advancement of Science. Washington, D.C., 137–189.

Carey, F. G., and J. M. Teal (1969a). Regulation of body temperature by the Bluefin Tuna. *Comp. Biochem. Physiol.* 28:205–213.

Carey, F. G., and J. M. Teal (1969b). Mako and Porbeagle: Warm-bodied sharks. *Comp. Biochem. Physiol.* 28:199–204.

Cena, K., and J. A. Clark (1972). Effect of solar radiation on temperatures of working honey bees. *Nature New Biol.* 236:222–23.

Cena, K., and J. A. Clark (1973). Thermographic measurements of the surface temperatures of animals. *J. Mammal.* 54:1003–7.

Chagnon, C. E., and C. E. Junge (1961). The vertical distribution of submicron particles in the stratosphere. *J. Meteorol.* 18:746–52.

Chapman, E. A., and D. Graham (1974). The effect of light on the triangular acid cycle in green leaves. *Plant Physiol.* 53:879–85.

Charles-Edwards, D. A. (1971). A simple kinetic model for leaf photosynthesis and respiration. *Planta* 101:43–50.

Charney, E., and F. S. Brackett (1961). Spectral dependence of scattering from a spherical alga and its implications for the state of organization of the light-accepting pigments. *Arch. Biochem. Biophys.* 92:1–12.

Chartier, P. (1966). Étude théorique de l'assimilation brute de la feuille. *Ann. Physiol. Vég.* 8:167–96.

Chartier, P. (1970). A model of CO_2 assimilation in the leaf. In I. Setlik (ed.), *Prediction and Measurement of Photosynthetic Productivity*. Wageningen: Pudoc.

Chartier, P., and J. Prioul (1976). The effects of irradiance, carbon dioxide and oxygen on the net photosynthetic rate of the leaf: A mechanistic model. *Photosynthetica* 10:20–24.

Clark, J. A., K. Cena, and J. L. Monteith (1973a). Measurements of the local heat balance of animal costs and human clothing. *J. Appl. Physiol.* 35:751–54.

Clark, J. A., K. Cena, and N. J. Mills (1973b). Radiative temperatures of butterfly wings. *Zeitschr. Angewandte Entomol.* 73:327–32.

Clark, L. R. (1949). Behavior of swarm hoppers of the Australian plague locust, *Chortoicetes terminifera* Walk., New South Wales Bull. Counc. Scient. Ind. Res., Australia, No. 245.

Clausen, J., D. D. Keck, and W. M. Hiesy (1940). *Experimental Studies on the Nature of Species, I: Effect of Varied Environments on Western North American Plants; Carnegie Inst. Wash. Pub. 520.*

Clements, E. S. (1904). The relation of leaf structure to physical factors. *Trans. Amer. Microsc. Soc.* 26:19–102.

Clements, H. F. (1938). Mechanisms of freezing resistance in the needles of *Pinus ponderosa* and *Pseudotsuga mucronata, State Col. Wash. Res. Stud.* 6:3–45.

Clum, H. H. (1926). The effect of transpiration and environmental factors on leaf temperature, I and II. *Amer. Jour. Bot.* 13:194–230.

Coblentz, W. W. (1913). The diffuse reflecting power of various substances. *Bull. Nat. Bur. Stand. Wash.* 9:283–325.

Cole, L. C. (1943). Experiments on toleration of high temperature in lizards with reference to adaptive coloration. *Ecology* 24:94–108.

Compton, J. T., and M. W. Garratt (1977). Leaf optical systems modeled as a stochastic process. *Appl. Opt.* 16:635–42.

Cook, D. B., and W. J. Hamilton (1942). Winter habits of white-tailed deer in central New York. *J. Wildl. Management* 6:287–91.

Cook, G. D., J. R. Dixon, and A. C. Leopold (1964). Transpiration: Its effects on plant leaf temperature. *Science* 144:546–47.

Coulson, K. L. (1966). Effects of reflection properties of natural surfaces in aerial reconnaissance. *Appl. Opt.* 5:1–13.

Coulson, K. L. (1975). *Solar and Terrestrial Radiation.* New York: Academic Press.

Cowles, R. B. (1940). Additional implications of reptilian sensitivity to high temperature. *Amer. Nat.* 74:542–61.

Cowles, R. B. (1946). Fur and feathers: A response to falling temperature? *Science* 103:74–75.

Cowles, R. B., and C. M. Bogert (1944). A preliminary study of the thermal requirements of desert reptiles. *Bull. Amer. Mus. Nat. Hist.* 83:265–96.

Cummings, N. W., and B. Richardson (1927). Evaporation from lakes. *Phys. Rev.* 30:527–34.

Curcio, J. A., and C. C. Petty (1951). The near infrared absorption spectrum of liquid water. *J. Opt. Soc. Amer.* 41:302–4.

Curcio, J. S., G. L. Knestrick, and T. H. Cosden (1961). Atmospheric scattering in the visible and infrared. NRL Rept. 5567, Nav. Res. Lab., Wash., D.C.

Curtis, O. F. (1936). Leaf temperatures and the cooling of leaves by radiation. *Plant Physiol.* 11:343–64.

Dahl, E., and E. Mork (1959). On the relationships between temperature, respiration, and growth in Norway spruce [*Picea abies* (L.) Karst.]. *Det Norske Skogforsoksvesen* 53:83–93.

Davies, O. L., ed. (1967). *The Design and Analysis of Industrial Experiments,* 2nd ed. Edinburgh: Oliver and Boyd.

Davis, L. B., and R. C. Birkebak (1975). Convective exchange transfer in fur. In

D. M. Gates and R. B. Schmerl (eds.), *Perspectives of Biophysical Ecology*. Chap. 29, pp. 525–48.

Dawson, W. R. (1958). Relation of oxygen consumption and evaporative water loss to temperature in the cardinal. *Physiol. Zool.* 31:37–48.

Dawson, W. R., and J. W. Hudson (1970). Birds. In G. C. Whittow (ed.), *Comparative Physiology of Thermoregulation*. Academic: New York. Vol. I, 223–310.

de Bary, E. (1964). Influence of multiple scattering of the intensity and polarization of diffuse sky radiation. *Appl. Optics* 3:1293–1303.

Decker, J. P. (1959). Comparative responses of carbon dioxide outburst and uptake in tobacco. *Plant Physiol.* 34:100–2.

Derby, R., and D. M. Gates (1966). The temperature of tree trunks, calculated and observed. *Amer. J. Bot.* 53:580–87.

Diermendjian, D., and Z. Sekera (1954). Global radiation resulting from multiple scattering in a Rayleigh atmosphere. *Tellus* 6:382–98.

Digby, P. B. (1955). Factors affecting the temperature excess of insects in sunshine. *J. Exp. Biol.* 32:279–98.

Dinger, J. E. (1941). The absorption of radiant energy in plants. Ph.D. thesis, Iowa State College, Ames, Iowa.

Dirmhirn, I. (1964). *Das Strahlungsfeld im Lebensraum*. Akad. Verlagsgessellschaft Frankfurt am Main.

Downes, R. W. (1969). Differences in transpiration rates between tropical and temperate grasses under controlled conditions. *Planta* 88:261–73.

Drummond, A. J. (1958). Notes on the measurement of natural illumination, II: Daylight and skylight at Pretoria, the luminous efficiency of daylight. *Archiv. Meteorol. Geophysik. Bioklim. Ser. B* 9:149–63.

Eaton, F. M., and G. O. Belden (1929). Leaf temperatures of cotton and their relation to transpiration, varietal differences and yields. *USDA Tech. Bull.* 91:1–39.

Edwards, R. L. (1960). Relationship between grasshopper abundance and weather conditions in Saskatchewan, 1930–1958. *Canad. Entomol.* 92:619–24.

Eglinton, G., and R. J. Hamilton (1967). Leaf epicuticular waxes. *Science* 156:1322–35.

Ehrler, W. L., and C. H. M. van Bavel (1968). Leaf diffusion resistance, illuminance, and transpiration. *Plant Physiol.* 43:208–14.

Ehleringer, J., and O. Bjorkman (1977). Quantum yields for CO_2 uptake in C_3 and C_4 plants. *Plant Physiol.* 59:86–90.

Ehleringer, J. R., and P. Miller (1975). A simulation model of plant water relations and production in the alpine tundra, Colorado. *Oecologia* 19:177–94.

Ehlers, J. H. (1915). The temperatures of leaves of *Pinus* in winter. *Amer. J. Bot.* 2:32–70.

Ellyard, P. W., and M. Gibbs (1969). Inhibition of photosynthesis by oxygen in isolated spinach chloroplasts. *Plant Physiol.* 44:1115–21.

Elterman, L. (1964). *An Atlas of Aerosol Attenuation and Extinction Profiles for the Troposphere and Stratosphere*. Environ. Res. Paper No. 241, Air Force Cambridge Research Labs.

Emerson, R. (1958). The quantum yield of photosynthesis. *Annual Rev. Plant Physiol.* 9:1–24.

Emerson, R., and C. Lewis (1941). Carbon dioxide exchange and the measurement of the quantum yield of photosynthesis. *Amer. J. Bot.* 28:789–804.

Erskine, D. J., and J. R. Spotila (1977). Heat energy budget analysis and heat transfer in the large mouth black bass (*Micropterus salmoides*). *Physiol. Zool.* 50:157–69.

Farlow, J. O., C. V. Thompson, and D. E. Rosner (1976). Plates of the dinosaur Stegosaurus: Forced convection heat loss fins? *Science* 192:1123–25.

Farrar, J. F., and O. P. Mapunda (1977). Optical properties of some African crop plants. *Appl. Optics* 16:248–51.

Farquahar, G. D., and I. R. Cowan (1974). Oscillations in Stomatal Conductance. *Plant Physiol.* 54:769–772.

Forrester, M. L., G. Krotkov, and C. D. Nelson (1966). Effect of oxygen on photosynthesis, photorespiration, and respiration in detached leaves, I: Soybean. *Plant Physiol.* 41:422–27.

Frisch, von K. (1950). *Bees: Their Vision, Chemical Senses, and Language.* Ithaca, N.Y.: Cornell Univ. Press.

Fritts, H. (1966). Growth rings of trees: Their correlation with climate. *Science* 154:973–79.

Fritz, S., and T. H. MacDonald (1949). Average solar radiation in the United States. *Heating and Ventilating* 46:61–64.

Fuchs, S., and C. B. Tanner (1967). Evaporation from a drying soil. *J. Appl. Meteorol.* 6:852–57.

Gaastra, P. (1959). Photosynthesis of crop plants as influenced by light, carbon dioxide, temperature, and stomatal diffusion resistance. *Lab. Plant Physiol. Res. Agri. Univ. Wageningen* 59(13):1–68.

Gaertner, E. E. (1964). Tree growth in relation to the environment. *Bot. Rev.* 30:393–436.

Gates, D. M. (1956). Infrared determination of precipitable water vapor in a vertical column of the earth's atmosphere. *J. Meteorol.* 13:369–75.

Gates, D. M. (1960). Near infrared atmospheric transmission to solar radiation. *J. Opt. Soc. Amer.* 50:1299–1304.

Gates, D. M. (1961). Winter thermal radiation studies in Yellowstone Park. *Science* 134:32–35.

Gates, D. M. (1962). *Energy Exchange in the Biosphere.* New York: Harper and Row.

Gates, D. M. (1963). Leaf temperature and energy exchange. *Archiv. Meteor. Geophys. Bioklim. Ser. B* 12:321–36.

Gates, D. M. (1966a). Transpiration and energy exchange. *Quart. Rev. Biol.* 41:353–64.

Gates, D. M. (1966b). Spectral distribution of solar radiation at the earth's surface. *Science* 151:523–29.

Gates, D. M. (1968). Sensing biological environments with a portable radiation thermometer. *Appl. Opt.* 7:1803–9.

Gates, D. M. (1969a). The ecology of an elfin forest in Puerto Rico, 4: Transpiration rates and temperatures of leaves in cool humid environment. *J. Arnold Arboretum* 50:93–98.

Gates, D. M. (1969b). Infrared measurement of plant and animal surface temperature and their interpretation. In P. L. Johnson (eds.), *Remote Sensing in Ecology.* Athens, Ga.: Univ. Ga. Press. Chap. 6, pp. 95–107.

Gates, D. M. (1970). Animal climates: Where animals must live. *Environ. Res.* 3:132–44.

Gates, D. M. (1972). *Man and His Environment: Climate.* New York: Harper and Row.

Gates, D. M. (1974). Review of "Life's Color Code" by W. J. Hamilton. *Bioscience* 24:120.

Gates, D. M., and C. M. Benedict (1963). Convection phenomena from plants in still air. *Amer. J. Bot.* 50:563–73.

Gates, D. M., and W. J. Harrop (1963). Infrared transmission of the atmosphere to solar radiation. *Appl. Opt.* 2:887–98.

Gates, D. M., and R. Janke (1965). The energy environment of the alpine tundra. *Oecol. Plant.* 1:39–62.

Gates, D. M., and L. E. Papian (1971). *Atlas of Energy Budgets of Plant Leaves.* New York: Academic Press.

Gates, D. M., and W. Tantraporn (1952). The reflectivity of deciduous trees and herbaceous plants in the infrared to 25 microns. *Science* 115:613–16.

Gates, D. M., W. M. Hiesey, H. W. Milner, and M. A. Nobs (1964). Temperatures of *Mimulus* leaves in natural environments and in a controlled growth chamber. *Carnegie Inst. Wash. Yearb.* 63:418–30.

Gates, D. M., H. J. Keegan, J. C. Schleter, and V. R. Weidner (1965). Spectral properties of plants. *Appl. Opt.* 4:11–20.

Gates, D. M., R. Alderfer, and S. E. Taylor (1968). Leaf temperatures of desert plants. *Science* 159:994–95.

Gates, D. M., H. B. Johnson, C. S. Yocum, and P. W. Lommen (1969). Geophysical factors affecting plant productivity. Proc. Internat. Symp. on the Productivity of Photosynthetic Systems, part II (Theoretical foundations of optimization of the photosynthetic productivity). Moscow, USSR.

Gates, F. C. (1930). Aspen association in northern lower Michigan. *Bot. Gaz.* 40:233–59.

Gausman, H. W., and Allen, W. A. (1973). Optical parameters of leaves of 30 plant species. *Plant Physiol.* 52:57–62.

Gausman, H. W., and R. Cardenas (1968). Effect of pubescence on reflectance of light. Proc. 5th Symp. on Remote Sensing of Environment, Univ. Mich., Ann Arbor, pp. 291–97.

Gausman, H. W., W. A. Allen, and R. Cardenas (1969). Reflectance of cotton leaves and their structure. In *Remote Sensing of Environment.* New York: American Elsevier. Vol. 1, pp. 19–22.

Geiger, R. (1965). *The Climate Near the Ground.* Cambridge: Harvard University Press.

Gessaman, J. A., ed. (1973). *Ecological Energetics of Homeotherms; Monogr. Ser.* 20:1–155. Logan, Utah: Utah State Univ. Press.

Grace, J. (1978). *Plant Response to Wind.* New York: Academic Press.

Graner, E. A. (1942). Genetics of manihot, I: Inheritance of leaf form and color of the outer root skin in *Manihot utilissima.* Pohl. Bragantia 2:13–22.

Green, G. W. (1968). In W. P. Lowry (ed.), *Weather and Insects in Biometeorology.* Corvalis, Oreg.: Oreg. State Univ. Press. Pp. 81–112.

Hafez, E. S. (1968). *Adaptation of Domestic Animals.* Philadelphia: Lea and Febiger.

Hall, A. E. (1971). A model of leaf photosynthesis and respiration. *Carnegie Inst. Wash. Yearb.* 70:530–40.

Hall, A. E., and O. Björkman (1975). Model of leaf photosynthesis and respiration. In D. M. Gates and R. B. Schmerl (eds.), *Perspectives of Biophysical Ecology.* Chap. 4, pp. 55–73.

Hall, A. E., and M. R. Kaufman (1975). Regulation of water transport in the soil-plant-atmosphere continuum. In D. M. Gates and R. B. Schmerl (eds.), *Perspectives of Biophysical Ecology*. New York: Springer-Verlag. Chap. 11.

Hamilton, W. J. (1973). *Life's Color Code*. New York: McGraw-Hill.

Hamilton, W. J., and F. Heppner (1967). Radiant solar energy and the function of black homeotherm pigmentation: An hypotheses. *Science* 155:196–97.

Hammel, H. T., F. T. Caldwell, and R. M. Abrams (1967). Regulation of body temperature in the blue-tongued lizard. *Science* 156:1260–62.

Hanawalt, P. C. (1966). The u.v. sensitivity of bacteria; its relation to the DNA replication cycle. *Photochem. Photobiol.* 5:1–12.

Hanson, H. C. (1917). Leaf-structure as related to environment. *Amer. J. Bot.* 4:533–60.

Haurwitz, B. (1934). Daytime radiation at Blue Hill Observatory in 1933 with applications to turbidity in American air masses. *Harvard Meteorol. Stud.* 1.

Heath, J. E. (1964). Head-body temperature differences in horned lizards. *Physiol. Zool.* 37:273–79.

Heath, J. E. (1965). Temperature regulation and diurnal activity in horned lizards. *Univ. Calif. Berkeley Publ. Zool.* 64:97–136.

Heath, O. V. S. (1969). *The Physiological Aspects of Photosynthesis*. Stanford, Calif.: Stanford Univ. Press.

Heath, O. V. S., and H. Meidner (1957). Effects of carbon dioxide and temperature on stomata of *Allium sepa* L. *Nature* 180:181–82.

Heath, O. V. S., and B. Orchard (1957). Temperature effects on the minimum intercellular space carbon dioxide concentration. *Nature* 180:180–81.

Heinrich, B. (1971). Temperature regulation of the sphinx moth, *Manduca sexta,* I. *J. Exp. Biol.* 54:141–52.

Heinrich, B. (1974). Thermoregulation in endothermic insects. *Science* 185: 747–56.

Heller, H. C. (1971). Altitudinal zonation of chipmunks (*Eutamias*): Interspecific aggression. *Ecology* 52:312–19.

Heller, H. C., and D. M. Gates (1971). Altitudinal zonation of chipmunks (*Eutamias*): Energy budgets. *Ecology* 52:424–53.

Hesselbring, H. (1914). The effect of shading on the transpiration and assimilation of the tobacco plant in Cuba. *Bot. Gaz.* 57:257–86.

Hew, C. S., G. Krotkov, and D. T. Canvin (1969). Determination of the rate of CO_2 evolution by green leaves in light. *Plant Physiol.* 44:662–70.

Hilpert, R. (1933). Wärmeabgabe von geheizten Drähten und Rohren im Luftstrom. *Forsch. Geb. Ing. Wesen.* 4:215–24.

Hofstra, G., and J. D. Hesketh (1969). Effects of temperature on the gas exchange of leaves in the light and dark. *Planta* (Berlin) 85:228–37.

Holdridge, L. R. (1947). Determination of world plant formations from simple climatic data. *Science* 105:367–68.

Holdridge, L. R., W. C. Grenke, W. H. Hatheway, T. Liang, and J. A. Tasi (1971). *Forest Environments in Tropical Life Zones: A Pilot Study*. New York: Pergamon Press.

Holmgren, P., and P. G. Jarvis (1967). Carbon dioxide efflux from leaves in light and darkness. *Physiol. Plant.* 20:1045–51.

Holmgren, P., P. G. Jarvis, and M. S. Jarvis (1965). Resistances to carbon dioxide and water vapor transfer in leaves of different plant species. *Physiol. Plant.* 18:557–73.

Hooper, E. T. (1941). Mammals of the lava fields and adjoining areas in Valencia County, New Mexico. *Misc. Publ. Museum Zool. Univ. Mich.* 51:1–47.

Hopmans, P. A. M. (1971). *Rhythms in Stomatal Opening of Bean Leaves.* Wageningen: H. Veenman & Zonen.

Howard, J. A. (1966). Spectral energy relations of isobilateral leaves. *Austr. J. Biol. Sci.* 19:757–66.

Huber, B. (1935). Der Wärmehaushalt der Pflanzen. *Naturwiss. Landwirt* 17:1–148.

Huber, B. (1937). Mikroklimatische und Pflanzentemperaturregisrierungen mit dem Multithermographen von Hartmann und Braun. *Jahrb. Wiss. Bot.* 84: 671–709.

Huey, R. B. (1974). Behavioral thermoregulation in lizards: Importance of associated costs. *Science* 184:1001–3.

Hutchinson, V. H., and J. L. Larimer (1960). Reflectivity of the integuments of some lizards from different habitats. *Ecology* 41:199–209.

Irvine, J. E. (1970). Evidence for photorespiration in tropical grasses. *Physiol. Plant.* 23:607–12.

Irvine, W. M., and J. B. Pollack (1968). Infrared optical properties of water and ice spheres. *Icarus* 8:324–60.

Irving, L., H. Krog, and M. Monson (1955). The metabolism of some Alaskan animals in winter and summer. *Physiol. Zool.* 28:173–185.

Jackson, W. A., and R. J. Volk (1970). Photorespiration. *Ann. Rev. Plant Physiol.* 21:385–432.

Janke, R. A. (1970). Transpiration resistance in *Vaccinium myrtillus. Amer. J. Bot.* 57:1051–54.

Jarvis, P. G. (1975). Water transfer in plants. In D. A. de Vries and N. H. Afgan (eds.), *Heat and Mass Transfer in the Biosphere.* Washington, D.C.: Scripta Vol. I, 369–394.

Jerison, H. J. (1971). More on why birds and mammals have big brains. *Amer. Nat.* 105:185–89.

Johnson, C. G., and L. P. Smith (1965). *The Biological Significance of Climatic Changes in Britain.* London and New York: Academic Press.

Johnson, F., H. Eyring, and M. Polissar (1954). *The Kinetic Basis of Molecular Biology.* New York: John Wiley and Sons.

Johnson, F. H. (1957). *Influence of Temperature on Biological Systems.* Amer. Physiol. Soc., Wash., D.C.

Johnson, F. S. (1954). The solar constant. *J. Meteorol.* 11:431–39.

Johnson, H. B. (1975). Gas exchange strategies in desert plants. In D. M. Gates and R. B. Schmerl (eds.), *Perspectives of Biophysical Ecology.* New York: Springer-Verlag.

Joliffe, P., and E. Tregunna (1973). Environmental regulation of the oxygen effect on apparent photosynthesis in wheat. *Can. J. Bot.* 51:841–853.

Jones, H., and R. Slatyer (1972). Estimation of the transport and carboxylation components of the intracellular limitation to leaf photosynthesis. *Plant Physiol.* 50:283–88.

Kanemasu, E. T., G. W. Thurtell, and C. B. Tanner (1969). Design, calibration, and field use of a stomatal diffusion porometer. *Plant Physiol.* 44: 881–85.

Kaplan, L. D. (1960). The influence of carbon dioxide variations on the atmospheric heat balance. *Tellus* 12:204–8.

Kavanau, J. L., and K. S. Norris (1961). Behavior studies by capacitance sensing. *Science* 134:730–32.

Keeling, C. D., R. B. Bacastow, A. E. Bainbridge, C. A. Ekdahl, P. R. Guenther, L. S. Waterman, and J. F. S. Chin (1976). Atmospheric carbon dioxide variations at Mauna Loa Observatory, Hawaii. *Tallus* 28:538–64.

Kelly, C. F., T. E. Bond, and N. R. Ittner (1957). Sky temperatures in the Imperial Valley of California. *Amer. Geophys. Union Trans.* 38:308–13.

Kendeigh, S. C. (1944). Effect of air temperature on the rate of energy metabolism in the English Sparrow. *J. Exp. Zool.* 96:1–16.

Kendeigh, S. C. 1974. *Ecology*. Englewood Cliffs, N.J.: Prentice-Hall.

Kennedy, J. S. (1939). The behavior of the desert locust (*Schistocerca gregaria*, Forsk.) (Orthoptera) in an outbreak center. *Roy. Entomol. Soc. Trans. London* 89:385–542.

Kenny, P., and P. J. McGruddy (1972). A circuit for a self-timing stomatal diffusion porometer. *Agr. Meteorol.* 10:393–99.

Kestin, J. (1966). The effect of free-steam turbulence on heat transfer. In T. F. Irvine and J. P. Harnett (eds.), *Advances in Heat Transfer*, vol. 3. New York: Academic Press.

Kimball, H. H. (1919). Variations in the total and luminous solar radiation with geographical position in the United States. *Month. Weath. Rev.* 47:769–98.

Kimball, H. H. (1924). Records of total solar radiation intensity and their relation to daylight intensity. *Month. Weath. Rev.* 52:473–79.

Kimball, H. H. (1938). The duration and intensity of twilight. *Month. Weath. Rev.* 66:279–86.

Kimball, M. H., and F. A. Brooks (1959). Plant climates of California. *Calif. Agr.* 13:7–12.

Klauber, L. M. (1939). Studies of reptile life in the arid southwest, II: Speculations of protective coloration and protective reflectivity. *Bull. Zool. Soc.* 14:65–79.

Kleiber, M. (1961). *The Fire of Life: An Introduction to Animal Energetics*. New York: John Wiley & Sons.

Kleiber, M. (1974). Linearized heat transfer relations in biology. *Science* 183:978.

Knipling, E. B. (1969). Physical and physiological basis for the reflectance of visible and near-infrared radiation from vegetation. Proc. Symp. Inform. Processing, Purdue Univ., pp. 732–41.

Knoerr, K. R., and L. W. Gay (1965). Tree leaf energy balance. *Ecology* 46:17–24.

Knutson, R. M. (1974). Heat production and temperature regulation in eastern skunk cabbage. *Science* 186:746–47.

Kondratyev, K. Y. (1969). *Radiation in the Atmosphere*. New York: Academic Press.

Koppen, W. (1900). Versuch einer Klassifikation der Klimate, vorzugsweise nach ihren Beziehungen zur Pflanzenwelt. *Geogr. Zeitschr.* 6:593–611, 657–79.

Koppen, W. (1918). Klassifikation der Klimate nach Temperatur, *Niederschlag Jahresverlauf.* 64:193–203, 243–48.

Koriba, K. (1943). Über die Konvektion und Verdunstung als physikalische Grundlage der Transpiration. *Japan. J. Bot.* 13:1–260.

Kovarik, M. (1964). Flow of heat in an irradiated protective cover. *Nature.* 201:1085–87.

Kreith, F. (1966). *Principles of Heat Transfer.* Scranton, Pa.: International Textbook Co.

Krog, J. (1955). Notes on temperature measurements indicative of special organization in Arctic and sub-Arctic plants for utilization of radiated heat from the sun. *Physiol. Plant.* 8:836–39.

Ku, S., and G. Edwards (1977). Oxygen inhibition of photosynthesis, I: Temperature dependence and relation to O_2/CO_2 solubility ratio. *Plant Physiol.* 59:986–90.

Kuiper, P. J. (1961). The effects of environmental factors on the transpiration of leaves, with special reference to stomatal light response. *Lab. Plant Physiol. Res. Agr. Univ. Wageningen* 61:1–49.

Kumar, R., and L. Silva (1973). Reflectance model of a plant leaf. The Laboratory for Applications of Remote Sensing. Inform. Note 022473, Purdue Univ., West Lafayette, Ind.

Ladurie, E. L. (1971). *Times of Feast, Times of Famine: A History of Climate since the year 1000.* New York: Doubleday and Co.

Laing, W., W. L. Ogren, and R. Hageman (1974). Regulation of soybean net photosynthesis CO_2 fixation by the interaction of CO_2, O_2 and ribulose 1,5-diphosphate carboxylase. *Plant Physiol.* 54:678–85.

Landsberg, H. E. (1961). Solar radiation at the earth's surface. *Solar Energy* 5:95–98.

Lange, O. L. (1959). Untersuchungen über Wärmehaushalt und Hitzeresistenz mauretanischer Wüsten-und Savannenpflanzen. *Flora* 147:595–651.

Lange, O. L. (1965). The heat resistance of plants, its determination and variability, methodology of plant eco-physiology. Proceedings of the Montpellier Symposium, UNESCO.

Lange, O. L. (1967). Investigations on the variability of heat-resistance in plants. In A. S. Troshin (ed.), *The Cell and Environmental Temperature.* New York: Pergamon Press. Pp. 131–41.

Lange, O. L., R. Lösch, E. D. Schulze, and L. Kappen (1971). Responses of stomata to changes in humidity. *Planta* 100:76–86.

Lange, O. L., E. D. Schulze, L. Kappen, U. Buschbom, and M. Evenari (1975). Photosynthesis of desert plants as influenced by internal and external factors. In D. M. Gates and R. B. Schmerl (eds.), *Perspectives of Biophysical Ecology.* New York: Springer-Verlag. Chap. 8.

Larcher, W. (1954). Die Kalteresistenz Mediterraner Immergruner und ihre Beeinflussbarkeit. *Planta* 44:607–35.

Larsen, P. (1966). Light requirements in plant production and growth regulation. *Acta Agr. Scand. Suppl.* 16:161–72.

Laude, H. M. (1953). The nature of summer dormancy in perennial grasses. *Bot. Gaz.* 114:284–92.

Laue, E. G., and A. J. Drummond (1968). Solar constant: First direct measurements. *Science* 161:888–91.

LaRue, C. D. (1930). The water supply of the epidermis of leaves. *Mich. Acad. Sci.* 13:131–39.

Lee, R., and D. M. Gates (1964). Diffusion resistance in leaves as related to their stomatal anatomy and micro-structure. *Amer. J. Bot.* 51:963–75.

Leith, H. (1969). Predicted annual fixation of carbon for the land masses and oceans of the world. In *Analysis of Temperate Forest Ecosystems*. New York: Springer-Verlag.

Leith, H. (1972). Über die Primärproduktion die Pflanzendecke der Erde. *Angew. Bot.* 46:1–37.

List, R. J., ed. (1963). *Smithsonian Meteorological Tables*, 6th ed. Smithsonian Inst., Washington, D.C.

Liu, B. Y., and R. C. Jordan (1960). The interrelationship and characteristic distribution of direct, diffuse, and total solar radiation. *Solar Energy* 4:1–19.

Lommen, P. W., C. R. Schwintzer, C. S. Yocum, and D. M. Gates (1971). A model describing photosynthesis in terms of gas diffusion and enzyme kinetics. *Planta* 98:195–220.

Lommen, P. W., Smith, S. K., Yocum, C., and D. M. Gates (1975). Photosynthetic Model. In D. M. Gates and R. B. Schmerl (eds.), *Perspectives of Biophysical Ecology*. New York: Springer-Verlag.

Loomis, R. S., and W. A. Williams (1963). Maximum crop productivity: An estimate. *Crop Sci.* 3:67–72.

Lumb, F. E. (1964). The influence of clouds on hourly amounts of total solar radiation at the sea surface. *Quart. J. Roy. Meteorol. Soc.* 90:43–56.

Lustick, S. (1969). Bird energetics: Effects of artificial radiation. *Science* 163:387–90.

Lutova, M. I., and I. G. Zavadskaya (1966). Effects of the plant keeping duration at different temperatures on the cell heat resistance. *Tsitologiya* 8:484–93.

MacBryde, B., R. L. Jefferies, R. Alderfer, and D. M. Gates (1971). Water and energy relations of plant leaves during periods of heat stress. *Oecol. Plant.* 6:151–62.

McCullough, E. C., and W. P. Porter (1972). Computing clear day solar radiation spectra for the terrestrial ecological environment. *Ecology* 52:1008–15.

McIntosh, R. P. (1974). "Plant ecology 1947–72" *Ann. Mo. Bot. Garden* 61: 132–65.

McMahon, T. (1973). Size and shape in biology. *Science* 179:1201–4.

McNab, B. K. (1970). Body weight and the energetics of temperature regulation. *J. Exp. Biol.* 53:329–48.

McNab, B. K., and P. Morrison (1963). Body temperature and metabolism in subspecies of *Peromyscus* from arid and mesic environments. *Ecol. Monogr.* 33:63–82.

Maloiy, G. M., ed. (1972). *Comparative Physiology of Desert Animals*. New York: Academic Press.

Marr, J. W. (1961). *Ecosystems of the East Slope of the Front Range in Colorado* (Univ. Colo. Ser. Biol. No. 8). Boulder, Colo.: Univ. Colo. Press.

Marsh, H., J. Galmiche, and M. Gibbs (1965). Effect of light on the tricarboxylic acid cycle in *Scendesmus*. *Plant Physiol.* 40:1013.

Martin, E. V. (1943). *Studies of Evaporation and Transpiration Under Controlled Conditions*. Carnegie Inst. Wash. Publ., Washington, D.C.

Mazur, P. (1969). Freezing injury in plants. *Ann. Rev. Plant. Physiol.* 20: 419–48.

Meidner, H., and T. A. Mansfield (1965). Stomatal responses to illumination. *Biol. Rev.* 40:483–509.

Mellor, R. S., F. B. Salisbury, and K. Raschke (1964). Leaf temperatures in controlled environments. *Planta* 61:56–72.

Merriam, C. H. (1894). Laws of temperature control of the geographic distribution of terrestrial animals and plants. *Nat. Geog. Mag.* 6:229–38.

Meyers, V. I., and W. A. Allen (1968). Electrooptical remote sensing methods as nondestructive testing and measuring techniques in agriculture. *Appl. Opt.* 7:1819–38.

Miller, A. A. (1961). *Climatology.* London: Methuen and Co.

Miller, E. C., and A. R. Saunders (1923). Some observations on the temperature of the leaves of crop plants. *J. Agr. Res.* 26:15–43.

Miller, P., and D. M. Gates (1967). Transpiration resistance of plants. *Amer. Midl. Nat.* 77:77–85.

Minnaert, M. (1954). *The Nature of Light and Color in the Open Air* (revised by K. E. Brian Jay). New York: Dover Publications.

Mitchell, J. W. (1976). Heat transfer from spheres and other animal forms. *Biophys. J.* 16:561–69.

Moen, A. M. (1968a). Energy exchange of white-tailed deer, western Minnesota. *Ecology* 49:676–82.

Moen, A. M. (1968b). Surface temperatures and radiant heat loss from white-tailed deer. *J. Wildl. Management* 32:338–44.

Monteith, J. L. (1965a). Evaporation and environment. In *The State and Movement of Water in Living Organisms* (19th Symp. Soc. Exp. Biol.). New York: Academic Press.

Monteith, J. L. (1965b). Evaporation and environment. *Proc. Symp. Soc. Exp. Biol.* 19:205–234.

Monteith, J. L. (1973). *Principles of Environmental Physics.* London: Edward Arnold Publications.

Monteith, J. L., ed. (1976). *Vegetation and the Atmosphere,* vols. 1 and 2. London: Academic Press.

Monteith, J. L., ed. (1975/76). *Vegetation and the Atmosphere,* vol. 1: Principles (1975) and 2: Case Studies (1976). New York: Academic Press.

Monteith, J. L., and T. A. Bull (1970). A field porometer, II: Theory, calibration, and performance. *J. Appl. Ecol.* 7:623–38.

Mooney, H. A., and W. D. Billings (1961). Comparative physiological ecology of arctic and alpine populations of *Oxyria digyna. Ecol. Monogr.* 31:1–29.

Mooney, H. A., and A. W. Johnson (1965). Comparative physiological ecology of an arctic and an alpine population of *Thalictrum alpinum* L., *Ecology* 46:721–27.

Morhardt, S. S. (1975). Use of climate diagrams to describe microhabitats occupied by Belding ground squirrels and to predict rates of change of body temperature. In D. M. Gates and R. B. Schmerl (eds.), *Perspectives of Biophysical Ecology.* New York: Springer-Verlag.

Morhardt, S. S., and D. M. Gates (1974). Energy exchange analysis of the Belding ground squrrel and its habitat. *Ecol. Monogr.* 44:17–44.

Moss, R. A., and W. E. Loomis (1952). Absorption spectra of leaves, I: The visible spectrum. *Plant Physiol.* 27:370–91.

Nagy, K. A., D. K. Odell, and R. S. Seymour (1972). Temperature regulation by the inflorescence of *Philodendron. Science* 178:1195–97.

Nobel, P. S. (1974). *Introduction to Biophysical Plant Physiology.* San Francisco: W. H. Freeman and Co.

Norris, K. S. (1967). Color adaptation in desert reptiles and its thermal relationships. In W. W. Milstead (ed.), *Lizard Ecology, A symposium*. Columbia, Mo.: Univ. Mo. Press.

Ohmart, R. D., and R. C. Lasiewski (1970). Road runners: Energy conservation by hypothermia and absorption of sunlight. *Science* 172:67–69.

O'Leary, J. W. (1975). Environmental influence on total water consumption by whole plants. In D. M. Gates and R. B. Schmerl (eds.), *Perspectives in Biophysical Ecology*. New York: Springer-Verlag.

Packard, G. C., C. R. Tracy, and J. J. Roth (1977). The physiological ecology of reptilian eggs and embryos, and the evolution of viviparity within the class reptilia. *Biol. Rev.* 52:71–105.

Parker, J. (1961). Seasonal changes in cold resistance of some northeastern woody evergreens. *J. Forestr.* 59:108–11.

Parker, J. (1963). Cold resistance in woody plants. *Bot. Rev.* 29:123–201.

Parker, R. F., and R. L. Thompson (1942). The effect of external temperature on the course of infectious myxomatosis of rabbits. *J. Exp. Med.* 75:567–73.

Parkhurst, D. F., and D. M. Gates (1966). Transpiration resistance and energy budget of *Populus sargentii* leaves. *Nature* 210:172–74.

Parkhurst, D. G., and O. L. Loucks (1972). Optimal leaf size in relation to environment. *J. Ecol.* 60:505–37.

Parkhurst, D. F., P. R. Duncan, E. Kreith, and D. M. Gates (1968a). Convection heat transfer from broad leaves of plants. *J. Heat Trans.* 90:71–76.

Parkhurst, D. F., P. R. Duncan, D. M. Gates, and F. Kreith (1968b). Wind tunnel modeling of convection of heat between air and broad leaves of plants. *Agr. Meteorol.* 5:33–47.

Parlange, J. Y., P. E. Waggoner, and G. H. Heichel (1971). Boundary layer resistance and temperature distribution on still and flapping leaves. *Plant Physiol.* 48:437–42.

Parry, D. A. (1951). Factors determining the temperatures of terrestrial arthropods in sunlight. *J. Exp. Biol.* 28:445–62.

Pearman, G. I. (1966). The reflection of visible radiation from leaves of some western Australian species. *Austr. J. Biol. Sci.* 19:97–103.

Pearman, G. I., H. L. Weaver, and C. B. Tanner (1972). Boundary layer heat transfer coefficients under field conditions. *Agr. Meteorol.* 10:83–92.

Pearson, O. P. (1954). Habits of the lizard *Liolaemus multiformis multiformis* at high altitudes in southern Peru. *Copeia* 2:111–16.

Peisker, M. (1974). A model describing the influence of oxygen on photosynthetic carboxylation. *Photosynthetica* 8:47–50.

Peisker, M. (1976). Ein Modell der Sauerstoffabhangigkeit des photosynthetischen CO_2-Gaswechsels von C_3 Pflanzen. *Kulturpflanze* 24:221.

Penman, H. L. (1948). Natural evaporation from open water, bare soil, and grass. *Proc. Roy. Soc. London Ser. A* 193:120–45.

Penman, H. L., and R. K. Schofield (1951). Some physical aspects of assimilation and transpiration. *Symp. Soc. Exp. Biol.* 5:115–129.

Penndorf, R. (1954). The vertical distribution of Mie particles in the troposphere. Geophys. Res. Paper No. 25, Air Force Cambridge Research Center.

Perttu, K. (1971). Factors affecting needle and leaf temperatures. Dept. of Reforestation, Royal College Forestry, Stockholm, No. 25.

Plass, G. N. (1956). The carbon dioxide theory of climate change. *Tellus* 8:140–54.

Plass, G. N. (1959). Carbon dioxide and climate. *Sci. Amer.* 201:41–48.

Porter, W. P. (1967). Solar radiation through the living body walls of vertebrates with emphasis on desert reptiles. *Ecol. Monogr.* 37:273–96.

Porter, W. P. (1969). Thermal radiation in metabolic chambers. *Science* 166:115–17.

Porter, W. P., and D. M. Gates (1969). Thermodynamic equilibria of animals with environment. *Ecol. Monogr.* 39:245–70.

Porter, W. P., J. W. Mitchell, W. A. Beckman, and C. B. DeWitt (1973). Behavioral impliations of mechanistic ecology. *Oecologia* 13:1–54.

Precht, H., J. Christopherson, H. Hensel, and W. Larcher, eds. (1973). *Temperature and Life*. New York: Springer-Verlag.

Prioul, J. L., and P. Chartier (1977). Partitioning of transfer and carboxylation components of intracellular resistance to photosynthetic CO_2 fixation: A critical analysis of the methods used. *Ann. Bot.* 41:789–800.

Pruitt, W. O., and D. E. Angus (1961). Comparisons of evapotranspiration with solar and net radiation and evaporation from water surfaces. In *Investigation of Energy and Mass Transfers Near the Ground, Including Influences of the Soil-Plant-Atmosphere System*, First Annual Report, U.S. Army Electronics Proving Grounds Technical Program, Univ. Calif., Davis. Chap. 6.

Rabinowitz, E. (1951). Photosynthesis and related processes. Vol. II, Part I. New York: Interscience.

Raschke, K. (1954). Ein Verfahren der Transpirationsbestimmung an Blattpunkten, Blatteilen und ganzen Blättern in situ. *Naturwissenschaften:* 41:308.

Raschke, K. (1956a). Micrometeorologically measured energy exchanges of an Alocasia leaf. *Arch. Meteorol. Geophys. Bioklimatol.* B7:240–68.

Raschke, K. (1956b). The physical relationships between heat-transfer coefficients, radiation exchange, temperature, and transpiration of a leaf. *Planta* 48:200–238.

Raschke, K. (1960). Heat transfer between the plant and the environment. *Ann. Rev. Plant Physiol.* 11:111–26.

Raschke, K. (1975). Stomatal action. *Ann. Rev. Plant Physiol.* 26:309–40.

Raunkiaer, C. (1934). *The Life Forms of Plants and Plant Geography*. Oxford: Clarendon Press.

Raven, J. (1972). Endogenous inorganic carbon sources in plant photosynthesis, II: Comparison of total CO_2 production in the light with measured CO_2 evolution in the light. *New Phytol.* 71:995–1014.

Redmond, D. R. (1955). Studies in forest pathology, XV: Rootlets, mycorrhiza, and soil temperatures in relation to birch dieback. *Canad. J. Bot.* 33: 595–627.

Reed, K. L., E. R. Hamerly, B. E. Dinger, and P. G. Jarvis (1976). An analytical model for field measurement of photosynthesis. *J. Appl. Ecol.* 13:925–42.

Regal, P. J. (1975). The evolutionary origin of feathers. *Q. Rev. Biol.* 50:35–66.

Reichle, D. E. (1970). *Analysis of Temperate Forest Ecosystems*. New York: Springer-Verlag.

Renck, L. E. (1972). Light scatter in biological materials. M.S. thesis, Purdue Univ., West Lafayette, Ind.

Revfeim, K. J. (1978). A simple procedure for estimating global daily radiation on any surface. *J. Appl. Meteorol.* 17:1126–31.

Riechert, S. E., and C. R. Tracy (1975). Thermal balance and prey availability:

Bases for a model relating web-site characteristics to spider reproductive success. *Ecology* 56:265–84.

Robinson, N., ed. (1966). *Solar Radiation*. Amsterdam: Elsevier Publ. Co.

Rose, A. H., ed. (1967). *Thermobiology*. London and New York: Academic Press.

Rosenberg, N. J. (1974). *Microclimate: The Biological Environment*. New York: John Wiley and Sons.

Sakai, A., and C. J. Weiser (1973). Freezing resistance of trees of North America with reference to tree regions. *Ecology* 54:118–26.

Salisbury, F. B., and C. W. Ross (1969). *Plant Physiology*. Belmont, Calif.: Wadsworth Publ. Co.

Salisbury, F. B., and G. G. Spomer (1964). Leaf temperatures of alpine plants in the field. *Planta* 60:497–505.

Saville, D. B. (1972). Arctic adaptations in plants. Monograph 6, Canad. Dept. Agr., Ottawa.

Sayre, J. D. (1926). Physiology of stomata of *Rumex patienta*. *Ohio J. Sci.* 26:233–66.

Schmidt-Nielsen, K. (1964). *Desert Animals: Physiological Problems of Heat and Water*. New York: Oxford Univ. Press.

Schmidt-Nielsen, K. (1972). *How Animals Work*. Cambridge: Cambridge Univ. Press.

Schmidt-Nielsen, K., T. J. Dawson, H. T. Hammel, D. Hinds, and D. C. Jackson (1965). The jack-rabbit: A study in its desert survival. *Hvalrodets Skrifter* 48:125–42.

Schneider, S. H. (1975). On the carbon dioxide–climate confusion. *J. Atmos. Sci.* 320:2060.

Schulze, E. D., O. L. Lange, U. Buschbom, L. Kappen, and M. Evenari (1972). Stomatal responses to changes in humidity in plants growing in the desert. *Planta* 108:259–70.

Schulze, E. D., O. L. Lange, M. Evenari, L. Kappen, and U. Buschbom (1974). The role of air humidity and leaf temperature in controlling stomatal resistance of *Prunus armeniaca* L. under desert conditions, I: A simulation of the daily course of stomatal resistance. *Oecologia* 17:159–70.

Schulze, R. (1970). *Strahlenkima der Erde*. Darmstadt: Dr. Dietrich Steinkopff Verlag.

Shields, L. M. (1950). Leaf xeromorphy as related to physiological and structural influences. *Bot. Rev.* 16:399–447.

Shreve, E. B. (1919). A thermo-electrical method for the determination of leaf temperature. *Plant World* 22:172–80.

Shreve, F. (1924). Soil temperature as influenced by altitude and slope exposure. *Ecology* 5:128–36.

Shull, C. A. (1929). A spectrophotometric study of reflection of light from leaf surfaces. *Bot. Gaz.* 87:583–607.

Shull, C. A. (1930). The mass factor in the energy relations of leaves. *Plant Physiol.* 5:279–82.

Simpson, G. G., A. Roe, and R. C. Lewontin (1960). *Quantitative Zoology*, rev. ed. New York: Harcourt, Brace.

Sinclair, T. R. (1968). Pathway of solar radiation through leaves. M.S. thesis, Purdue Univ., West Lafayette, Ind.

Sinclair, T. R., M. M. Schreiber, and R. M. Hoffer (1973). A diffuse reflectance hypothesis for the pathway of solar radiation through leaves. *Agron. J.* 65:276–83.

Skuldt, D. J., W. A. Beckman, J. W. Mitchell, and W. P. Porter (1975). Conduction and radiation in artificial fur. In D. M. Gates and R. B. Schmerl (eds.), *Perspectives of Biophysical Ecology.* New York: Springer-Verlag.

Slatyer, R. O. (1964). Efficiency of water utilization by arid zone vegetation. *Ann. Arid Zone* 3:1–12.

Slatyer, R. O., and J. F. Bierhuizen (1964). Transpiration from cotton leaves under a range of environmental conditions in relation to internal and external diffusive resistances. *Austral. J. Biol. Sci.* 17:115–30.

Slatyer, R. O., and I. C. McIlroy (1961). *Practical Micrometeorology.* Paris: UNESCO.

Smith, E. (1937). The influence of light and carbon dioxide on photosynthesis. *Gen. Physiol.* 20:807–30.

Smith, E. (1938). Limiting factors in photosynthesis: Light and carbon dioxide. *Gen. Physiol.* 22:21–35.

Sogin, H. H., and V. S. Subramanian (1961). Local mass transfer from circular cylinders in cross flow. *J. Heat Trans.* 83:483–93.

Spomer, G. G., and F. B. Salisbury (1968). Eco-physiology of *Geum turbinatum* and implications concerning alpine environments. *Bot. Gaz.* 129:33–49.

Spotila, J. M., O. Soule, and D. M. Gates (1972). The biophysical ecology of the alligator: Heat energy budgets and climate spaces. *Ecology* 53:1094–1102.

Spotila, J. R., P. W. Lommen, G. S. Bakken, and D. M. Gates (1973). A mathematical model for body temperatures of large reptiles: Implications for dinosaur ecology. *Amer. Nat.* 107:391–404.

Stair, R. (1951). Ultraviolet spectral distribution of radiant energy from the sun. *J. Res. Nat. Bur. Stand.* 46:353–57.

Stair, R. (1952). Ultraviolet radiant energy from the sun observed at 11,190 feet. *J. Res. Nat. Bur. Stand.* 49:227–34.

Stair, R., and H. T. Ellis (1968). The solar constant based on new spectral irradiance data from 310 to 530 nm. *J. Appl. Meteorol.* 7:635–44.

Stälfelt, M. G. (1961). The effect of the water deficit on the stomatal movements in a carbon dioxide-free atmosphere. *Physiol. Plant.* 14:826–43.

Stanhill, G. (1961). The accuracy of meteorological estimates of evapotranspiration in acid climates. *J. Inst. Water Eng.* 15:477–82.

Steen, J. (1958). Climatic adaptation in some small northern birds. *Ecology* 39:625–629.

Stefan, J. (1881). Über die Verdampfung aus einem kreisförmig oder elliptisech begrentzen Becken. *Sitzungsber. Kais. Acad. Wiss. Wien. Mathem.-Naturwiss. (Classe II Abt.)* 83:943–54.

Stigter, C. J. (1972). Leaf diffusion resistance to water vapor and its direct measurement, I: Introduction and review concerning relevant factors and methods. *Lab. Phys. Meteorol. Agr. Univ. Wageningen* 72:5–47.

Stigter, C. J., J. Birnie, and B. Lammers (1973). Leaf diffusion resistance to water vapor and its direct measurement, II: Design, calibration, and pertinent theory of an improved leaf-diffusion resistance meter. *Lab. Phys. Meteorol. Agr. Univ. Wageningen* 73:1–55.

Stiles, F. G. (1971). Time, energy, and territoriality of the Anna hummingbird (*Calypte anna*). *Science* 173:818–21.

Stoll, A. M., and J. D. Hardy (1955). Thermal radiation measurements in summer and winter Alaskan climates. *Trans. Amer. Geophys. Union* 36:213–25.

Stowe, L. G., and J. A. Teeri (1978). The geographic distribution of C_4 species of the dicotyledonae in relation to climate. *Amer. Nat.* 112:609–23.

Summer, F. B. (1921). Desert and lava-dwelling mice and the problem of protective coloration in mammals. *J. Mammal.* 2:75–86.

Swinbank, W. C. (1963). Longwave radiation from clear skies. *Quart. J. Roy. Meteorol. Soc.* 89:339–48.

Tageera, S. V., and A. B. Brandt (1960). A study of the optical properties of leaves depending on the angle of light incidence. *Biofizika* 5:308–17.

Talbert, C. M., and A. E. Holch (1957). A study of the lobbing of sun and shade leaves. *Ecology* 38:655–58.

Tanner, C. B., and M. Fuchs (1968). Evaporation from unsaturated surfaces: A generalized combination method. *J. Geophys. Res.* 73:1299–1304.

Tansey, M. R., and M. A. Jack (1975). Moonlight, mushrooms, and moulds. *J. Theor. Biol.* 51:403–7.

Taylor, S. E. (1971). Ecological implications of leaf morphology considered from the standpoint of energy relations and productivity. Ph.D. thesis., Washington University, St. Louis, Mo.

Taylor, S. E. (1975). Optimal leaf form. In D. M. Gates and R. B. Schmerl (eds.), *Perspectives of Biophysical Ecology*. New York: Springer-Verlag.

Taylor, E. S., and D. M. Gates (1970). Some field methods for obtaining meaningful leaf diffusion resistances and transpiration rates. *Oecol. Plant.* 5:103–11.

Teeri, J. A., and L. G. Stowe (1976). Climatic patterns and the distribution of C_4 grasses in North America. *Oecologia* 23:1–12.

Templeton, J. R. (1970). Reptiles. In G. C. Whittow (ed.), *Comparative Physiology of Thermoregulation*, Vol. I. New York: Academic Press.

Tenhunen, J. D., and D. M. Gates (1975). Light intensity and leaf temperature as determining factors in diffusion resistance. In *Perspectives of Biophysical Ecology*. New York: Springer-Verlag.

Tenhunen, J. D., C. S. Yocum, and D. M. Gates (1976). Development of a photosynthesis model with an emphasis on ecological applications, I: Theory. *Oecologia* 26:89–100.

Tenhunen, J. D., J. A. Weber, C. S. Yocum, and D. M. Gates (1976). Development of a photosynthesis model with an emphasis on ecological applications, II: Analysis of a data set describing the P_M surface. *Oecologia* 26:101–19.

Tenhunen, J. D., J. A. Weber, L. H. Filipek, and D. M. Gates (1977). Development of a photosynthesis model with an emphasis on ecological applications, III: Carbon dioxide and oxygen dependencies. *Oecologia* 30:189–307.

Tenhunen, J. D., J. D. Hesketh, and D. M. Gates (1980a). Leaf photosynthesis models. In *Predicting Photosynthate Production and Use for Ecosystem Models*. Cleveland, Ohio: CRC Press.

Tenhunen, J. D., J. D. Hesketh, and P. C. Harley (1980b). Modeling C_3 leaf respiration in the light. In *Predicting Photosynthate Production and Use for Ecosystem Models*. Cleveland, Ohio: CRC Press.

Thom, A. S. (1968). The exchange of momentum, mass, and heat between an artificial leaf and the airflow in a wind tunnel. *Quart. J. Roy. Meteorol. Soc.* 94:44–55.

Thompson, D. W. (1917). *On Growth and Form*. London: Cambridge Univ. Press.

Thornthwaite, C. W. (1931). The climates of North America according to a new classification. *Geogr. Rev.* 21:633–55.

Thornthwaite, C. W. (1933). The climates of the earth. *Geogr. Rev.* 28:433–40.

Tibbals, E. C., E. K. Carr, D. M. Gates, and F. Kreith (1964). Radiation and convection in conifers. *Amer. J. Bot.* 51:529–38.

Tikhomirov, B. A., V. F. Shamurin, and V. S. Shtepa (1960). The temperature of arctic plants. *Acad. Sci. USSR Biol. Ser.* 3:429–42.

Timoshenko, S. (1962). *Elements of Strength of Materials*. Princeton, N.J.: Van Nostrand.

Ting, I. P., and W. E. Loomis (1963). *Diffusion through stomates. Amer. J. Bot.* 50:866–72.

Tinkle, D. W., D. McGregor, and S. Dana (1962). Home range ecology of *Uta stansburiana stejnegeri*. *Ecology* 43:223–29.

Tracy, C. R. (1972). Newton's Law: Its applicability for expressing heat losses from homeotherms. *BioScience* 22:656–59.

Tracy, C. R. (1975). Tyrannosaurs: Evidence for endothermy? *Amer. Nat.* 110: 1105–6.

Tracy, C. R. (1977). Minimum size of mammalian homeotherms: Role of the thermal environment. *Science* 198:1034–35.

Tregear, R. T. (1965). Hair density, wind speed, and heat loss in mammals. *Amer. J. Appl. Physiol.* 20:796–801.

Tucker, C. J., and M. W. Garratt (1977). Leaf optical systems as a stochastic process. *Appl. Optics* 16:635–42.

Turner, J. S., and E. G. Brittain (1962). Oxygen as a factor in photosynthesis. *Biol. Rev.* 37:130–70.

Turner, N. C. (1974). Stomatal behavior and water status of maize, sorghum, and tobacco under field conditions. *Plant Physiol.* 53:360–65.

Turner, von H., and W. Tranquilini (1961). Die Strahlungsverhältnisse und ihr Einfluss auf die Photosynthese der Pflanzen. *Mitt. Forstl. Bundes-Versuchsanstalt Mariabrunn* 59:69–103.

Turrell, F. M., S. W. Austin, and R. L. Perry (1962). Nocturnal thermal exchange of citrus leaves. *Amer. J. Bot.* 49:97–109.

Tyler, J. E., and R. W. Preisendorfer (1962). Transmission of energy within the sea, VIII: Light. In M. N. Hill (ed.), *The Sea,* Vol. I. New York: John Wiley.

Underwood, C. R., and E. J. Ward (1966). The solar radiation area of man. *Ergonomics* 9:155–68.

Urbach, F., ed. (1969). *The Biologic Effects of Ultraviolet Radiation*. Oxford: Pergamon Press.

Uvarov, B. P. (1931). Insects and climate. *Trans. Entomol. Soc. London* 79:1–247.

van Bavel, C. H. M., F. S. Nakayama, and W. L. Ehrler (1965). Measuring transpiration resistance of leaves. *Plant Physiol.* 40:535–40.

Vegis, A. (1963). In L. T. Evans (ed.), Climatic control of germination, bud break, and dormancy. *Environmental Control of Plant Growth*. New York: Academic Press. Pp. 265–285.

Verduin, J. (1947). Diffusion through multiperforate septa. In J. Frank and W. E. Loomis (eds.), *Photosynthesis in Plants*. Ames, Iowa: Iowa State College.

Vogel, S. (1968). Sun leaves and shade leaves: Differences in convective heat dissipation. *Ecology* 49:1203–4.

Vogel, S. (1970). Convective cooling at low airspeeds and the shapes of broad leaves. *J. Exp. Bot.* 21:91–101.

Volz, F. E., and K. Bullrich (1961). Scattering function and polarization of skylight in the ultraviolet to the near infrared region, with haze of scattering type 2. *J. Meteorol.* 18:306–18.

Waggoner, P. E., and R. H. Shaw (1952). Temperature of potato and tomato leaves. *Plant Physiol.* 23:710–24.

Walter, H. (1926). Die Verdunstung von Wasser in bewegter Luft und ihre Abhängigkeit von der Grösse der Oberfläche. *Z. Bot.* 18:1–47.

Waterman, H. (1955). Polarized light and animal navigation. *Sci. Amer.* 193:88–94.

Wathen, P. M., J. W. Mitchell, and W. P. Porter (1971). Theoretical and experimental studies of energy exchange from jack-rabbit ears and cylindrically shaped appendages. *Biophys. J.* 11:1030–47.

Watson, A. N. (1933). Preliminary study of the relation between thermal emissivity and plant temperatures. *Ohio J. Sci.* 33:435–50.

Watts, W. R. (1971). Role of temperature in the regulation of leaf extension in *Zea mays*. *Nature* 229:46–47.

Weathers, W. W. (1970). Physiological thermoregulation in the lizard, *Dipsosaurus dorsalis*. *Copeia* 3:549–57.

Welch, W. R., and C. R. Tracy (1977). Respiratory water loss: A predictive model. *J. Theor. Biol.* 65:253–65.

Welkie, G. W., and M. Caldwell (1970). Leaf anatomy of species in some dicotyledon families as related to the C_3 and C_4 pathways of carbon fixation. *Canad. J. Bot.* 48:2135–46.

Wellington, W. G. (1974). Bumblebee ocelli and navigation at dusk. *Science* 183:550–51.

Went, F. W. (1949). Ecology of desert plants, II: The effect of rain and temperature. *Ecology* 30:1–13.

Went, F. W. (1957). *The Experimental Control of Plant Growth*. New York: Ronald Press.

Went, F. W. (1961). Temperature: Survey, thermoperiodicity. In *Handbuch der Pflanzenphysiologie*, Vol. 16. Berlin: Springer-Verlag.

West-Eberhard, M. J. (1973). Size variation and the distribution of hemimetabolous aquatic insects: Two thermal equilibrium hypotheses. *Science* 200:444–446.

Westlake, D. F. (1966). The light climate for plants in rivers. In R. Bainbridge, G. C. Evans, and O. Rackham (eds.), *Light as an Ecological Factor*. New York: John Wiley.

Whitehead, F. H. (1962). Experimental studies of the effect of wind on plant growth and anatomy, II: *Helianthus annuus*. *New Phytol.* 61:59–62.

Whiteman, P. C., and D. Koller (1967). Interactions of carbon dioxide concentration, light intensity, and temperature on plant resistances to water vapor and carbon dioxide diffusion. *New Phytol.* 66:463–73.

Whittow, G. C., ed. (1970). *Comparative Physiology, Vol. I: Invertebrates and Nonmammalian Vertebrates*. New York: Academic Press.

Wielgolaski, F. E. (1966). The influence of air temperature on plant growth and development during the period of maximal stem elongation. *Oikos* 17:121–41.

Wigley, G., and J. A. Clark (1974). Heat transport coefficients for constant energy flux models of broad leaves. *Boundary-Layer Meteorol.* 1:123–456.

Williams, W. T. (1950). Studies in stomatal behavior, IV: The water-relations of the epidermis. *J. Exp. Bot.* 1:114–31.

Willstätter, R., and A. Stoll (1918). Untersuchunger über die Assimilation der Kohlensäure. Berlin: Springer-Verlag.

Wilson, J. W. (1957). Observations on the temperatures of arctic plants and their environments. *J. Ecol.* 45:499–531.

Wilson, J. W. (1966). Effects of temperature on net assimilation rates. *Ann. Bot.* 30:753–61.

Wong, C. L., and W. R. Blevin (1967). Infrared reflectances of plant leaves. *Austral. J. Biol. Sci.* 20:501–8.

Woolley, J. T. (1971). Reflectance and transmittance of light by leaves. *Plant Physiol.* 47:656–62.

Wuenscher, J. E., and T. T. Kozlowski (1971). The response of transpiration resistance to leaf temperature as a desiccation resistance mechanism in tree seedlings. *Physiol. Plant.* 24:254–59.

Yeats, N. T. (1967). The heat tolerance of sheep and cattle in relation to fleece or coat character. In S. W. Tromp and W. H. Weihe (eds.), *Biometeorology (Proc. Third Internat. Biometeorol. Congr.)*. Oxford: Pergamon Press. Pp. 464–470.

Index